WHERE TO WATCH BIRDS IN
SOUTHERN & WESTERN SPAIN

ERNEST GARCIA
WITH ANDREW PATERSON

HELM
LONDON · OXFORD · NEW YORK · NEW DELHI · SYDNEY

HELM
Bloomsbury Publishing Plc
50 Bedford Square, London, WC1B 3DP, UK
29 Earlsfort Terrace, Dublin 2, Ireland

BLOOMSBURY, HELM and the Helm logo are
trademarks of Bloomsbury Publishing Plc

First published in the United Kingdom 1994
This edition published 2019

Copyright © Ernest Garcia, 2019

Ernest Garcia has asserted his right under the Copyright, Designs
and Patents Act, 1988, to be identified as Author of this work

All rights reserved. No part of this publication may be reproduced or
transmitted in any form or by any means, electronic or mechanical,
including photocopying, recording, or any information storage or retrieval
system, without prior permission in writing from the publishers

A catalogue record for this book is available from the British Library

ISBN: PB: 978-1-4729-5184-7; ePDF: 978-1-4729-8678-9; ePub: 978-1-4729-5183-0

2 4 6 8 10 9 7 5 3

Line drawings by Stephen Message
Maps by Brian Southern

Typeset by Mark Heslington Ltd, Scarborough, North Yorkshire
Printed and bound in India by Replika Press Pvt. Ltd.

MIX
Paper from
responsible sources
FSC® C016779

To find out more about our authors and books visit www.bloomsbury.com
and sign up for our newsletters

Front cover (top): Rufous-tailed Scrub-robin © Carlos Bocos; (bottom): Black Stork
© Juan Luis Muñoz Roldan. Spine: Bonelli's Eagle © Juan Luis Muñoz Roldan.
Back cover (left): Red-rumped Swallow © Philip Croft; (middle): Squacco Heron
© Philip Croft; (right): White-headed Duck © Erni/Shutterstock

CONTENTS

Foreword	7
Introduction to the Fourth Edition	9
How to use this book	16
Visiting southern and western Spain	19
Landscape and climate	29
Birds and other wildlife of southern and western Spain	33
Watching seabirds and the migration of raptors and storks	39

ANDALUCÍA 49

Huelva Province 49
H1	Guadiana estuary (Marismas de Isla Cristina)	51
H2	Piedras estuary (Marismas del Río Piedras)	53
H3	Laguna de El Portil	55
H4	Odiel estuary (Marismas del Odiel)	57
H5	Doñana National Park	62
H6	Sierra Pelada	72
H7	Western Sierra Morena	74

Cádiz Province 78
CA1	Tarifa beach (Playa de Los Lances)	79
CA2	Central Strait of Gibraltar	82
CA3	Punta Secreta	88
CA4	La Janda	90
CA5	Bolonia, Zahara and Sierra de la Plata	93
CA6	The corkwoods (Los Alcornocales)	96
CA7	Palmones estuary	103
CA8	Guadiaro estuary (Sotogrande)	104
CA9	Cape Trafalgar	106
CA10	Barbate pinewoods and estuary	107
CA11	Sancti Petri marshes	110
CA12	Cádiz Bay (Bahía de Cádiz)	112
CA13	Lagunas de Puerto Real	116
CA14	Lagunas del Puerto de Santa María	117
CA15	Laguna de Medina	119
CA16	Chipiona coast	122
CA17	East Bank of the lower Guadalquivir	124
CA18	Arcos and district	127
CA19	Lagunas de Espera	129
CA20	Sierra de Grazalema	131
CA21	Peñon de Zaframagón	133

Sevilla Province 135
SE1	Eastern Guadalquivir Delta. Brazo del Este	136
SE2	Western Guadalquivir Delta	140
SE3	Dehesa de Abajo	142

SE4	Western Guadalquivir Valley Farmlands	144
SE5	Lagunas de La Lantejuela	147
SE6	Sierra Norte	149

Córdoba Province — **151**

CO1	Laguna de Zóñar	153
CO2	Lagunas Amarga and Dulce	155
CO3	Lagunas del Rincón and de Santiago	156
CO4	Laguna de Tíscar	158
CO5	Laguna del Salobral	159
CO6	Cordobilla reservoir	160
CO7	Malpasillo reservoir	161
CO8	Sierras Subbéticas	163
CO9	Sierra de Hornachuelos	166
CO10	Sierra de Cardeña y Montoro	169
CO11	Alto Guadiato farmlands	172
CO12	Zújar and Guadamatilla Valleys	174
CO13	Puente Nuevo reservoir	176
CO14	Río Guadalquivir at Córdoba	177

Málaga Province — **180**

MA1	Laguna de Fuente de Piedra	181
MA2	Lagunas de Campillos	185
MA3	Teba: sierra and gorge	186
MA4	El Chorro and the Abdalajís valley	188
MA5	El Torcal	191
MA6	Sierra de Camarolos	193
MA7	Málaga hills (Montes de Málaga)	194
MA8	Guadalhorce estuary	196
MA9	Juanar (Sierra Blanca)	199
MA10	Sierras de las Nieves National Park	201
MA11	Sierra Crestellina	204
MA12	Sierra Bermeja and Central Serranía de Ronda	207
MA13	Upper Guadiaro valley	210
MA14	Sierra Tejeda	212
MA15	Vélez estuary	214
MA16	Alpujata migration watchpoint (Mirador de las Águilas)	216
MA17	Punta Calaburras	217

Granada Province — **220**

GR1	Sierra Nevada	221
GR2	The Alpujarras	225
GR3	La Ragua pass (Puerto de la Ragua)	228
GR4	Laguna de Padul	230
GR5	Alhama Reservoir (Pantaneta de Alhama de Granada)	232
GR6	Sierra de Huétor	234
GR7	Guadix basin (Hoya de Guadix)	235
GR8	Sierra de Baza	238
GR9	Baza basin (Hoya de Baza)	239
GR10	Suárez Wetland (Charca de Suárez)	241

Jaén Province		**244**
J1	Sierras de Cazorla and Segura	246
J2	Despeñaperros and Aldeaquemada	250
J3	Sierra de Andújar	253
J4	Sierra Mágina	256
J5	Guadalén and Giribaile Reservoirs	259
J6	Laguna Grande de Baeza	262
J7	Upper Guadalquivir Reservoirs	264
Almería Province		**266**
AL1	Cabo de Gata Natural Park	267
AL2	Punta Entinas to Roquetas de Mar	272
AL3	Las Norias lakes (Cañada de las Norias)	275
AL4	Adra lakes (Albufera de Adra)	278
AL5	Tabernas desert	280
AL6	Sierra Alhamilla	282
AL7	Sierra de María	284
AL8	Antas estuary	286
AL9	Almanzora estuary	287

EXTREMADURA 290

Badajoz Province		**294**
BA1	Dehesas and reservoirs of Southwestern Badajoz	295
BA2	The Western Sierras. Sierra de San Pedro	297
BA3	Southern Badajoz	300
BA4	Sierra de Hornachos	303
BA5	River Guadiana at Mérida	305
BA6	Cornalvo reservoir	307
BA7	Los Canchales reservoir	308
BA8	Sierra de Tiros	310
BA9	Plains of La Serena	312
BA10	Orellana and other eastern reservoirs	315
BA11	Cíjara Game Reserve (Reserva Regional de Cíjara)	317
BA12	River Guadiana at Badajoz City	319
BA13	Lagoons and farmlands of La Albuera	321
BA14	Alqueva reservoir inlet	323
Cáceres Province		**325**
CC1	Plains of Western Cáceres	326
CC2	Steppes of Malpartida de Cáceres	328
CC3	Cáceres–Trujillo steppes	330
CC4	Cuatro Lugares steppes	333
CC5	Sierras de Las Villuercas	334
CC6	Monfragüe National Park	337
CC7	Sierra de Gata and Borbollón reservoir	341
CC8	Jerte valley	343
CC9	Arrocampo reservoir and Campo de Arañuelo	347
CC10	Sierra Brava reservoir	350
CC11	Guadiana Valley ricefields	351
CC12	Alcollarín reservoir	354

GIBRALTAR	**357**
Status list of the birds of southern and western Spain	368
Appendices	**399**
Appendix 1. Scientific names of plant species mentioned	399
Appendix 2. Glossary of local geographical terms	399
Further reading	400
Site index	401
Species index	405

FOREWORD

Visitors to Spain in search of birds naturally gravitate to the south and west, to Andalucía and Extremadura. They are attracted by the spectacular assemblages of breeding and migrant raptors, by the exotic wetland species and by the colourful passerine communities of the Mediterranean. But there is much more on offer. Here are some of the best unspoilt wildlife habitats surviving in Europe, supporting a wide range of species that are uncommon or absent elsewhere. Nowhere else in Europe can you find Rüppell's Vulture, Red-knobbed Coot, White-rumped Swift, Little Swift or Trumpeter Finch. Here too there are excellent and, in many cases, unparalleled opportunities for seeing a large range of Iberian and Mediterranean species. The Marbled and White-headed Ducks, Balearic Shearwater, Squacco Heron, Lesser Flamingo, Black-winged Kite, Purple Swamphen, Great Bustard, Pin-tailed Sandgrouse, Collared Pratincole, Audouin's Gull, Gull-billed Tern, Great Spotted Cuckoo, Red-rumped Swallow, Rufous-tailed Scrub-robin, Black Wheatear, Western Olivaceous Warbler, Penduline Tit, Southern Grey Shrike, Spotless Starling and Spanish Sparrow come rapidly to mind as examples, but there are many others. A major fraction of the world populations of the Spanish Imperial Eagle and Iberian Azure-winged Magpie inhabits parts of the region. The proximity to Africa, the migration routes, the combination of mountain, steppe and maritime habitats and, of course, the year-round superb climate are all extra ingredients that make a birding visit to southern Spain memorable and rewarding.

The Mediterranean countries still do not have a good name in conservation circles. However, matters are improving and the former hordes of callous hunters, intent on blasting or netting everything that moves, are on the wane, except in a few notorious blackspots such as Malta, Cyprus and Lebanon. In Spain at least, there have been great improvements over recent years accompanied by a widespread and ever-growing interest and awareness among the Spaniards themselves.

A key driving force in encouraging interest in birds in Spain and in bringing about their conservation is the Spanish Ornithological Society and Birdlife International partner, the Sociedad Española de Ornitología (SEO/Birdlife). If you are a regular birding visitor to Spain, or are resident there, you can best contribute to their efforts by joining the Society. You will find that their twice-yearly magazine *Aves y Naturaleza* is a dynamic and informative update on anything to do with birds in Spain, including recent observations and some wonderful photographs. SEO/Birdlife also publish the scientific journal *Ardeola*, nearly all of which is in English. You can contact the Society via their website *www.seo.org*.

We are all fortunate that there is still a great deal left to conserve in Spain. Conservation, though, has its price and the case for preserving important areas is strengthened if they can be seen to attract tourists as well. Fortunately, national and regional authorities in Spain have become well aware of the benefits of ecotourism – witness the number of Spanish stands at the British Birdfair at Rutland Water for example – and they have invested a great deal of effort and money not merely on publicity but also on improving the infrastructure and management of their many nature reserves. This book has always been intended as a contribution to conservation of Spanish birds and their habitats by encouraging visits to areas that deserve to be protected. Do tell hoteliers and other local people why you are visiting. By doing so you will be helping to ensure that Spanish wildlife remains the spectacle that it is.

Enjoy your stay.
¡ *Buen viaje* !

INTRODUCTION TO THE FOURTH EDITION

This book was first published in 1994, with two subsequent editions in 2001 and 2008. It is the sister volume to *Where to Watch Birds in Northern and Eastern Spain*, by Ernest Garcia and Michael Rebane, which covers the rest of Spain. The original work was a joint effort between me and Andy Paterson. Andy and I are still going strong – well, going at least – but Andy is now busy with other interests, so this revision is my sole responsibility. The entire text has been fully revised and updated. I have added a further eight completely new major sites and expanded the treatment of many of the original locations, introducing 49 new sub-sites.

It is over a decade since the appearance of the third edition, during which the southern Spanish natural environment and its wildlife have experienced major changes. Here I draw your attention to some of the most important and interesting ones. Some are developments of conservation significance, both positive and negative, and others involve changes in the numbers and geographical distribution of certain species.

Conservation Issues

The good news remains very real. Interest in Spanish birds and wildlife, and in the environment generally, continues to grow strongly, as evidenced by the rising membership of the Sociedad Española de Ornitología (Spanish Ornithological Society – also known as SEO/Birdlife) and other organisations. When SEO/Birdlife was founded in 1954 the number of Spaniards with a serious interest in birds (hunting excluded) probably did not exceed a couple of dozen. Now, over 60 years later, there are many thousands and you can expect to meet numbers of Spanish birders at key sites and in the wider countryside. The popularity of digital photography has also given a boost to conservation since so many people have found that wildlife subjects, birds included, provide interesting and rewarding challenges. School parties are taken routinely to visit Doñana, Monfragüe and other places where there are information centres and suitable facilities: you will hear such groups before you see them, since Spanish children tend towards voluble exuberance, but do tolerate them patiently. It is vital that younger generations develop a sympathy for their natural heritage.

There has been welcome expansion in the network of local wildlife groups, in particular the provincial and regional groups of the SEO – who undertake much survey work – which has already included three breeding bird atlases, a winter atlas, a migration atlas, national censuses of a wide range of species and much more. The various local authorities have taken note of this interest by establishing visitors' centres in many key places and by producing a great deal of printed informative material, for the benefit of both the Spanish public and also of the ever-increasing number of foreign visitors who come to southern and western Spain in search of its rich wildlife. Ecotourism is clearly here to stay as an important dimension of Spain's all-important and massive tourist industry.

The network of protected areas established and maintained by official bodies and, to a lesser extent, by voluntary organisations, has continued to expand, especially under the encouragement of the European Union. An increasingly complete network of Important Bird Areas (IBAs), or ZEPAs (Zonas de Especial Protección para las Aves) as they are known in Spain, is now in place both in Andalucía and Extremadura, and indeed elsewhere. The IBAs identify key regions and provide a basic level of regulation of the land use there and are invaluable in providing essential protection to key habitats and their species. Many of the ZEPAs are also included in the broader network of Special Protection Areas or SPAs (LICs: Lugares de Importancia Comunitaria in Spain), which was set up under the Natura 2000 initiative and provides some protection of the full range of biodiversity, not just birds. All these EU-inspired measures, together especially with some adjustments to the Common Agricultural Policy, go some way to reduce and perhaps eventually reverse the harmful effects of some EU policies on the natural environment. Designating a site as protected is, of course, very far from ensuring its conservation, but it is an essential first step. Many of the ZEPAs/LICs enjoy higher levels of security under the Spanish network of protected areas. There are four principal categories of these, defined as follows:

Reserva Natural (Nature Reserve) The lowest level of protection and roughly equivalent to that of an SSSI (Site of Special Scientific Interest) in Britain. Nature Reserves are intended to protect specific ecosystems or communities. These are mainly small sites, many of them lagoons. Only activities compatible with the survival of the sites are permitted there, for example, fish farming is banned at wetland sites. Visitors are encouraged and information centres and hides may be provided.

Paraje Natural (Natural Locality) A similar status to Nature Reserve, protecting mainly restricted areas of general scenic or biological interest. Traditional activities (not hunting) are permitted to continue.

Parque Natural (Natural Park) An extensive area, often of sierras and woodlands, which offers well-preserved natural or semi-natural habitats. The designation protects the area from further unsuitable development, while still allowing compatible traditional activities to continue. A feature of a Natural Park is that it aims to provide educational and recreational facilities for the general public. The protection of traditional architecture and cultural aspects is included. Many of the Parks have information centres that offer factsheets, maps, displays and other information about the region.

Parque Nacional (National Park) A National Park enjoys the highest level of protection. The designation protects extensive areas of international importance within which all human activities are strictly controlled. Access too is limited where this is in the interests of the fauna and flora. There are 15 National Parks in Spain: three of them, Doñana (H5), the Sierra de las Nieves (MA10) and the Sierra Nevada (GR1), in Andalucía, and one, Monfragüe (CC6), in Extremadura.

All this notwithstanding, the natural environment remains under grave threat from a diversity of causes. Changes in land use often have implications for the environment that give cause for concern. A particular worry that has come to the fore in recent years is the enormous expansion of intensive olive cultivation especially, but by no means exclusively, within the Guadalquivir valley. The new olive groves have swallowed up great expanses formerly devoted to cereal crops and pastures. The cultivated olive finds the warm climate of the Upper Guadalquivir highly amenable and a great proportion of the lowland farmlands, and an increasing fraction of the lower slopes of the hills, has been given over to olive groves. Jaén province alone is apparently the origin of some 40% of the global production of olive oil. Traditional olive groves are untidy orchards of old trees, intermixed with a rich variety of herbaceous and shrubby plants. Such places are the home of the Rufous-tailed Scrub-robin, Turtle Dove, Woodchat Shrike and many other birds and other animals. Unfortunately, most olive groves are now intensively managed, the ground between the trees scrubbed bare of vegetation, apparently 'to reduce competition'. The result is a monoculture of minimal biodiversity, effectively birdless over great areas. To counter this, SEO/Birdlife is coordinating a conservation project named 'Olivares Vivos' aimed at promoting wildlife-friendly, sustainable olive cultivation (see English-language website *http://www.olivaresvivos.com/en/*). This has attracted widespread support within the farming community and augurs well for the future.

The extension of irrigated areas in Extremadura and elsewhere may have incidentally provided additional wetland habitats, for example, by the ongoing expansion of rice cultivation, but some of this has involved the loss of steppe habitats that formerly accommodated bustards and other threatened birds. The grubbing-up of the evergreen oak woodlands of Extremadura in particular is causing significant loss of habitats for wintering Cranes and many breeding species. Cork Oak forests, vital habitats for some of the best of the area's wildlife, may also be threatened if demand for cork continues to decline, something to remember when choosing wine: screw-tops are definitely not 'green'. Recent changes to the Common Agricultural Policy may go some way to limit or correct damage caused by insensitive changes in land use but the problem is likely to remain severe for the foreseeable future.

The loss of green land to such developments as urban expansion – especially to tourist apartments and villas on the coasts – and new motorways and roads continues to have serious effects in many places. Among other infrastructures, wind farms are a controversial, intrusive and increasingly widespread feature of our area. Wind-power has been fêted for its environment-friendly credentials: wind is pollution-free after all. However, a visit to the Strait of Gibraltar makes the concept of visual pollution starkly real since hundreds upon hundreds of wind turbines now disfigure the hillsides in what used to be an area of outstanding, nay exceptional, natural beauty. The turbine blades chop up birds too but we do not yet know whether this is having an impact at the population level, although the few Egyptian Vulture pairs that live near the Strait have suffered significant losses. Wind farms have expanded in the Strait area (CA2), on the plain of La Janda (CA4), near the Sierra Crestellina (MA11) and more locally elsewhere. Whereas it seems highly improbable that wind farms can make a noticeable contribution to ameliorating any human-induced climatic change, it is certain that, if we must have them, they should be sited more sensitively and only if validated by Environmental Impact Assessments.

Climate change may itself be having a broader impact on the regional ecology. The most serious impact is probably water shortage. Much of southern Spain is relatively arid, with very dry years occurring from time to time.

However, years of poor rainfall seem to be on the increase. Indeed, 2005 saw the most severe drought in the region for 125 years, with many wetlands drying up completely for much longer than the usual mid-summer period and forest fires becoming exceptionally widespread and damaging. Much the same happened in 2017. The competing demands for ever-increasing quantities of water from tourism (think swimming pools and golf courses), agriculture, horticulture and direct human consumption put great pressure on the water supplies available for such places as natural wetlands. The overall response of the authorities has been to build more reservoirs but water problems are likely to continue for the forseeable future and very probably will become more acute.

A number of other problems deserve mention. The disturbance of the countryside by noisy visitors is a problem in some areas where quad bikes have become popular, allowing people access to formerly quiet and remote forest tracks. Such vehicles are prohibited from many protected places but the law is very poorly enforced. Sundays are worst; indeed the people involved are known locally as 'domingueros': the Sunday-trippers, a point to bear in mind when planning your birding trips.

More insidious and pervasive is the effect of poison baits laid for the control of foxes and other 'vermin', especially on large estates where the practice is hard to detect. These, and the effects of some poorly designed electricity pylons, which electrocute birds, have had severe impacts on the populations of many raptors. SEO has campaigned actively against both threats, with some notable success, but the dangers are far from over.

Changes in the Avifauna

The bird populations of so large a region are bound to fluctuate in the long term but even the short period of the last few decades has seen significant changes.

The Spanish White Stork population was at a record low of 6,753 pairs in 1984 but the 2004 census found a record 33,217 pairs, with large increases in the traditional core areas of Extremadura and western Andalucía. Such increases have continued but more recent census data is lacking. They have still not made any significant inroads into eastern Andalucía, which is probably too arid for them. The storks have suffered local setbacks from unsympathetic local authorities, such as those that removed the 40+ nests from the historical core of Cáceres city or the 25 nests on Plasencia cathedral. In the latter case a public outcry prevented the installation of devices to dissuade the storks, which promptly rebuilt 18 nests there. White Storks continue to benefit from the food hand-outs at landfill garbage dumps, to the extent that many have become virtually resident in Spain and no longer risk the hazards of the annual return trip to Africa.

The Griffon Vulture population burgeoned from 3,240 pairs in 1979 to over 17,000 pairs in 1999 and had increased further to 25,000 pairs by 2008. The population has remained at least stable since then but further large increases have been reported from some key sites. At Monfragüe, for example, the nesting population increased from 530 pairs in 2013 to 681 in 2017. Vultures, like White Storks, are also beneficiaries of the sloppiness of human waste-disposal but they have also benefited from a reduction of direct persecution, now that raptor-shooting is universally prohibited in Spain. Many of the key colonies are directly protected. Black Vultures too have increased greatly in Spain, from 365

White-rumped Swifts and Little Swift

pairs in 1986 to some 3,000 pairs in 2018, and this species has never been easier to find nor more abundant in its regular haunts in Extremadura and the Sierra Morena. The Lammergeier, the key target species at what was its last Andalucían redoubt at Cazorla (J1), has benefited from a reintroduction programme there, which saw the first releases of captive-bred juveniles in 2006 and the first successful nesting in 2015. Introduction of Ospreys to reservoirs and estuarine habitats in western Andalucía and on the Guadiana River in Extremadura had resulted in a nesting population of some 20 pairs by 2018, with every indication of further increase.

A particularly welcome development in recent years has been the ongoing recovery of the Spanish Imperial Eagle, whose population declined to only 131 pairs in 1999 but numbered well over 400 pairs by 2018. It is increasing steadily, reoccupying some former haunts as well as new locations. Specific conservation measures coordinated by SEO, including agreements with landowners and the remedying of unsafe pylons, have played a large part in making this welcome trend possible.

Other breeding species showing significant and ongoing increases in numbers and, in some cases, geographical spread within our area include Great Cormorant, Great White Egret, Grey Heron, Black Stork, Glossy Ibis, Black-winged Kite, Marsh and Montagu's Harriers, Short-toed, Booted, Bonelli's and Golden Eagles, Purple Swamphen, Black-headed Gull, Yellow-legged Gull (alas!), Little and White-rumped Swifts and Red-rumped Swallow.

The good health of the Montagu's Harrier population owes a great deal to human intervention, as indeed do the declines that it has suffered since during the late 20th century. A big effort to locate Montagu's Harrier nests is made annually by dedicated groups both in Andalucía and Extremadura, with nests threatened with destruction by mowing being removed and the young safely released into the wild in due course. This is no small task. For example, in 2016 460 of 693 nests located in Andalucía were given some form of direct protection. Such efforts have meant that the population of Montagu's Harrier in Andalucía numbers some 800 pairs, with a further 500 pairs in Extremadura, and these enjoy high breeding success, fledging around two young per successful nest.

The tendency for some species of African origin to appear in southern Iberia and eventually to colonise the region appears to be continuing. The White-rumped and Little Swifts are examples of relatively recent successful colonisations but since then there have been several instances of breeding by Cream-coloured Coursers and Lesser Flamingos. Common Bulbuls nested successfully at Tarifa in 2014 and Laughing Doves raised young near Sevilla in 2018. Whether or not these events prove to be forerunners of longer-term developments, they serve to remind us to be prepared for the unexpected – it seems to happen more often than we might once have thought, especially in southern Spain.

A number of species have shown an increasing trend to winter within our area. White Storks undoubtedly overwinter to a much greater extent than used to be the case, and they have been joined by increasing numbers of Black Storks. There are increasing winter records of a number of raptor species, such as Black Kites, Egyptian Vultures and Short-toed Eagles. In particular, Booted Eagles, which were formerly regarded as fully migratory, may now be seen all winter in favoured localities, especially near the coast. Wintering Ospreys are reported ever more frequently and a few Pallid Harriers have recently begun to winter in the region. Post-breeding dispersal from North Africa brings Long-legged Buzzards and Lanners to the south of our region increasingly frequently.

There is a debit list too, unfortunately. Some species have shown worrying recent declines that appear to be continuing. Hardest hit are the birds of the farmed countryside, especially, but not only, the steppic species, victims of agricultural intensification and land use changes. The tendency to do away with traditional rotations that feature short-term and long-term fallows, the reduction in the extent of cereals and other dry crops in favour of irrigated crops and olive groves and, of course, the overuse of agricultural chemicals have all had very serious effects on some bird populations. The Little Bustard has suffered particularly; what was once a common bird of open country has now become distinctly hard to find, with numbers falling by 80%+ even in its former strongholds. Both sandgrouse have also declined markedly, perhaps by 50%. The Lesser Kestrel suffered a sharp decline from the 1970s onwards that was reversed in the late 1980s but has since resumed. Lesser Kestrel numbers in Extremadura suffered a severe drop, some 70%, between 2017 and 2018, with many colonies abandoned in 2018. Even such ubiquitous and abundant species as the Barn Swallow, House Sparrow and Corn Bunting have been found by Spanish common bird censuses to have suffered significant recent declines.

Among the raptors, a particularly serious decline has affected the Egyptian Vulture, which seems to have been especially hard-hit by poison baits laid for mammalian predators. In 2018 only 23 pairs could be found in the whole of Andalucía, where once there were hundreds.

Waterbirds in general have benefited greatly by the widespread protection afforded to wetlands. However, a wetland is nothing without water and drought years have serious impacts within the region. The resurgent population of the White-headed Duck is holding its own but two other iconic aquatic species of southern Spain, the Marbled Duck and the Red-knobbed Coot, remain at a low ebb. Time will tell whether the climate is set to become drier in the long term, which would have profound effects on regional biodiversity.

Classification Status and Names

The whole issue of bird names, and especially the defining of species-boundaries, remains highly controversial. The sequences of species in lists and field guides is also in a state of flux. For the sake of simplicity, the taxonomic decisions, names and sequence adopted by the most recent version of the Western Palearctic List of the Association of European Records and Rarities Committees (Crochet & Joynt 2015) have been followed here. These names will be familiar to all users and also correspond with those used in the regional avifauna, *The Birds of the Iberian Peninsula* (De Juana & Garcia 2015), in the most recent List of the Birds of Spain (Gutiérrez *et al.* 2012) and to a large extent in the *Collins* bird guide (Svensson *et al.* 2009). The chief divergence from the WP List is in referring to the Iberian Green Woodpecker, drawing attention to the distinctive endemic form, *Picus viridis sharpei*, which is treated as a separate species by some authorities. Qualifiers (e.g. 'Common') are given in the species accounts only for species such as coots, cuckoos and swallows that are represented in the area by more than one species. The full English, Spanish and scientific names are given for all species in the Status List chapter.

Ernest Garcia

HOW TO USE THIS BOOK

The core of the book is the Site Accounts but the preliminary chapters should help you to plan your visits. They describe the area and its birds. The Status List (pages 368–98) gives the status of all those that occur with any regularity in the region as well as records of rarities. It is the principal reference to those widespread species that are seldom mentioned in the Site Accounts. The Species Index is cross-referenced with the Site Accounts to help you to track down those species that are of particular interest to you. For example, to see White-headed Ducks you can look up that species in the Species Index and find a reference to all those sites where they regularly occur. The Site Index will help you locate particular Site Accounts.

In many ways, southern Spain is one big birding site. Certainly you do not have to travel far to find a great deal of interest, especially if the Mediterranean region is new to you. A striking feature is the sheer numbers of larks, finches, sparrows, buntings and other common birds in the countryside, a welcome contrast with some more northern countries where farmland especially is impoverished birdwise. Birds of prey are everywhere: Montagu's Harriers quarter the fields and soaring Griffon Vultures and other large raptors are omnipresent. The towns and cities offer hordes of screaming swifts in spring and summer, which share the rooftops, especially in the west, with White Storks and Lesser Kestrels. Still, there is a lot more to see: in the woodlands, forests, coasts, wetlands and mountains. This book provides a dossier of the principal sites of the region, selecting those that are readily accessible and can be relied upon to provide the main regional specialities. Many less well-known areas that often repay closer investigation are also included. Southern Spain is both fun and rewarding to explore and there remains a great deal of scope for birding in little-known corners of both Andalucía and Extremadura.

PROVINCE CHAPTERS

Andalucía and Extremadura are two of the autonomous regions of the Spanish state. Each region has its own administration and is subdivided into a number of provinces. The Site Accounts are grouped by provinces into ten chapters. Each province chapter has an introductory section giving, in turn, a brief description of the province, a list of the sites with their code numbers, general information and useful addresses, and an outline of access to the province. Some suggestions on where to stay are also made in this section or, for Extremadura, in the introduction to the region. A province map shows the position of individual sites within the province, in relation to main towns and access roads, for ease of location.

The sites themselves are coded by letters and numbers. The former are the national provincial abbreviations, which are:

Andalucía (8 provinces): Huelva (H), Sevilla (SE), Cádiz (CA), Córdoba (CO), Málaga (MA), Granada (GR), Jaén (J) and Almería (AL).
Extremadura (2 provinces): Cáceres (CC) and Badajoz (BA).
Gibraltar is treated separately (pages 357–67).

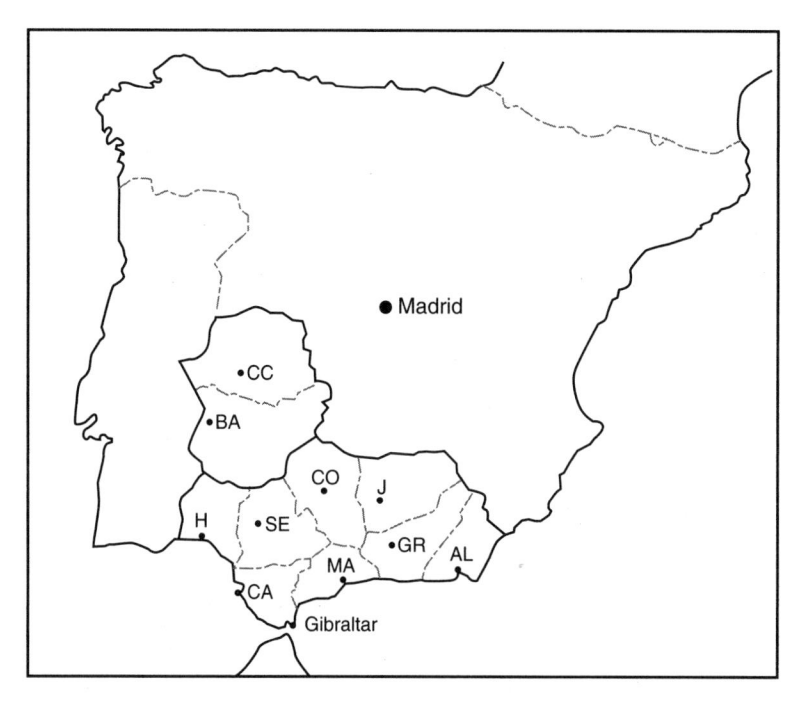

Map 1: Provinces of Andalucía and Extremadura with Gibraltar:
AL (Almería), BA (Badajoz), CA (Cádiz), CC (Cáceres), CO (Córdoba), GR (Granada), H (Huelva), J (Jaén), MA (Málaga), SE (Sevilla)

Map 1 shows the locations of the provinces of Andalucía and Extremadura in relation to the Iberian Peninsula as a whole.

SITE ACCOUNTS

Each site account gives the following information.

Site name and reference number

Status The conservation status of the site, including the official names of any protected areas where these differ from the site name. The latter are useful when seeking further information from websites.

Site map Showing the main features of the site and the access roads and paths. The map scales should be noted carefully. The smaller wetland sites have smaller-scale maps. The extensive sites covering the plains and mountains (sierras) are necessarily drawn on a larger scale.

Site description The principal structural and botanical features of the area. Any available visitor facilities are mentioned.

Most major sites have visitors' centres, which usually include comprehensive displays explaining the topography, geology, history and wildlife of the region. The centres are also invaluable sources of printed guides (in several languages) to walking trails and other points of interest. Opening hours vary, sometimes seasonally, but in general such centres (but not the sites themselves) are often closed on one day per week (often Mondays) and close for an afternoon lunchbreak (often 14.00–16.00 hours) and on Sunday afternoons.

Species   The major bird species typical of the site. Species of other wildlife are also mentioned briefly where appropriate.

Timing   The best times of year to visit and any other factors to consider.

Access   How to get to, enter and explore the site. The wheelchair symbol (♿) highlights a brief statement of the extent to which a site is accessible to those who use wheelchairs or who have difficulty in walking for extended periods.

Calendar   The main bird species are listed under all or some of the following headings: All year, Breeding season, Winter, Passage periods.

'All Year' often means 'Resident' but for some species (Red Kite, Crag Martin etc.) there is actually a great deal of turnover among the populations involved. The lists are sometimes extensive but they are not exhaustive. In particular, they usually omit such species as Barn Swallow, Stonechat and Corn Bunting, which are widespread and common throughout Spain. Common species are included where they help to define the ornithological flavour of a site.

KEY TO SITE MAPS

Symbol	Meaning	Symbol	Meaning
	town / village		marsh
	motorway or major road		reeds
	secondary road		sierra
	minor road		bridge
	track		viewpoint
	path		monastery / church
	regional boundary		mountain refuge
	railtrack		information centre
	river		hide
	lake		car park
	extent of reserve		lighthouse
	major peaks		

VISITING SOUTHERN AND WESTERN SPAIN

PLANNING YOUR VISIT

This book can be used in a variety of ways, depending on the nature of your visit or, indeed, whether you live in the region. As far as visitors are concerned, there are two main options:

Touring
The best option for a birding holiday. Most touring birders fly to a suitable airport (see below) and then travel widely in a hired car, staying at a variety of hotels. Clearly, the amount of time available will dictate what is feasible. An ideal two-week trip, with a good chance of producing 200 or more species and all or most of the regional specialities, would involve visiting western Andalucía from mid-April to early May. From here you would spend four or five days visiting sites in Cádiz, including at least one day watching raptor migration at the Strait of Gibraltar, a further four or five days based around Doñana and visiting that area, and then five or six days in Extremadura visiting the best sites of both provinces.

Key sites to include in such an itinerary are listed below. Those that can feasibly be visited on the same day are grouped together. Taken together these sites provide a good introduction to the region but there are of course many alternatives and the time of year has a big influence on when particular places are at their best.

Gibraltar: Allow one day.
Cádiz Province:
  Tarifa (CA1 & CA2), Sierra de la Plata (CA5) and La Janda (CA4).
  Laguna de Medina (CA15) and East Bank of the Río Guadalquivir (CA17).
  Grazalema (CA20).
Sevilla Province:
  Brazo del Este (SE1 – whatever you do don't miss this one!). Allow at least one day.
Huelva Province:
  Doñana (H5). Allow at least two days.
  Odiel (H4).
  From most likely starting points, allow half a day to drive from western Andalucía to Extremadura, or vice versa.
Badajoz Province:
  La Serena (BA9).
Cáceres Province:
  Cáceres–Trujillo steppes (CC3), Arrocampo (CC9), Sierra Brava (CC10) and Alcollarín (CC12) reservoirs.
  Guadiana Valley ricefields (CC11).
  Monfragüe (CC6) (also not to be missed). Allow at least one day.

Centre-based visits

This is the option for those who have a family in tow or who simply don't want to drive very far. Most typical family holidays will be based on the coastal resorts but many of these are well placed for visiting a wide range of sites.

Almería can be very interesting and Roquetas de Mar makes a very good centre, featured by many package tours. From Roquetas you are within three hours of all the Almería sites as well as the eastern end of the Sierra Nevada (Granada). The centre is accordingly well placed for seeing waders, seabirds and steppe species, including the 'desert' specialities such as Dupont's Lark and the Trumpeter Finch.

The section of the Costa del Sol between Málaga and Almería, for example, the resort of Nerja, is not recommended as a base, although it provides good access to the Sierra Nevada. Wetlands are particularly lacking here.

The western Costa del Sol, from Torremolinos south to Gibraltar, is much better. Even from Torremolinos you are within a two-hour motorway drive of parts of Cádiz, such as Tarifa (CA1) and La Janda (CA4), the southern Córdoba lakes and most of the Málaga sites. Inland resorts to consider in this sector are Ronda, Benaoján, Grazalema and Jimena, all of which are good centres for walking holidays.

Gibraltar itself is an excellent base and provides easy access to all the sites in Cádiz province and most of those in Málaga. The border can be an obstacle since queues may build up there during busy periods. However, leaving Gibraltar in the mornings and returning in the evenings is usually straightforward: the commuters are heading in the opposite direction then. Gibraltar is particularly good for raptor and seabird watching and many visitors spend their entire holidays there. April and September are the best months for seeing a large diversity of species. However, it is unsuitable for most family holidays, particularly since the beaches are small and crowded. (Gibraltar is also the ONLY place in our entire region where you can get a decent cup of tea; be warned, the Spanish offerings in this respect are unspeakably awful – tea addicts, bring your own with you, and a travel kettle!)

The Atlantic coast of Cádiz has comparatively few hotels but has become increasingly popular. A base at Zahara de los Atunes has all the essentials for a sun, sea and sand family holiday in abundance (and no raucous night life) and is on the shore of the Strait itself, within easy striking range of all of Cádiz province and Sevilla as well. The airports at Gibraltar, Jerez and Sevilla provide access to the area.

The Huelva coast has more hotels and apartments and naturally the Doñana area is an obvious base for a birding holiday. Matalascañas offers the usual seaside amenities but is also an excellent centre for touring the area. Fly to Sevilla or to Faro in adjacent Portugal.

It is obviously possible to establish a base-camp at any of the inland localities but none of these would be ideal for general family holidays. A single-base visit to Extremadura could be organised around a hotel in Mérida, Cáceres or, best of all, Trujillo. You would need to fly to Madrid or Sevilla first.

Where to stay

Finding somewhere to stay in Andalucía and Extremadura has never been easier, thanks to the well known internet booking services. Just search for 'Hotels in...' and plenty of options appear. There are hotels in all cities and towns, and in many villages, as well as along all the main trunk roads. Apartments booked through Airbnb or letting agencies are another good option, especially if you are staying for more than a week or two. It is simply a matter of personal preferences and budgets. Coastal locations are particularly well served but there are ample tourist facilities serving popular tourist sites inland, such as the Sierra Nevada and Monfragüe National Parks, Cazorla and the Sierra Morena.

Access by birders with disabilities

Very few of the sites are conveniently accessible by public transport so the advice provided assumes that birders with disabilities will have the use of a car. The information on 'Access' in the Site Accounts should allow those who have limited mobility to gauge the suitability or otherwise of particular sites. The wheelchair symbol (♿) highlights a brief explanation of the extent to which a site is accessible to those who use wheelchairs or who have difficulty in walking for extended periods. Hides and other watchpoints usually have parking fairly close by. Modern visitors' centres are generally designed with the needs of all visitors in mind and so are many of the hides, and the situation is improving. In particular, universal access trails are a feature at an increasing number of visitors' centres and these allow all birders, including wheelchair users, to visit the habitats there.

Many sites in open country, for example in the Extremaduran steppelands, involve scanning large areas from quiet roadsides; such places should be accessible to all. It is only really the rougher and steeper trails that may prove an insurmountable hurdle to some birders, including many people who are not actually disabled! However, there are always plenty of other easy-to-reach wonderful places to visit instead.

INTERNET BIRDING IN SOUTHERN AND WESTERN SPAIN

Websites have proliferated to the extent that a search for any of the locations in this book will provide a selection of 'hits', some of which are useful for last-minute information such as the appearances of rarities. The majority of sites and reserves have their own websites, generally giving details of access, species lists, opening hours of visitors' centres and maps. It is a good idea to Google those sites that you are likely to visit. Most are in Spanish but the major ones often feature an English version. The following general websites are often useful.

Rare birds in Spain (www.rarebirdspain.net/home.htm). Essential reading for news and often excellent photographs of recent rarities from throughout Spain. In English and updated weekly, usually on Mondays. One to check before visiting the region to ensure that you don't miss anything special. The site also offers, under 'Files' in the right-hand column, the latest version of the Spanish national bird list, including both Spanish and English names, as well as English and Spanish language versions of the form for submitting descriptions of rarities. Former rarities committee member Ricard Gutiérrez maintains this site.

Sociedad Española de Ornitología/SEO/Birdlife (www.seo.org). The website of the Spanish Ornithological Society provides a great deal of information about the Society and its activities as well as links to kindred organisations and to its own local groups. Some entries are bilingual (Spanish/English). The postal address of the Society is SEO/Birdlife, C/Melquiades Biencinto 34, 28053, Madrid. (Email: *seo@seo.org*). Particularly useful pages on this site, under *Quiénes somos* (Who we are), include:

'**Comité de Rarezas**'; **the Rarities Committee site** (https://www.seo.org/comite-de-rarezas/): including the proforma 'rellenar ficha' (at the bottom of the title page) for submitting details of rare birds. Such reports are best submitted online to: rarezas@seo.org. The English equivalent of the rarities form is acceptable. Both forms are also available on the 'Rare birds in Spain' website.

Grupos Locales; SEO Local Groups
Click on the map for further details of local groups in those regions (most of them) where they exist. Some groups have their own websites and these are an excellent source of local information.

ADENEX (www.adenex.org). Website of La Asociación para la Defensa de la Naturaleza y los Recursos de Extremadura. Extensive and informative.

Aves de Extremadura (*http://aves-extremadura.blogspot.com/*). Blog of Javier Prieta. Features summaries of published information relevant to the region.

Gibraltar Ornithological and Natural History Society (*www.gonhs.org*). Information on birds and other wildlife of Gibraltar, the Gibraltar Bird Report and other publications, and the Society's many activities. Bird News is updated regularly. Includes details of how to book accommodation in the Upper Rock Nature Reserve and bird observatory.

Colectivo Ornitológico Cigüeña Negra (COCN) (*http://cocn.tarifainfo.com/central.html*). Website of local group mainly studying migration at the Strait of Gibraltar. Includes image gallery and recent records. In Spanish.

Fundación Migres (*http://www.fundacionmigres.org/*). Group funded by the Andalucían Environment Agency to monitor raptor and seabird passage at the Strait of Gibraltar. Has Spanish and English versions.

RAM (Red de observación de Aves y Mamíferos marinos) (*http://redavesmarinas.blogspot.co.uk/p/whats.html*). Pan-Iberian seabird and marine mammal monitoring network. Organises coordinated watches at key points all around the Iberian Peninsula. Bilingual. Includes data from the Andalucían coast, from such watchpoints as Cape Trafalgar, Tarifa, Cabo de Gata and Punta Calaburras.

Andalucía Bird Society (*http://www.andaluciabirdsociety.org/*). English-language website of society that caters principally for resident and visiting expatriates. The Society has its own publications and organises excursions.

Birding Cádiz Province (*https://birdingcadizprovince.weebly.com/*). Excellent site maintained and regularly updated by John Cantelo. Features detailed site information, news of recent observations and much more. Highly recommended if you are visiting the province.

Ventana del Visitante de los Espacios Naturales (*http://www.juntadeandalucia.es/medioambiente/servtc5/ventana/entrar.do*). Official website of the protected spaces of Andalucía. Bilingual (Spanish/English). Lists all the Andalucían sites and offers downloadable site maps, visitors' guides and visitors' centre opening timetables.

MAPS

Satellite navigation aids are extremely useful, especially for finding out where you are in the absence of any signposting and of course for guiding you to desired locations. However, the site maps are designed to guide you to and around the various sites. You may need to inspect the header map, at the start of each province chapter, ideally in conjunction with a published road map, to establish the exact location of the sites covered. A large-scale road map (Firestone, Michelin or equivalents) is essential to give an overview of your journey. Best of all is the annual 1:300,000 (1cm = 3km) official road atlas – Mapa Oficial de Carreteras – published by the Spanish Ministry of Information. The latest version (see the Ministry of Information website: www.fomento.es) can be purchased from any bookshop (Librería), motorway service station or large newsagent anywhere in Spain. It costs around €25, including an interactive DVD. The atlas maps are excellent and up-to-date and there are legends in English and French. The atlas also includes street maps of all the provincial capitals.

WHEN TO GO

All times of year are of interest but there are important seasonal differences.

Spring (March–May) is generally best for seeing a wide range of species. Visits in early spring (March–April) allow you to find many of the wintering species as well as some of the arriving migrants. The attractions of springtime include the birdsong, which can be spectacular, and also makes it much easier to find passerines, the displays of raptors and bustards, and the splendid displays of wild flowers.

Summer (June–August) has the disadvantages of high temperatures, dried-up lakes and large numbers of tourists along the coasts but there is still plenty to see at most sites.

Autumn (September–November) is a pleasant season, especially once the rains arrive. The rains may start any time after mid-September but this varies considerably from year to year. Persistent heavy rain is not usually expected until November. There are plenty of birds, including migrants, and many winter visitors arrive from October onwards.

Winter (December–February) has the attraction of the wintering species, including great numbers of Cranes, raptors, waterfowl and waders, although summer visitors are naturally then absent.

Lesser Kestrels

The amount of daylight available varies significantly according to the season and is of obvious importance. Spain and Gibraltar run on Central European Time: GMT plus 1 hour in winter and GMT plus 2 hours in summer. In summer, birding is possible from about 06.00–22.00 hours local time, although the heat of the day is best avoided. In winter the days are noticeably longer than in northern Europe; birding is possible from about 08.00–18.00 hours local time at the solstice.

WHAT TO WEAR

We realise that birders are not trend-setters in the world of sartorial elegance. What you wear is largely up to you but sombre colours are an obvious good idea. Newcomers to the region may need to beware assuming it is always sweltering in southern Spain. It often is but, then again, it is often literally freezing cold.

In summer you require light clothing, although something warmer and windproof is useful when seawatching or when visiting high mountains. At other times it is necessary to pack some warm clothing, especially when visiting inland and mountain sites. Winter can be very cold, with frost and snow in many areas away from the coast. Hence, waterproof and windproof layers will be invaluable, not least because the cooler times of year are often very wet as well. In general, you can rely on hot weather anywhere between June and September but you should be prepared for something cooler at other times of year. Sunglasses are essential to counter the intense glare of the Mediterranean light and are useful year-round, especially when driving. They are also extremely helpful for detecting raptors against bright clouds.

GETTING THERE

By air
Many visitors travel by air, either to take advantage of one of the many relatively inexpensive one-centre package holidays or as a preliminary to touring the country in a hire car. Package holidays are mainly centred on Málaga (for

the Costa del Sol) and Almería. Flights by major carriers and low-cost airlines are also available to these destinations as well as to Sevilla, Jerez, Granada and Gibraltar. Faro airport, in the Algarve, Portugal, also offers good access to the southwestern Spanish provinces. Madrid is a handy starting point for a visit to Extremadura, usually well under four hours' drive from Madrid airport. Flights to Madrid originate from many parts of Europe, including all the main British airports. Sevilla (Seville) airport also has excellent motorway access to Extremadura and western Andalucía and it benefits from much simpler and quieter traffic links than Madrid airport.

By car
A car is not absolutely essential for enjoying southern Spain; you can spend enjoyable visits walking in the Serranía de Ronda or just sitting and watching migration at Gibraltar, for example. It is, however, impossible to visit many of the sites in this book without using a vehicle and you would be severely limited in your options without one.

Flying out and hiring a car is probably the best strategy for the short-term visitor. Car hire is available at all the airports and resorts but it is advisable and usually cheaper to make arrangements beforehand through a travel agency or your airline. Unlimited-mileage rental is the norm and is essential; you will probably cover many hundreds of miles in a week or so. Hiring a car avoids the cost of travelling to Spain by road and saves your own vehicle from the

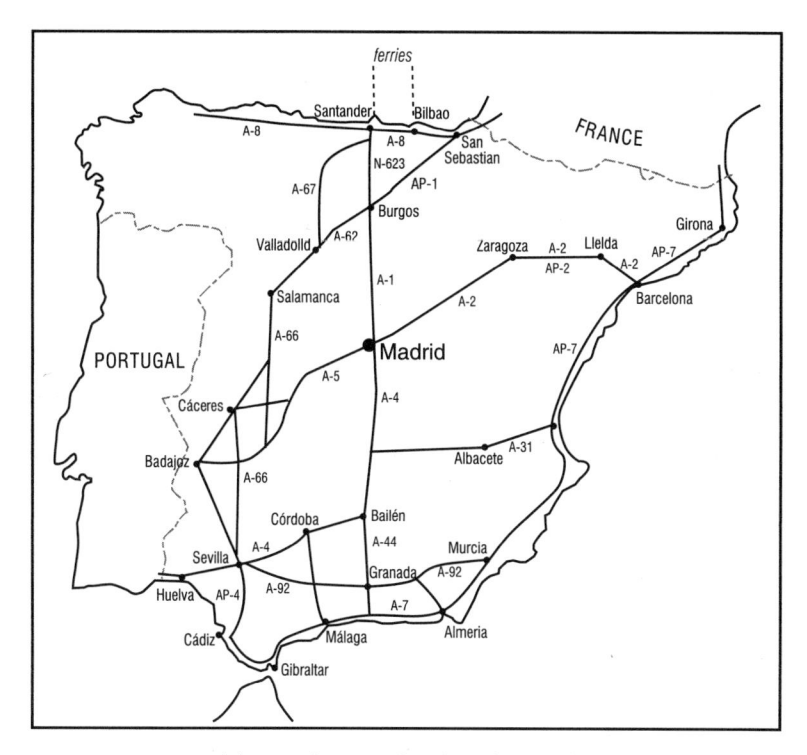

Map 2: Routes to Andalucía and Extremadura from the rest of Spain

undoubted wear and tear of a birding holiday. British visitors will also avoid the inconvenience of driving a right-hand drive car on the Continent.

If you do come in your own vehicle, several choices are on offer. Travellers from Britain will find that the easiest option is to take a ferry from Plymouth to Santander or from Portsmouth to Bilbao or Santander. Brittany Ferries was the sole operator in 2018. These crossings take up to 30 hours but there are excellent opportunities for seeing seabirds and cetaceans during the journey, especially during the summer and autumn. Otherwise you can drive down through France, taking in the Pyrenees on the way if desired. Map 2 provides a route guide for access to the region.

For travellers from the north coast, the quickest way to reach Extremadura is via Madrid, taking the N-623 and A-1 from Santander or the AP-68 and A-1 from Bilbao. Take the A-5 Autovía de Extremadura from Madrid to both Cáceres and Badajoz provinces. For direct access to Andalucía from Madrid take the A-4 south as far as Córdoba and then Sevilla, or turn off at Bailén on to the A-44 for Málaga and the eastern provinces. Travellers following the Mediterranean coast from France will usually reach Andalucía on the AP-7/A-7, which provides good access to all the coastal provinces, following the shoreline as far west as Cádiz.

DRIVING

Roads

Cities apart, the area offers pleasant and varied driving on often traffic-free roads. The roads themselves are often excellent. Nowadays, even most minor roads have been very much improved, often with the assistance of extravagant European Community funding. However, when travelling off the beaten track you should be aware that such roads are often not at all good and driving on the dirt tracks mentioned in many Site Accounts is an acquired art. Common sense should suffice but remember that unsurfaced roads are gritty and slippery, especially when wet. They are best avoided altogether in poor weather unless you have an all-terrain vehicle. All the sites mentioned in this book are safely accessible, given reasonable care, but naturally no liability is accepted for any mishaps that may occur.

Road numbers

Changing the road numbers seems to be an ongoing pastime for the relevant authorities in Spain so pre-2018 maps are now out of date in this respect. The site maps are faithful to the official road atlas (Mapa Oficial de Carreteras), 2018 edition, but some numbers are liable to change again. Many major roads have two or three numbers, often shown together, e.g. E-15/A-7/N-340. In particular, a road often carries an E- number under the EU scheme alongside a national designation, e.g. E-902/N-323. *All the 'E-' numbers are ignored for the sake of simplicity.* 'N-' roads (Carreteras Nacionales) are major trunk roads, often dual carriageways. 'A-' roads are usually (but not always!) motorways (Autovías). 'AP' roads are toll motorways (Autopistas de Peáje), the best choice where available for getting from A to B since they are largely avoided by lorries and carry relatively light traffic; the Mediterranean coastal motorway is an unfortunate exception.

Kilometre posts
Irrespective of the vagaries resulting from official tampering with road numbers, main roads (N- & A- roads generally) benefit from marker posts. These are green and indicate kilometres from the point of origin chosen for the numbering (and so read 'backwards' if you are going the other way). They are separated by white unmarked 250m posts or by marked (1, 2, 3, etc.) 100m posts. This system is very helpful for indicating locations and turn-offs with some precision. References in the Site Accounts to km-'X' accordingly refer to the roadside marker posts – or to motorway junctions, which are indicated by kilometre and not by sequential numbers.

Traffic densities
Traffic is refreshingly light by the sorry standards of northern Europe and southeast England in particular. Nevertheless, cities and their outskirts are often busy at peak periods and trunk roads may become painfully congested at holiday times, notably during the four days (Thursday–Sunday) of the Easter weekend, at Christmas, at New Year and at weekends during the summer months (July–September). The holiday travel periods, when Spaniards flock to and from the coasts, are also best avoided: they comprise the middle of July and the beginning and end of August especially. Traffic problems occur along the main trunk roads; country roads are not greatly affected but congestion occurs around particular towns and villages during their local festivals (fiestas). The 'mother of all fiestas' is probably that of El Rocío in Doñana (H5); over a million people descend on the local church in a massive outburst of religious fervour and festivities. This occurs on the seventh weekend after Easter, at Pentecost (Whitsun). Birding then is impossible.

Signs
The following signs are likely to be noticed during a birding holiday.

Entrada Prohibida/Paso Prohibido	No Entry
Incendio Prohibido	No Fires
Basura Prohibida	No dumping of rubbish
Privado/Particular	Private
Camino Privado/Particular	Private Road (i.e. keep out)
Coto Privado de Caza	Hunting Rights Reserved; usually also indicated by little rectangular signs, diagonally split black and white
Ganado Bravo	Fighting Cattle (see below)

Protected areas generally have green signs headed CMA or AMA (Consejería or Agencia de Medio Ambiente – Environment Agency) and giving the status of the site, e.g. Reserva Natural.

NATURAL HAZARDS

The Spanish countryside is a great deal wilder than that of northern Europe and certain hazards need to be kept in mind.

Bulls: Southwest Andalucía and parts of Extremadura are the land of the fighting bull. These beasts are big (half a ton or more), remarkably rapid for their size, and boast a set and spread of horns equipped for dealing with the unexpected (you). A set of these in the wrong place could really ruin your day and would be the ultimate catharsis. Fighting cattle are not necessarily black. The cows of the breed are every bit as nasty as the bulls, particularly if they have calves at heel. Therefore, a strong word of warning: these animals are killers and they will kill you if they get hold of you. Never enter fenced fields containing cattle, especially if there are 'GANADO BRAVO' signs. Quite a few of the local varieties of 'ordinary' cattle also have impressive horns and should also be treated with respect, again especially if they are with their calves.

Mosquitoes: These are present in many areas throughout the whole year and can be a nuisance. Fortunately malaria is not a hazard in Spain. If you are susceptible to insect bites you would be well advised to come equipped with your favourite repellent and anti-histamine cream. Anti-mosquito 'plugs', which fit into an electrical socket and burn a repellent tablet, are also helpful.

Dogs: The sheep flocks in Extremadura are accompanied by equally woolly, large, white sheep dogs, which seem to be particularly unfriendly. Elsewhere the numerous strays are generally too emaciated to pose any direct threat but they are a definite menace on the roads. They show a total lack of road sense and have a suicidal knack of turning up in the middle of the carriageway even on the busiest stretches. Drivers beware! Main roads are littered with dog corpses in various stages of mummification but there always seem to be some live ones left.

Sunburn: Sunny Spain often lives up to its name, even in mid-winter. Always wear a hat and use a good sunblock to prevent sunburn or worse. Remember that the cooling effect of coastal sea breezes may well disguise the onset of sunburn; the painful reckoning will come later. Sunglasses are a boon when driving, in snowy conditions, when birding near water and, especially, when scanning high clouds for raptors.

SECURITY

Tourist areas worldwide are a happy hunting ground for thieves but the area is no worse in this respect than most other similar regions. The cities and coastal resorts are the higher-risk areas, the countryside usually being quite safe, but common sense precautions to guard your belongings are always advisable. Pickpockets are rife in towns, especially those on the tourist beat, and theft from cars can be a hazard in built-up areas. The cities of Sevilla and Córdoba are notorious blackspots in this respect. The obvious rule is never to leave valuables unattended in vehicles, especially in full view.

LANDSCAPE AND CLIMATE

Andalucía and Extremadura together include a broad range of terrain, habitats and climatic regimes, allowing them to accommodate an impressive range of bird species, including some not found elsewhere in Europe. Habitat characteristics are an important part of the site descriptions but general landscape features are summarised here.

LANDSCAPE

The Andalucían mountains
Andalucía is dominated by the mountains of the Betic Cordillera, running northeast from the Strait to and beyond Almería. They include the Sierra Nevada, whose principal peak (Mulhacén, 3,478m) is the loftiest in Iberia. In the west many of these ranges are heavily wooded, with extensive forests of Iberian Holm Oak, Cork Oak and pine. The harvesting of cork is of great importance both economically and from the point of view of conservation. The trunks are stripped every seven years or so, revealing the bright red inner bark. Between times the forests are largely undisturbed and shelter a thriving community of passerines, raptors and other wildlife. The drier eastern ranges are relatively barren, as are the highest tops. The many cliffs and the gorges along water courses provide secure nest sites for raptors and others. To the north of the Guadalquivir, the older and less rugged mountains of the Sierra Morena, also extensively wooded, separate Andalucía from Extremadura and Castilla-La Mancha.

The Atlantic coastlands
The coast running from Gibraltar westwards to the Portuguese Algarve is the Costa de la Luz (the Light Coast – and the ambient brilliance really is special). The Atlantic beaches are broad and sandy, often with a backing of large and sometimes massive dunes. The tides sustain well-developed salt marshes in the river mouths and there are extensive areas of active and abandoned salt pans, important as a habitat for waders. This coast has a succession of noteworthy wetland habitats, ranging from scattered small freshwater lagoons to the large inlets and marismas of the Huelva coast, most notably those of the Coto Doñana. All these are of the greatest interest and importance ornithologically. Forests of Maritime Pines and Stone Pines are characteristic and there are areas of heathland and low scrub, including the palmetto scrub dominated by the dwarf Fan Palm. Much of the hinterland is cultivated; orange and olive groves are characteristic of the region. The Costa de la Luz remains better preserved than the Costa del Sol despite having superior beaches. Until quite recently access difficulty and the frequency of strong winds discouraged greater touristic exploitation but clusters of holiday apartments and other developments have proliferated in many areas since the 1990s.

The Mediterranean coastlands
The Costa del Sol (Sun Coast) runs from Gibraltar eastwards to Cabo de Gata (Almería), where you turn the corner into the southern end of the Costa Blanca. The coastal strip is familiar as a notorious tourist trap and there are certainly lengthy stretches of extensive development ranging from the tasteful to the tacky. Until the 1960s the beaches were largely inhabited by Kentish Plovers but these have had to give way to the hordes of visitors and new residents who have colonised the area.

The mountains sweep down very close to the coast but there are stretches where alluvial deposits in the river mouths have produced the fertile and intensively cultivated, if narrow, coastal plain. The central portion of the Costa del Sol lacks the coastal plain and large beaches. Here the mountains reach the sea as high cliffs.

Conditions become progressively more arid as one moves east until in Almería there are areas of true steppe and semi-desert. Tidal activity is limited in the Mediterranean and the beaches are relatively narrow. Unlike on the Atlantic shore, wetland habitats are relatively few but there are some interesting small estuaries and major expanses of salt pans.

The river valleys
Three major rivers, the Tajo (Tagus), Guadiana and Guadalquivir cross the area. The Guadalquivir valley is the main feature of the plain of Andalucía, the broad fertile lowlands that flank the river from the Atlantic eastwards across the provinces of Sevilla, Córdoba and Jaén. Water is at a premium in Spain and the valleys are intensively cultivated. Irrigation schemes extend the influence of the rivers across wider regions and natural habitats are limited. Nevertheless, dams have created new wetlands, especially in Extremadura, and a number of important freshwater lagoons also occur. The region has seen a great expansion of rice cultivation in recent years and enormous areas of rice paddies are now a feature of the lower Guadalquivir valley in Sevilla and the Guadiana valley in Extremadura, and more locally elsewhere. The ricefields have provided valuable habitat for aquatic birds in areas where wetlands were previously limited or absent.

Minor watercourses abound although many of these, especially in eastern Andalucía, are seasonal and completely dry for months on end. Heavy rains in the mountains, including summer thunderstorms, can produce unexpected and sometimes dangerous flash flooding in even the driest-looking stream beds. The watercourses often hold concentrations of breeding birds, notably colonies of egrets and Bee-eaters and numbers of Nightingales, Cetti's Warblers, Olivaceous Warblers and other passerines. It is often rewarding to stop at bridges to look for aquatic birds and those attracted by the riparian vegetation. The bridges themselves often harbour nesting birds, notably Red-rumped Swallows, White-rumped Swifts, Crag Martins and Rock Sparrows.

The Extremaduran plateau
Extremadura is the western part of the central Spanish plateau. Much of the area is gently undulating country with more mountainous areas to the east (Sierras de Guadalupe) and to the north (Sierra de Gredos, mainly outside the region). Cultivation of the Iberian Holm Oak or encina is characteristic. The trees are a source of animal fodder, the acorns especially being a staple of the

numerous pig herds. The grazing woodlands of well-spaced encinas and Cork Oaks, with their limited understorey, comprise a distinctive, savanna-like habitat, the dehesa, which covers vast areas. Some dehesas are interplanted with cereal crops that often harbour rodents, which in turn attract Black-winged Kites.

The large expanses of open country are steppe-like, with huge areas of rough pasture interspersed with cereal fields. The remaining natural vegetation is, strictly speaking, a garrigue or low Mediterranean scrub. True steppe, such as is found in Almería and other areas of minimal rainfall (below 300mm annually), is lacking in Extremadura but the plains still support a thriving community of bustards, sandgrouse, larks and other steppic birds.

CLIMATE

Southern Spain has a Mediterranean climate with hot, dry summers and mild, wet winters, but there are important variations in temperature and rainfall across the region. In addition, the rains themselves are not always reliable and some recent years have seen prolonged droughts (sequías) with only a fraction of the normal rainfall and serious consequences for wetland habitats especially (not to mention for crops and drinking water supplies). The rainfall regime can be extremely variable. For example, 2017 was a severe drought year, especially during the autumn and early winter, when reservoir levels fell to exceptionally low levels. Nevertheless, the first half of 2018 was extraordinarily cool and wet, with record rains in March, leading to optimal and in some cases excessive water levels at wetlands.

The Atlantic coast has the mildest temperature regime: summer daytime temperatures on the Huelva and Cádiz coasts reach 28°C, cooling by a few degrees at night. The Mediterranean coastlands are several degrees warmer than the Atlantic shores in summer. In winter, all the coastal regions are extremely mild with average daytime temperatures above 10°C and a virtual absence of frost or snow. Almería is the warmest part of southern Spain in winter, with mean temperatures averaging 18°C. Inland in Andalucía temperatures are more extreme. Sevilla and Córdoba in the Guadalquivir valley are notoriously hot in summer, regularly achieving maximum temperatures over 40°C. Away from the coasts in winter, the days are cool and frosts are not uncommon, especially in eastern Andalucía.

Most rain arrives from the Atlantic and is deposited on the western Andalucían mountains. The result is the relative lushness of Huelva and Cádiz provinces and the extreme dryness of eastern Andalucía, with semi-arid conditions in southeastern Almería. Grazalema (Cádiz), in the western part of the Serranía de Ronda, has a massive annual rainfall of 2,000mm (nearly 80 inches), the highest in Spain. In contrast, Cabo de Gata (Almería), in the rainshadow of the Sierra Nevada, receives only some 200mm (under 8 inches), the lowest annual rainfall of anywhere in Europe.

Extremadura is drier than western Andalucía and often hotter in summer, when daytime temperatures regularly exceed 35°C and occasionally rise above 40°C. Nights are some 10°C cooler. Winters in Extremadura average below 10°C during the day and nights are cold, with frequent frosts.

Temperatures drop rapidly with altitude and the higher mountain ranges are snow-capped in winter. Snow also affects intermediate ranges, such as the

Serranía de Ronda, during cold snaps. Some snow lingers in hollows on the north side of the Sierra Nevada peaks throughout the summer.

CLIMATE AND BIRDS

The climate of southern Spain (and the relatively limited application of pesticides) encourages invertebrate life and the wide range of insectivorous birds dependent on it. Spring comes early in southern Andalucía, where Swallows lay their first clutches in February. The hot, dry summers do affect birds, however, especially aquatic species that have to fledge their young and move on before their breeding lagoons dry up in the annual drought. Very dry years may mean a complete failure of nesting by the Flamingos at Fuente de Piedra, for example.

The mild winters of the coastal lowlands make them attractive to many species at that time. There are enough flying insects to support Crag Martins, and even some swallows and House Martins, in winter in western Andalucía. Large numbers of larks, pipits, thrushes, warblers and finches winter in southern Spain and there are notable populations then of Greylag Geese, other waterfowl, raptors, Cranes and waders.

Wind has an important influence on birds, especially in the coastal regions. It must be taken into account by anyone hoping to see the migrations of storks and raptors in the Gibraltar area: in general, these happen at the downwind end of the Strait. Broadly speaking, westerly winds (vientos de poniente) dominate in winter and easterlies (vientos de levante) in summer. Westerly and southwesterly gales bring seabirds onshore along the Huelva coast, especially in winter and early spring. They also force seabirds into the Mediterranean through the Strait. The easterly winds blow down the Mediterranean coast usually as a result of anticyclonic weather over Iberia. These easterlies get stronger as they approach the Strait where they can blow with some ferocity (reaching force 10+) for days on end. Levanter conditions during migration periods often cause passerines and other migrants to descend in droves along the coasts. These falls of migrants are especially likely if easterly winds are accompanied by rain or thunderstorms.

Flamingos & Garganey

BIRDS AND OTHER WILDLIFE OF SOUTHERN SPAIN

BIRDS

Southern Spain offers many special bird species. It is a key place – and in some cases the only place – in Europe in which to find Marbled and White-headed Ducks, Barbary Partridges, Bald Ibises (introduced), Black-winged Kites, Rüppell's Vultures, Spanish Imperial Eagles, Purple Swamphens, Red-knobbed Coots, Lesser Crested Terns, Black-bellied Sandgrouse, Red-necked Nightjars, White-rumped and Little Swifts, Iberian Green Woodpeckers, Black Wheatears, Iberian Chiffchaffs, Iberian Azure-winged Magpies, Spotless Starlings, Common Waxbills, Red Avadavats and Trumpeter Finches. In addition, the region provides ample opportunities to see Squacco Herons, Black Storks, Glossy Ibises, Black Vultures, Little and Great Bustards, Slender-billed and Audouin's Gulls, Gull-billed Terns, Pin-tailed Sandgrouse, Great Spotted Cuckoos; Dupont's, Lesser Short-toed and Thekla Larks, Rufous-tailed Scrub-robins, Olivaceous Warblers and Southern Grey Shrikes, all of which have a restricted range in other parts of Europe.

The rich bird communities of the region result from the combination of the strategic location and extensive favourable habitats. Summary information on all the species that occur is in the Status List (pages 368–98). A much more detailed account of the avifauna is available in *The Birds of the Iberian Peninsula* (see Further Reading). Many species can be seen all year round but there is a great deal of movement affecting the populations of some of these nominally resident birds. Waterbirds, such as Greater Flamingos and White-headed Ducks, move around as and when their regular haunts dry up. Montane raptors, such as Bonelli's and Golden Eagles, often descend to the lowlands in winter. Steppe birds, such as Little Bustards and Calandra Larks, are among a range of species that perform often complex movements within the Iberian Peninsula. There are some species as diverse as Common Buzzards, Lapwings and Blackcaps whose breeding populations are greatly increased by individuals from northern Europe during migration times and in winter. In addition, there are a number that make the usual definitions of 'resident' or 'winter visitor' more or less meaningless. Such itinerant birds typically leave southern Spain for Africa as soon as they finish nesting but return with the arrival of the winter rains. White Storks are a classic example of this: many cross the Strait to Africa in July and August but they return in numbers from November onwards. Similar movements are carried out by Lesser Kestrels, Great Spotted Cuckoos and significant numbers of Barn Swallows and House Martins. Among the conventional migrants, some return very early in the year: Black Kites, Pallid Swifts and Subalpine Warblers seen in January or February are usually new arrivals and not wintering birds. The same applies to some Egyptian Vultures, Short-toed Eagles and Booted Eagles but small numbers of these species increasingly overwinter in the region.

Resident species
Typical residents include Little and Cattle Egrets, many White Storks, Spoonbills, Greater Flamingos, White-headed Ducks, 14 raptor species among which Griffon and Black Vultures are especially obvious, Eagle Owls and numerous passerines, among which special mention can be made of Calandra and Thekla Larks, Crag Martins, Stonechats, Black Wheatears, Blue Rock Thrushes, Cetti's Warblers, Zitting Cisticolas, Crested Tits, Short-toed Treecreepers, Southern Grey Shrikes, Azure-winged Magpies, Red-billed Choughs, Ravens, Spotless Starlings, Spanish and Rock Sparrows, Serins and Cirl, Rock and Corn Buntings. The steppe-like habitats of Extremadura have a typical community that includes Great and Little Bustards, Stone-curlews and Black-bellied and Pin-tailed Sandgrouse, with Dupont's Larks and Trumpeter Finches in the true steppes of Almería.

Wintering species
In winter (loosely defined as November to February) there are notable concentrations of waterfowl, including internationally significant numbers of Greylag Geese in Doñana. Resident raptors are joined by Hen Harriers and Merlins, as well as numbers of continental Red Kites and Common Buzzards and, locally, Ospreys. Waders and gulls winter in large numbers along the coasts where there are also significant local concentrations of Common Scoters, Caspian Terns and Sandwich Terns. Extremadura is a major wintering zone for Cranes: over 150,000 are regularly present then. The conspicuous wintering passerines include very large numbers of Skylarks, Meadow Pipits, White Wagtails, Song Thrushes, Blackcaps, Common Chiffchaffs, Common Starlings and finches. Penduline Tits are also widespread in this season.

Summer residents
These breed in the region but spend the winter in Africa. Some, such as Black Kites and Barn Swallows, return as early as February. Most are back by April but a few, such as White-rumped Swifts, may not appear until May. Departures begin from July onwards and few individuals are seen after late October, although there are quite a number of records of occasional wintering by such species. The typical 'summer' visitors include Little Bitterns, Purple Herons, Black Storks, Black Kites, Short-toed Eagles, Montagu's Harriers, Booted Eagles, Hobbies, Quails, Collared Pratincoles, Little Ringed Plovers, most terns, Turtle Doves, Cuckoos, Scops Owls, nightjars, swifts, Bee-eaters, Rollers, Greater Short-toed Larks, Tawny Pipits, Rufous-tailed Scrub-robins, Nightingales, Black-eared Wheatears, Rufous-tailed Rock Thrushes, many warblers, flycatchers, Golden Orioles, Woodchat Shrikes and Ortolan Buntings.

Passage migrants
Bird migration is a prominent feature of southern Spain and it occurs in some form in every month of the year. Northern European populations wintering in Africa cross the region twice annually and seabird movements are prominent along the coasts and through the Strait. The enormous numbers of birds that occur as summer or winter visitors also contribute prominently to the migratory movements of the region. Visible migration is the highlight of the Gibraltar area but is often seen elsewhere. Wetlands, including ricefields, especially attract a steady turnover of ducks and waders on migration.

Apart from those species that are represented by breeding or wintering populations, the typical migrants include Honey-buzzards, Little Stints, Curlew Sandpipers, phalaropes, Pomarine Skuas, Lesser Crested Terns, Tree Pipits, Whinchats and Grasshopper, Sedge, Wood and Willow Warblers. In addition, there is an extensive and ever-lengthening list of vagrants (see the Status List, pages 368–98).

THE BIRDWATCHING CALENDAR FOR ANDALUCÍA AND EXTREMADURA

January Strait: northward passage of White Storks. Wintering seabirds present. Peak numbers of most wintering species generally. First Barn Swallows arrive.

February Strait: northward passage begins of Black Kites and Short-toed Eagles. Hirundines and other passerines move north. Raptors displaying over breeding sites. Cranes leave Extremadura. Waterfowl numbers decline.

March Strait: northward passage of Black Kites, Short-toed Eagles and other raptors. Migrants generally obvious. Cory's (Scopoli's) Shearwaters and Audouin's Gulls enter Mediterranean. Gannets and other seabirds leaving winter quarters. Main arrivals of Pallid Swifts.

April Strait: large variety of raptors and Black Storks on passage north. Best month for falls of passerine migrants. Common Swifts arrive. Breeding season in full swing for most species.

May Strait: concentrated arrival of Honey-buzzards and passage of Griffon Vultures. Passage of Melodious Warblers, flycatchers and other late migrants. Arrival of White-rumped Swifts, Rufous-tailed Scrub-robins and Olivaceous Warblers. Lesser Crested Terns enter Mediterranean. Breeding season in full swing generally.

June Strait: Griffon Vultures and small numbers of other raptors still arriving. Balearic Shearwaters leave Mediterranean. Breeding birds the main attraction everywhere.

July Tarifa: White Stork southward migration begins. Strait: westward movement of Balearic Shearwaters, Audouin's and Mediterranean Gulls and Gull-billed Terns. Passage waders.

August Strait: southward passage of White Storks and raptors: mainly Black Kites. Westward passage of Audouin's Gulls. Black Tern passage along coasts. Black Stork flocks at Extremaduran reservoirs.

September Strait: main passage month for raptors and Black Storks. Migrants widespread generally. First wintering seabirds arrive.

October Strait: passage of Griffon Vultures and small numbers of Red Kites, Hen Harriers and Common Buzzards, as well as finches and other passerines.

Balearic Shearwaters re-enter Mediterranean. White-rumped Swifts depart. Waterfowl numbers increase. Lesser Crested Terns leave Mediterranean.

November Strait: Griffon Vultures and small numbers of other raptors crossing south. Passerine migration continues. White Storks begin to return north. Wintering seabirds obvious. Cory's (Scopoli's) Shearwaters leave Mediterranean. Cranes arrive in Extremadura. Wintering waterfowl numbers high.

December Strait: northward passage of White Storks. Wintering seabirds present. Peak numbers of most wintering species generally.

OTHER ANIMALS

Invertebrates

The region boasts a huge diversity of terrestrial and aquatic invertebrates. They range in form from the numerous dragonflies and damselflies, to the bizarre Praying Mantises, the lumbering Rhinoceros and Dung Beetles, the alarming (though innocuous) giant black Carpenter Bee *Xylocapa violacea*, an assemblage of often-large spiders, the cacophonous cicadas, the hordes of grasshoppers and crickets and the ever-present ants and flies. Brock (2017) is a good introduction to the more conspicuous insects.

Special mention must be made of the butterflies. The whole region is a paradise for butterflies and Andalucía itself constitutes a significant reservoir for species, such as the Clouded Yellow *Colias crocea*, which migrate to northern Europe. The variety of available habitats ensures a pleasing diversity of species and there are butterflies in all places and, at least in the warmer areas, at all seasons. The list is a long one; over 130 species and including some 30 Blues alone! The many striking species that are readily noticed include the Swallowtail *Papilio machaon*, Scarce Swallowtail *Iphiclides podalirius*, Plain Tiger *Danaus chrysippus*, Monarch *Danaus plexippus*, Spanish Festoon *Zerynthia rumina*, Moroccan Orange Tip *Anthocharis belia*, Cleopatra *Gonepteryx cleopatra* and, perhaps the most impressive of all, the Two-tailed Pasha *Charaxes jasius*. The Painted Lady *Vanessa cardui* is common and there are enormous invasions in some springs, after suitable winters in Morocco, when great numbers cross the Strait and Sea of Alborán and move north.

Amphibians and Reptiles

Members of these two groups are not only interesting in their own right but also are important as food for many storks, herons and raptors. Waterside margins everywhere are home to quantities of frogs, notably the Iberian Water Frog *Rana perezi*, whose strident croaking commands attention. Two species, the green Common Tree Frog *Hyla arborea* and the Stripeless Tree Frog *Hyla meridionalis*, inhabit the taller vegetation, in which they climb adeptly. A range of newts, toads and salamanders completes the amphibian list.

Reptiles are much more widespread than amphibians, unsurprisingly given the aridity of many areas. Lizards are everywhere. They range from the almost iguana-like Ocellated Lizard *Timon lepidus*, up to 75cm long, down to the abundant wall lizards and geckoes. The latter are obvious at night when they emerge to haunt balconies and walls, especially those that are illuminated by electric lights and so attract insects. The southern provinces of the area are the haunt of the strictly protected Mediterranean Chameleon *Chamaeleo chamaeleon*. Waterside margins are commonly covered in the basking bodies of terrapins: both the European Pond Terrapin *Emys orbicularis* and the Spanish Pond Terrapin *Mauremys leprosa* occur, splashing noisily into the water when approached. Marine turtles occur offshore but are seldom seen from land.

Snakes are common, as evidenced perhaps by the numbers of Short-toed Eagles. The eagles have eight species to choose from, ranging from the familiar Mediterranean Grass Snake *Natrix astreptophora* to the venomous Lataste's Viper *Vipera latasti* and the Montpellier Snake *Malpolon monspessulanus*, the latter occasionally reaching impressive lengths of over two metres.

Mammals

The region is a stronghold of that iconic feline, the endemic Iberian Lynx *Lynx pardina*. Lynx numbers have been boosted by conservation measures including captive breeding and it is now quite possible to encounter the animals, especially in the Sierra Morena (see J3). Sightings at Doñana are also a possibility if you are lucky. In contrast, although the Wolf *Canis lupus* is doing well in northwestern Iberia it still has only a relict presence in the south. The Otter *Lutra lutra* is common in many rivers and seems to be increasing in Extremadura. Small carnivores are quite frequently picked up by car headlights as they cross roads at night: the spotted ones are the Genet *Genetta genetta* and the brown ones with a tuft on the tail are the Egyptian Mongoose *Herpestes ichneumon*. Other interesting widespread carnivores of the region include the Beech Marten *Martes foina* and the Wild Cat *Felis sylvestris*.

Herbivores are locally abundant and both Red Deer *Cervus elaphus* and Fallow Deer *Dama dama* are especially readily observed in Doñana, together with the Wild Boar *Sus scrofa*. Red Deer ranches are also a feature of the Sierra Morena. The Spanish Ibex *Capra pyrenaica* is quite easily seen in the Andalucían mountains, notably above Grazalema (CA20), in the Sierra de las Nieves (MA10), the Sierra Bermeja (MA12) and at Cazorla (J1). The Mouflon or Wild Sheep *Ovis musimon* has been introduced locally, originally at Cazorla.

The many smaller mammals include the tiniest known, the Etruscan Shrew *Suncus etruscus*; a big one is 50mm long, plus 30mm of tail, and 2g in weight. The Vagrant or Algerian Hedgehog *Atelerix algirus* occurs along the Mediterranean coast, as does the more widespread Common Hedgehog *Erinaceus europaeus*. In both cases, most specimens seen are those which have failed to make a road crossing. There are moles in areas that have soil and their hummocks can occasionally be seen. The local species is the Iberian Mole *Talpa occidentalis*.

The Rabbit *Oryctolagus cuniculus* and the Iberian Hare *Lepus granatensis* are both reasonably common, despite heavy hunting both by people and numerous wild predators. The Red Squirrel *Sciurus vulgaris* occurs locally in mountain pinewoods in eastern Andalucía, for example in Cazorla (J1). The days when it was said that a squirrel could go from the Pyrenees to Tarifa without touching the ground are long gone; the destruction by man of the original Mediterranean forests was achieved centuries ago.

Bats are common, even numerous at times in some areas. The 18 species of the area fall into three general groups: the small or medium-sized pipistrelle types, the long-eared ones and the big, high-flying ones, usually Serotines *Eptesicus serotinus*, which appear in the summer months. Two even larger bat species of the region also merit mention. They are the European Free-tailed Bat *Tadarida teniotis* and the even larger Giant Noctule *Nyctalus lasiopterus* (wing span up to 46cm), the latter notorious for its habit of including small nocturnal migrating birds in its diet.

Cetaceans are well represented offshore and are a particular feature of the Strait and Bay of Gibraltar. Large pods of dolphins, chiefly Common Dolphins *Delphinus delphis* but also Striped Dolphins *Stenella coeruleoalba* and Bottlenose Dolphins *Tursiops truncatus*, are readily seen by seabird watchers from Europa Point (Gibraltar) and similar vantage points. A pod of Killer Whales *Orcinus orca* frequents the northwestern waters of the Strait, attendant on the tuna fishery there, and occasionally appears elsewhere in the area. The waters of the Strait also produce regular sightings of pods of Long-finned Pilot Whales *Globicephala melas* and quite often such larger species as the Fin Whale *Balaenoptera physalus*, Sperm Whale *Physeter macrocephalus* and Humpback Whale *Megaptera novaeangliae*.

PLANTS

The Iberian Peninsula is one of the major plant-hunting regions of Europe and the southern portions include quantities of endemic species in such regions as the Sierra Nevada (GR1) and Cazorla (J1). Only the most single-minded of birders would fail to remark upon the spectacle of the spring flowers and many will want to identify at least some of the species involved. Although the main vegetational features of most sites are outlined in the site descriptions, interested visitors will require a suitable field guide. Enthusiasts will find those by Thorogood (2016) and Hall (2017) very helpful.

WATCHING SEABIRDS AND THE MIGRATION OF RAPTORS AND STORKS

General birding apart, southern Spain offers particular opportunities in two specific areas: seabird watching and seeing the impressive migration of soaring birds. The general advice here will help you to make the most of these.

WHERE TO WATCH SEABIRDS

A large part of this guide is obviously devoted to land birds but there are also some very interesting seabirds to be observed along both the Atlantic and Mediterranean coasts of Andalucía, some of which are extremely rare in other parts of Europe. A few species of what are technically seabirds also winter in significant numbers inland, including in Extremadura.

Seawatching has come into its own in Spain, and indeed in the Iberian Peninsula as a whole. In particular, the Iberian Seabird Network (RAM: *http://redavesmarinas.blogspot.co.uk/p/whats.html*) coordinates regular watches and counts from key headlands on the coast within our region, including Chipiona lighthouse, Cape Trafalgar, Tarifa island, Punta Calaburras and Cabo de Gata. You can, of course, seawatch from any headland on the coast between Huelva and Almería but some of the bays are also good for watching resting gulls and terns. The estuaries, especially those of the Odiel and Guadalquivir, exercise a strong attraction, as do the few areas of standing fresh water along the coast and also the saltpans. This section highlights the more productive sites (see Map 3); details of each are given in the relevant province chapter. More specific information on each species is given in the Status List (pages 368–98) and by de Juana & Garcia (2015).

At sea
There has been relatively little observation carried out at sea either off the Atlantic or Mediterranean shores of Andalucía, except in the western approaches to the Strait of Gibraltar and within the Strait itself. Several seabird species occur offshore far more commonly than shore-based observations suggest, especially in Atlantic waters. The more regular of these include European, Leach's and Wilson's Storm-petrels, Great and Sooty Shearwaters, all four skua species, Sabine's Gull, Kittiwake, Arctic and Roseate Terns and Puffin.

Ferry crossings between Spain and North Africa provide opportunities for making observations at sea. Anybody crossing the Strait from Gibraltar or Algeciras to Tangier or Ceuta would be well advised to keep a weather eye open, both for seabirds and for landbirds overhead. For birding purposes, the diagonal crossings of the Strait, Algeciras/Tangier, are best since the trip takes about 2½ hours; the Ceuta trips are only one hour.

Ferries also serve the Spanish north-African enclave of Melilla from Málaga and Almería. These are much longer crossings, eight hours or so by slow boat.

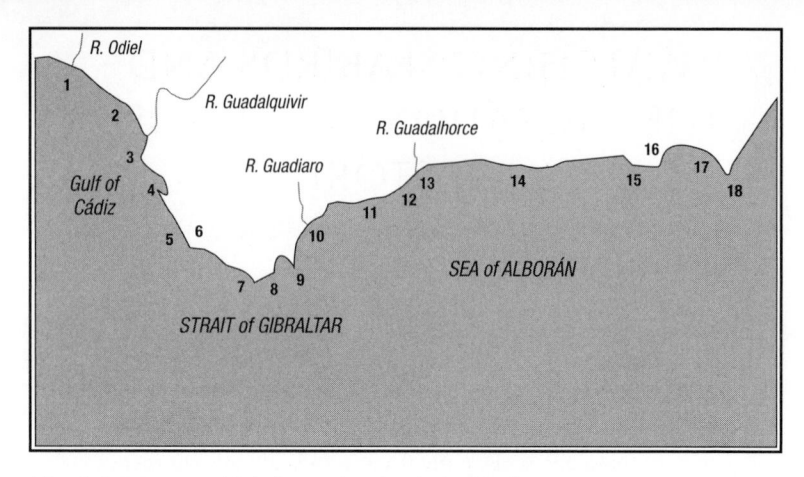

Map 3: Important seabird observation sites in Andalucía

1 Odiel estuary
2 Huelva coast
3 Chipiona lighthouse
4 Cádiz Bay
5 Cape Trafalgar
6 Barbate estuary
7 Tarifa
8 Punta Secreta
9 Europa Point (Gibraltar)
10 Sotogrande
11 Punta Calaburras
12 Benalmádena
13 Guadalhorce estuary
14 Granada coast
15 Punta Entinas
16 Salinas de Roquetas
17 Salinas de Cabo de Gata
18 Cabo de Gata

Most of the available observations from here have been made in summer, when usually the only species to be seen are a few Cory's and Balearic Shearwaters, European Storm-petrels – which are usually around fishing vessels or near cetaceans – and occasional Great Skuas, Yellow-legged Gulls and Audouin's Gulls. Winter crossings may well prove more interesting here.

Watching from ships is an acquired art. It is usually best to find an observation point towards the front of the ferry and as high up as possible. Conventional ferries are a must. Some services are provided by jetfoils, which are next to useless for seabird watching since they are too fast, too noisy, raise a lot of spray and have most of the passenger accommodation enclosed.

The ideal way to find and watch seabirds at sea is to embark on a 'pelagic' trip, a seabirding excursion on a chartered vessel. Private arrangements apart, the opportunities for such trips off the Andalucían coast are still quite limited, although bound to increase in future. Seabirding excursions in Cádiz Bay, the Huelva coast and within the Guadalquivir estuary can be arranged by Chipiona Charter (*http://www.chipionacharter.com/*). Prior contact with the company as far in advance of a proposed trip as possible is essential. Offshore seabirding in their area is likely to be most productive in summer and early autumn (July–October). It is worth noting that there are enterprises in both Tarifa and Gibraltar that take people out on whale- and dolphin-watching trips and an expedition on one of these boats might well prove fruitful seabirdwise at any time.

The Atlantic coast

The entire coast of Huelva province is most interesting, and all the region between the Portuguese border and the mouth of the Guadalquivir is worth visiting. The Odiel estuary (H4) is a notable site. In particular, the causeway (El Espigón) that runs along the west side of the river opposite Huelva city attracts thousands of gulls as well as large numbers of terns and waders, both in winter and during migration periods. The wintering community includes abundant Great Cormorants and Black-necked Grebes and small numbers of Caspian Terns and Slender-billed Gulls. Black Terns are often numerous on passage in spring and autumn. The tip of El Espigón offers an obvious vantage point for seawatching but access here has become difficult in recent years. Wintering Caspian Terns are also a feature of the Guadiana estuary in the far west (H1), which also attracts a good range of seabird species.

Further east, the beach at Doñana (H5) holds large numbers of wintering and migrant gulls and terns, including Audouin's Gulls. Offshore there are regular winter concentrations of up to a few thousand Common Scoters. The scoter flocks are worth close inspection since other sea ducks occur occasionally.

Within the Doñana region itself there are breeding populations of Gull-billed and Whiskered Terns and Slender-billed Gulls, these last being more or less resident. These species also occur commonly across the river in the Bonanza and Trebujena saltpans (CA17), which also attract migrant and wintering Caspian Terns and large mixed gull flocks.

The lighthouse at Chipiona (CA16), at the mouth of Guadalquivir, gives excellent views across the river and to seaward. Mixed gull and tern flocks are a feature here and Gannets and shearwaters, especially Balearic Shearwaters, are often present offshore. The potential for seawatching is high but avoid afternoons and evenings, when you will be facing into the sun.

From the mouth of the Guadalquivir southwards to Tarifa, the beaches are heavily used by migrant and wintering gulls, notably Yellow-legged and Lesser Black-backed Gulls and, to a lesser extent, Black-headed Gulls. There are always some wintering Mediterranean Gulls here and there is the possibility of finding scarce wintering species such as Common and Great Black-backed Gulls, particularly between Chipiona and Rota (CA16).

The Cádiz Bay salt pans and wetlands (CA11/CA12) harbour breeding colonies of Black-headed and Yellow-legged Gulls, as well as Little Terns. The bay itself is one of the very few areas in Andalucía where there is a good possibility, in winter, of seeing divers, most often Great Northern Divers. Hundreds of Great Crested Grebes regularly winter in the bay. Cape Trafalgar (CA9) has some potential for seawatching even though the Cape itself is rather low and the light is against you in the afternoons. Audouin's Gulls occur regularly. Further south, at Barbate (CA10), there is a large colony of Yellow-legged Gulls on the cliffs, and the beaches and the tidal mudflats along the river, together with the rubbish tip to the south of the town, attract enormous numbers of gulls, especially in winter.

Audouin's Gulls turn up with ever-increasing frequency near the Strait. The best site on the Atlantic coast to see them is Tarifa beach (CA1), which is a must at any time of year. This same site is also one of the likely places to see Lesser Crested Terns, which may appear in May or at any time between August and late October or early November, especially after easterly winds. The Royal Tern is also recorded here very occasionally in late summer or early autumn. The beach holds a considerable migrant and wintering population of Sandwich Terns, with some non-breeders remaining all summer. Grey Phalaropes may occur onshore in winter after westerly gales.

The Strait of Gibraltar

The northern shore of the Strait between Tarifa and Punta Secreta, formerly off-limits as a military zone, is now accessible, chiefly via the coastal footpath. The prime locations for seawatching remain the harbour wall at Tarifa (CA2) as well as the headlands of Punta Carnero and Punta Secreta (CA3). Best of all is Tarifa island, but birding access here is generally restricted to those participating in seabird counts organised by Migres (see CA2). Audouin's Gulls may be seen at Punta Secreta during most of the year but especially in late summer and autumn, when flocks roost on the coastal rocks. Lesser Crested Terns also appear here sometimes in autumn.

Gibraltar itself offers much for the seabirder. Apart from the very substantial colony of Yellow-legged Gulls (several thousand pairs) and the very few pairs of Shags, its claim to fame is the excellent seawatch site at Europa Point. Here you can obtain very close views at times of such species as Balearic, Levantine and Cory's Shearwaters. In particular, it can be very good for the observation of Audouin's Gulls from July to September (several thousand pass) and it offers the whole range of migrant and wintering seabirds, as well as the possibility of a vagrant species if you are exceptionally lucky: such unexpected species as Cape Petrel, Iceland Gull, Grey-headed Gull and Little Auk have turned up here.

The Sea of Alborán

The Alborán basin is the part of the Mediterranean off the Costa del Sol, from Cabo de Gata (Almería) in the east to the Strait of Gibraltar in the west. The Guadiaro estuary at Sotogrande (CA8) attracts a diversity of gulls and terns. Razorbills and occasional grebes also occur in winter. Yellow-legged Gulls are always present, along with some Lesser Black-backed and Audouin's Gulls during most of the year. In winter there are usually Sandwich Terns. This is another of the better places to see Lesser Crested Terns, which may appear between August and late October. Caspian Terns also occur during migration periods.

Punta Calaburras (MA17) is a popular and recommended watchpoint and a good site from which to see shearwaters, as well as Gannets, skuas, gulls and terns.

The western side of Málaga Bay, between Benalmádena and the River Guadalhorce, is a very productive area that has been extensively watched by Andrew Paterson. As a result it boasts a very fair share of the seabird rarities seen in Andalucía, including Masked and Brown Boobies, three Nearctic Gull species (Franklin's, Laughing and Ring-billed), Macaronesian (Little) Shearwater and all four skua species. Watching from the harbour wall of the Puerto Deportivo (marina) at Benalmádena is often particularly fruitful for Gannets, fly-by gulls and, at times, for close views of Balearic Shearwaters and occasional Levantine Shearwaters. The attractiveness of this area is probably a result of a combination of the change in coastal direction and the presence of fresh water at the nearby Guadalhorce estuary.

The river mouth and ponds of the Río Guadalhorce (MA8) are always interesting and often host a good variety of gulls and terns. The presence of fresh water attracts significant numbers of non-breeding Audouin's Gulls in late spring and early summer, as well as Mediterranean Gulls and all the common gull species, and sometimes small flocks of Slender-billed Gulls in May–August. Caspian and Lesser Crested Terns occur on passage, most frequently in autumn.

The Royal Tern has occurred on several occasions. Common and Little Terns appear on migration alongside large numbers of migrant marsh terns, notably Black Terns, with some Whiskered Terns in spring, and Black Terns in late summer and early autumn. Occasional White-winged Black Terns occur in autumn after prolonged easterlies.

The eastern Málaga coast from Rincón de la Victoria to Nerja attracts large numbers of Yellow-legged Gulls all year and numerous Mediterranean, Black-headed and Lesser Black-backed Gulls in winter, especially around the Vélez estuary (MA15). There appears to be little of particular interest off the Granada coast, apart from a few pairs of breeding Yellow-legged Gulls, although numbers of gulls and terns occur along the shores at the appropriate times of year. The river mouth to the west of Motril, the Suárez wetland (GR10) and the fishing port at that town, may repay some attention.

In Almería there are records of large numbers of gulls and terns at various sites, notably at Punta Entinas and the saltpans at Roquetas (AL2) and, especially, Cabo de Gata (AL1). The most interesting species at this last site is again Audouin's Gull and flocks of up to 1,000 have been recorded there. Seawatching from Cabo de Gata can be good during migration periods for Cory's and Balearic Shearwaters as well as migrant and wintering gulls and terns.

Inland sites

Several inland sites attract gulls and terns. The whole of the Doñana region (H5) and the Guadalquivir valley as far north as Sevilla offer interesting species, as do the ricefields of La Janda (CA4), the wetlands of southern Córdoba (CO1–7) and the Laguna de Fuente de Piedra (MA1). The last attracts migrant marsh terns and there are sizable breeding numbers of Black-headed Gulls and Gull-billed Terns in suitable years, with Slender-billed Gulls nesting occasionally. These sites are of particular interest during passage periods but they may also attract seabirds in winter: the occasional occurrence of large numbers of Mediterranean Gulls at the Laguna de Medina (CA15) and the Embalse de la Viñuela (Málaga) and the regular gatherings of several thousand Lesser Black-backed Gulls at Fuente de Piedra (MA1) are cases in point.

Seabirds are obviously less prominent in Extremadura but wintering Great Cormorants, Black-headed Gulls and Lesser Black-backed Gulls are locally abundant, especially at the larger reservoirs and along the course of the River Guadiana. Marsh Terns occur on passage and Gull-billed Terns breed at the Orellana (BA10) and Sierra Brava reservoirs (CC10) and at several other localities.

WATCHING RAPTOR AND STORK MIGRATION

Soaring birds migrating in large numbers are an unforgettable spectacle and one that can only be observed regularly at a limited number of localities worldwide. Flocks of raptors and storks on the move may be encountered anywhere in southern Spain but the Strait of Gibraltar is where the flypaths converge and the largest assemblages take place. The Strait offers the shortest sea crossing between Africa and western Europe, only 16km at its narrowest point. As such it is useful to those species, notably storks and broad-winged raptors, which rely on soaring flight, since thermals and updrafts are largely lacking over the sea.

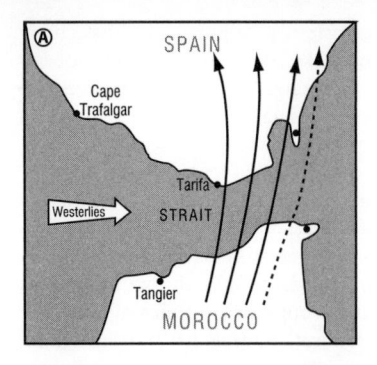

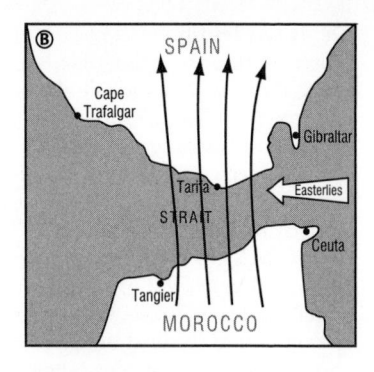

Map 4A. Routes of northbound soaring birds crossing the Strait:
a) in westerly winds, b) in easterly winds.

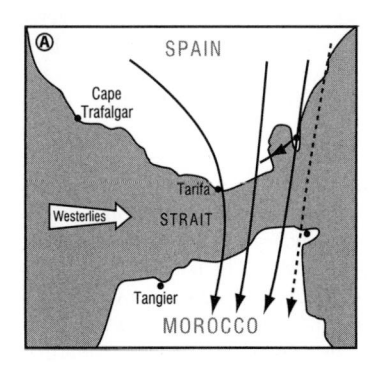

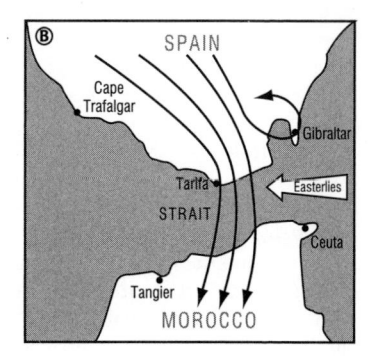

Map 4B. Routes of southbound soaring birds crossing the Strait:
a) in westerly winds, b) in easterly winds.

When to watch

The migration season at the Strait is a long one. Significant movements of northbound raptors occur from mid-February until June. Some non-breeding, immature raptors, chiefly Honey-buzzards and Short-toed Eagles, continue to arrive throughout July. The southbound raptor passage extends from late July until November. Different species are involved at different stages of both seasons (Table 2). White Storks are on the move from November to May (northbound) and July to September (southbound). Black Storks pass from February to April and again from September to late October mainly. In all, there is seldom a time when passage cannot be observed and no visit to southern Andalucía can be regarded as complete if it has not included a day or two watching the activity across the Strait. Residents in the area will need no encouragement: raptor watching is an addictive fixture that claims many hours from those fortunate enough to live close to the action.

Where to watch

Watching the migration at the Strait requires a modicum of planning and it is absolutely essential to pay due regard to the wind direction. A place that is overflown by streams of raptors on one occasion may be useless the next day, or even later the same day, if the wind changes. At the Strait the winds have generally got either an easterly or a westerly component. The rule of thumb is that movements take place at the downwind end of the Strait (Table 1).

Soaring birds are prone to lateral drift by the wind, although they may make some allowance for it. As a result, westerly winds will drift birds towards the eastern (Gibraltar) end of the Strait. Easterly winds will drift them towards the western (Cape Trafalgar) end. The effect is most marked during the northward migration, i.e. in 'spring'. Then birds will have crossed the open sea from Africa to Europe and will have been fully exposed to any lateral effects of the wind. In the southward migration, i.e. in 'autumn', birds tend to accumulate at the downwind end of the Strait and they will either cross there or move into the wind towards the shortest crossing point near Tarifa. If the winds are very strong, however, the birds will overshoot Tarifa and return inland to await better conditions. Strong easterlies between August and October regularly result in large numbers of birds flying east from Tarifa, into the wind, even as far as Gibraltar, where they mill about for some time before virtually always returning northwards again.

Maps 4A and 4B, which show the principal flightpaths in the region of the Strait, and Table 1 will help you to decide where to watch. The site descriptions for CA1, CA2, CA3 and Gibraltar give details of access to the various vantage points that need to be used. Finally, there is no substitute for experience and local advice should be sought and heeded whenever possible. The Observatorio del Estrecho at Punta Camorro and the watchpoint at Cazalla are regularly manned at migration times. The Observatory at Jews' Gate (Gibraltar) is also manned during the spring migration and the observers are an indispensable source of up-to-the-minute information. Visitors are always welcome at all these sites.

The situation is never entirely straightforward. Honey-buzzards are the most determined migrants and are least affected by the wind. They do drift but they allow themselves to do so since they are powerful fliers that can employ flapping flight for long periods over the sea and so can afford to deviate far from the regular flightpaths across the Strait. Strong winds that will ground all storks and most raptors may produce Honey-buzzard movements into and from the Costa del Sol and into and from the Cape Trafalgar/Doñana region.

White Storks are strongly attracted to the short sea crossing in the centre of

WIND	SITES	SITE CODE
Strong westerly	Gibraltar. Punta Secreta.	Gibraltar. CA3
Light/moderate westerly	Gibraltar. Punta Secreta. El Bujeo. El Algarrobo.	Gibraltar. CA3. CA2F/G
Calm	Punta Camorro. Cazalla. Alto del Cabrito. El Bujeo. El Algarrobo.	CA2C–G
Light/moderate easterly	Punta Camorro. Cazalla. Alto del Cabrito.	CA2C–E
Strong easterly	Cazalla. Los Lances.	CA2C. CA1

Table 1. Where to watch raptor and stork migration in relation to wind direction.

the Strait and tend not to cross at all if they cannot do so there. Hence White Storks are comparatively irregular at Gibraltar itself in spring, and virtually absent in autumn, even during westerlies. Black Storks, on the other hand, are less reliant upon soaring flight and occur at Gibraltar together with raptors.

During the southward passage, Booted and Short-toed Eagles especially tend to avoid Gibraltar even during westerly winds. They follow the mountains down to Tarifa and use the short crossing there. In contrast, they are among the most likely raptors to turn up at Gibraltar during strong easterlies in autumn, having aborted an attempt to cross at Tarifa and having coasted instead towards the Rock. These occasions regularly result in the impressive sight of up to several hundred eagles in view at once circling over Gibraltar. During the northward passage both species are common at Gibraltar in westerlies, after crossing the Strait.

Watching migration

Despite the numbers involved, there is not a continuous flow of raptors across the Strait, even if you happen to have found the right spot to watch from. Some points to consider are:

- Large raptors: Short-toed Eagles and larger raptors, and storks, are not on the move until well after dawn and they settle to roost well before sunset. Smaller raptors, including Black Kites and, especially, Honey-buzzards may cross early in the morning and late in the evening. The peak times for watching are 10:00–18.00 hours local times, approximately.
- Gaps in the flow of up to a couple of hours are quite usual, especially at the extremes of the season. If the gaps are any longer check that the wind direction has not changed.
- Birds that have crossed the Strait have often lost a great deal (or all) of their height and may arrive flapping low over the sea. Many nonetheless arrive at great height, especially during hot and calm weather.
- Birds leaving the Spanish coast will often soar to a great height beforehand, in order to try and glide across.
- Raptors fly higher in hot weather. An 'empty' sky may reveal hundreds of birds when scanned carefully. Honey-buzzards sometimes pass south over Gibraltar so high as to be only just visible through binoculars to watchers on the summit of the Rock and the radar there has picked up raptors at altitudes over 3,000m. Most migration is probably low enough to be visible, however.
- Persistent heavy rain will stop movements but showery, blustery weather does not deter most species.
- Counting the movements is difficult since they are usually on a broad front and a large team of observers is necessary. Counting the movement over a particular point is possible but allow the birds to pass you before you count them and don't contemplate counting soaring flocks: wait until they level off into a stream.
- It is a good idea to scan ahead. Raptors can be seen some kilometres before they arrive and it is best not to be taken by surprise. Rear-views are always frustrating.
- The gulls at Gibraltar and elsewhere provide an early warning service. An explosion of alarm calls often means that raptors have arrived.

- Sunglasses are invaluable when looking for birds against the light and especially against broken cloud, when the glare can otherwise be intolerable.
- Porter *et al.* (1981) and Forsman (1999, 2016) are the best available guides to raptor identification. The two Forsman books complement each other.

Three organisations: Fundación Migres, Colectivo Ornitológico Cigüeña Negra (COCN) and the Gibraltar Ornithological and Natural History Society (GONHS) monitor the migration of raptors and other species at the Strait. See CA2 and GIBRALTAR for further details.

What to expect

The Status List (pages 368–98) gives more details of the species involved but a short summary is provided here. Veteran watchers, who took part in the pioneering counts in the 1960s and 1970s, remark on the enormous increases since then in the numbers of both stork species and also of Sparrowhawks, Black Kites, Short-toed and Booted Eagles and Griffon Vultures. Common Buzzards and Red Kites have become very scarce, probably reflecting changes in winter quarters. Egyptian Vultures are beginning to show some recovery in numbers following recent decreases in their population. Table 2 gives the passage periods and relative abundances of the principal species using the Strait of Gibraltar flyway.

Other regular migrants: Several falcons occur regularly on passage but they do not concentrate at the Strait and only small numbers are seen there. They are the Lesser Kestrel, Common Kestrel, Hobby and Peregrine Falcon.

SPECIES	NORTHBOUND	SOUTHBOUND	NUMBERS
White Stork	Nov–May	Jul–Sep	>200,000
Black Stork	Feb–Apr	Sep–Oct	>5,000
Honey-buzzard	Late Apr–May	Aug–Sep	>100,000
Black Kite	Feb–May	Jul–Sep	>150,000
Red Kite	Feb–Apr	Sep–Nov	<100
Egyptian Vulture	Feb–May	Aug–Sep	>4,000
Griffon Vulture	Feb–Jun	Oct–Nov	>6,000
Short-toed Eagle	Feb–May	Sep–Oct	>20,000
Marsh Harrier	Feb–May	Sep–Oct	<1,000
Hen Harrier	Feb–Apr	Sep–Nov	<100
Montagu's Harrier	Apr–May	Aug–Sep	>2,000
Sparrowhawk	Feb–May	Aug–Nov	>2,000
Common Buzzard	Feb–Apr	Sep–Nov	<200
Booted Eagle	Mar–May	Sep–Oct	>20,000
Osprey	Mar–May	Sep–Oct	<500

Table 2. Principal passage periods and approximate (very) seasonal totals of the regular soaring migrants at the Gibraltar flyway.

Scarce but annual migrants: A few individuals of the following are recorded annually. Black-winged Kite, Rüppell's Vulture, Black Vulture, Pallid Harrier, Goshawk, Long-legged Buzzard, Lesser Spotted Eagle, Spanish Imperial Eagle, Bonelli's Eagle, Merlin, Eleonora's Falcon, Lanner Falcon.

Vagrants: Lammergeier, White-backed Vulture, Bateleur, Greater Spotted Eagle, Steppe Eagle, Golden Eagle, Red-footed Falcon, Saker Falcon.

HUELVA PROVINCE

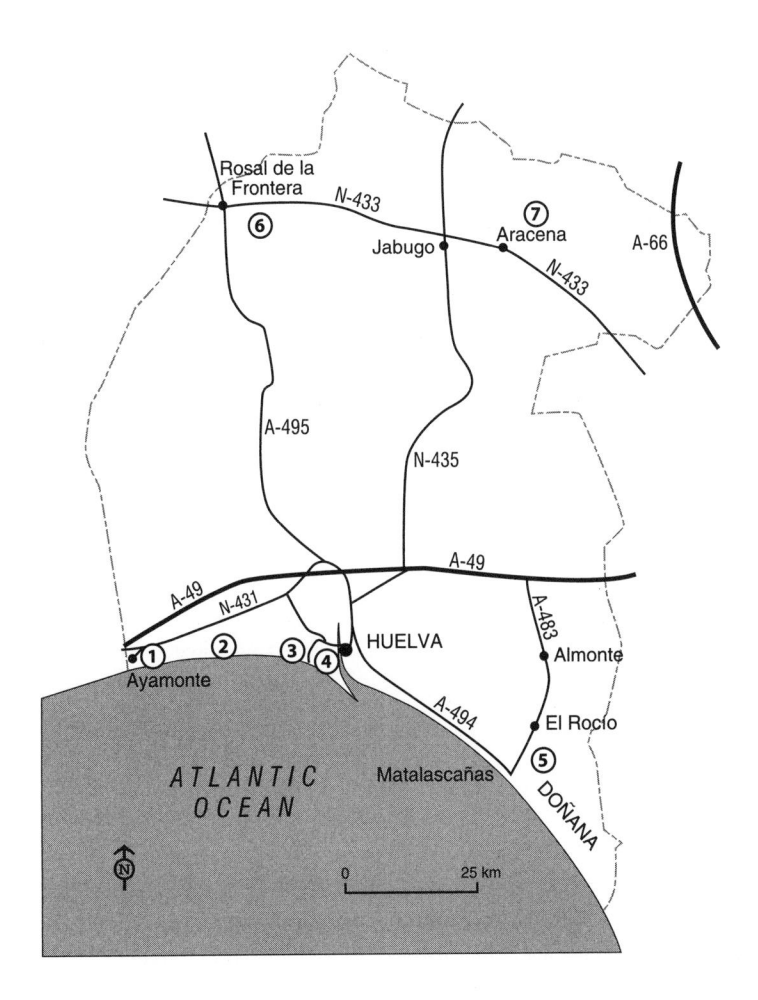

Huelva province is best known for the remarkable habitats and birds of the Coto Doñana (H5), but Doñana is only one of a series of wetland sites (H1–H4) that extend the entire length of the coastline. The province is accordingly of the greatest importance for waterbirds and waders of all kinds, and also for seabirds. The Odiel estuary (H4) is becoming increasingly popular among birders, being more compact than Doñana, more widely accessible and, in addition, offering better maritime habitats and opportunities to view seabirds. Odiel is not a substitute for Doñana but should be seen as an essential site to visit in its own right as well. Inland, Huelva offers the mountains of the western end of the Sierra Morena (H6, H7), a relatively unspoilt region of wooded hills and open country, supporting a good variety of raptors and other birds.

Sites in Huelva Province

H1 Guadiana estuary (Marismas de Isla Cristina)
H2 Piedras estuary (Marismas del Río Piedras)
H3 Laguna de El Portil
H4 Odiel estuary (Marismas del Odiel)
H5 Doñana National Park
H6 Sierra Pelada
H7 Western Sierra Morena

Getting there

Most of the sites listed are south of the A-49 Sevilla/Portugal motorway, which crosses the province east/west. Visitors from further east may travel via Sevilla using the A-381 Algeciras/Jerez motorway and the AP-4 Cádiz-Sevilla toll motorway (the AP-4 is scheduled to become toll-free from late 2019). The N-433 Sevilla–Portugal trunk road is the obvious access to the northern half of the province and connects with the A-66 motorway, linking western Andalucía with Extremadura and the north. The province is within easy reach of Sevilla and Jerez airports, and Faro airport in Algarve, Portugal.

Where to stay

Coastal Huelva has many tourist resorts and has seen ongoing proliferation of such developments as golf courses and their attendant blocks of holiday apartments and retirement villas, although much of the coastline, including the key wetlands, remains protected. The accommodation ranges from the Paradores at Ayamonte in the west and Mazagón in the east – both of them modern buildings and offering exemplary facilities and service – to a diversity of campsites and hotels. The coastal resort of Punta Umbría is a useful central location that offers a number of hotels.

The whole coastal area is within easy access of Matalascañas and El Rocío, which are the obvious bases for visitors to Doñana. Matalascañas has a range of hotels and campsites and there are more campsites just to the west at Mazagón. It is always a good idea to spend a night or two at El Rocío, whose lagoon never fails to provide an interesting spectacle. There are a good number of hotels there among which Hotel Toruño (2-star), which adjoins the lagoon, seems almost purpose built for the birders who appear to make up most of the clientele. There is a roof terrace offering panoramic views and an adjoining meeting room, ideal for small groups to gather to review their day's sightings. Pre-booking is advisable (*http://www.toruno.es/*) and you should request a first floor room with views over the lagoon.

Elsewhere in the province, there are small hotels in most of the towns. Visitors to the western Sierra Morena (H7) may find convenient accommodation in Aracena especially.

GUADIANA ESTUARY H1
(Marismas de Isla Cristina)

Status: Paraje Natural & ZEPA (2,498ha).

Site description
This is the southwesternmost corner of Andalucía, the River Guadiana marking the border with Portugal. Tidal activity has produced sand spits sheltering expanses of salt marsh and abandoned salt pans, dominated by glassworts. The xerophyte community is well developed. The low sand dunes have extensive covering vegetation, including areas of Stone Pine and Juniper woodland. Wide sandy beaches comprise the coastline and flank the principal mouth of the Guadiana. This is an excellent site in that it offers a range of well-preserved and relatively accessible habitats within a reasonably confined area, even though much of the littoral and dune habitat has been lost to the ever-expanding tourism developments.

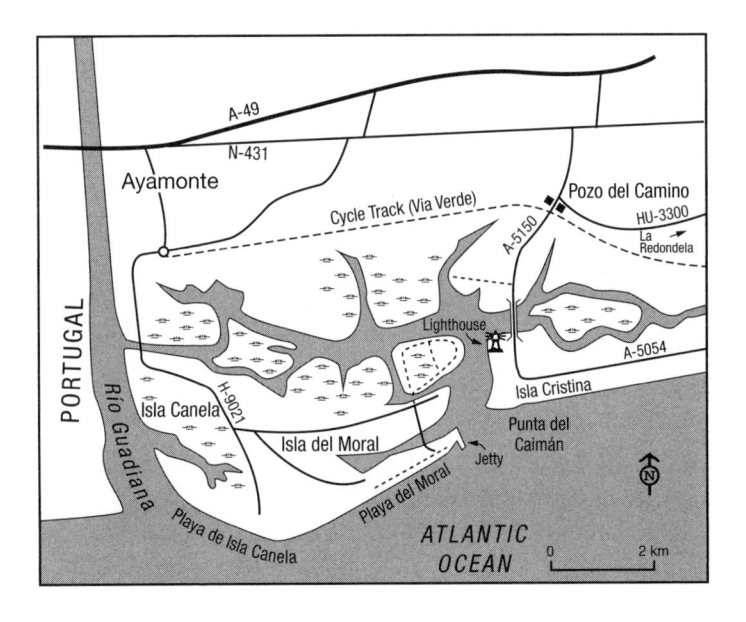

Species
The area is attractive all year round to a good variety of waders, gulls and terns. The largest numbers occur during passage periods and in winter but many species are represented even in summer. Thousands of Lesser Black-backed Gulls are present in winter, along with smaller numbers of Black-headed Gulls. As ever, the gull flocks offer the prospect of attracting rarer gull species. Caspian Terns are present in small numbers, even in summer when some non-breeders occur. Spoonbills occur in flocks of up to several hundred birds: many are visitors from the Marismas del Odiel (H4) but a breeding colony of up to 85 pairs

has been established here since the mid-1990s. Among the waders, Kentish Plovers occur in good numbers all year round; winter counts of over 300 are on record. The whole area is liable to hold significant numbers of a variety of passerine and other landbird species during passage periods.

Timing

Visits during passage periods and in winter are recommended but there are interesting birds here all year round. The sandy beaches, and the hotel complexes at Isla Cristina, Playa del Moral and Playa Canela, attract large numbers of tourists in summer when the area may get very congested. However, the salt marshes are not part of the regular tourist haunts. Visits in hot weather should be timed to avoid the heat of the day with its attendant glare and haze.

Access

A visit should include the salt marsh proper, the dune vegetation and the beaches. The coast road leading south from Ayamonte to Isla Canela gives access to the estuary and the beaches at Playa Canela. Driving eastwards the road rises near a stone tower, the Torre de Canela, affording views over the salt marsh. Continuing east towards Isla del Moral, a sandy track leads north immediately west of the village. The entrance is signposted Sendero Salinas del Duque. The track soon joins a causeway that allows exploration of the southern marshes. Entry is only possible on foot but there is limited parking available near the gate, which blocks the causeway to vehicles. The causeway follows the main creek of the salt marsh and skirts areas of dense glassworts and other xerophytes as well as giving close views over the abandoned saltpans, which attract waders.

The shore at Playa del Moral gives views across the marsh entrance. Sand banks and rocky breakwaters here attract gulls and terns especially. A stone jetty at the harbour entrance is very well placed for scanning for seabirds offshore.

The eastern side of the marsh is accessible from the Pozo del Camino/Isla Cristina road which crosses the creeks. A cycle and walking track (signposted Pista Via Verde del Litoral), following the former railway line between Pozo del Camino and Ayamonte, allows views across the salt marshes and an area of saltpans from the northern bank. The Via Verde also gives access to the eastern part of the protected area, which extends for some 2.5km beyond the mapped area, as far as the town of La Redondela. The western entrance to the Via Verde is at the roundabout at the southeast corner of Ayamonte town, where you can park.

A trail across the saltpans, signposted Salinas de Isla Cristina, is on the west side of the road just north of the bridge at Isla Cristina, at the bend in the A-5150. Parking is available at the entrance, where a boardwalk marks the start of the trail. It is also worth walking out onto the bridge from here, for views over the mudflats at low tide and to scan for gulls and terns. There is a sheltered walkway on both sides of the road.

A sand bar to seaward immediately south of Isla Cristina encloses a shallow inlet attractive at times to waders and also used by roosting gulls and terns. A rickety but safe wooden bridge crosses the inlet, giving access to the beach at Punta del Caimán. The inlet is reached by following signs to the faro (lighthouse) or playas (beaches) through Isla Cristina town. A promenade below the lighthouse and following the waterfront road is well placed for viewing waders in the inlet and the gull flocks roosting on the beach, and for scanning the main creek.

The Laguna del Prado, a freshwater lagoon with abundant marshland vegetation, is also worth visiting. The lagoon dries up in summer but when inundated it attracts wildfowl, herons and egrets and other waterfowl, including Purple Swamphens, as well as marshland passerines and migrants. Access is from La Redondela itself, from where you can walk westwards along the Via Verde from the old station platform. The Via Verde is reached from Calle de La Cerquilla, which follows the northern margin of the town and where you can park.

♿ Only the seaward sites, including especially the Isla Cristina promenade, are suitable for wheelchair access.

CALENDAR

All year: Little Egret, White Stork, Marsh Harrier, Purple Swamphen, Black-winged Stilt, Kentish Plover, Common Redshank, Yellow-legged Gull, Caspian Tern, Zitting Cisticola.	**Passage periods and winter:** Waterfowl, Spoonbill, waders: including Oystercatcher, Ringed and Grey Plovers, Sanderling, Little Stint, Dunlin, Black-tailed and Bar-tailed Godwits, Whimbrel, Curlew, Spotted Redshank, Greenshank, Green and Common Sandpipers, Turnstone, gulls and terns. Northern Gannets, Common Scoters and other seabirds offshore.
Breeding season: Montagu's Harrier, Collared Pratincole, Little Tern, Pallid Swift, Bee-eater, Hoopoe, Yellow Wagtail.	

PIEDRAS ESTUARY H2
(Marismas del Río Piedras)

Status: Paraje Natural & ZEPA 'Marismas del Río Piedras y Flecha del Rompido' (2,409ha.).

Site description
The curious sandy spit (La Flecha del Rompido) across the mouth of the Río Piedras shelters an area of salt marsh. Mudflats are exposed in the main creeks at low tide and along the landward side of La Flecha. The spit itself has exceptionally well-developed sand dune vegetation and is of considerable botanical importance. Coastal pinewoods are prominent to both east and west of the estuary.

Species
Waterfowl, waders, gulls and terns occur in some numbers, chiefly on passage and in winter. The flanking woodlands and the dune scrub hold good numbers of migrants at times.

Timing
Wader numbers are highest during passage periods and also in winter, when visits are recommended. There is considerable potential for seawatching off the

beach: the coastal dunes along the southern edge of La Flecha provide an elevated viewpoint. The beach gets busy in summer and is best avoided then.

Access

The site is reached from the N-431, driving south from Lepe (west bank) or Cartaya (east bank). On the western side, the main creek of the salt marsh is viewed from the fishing village of El Terrón. About 100m south from here a drivable if rough tarmac causeway leads across a minor creek and an area of salt flats right down to the beach (Playa El Terrón). The causeway doubles as a trail, Sendero de Nueva Umbría, and at quiet times of year it is worth walking rather than driving along it for views of waders on the salt flats. Once at the beach you can explore La Flecha on foot but it is advisable not to stray from the boardwalks, coastal track and the beach itself in the interests of conserving the fragile dune habitats.

On the eastern side there is good access to the central channel and also to extensive salt flats and coastal dune vegetation. Follow the wide signposted road 'Camino del Lancon' which leads west for 3.6km off the El Rompido/Cartaya road, just north of El Rompido. The entrance is almost directly opposite a Repsol petrol station, off a roundabout marked 'Rotunda de Lancon'. The first half of the road is good tarmac, serving a golf course, but the latter part is a sandy dirt track of indifferent quality and with some sticky patches during wet conditions. It is perfectly passable with care, nevertheless. Park at the end of the road to view the salt marsh fringes. Two footpaths (Sendero de la Turbera and Sendero de Río Piedras) are clearly signposted on the seaward side of the dirt road, giving access to the coastal pine belt, dune habitats, ephemeral freshwater pools and areas of salt pans and salt marsh. A third footpath loops through the salt marsh of the Marisma de San Miguel on the opposite (north) side of the road. All these areas are well worth exploring for waders and for migrant passerines in season.

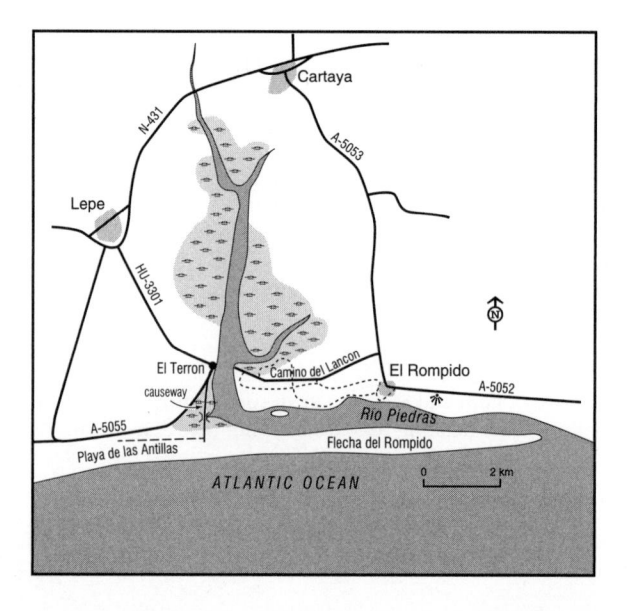

Good views over the main inlet to the marsh and La Flecha are obtainable from the village of El Rompido itself, where there are sizable mudflats. Parking for a shopping centre (Centro Comercial El Faro) is available off the roundabout where the coast road (A-5052) begins. Park here and walk down to the west end of the car park, where a boardwalk continues westwards along the riverside, giving excellent views of salt marshes to seaward and pinewoods and coastal scrub landwards. This boardwalk connects with the Río Piedras footpath mentioned above. The coast road itself rises east of El Rompido where a viewpoint (mirador) gives panoramic views of La Flecha and the mouth of the estuary.

&. All sites except for La Flecha itself are suitable for wheelchair access.

CALENDAR

All year: Marsh Harrier, Black-winged Stilt, Avocet, Little Ringed Plover, Kentish Plover, Common Redshank, Southern Grey Shrike, Dartford Warbler.

Breeding season: Spoonbill, Collared Pratincole, Little Tern, Hoopoe, Yellow Wagtail.

Passage periods and winter: Red-crested Pochard, Common Scoter, Red-breasted Merganser, Balearic Shearwater, Leach's Storm-petrel, Northern Gannet, Great Cormorant, Little Egret, Hen Harrier, Osprey, waders: including Knot, skuas, gulls: including Little Gull, terns: including Caspian Tern, Great Spotted Cuckoo, Bluethroat.

LAGUNA DE EL PORTIL   H3

Status: Reserva Natural (16ha, plus 1,300ha of peripheral protection).

Site description

A substantial freshwater lagoon, fringed with reeds and bulrushes and surrounded by a protected expanse of Stone Pine and Cork Oak woodland. It usually holds some water but levels can fall very low in summer, exposing expanses of mud. The lake is adjacent to the coast, at the eastern end of the resort village of El Portil.

Species

The lagoon attracts waterfowl and gulls, especially on passage and in winter, many species visiting occasionally from the adjacent Marismas del Odiel (H4). It is a reliable site for one or two Ferruginous Ducks, especially in winter, when White-headed Ducks and Red-knobbed Coots are also present occasionally. Waders, such as Little Stints, occur in small numbers when muddy areas are exposed. The fringing scrub has a notable population of the Mediterranean Chameleon.

Timing

Visits during passage periods and in winter are most likely to be productive but some waterfowl are always present. This is a site that you are bound to pass when travelling along the Huelva coast and it always deserves a look.

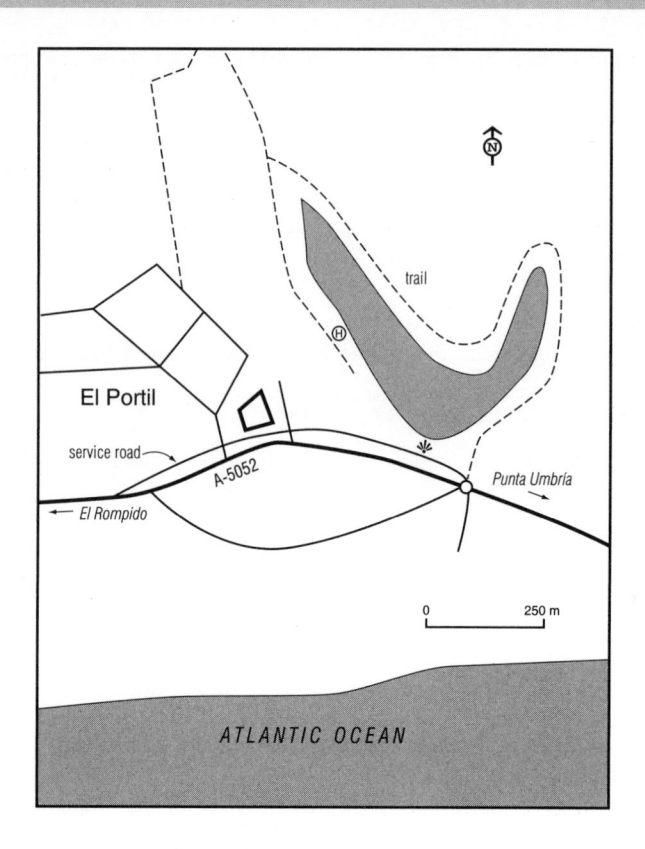

Access

The lake may be inspected quickly and easily from the service road that follows its southern boundary. Park by the wooden observation platform, which gives good, elevated views. A telescope is useful. The service road may only be entered from the roundabout at its eastern end, which is almost opposite the

White-headed Duck

lake itself. For closer scrutiny, park in the village and walk along the surrounding footpaths.

♿ Readily accessible at the roadside.

CALENDAR

All year: Mallard, Little and Great Crested Grebes, Purple Swamphen, Common Coot, Moorhen.

Breeding season: Little Bittern, Squacco and Purple Herons, Reed Warbler.

Passage periods and winter: Ducks: including Gadwall, Teal, Shoveler, Red-crested and Common Pochards, Ferruginous and White-headed Ducks, Black-necked Grebe, Grey Heron, Spoonbill, Red-knobbed Coot, gulls, waders including Little Stint and Ruff.

ODIEL ESTUARY H4
(Marismas del Odiel)

Status: Includes the Parajes Naturales of the Marismas del Odiel (7,185ha; also a ZEPA, UNESCO Biosphere Reserve and Ramsar site), Lagunas de Palos y Las Madres (693ha; Ramsar site), Estero de Domingo Rubio (343ha) and Enebrales de Punta Umbría (162ha).

Site description

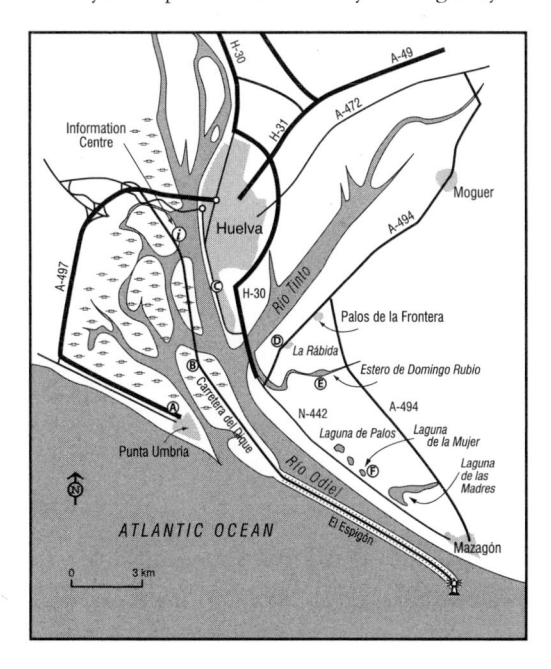

The provincial capital of Huelva lies at the confluence of the rivers Odiel and Tinto surrounded by a complex of broad tidal dykes fringed by salt flats and salt marshes. There are areas of fresh marsh and a number of small freshwater lagoons nearby. The spit of Punta Umbría encloses the marismas and is an extension of the sandy beaches of the central Huelva coast. An even longer spit, El Espigón, extends from the centre of the delta, southeastwards parallel to the coast, for some ten kilometres.

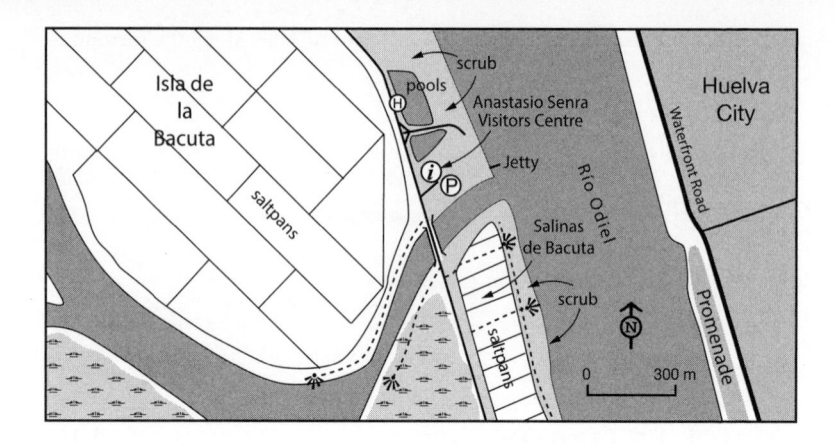

Species

The habitat diversity and strategic location of Odiel permits very high species diversity that includes many nesting and wintering landbirds as well as waterbirds, and a large range of passage migrants. The area is noteworthy for the large colonies of Spoonbills, over 500 pairs in good years, and it is indeed an excellent site to see this species. Spoonbills are scattered singly and in flocks across the whole region, foraging in the tidal creeks and salt pans as well as in the freshwater marshlands. They are confiding and give excellent views at close quarters. Spoonbills are very much a familiar part of the local landscape: the town of Punta Umbría even has a Spoonbill Street (Calle Espátula)!

Ospreys have long occurred regularly in winter, when 15 or more individuals may be present, and they also use the area on passage. However, there is also now a small breeding population, the result of a successful introduction project. Their nests are on purpose-built elevated platforms on the salt marshes.

A small breeding population of Common Shelducks is also noteworthy. Collared Pratincoles, Black-winged Stilts, Kentish Plovers and Common Redshanks are among the breeding waders and there is an important nesting population of Little Terns (up to 1,000 pairs).

The reedbed areas and freshwater lagoons hold good numbers of Purple Herons (20 pairs), among other nesting herons and egrets, as well as Little Bitterns, Purple Swamphens, Whiskered Terns and reedbed warblers.

Very large numbers of waders and up to 1,000 Greater Flamingos occur on passage and in winter. Scarce and rare wader species occur annually: recent records have included Broad-billed, Marsh and Purple Sandpipers and a Wilson's Phalarope. The area is attractive to gulls and flocks of thousands accumulate here. In summer these are mostly non-breeding Yellow-legged Gulls but a diverse range of species occurs at other seasons. There are excellent opportunities for scanning gull flocks and, in view of the numbers present, good chances of finding something unusual: the site has produced such scarce species as Great Black-backed and Common Gulls, and at least three Ring-billed Gulls. Terns are also attracted to the area and the wintering species include numbers of Sandwich Terns and small flocks of Caspian Terns. Hundreds of Cormorants and up to several thousand Black-necked Grebes are present in winter. There is potential for seawatching from Punta Umbría and, especially, from the end of El Espigón.

Timing
The site is interesting at all times of year. However, the resort areas, notably the beaches extending west from Punta Umbría, are crowded with human visitors in summer and during the Easter holidays. Visits on a rising tide offer the best prospect of seeing feeding waders at close quarters.

Access
The core of the Marismas del Odiel is strictly protected and not generally accessible, except on guided tours. The region still offers a number of points from which the key species of the area can be readily observed (Map 1).

A Punta Umbría The village and resort of Punta Umbría is reached from the coast road west of the estuary. It is possible to follow the road through the town for good views of the western channel and the islets beyond. Follow the signs marked playas (beaches). The beachfront road ends in a stone jetty, accessible only on foot, offering views of the estuary and the open sea. West of the town there are pinewoods and well-preserved dune vegetation, partly comprising the Paraje Natural of Los Enebrales de Punta Umbría. This site is worth searching for passerine migrants and holds numbers of the ubiquitous Azure-winged Magpies. It also gives northward views of the western salt pans and lagoons of Odiel, an area attractive to waterfowl and Flamingos especially.

♿ The coastal promenade is readily accessible.

B The central causeway This gives access to the core of the delta and is the best area for waders. The road is clearly marked at the roundabout at its northern end. It is signposted with the rather grand title of 'Carretera de las Islas. Dique Juan Carlos I, Rey de España'. The road travels down the centre of the reserve and beyond for a total of over 20km. It branches south from the Aljaraque/Huelva road (A-497) just west of the two road bridges across the Rio Odiel. The earlier part of this road passes through salt flats and salt pans before crossing the main western creek of the marshes and turning southeastwards to follow the southern bank of the river, opposite the city of Huelva. The whole road sees little traffic (except in summer) and it is usually safe to park alongside any area that is seen to hold birds, although not of course on the bridges.

The road continues through salt marshes and by shallow sandy lagoons, affording good views of the Odiel mudflats at low tide and of waders feeding in the salt pans at any time. An information centre for the Odiel marshes (Centro de visitantes Anastasio Senra, Ctra. del Dique Juan Carlos 1, km3, 21071, Huelva. Tel. 959 509 011) is on the left-hand side at km-3. Guided tours of the reserve areas may be booked here. There is ample parking at the centre from where several interesting options are available to you on foot (see Map 2). Visit all these sites if time permits. The nearest is a short jetty projecting into the Odiel river alongside the car park. Gulls and terns may be seen here, as well as waders on the banks and on the mudflats exposed at low tide. Three short footpaths are available nearby, all originating from the main road. The path marked Sendero de Calatilla de Bacuta, begins on the opposite side of the road from the centre and leads for 747m along the north side of a creek, giving good views of a series of salt pans as well as open water. It ends at a hide and viewpoint. Osprey nests on artificial platforms may be seen from here on the salt pans on the other side of the creek. A representative range of waterfowl, waders, gulls and terns may be expected and the low scrub along the path often shelters passerines. The 'Prohibido el Paso' sign at the start of the trail refers to vehicles and access on foot is permitted.

A similar but shorter path, also ending at a hide overlooking the salt marsh, is on the southern side of the creek. A third path network, on the eastern side of the road, allows you to inspect the traditional salt pans (Salinas de Bacuta) to the south of the bridge. A final area to visit is the pools on the eastern side of the road north of the information centre and adjacent buildings. Here too there is a hide looking over shallow water fringed by muddy margins and tamarisks; waterfowl and a good variety of waders are often present here.

Once you return to your car and continue southwards you soon cross a high bridge over the central creek. Beyond the road crosses an area of scrub and pinewoods that should be examined for passerine migrants in season. 'Anything' could turn up here but there will often be a good diversity of chats and warblers among others. There is parking available next to the principal woodland clump, on the west side of the road and 6km from the information centre. An 800m trail, El Acebuchal, traverses the area from here. The final 6km is a spectacular drive along the narrow and tapering spit, El Espigón, which stretches seemingly endlessly parallel to the coast, to arrive eventually at a lighthouse. The whole road allows close inspection of the Odiel mudflats and the sandy beaches to seaward. The end of El Espigón is a good site for seawatching, with seabirds passing close by during favourable, chiefly westerly, winds. Unfortunately, driving to the tip was prohibited from 2017 onwards 'for security reasons'. The best option is to stop at the last car park on the seaward side just before the spit proper. Here a short footpath (400m), marked Sendero de Cabeza Alta, gives access to a tidal inlet attractive to waders. There are also boardwalks to the seaward beaches, from where you can scan for seabirds offshore.

♿ Much excellent viewing is possible from or near your vehicle.

C Huelva city waterfront   The waterfront road follows the course of the Río Odiel, giving views over the river and its muddy fringes. The area attracts large numbers of gulls and some terns and waders, chiefly in winter and during passage periods. The broad esplanade that runs along much of the waterfront makes viewing easy and there is ample parking all the way. Visit in the morning to avoid staring into the sun.

♿ Viewing is possible from or near your vehicle.

D La Rábida Huelva displays abundant monuments celebrating the discoveries of Christopher Columbus and the Spanish contribution to the colonisation of the Americas. A large statue of Columbus himself dominates the point of confluence of the Río Odiel and Río Tinto. Other monuments, and a renowned monastery, are concentrated at Columbus's point of departure on the eastern bank of the Río Tinto at La Rábida, which accordingly is heavily signposted throughout the area. Follow signs to La Rábida and continue down to the river, where a jetty (Muelle de la Reina) gives good views over the channel and mudflats. Large flocks of gulls often gather here.

♿ Viewing is possible from or near your vehicle.

E El Estero de Domingo Rubio This is a creek that enters the estuary just southeast of La Rábida (see D). The fringing salt marsh attracts Spoonbills, herons, waterfowl and waders. The road to La Rábida follows the creek and gives good views over the western portions. The creek is tidal only along its lower reaches.

The upstream portions include a freshwater lagoon with fringing freshwater marshes, bordered by Stone Pine woods. Access to the freshwater sector is from the Palos/Mazagón road (A-494). There is a signposted walking track just south of the bridge, which crosses the creek itself, and parking is available under the trees on the west side of the road nearby. The track itself is one kilometre long, giving good views of the lagoon and marshes. It is also possible to walk upstream from the bridge. Breeding birds include Purple Herons, Little Bitterns, Purple Swamphens and Whiskered Terns. Wintering waterfowl include White-headed Ducks.

♿ Only the lower reaches are accessible.

F Lagunas de Palos y Las Madres Four lagoons (de Palos, de la Jara, de las Madres and de la Mujer) make up a reed-fringed complex along the N-442 coast road, immediately southeast of the oil refineries and industrial complexes south of La Rábida. The larger lagoons have permanent water. The lagoons are fringed by Stone Pine woodland and numerous, often boggy, hollows thickly vegetated by tamarisks. The whole area has been heavily encroached upon by plastic greenhouses, largely devoted to strawberry cultivation. However, the lagoons retain their natural marginal vegetation, which often holds numbers of migrants, and they attract waterfowl and waders. Little Bitterns, Night and Squacco Herons and Purple Swamphens breed. White-headed Ducks are often present and may breed. Access to the lagoons is via short, drivable, sandy tracks leading directly off the north (inland) side of the N-442: those at kilometres 11.1 (for Laguna de Palos) and 12 and 13 (for Laguna de la Mujer) are best. Two signposted trails visit the lagoons and a number of hides. The Laguna de Palos can also be reached from the service road (Polígono Nuevo) serving the industrial complex just north of the lake. If you drive north from the N-442 you will find a parking area on the right midway between the first two gantries that cross the road. From here a trail leads to the lake. The large Laguna de Las Madres is visible from the (busy) N-442; park opposite where the lake comes into view.

♿ Partial viewing is possible from or near your vehicle.

CALENDAR

All year: Common Shelduck, Gadwall, Mallard, Red-crested and Common Pochards; Little, Great Crested and Black-necked Grebes, Cattle and Little Egrets, Grey Heron, White Stork, Spoonbill, Greater Flamingo (sometimes breeds), Osprey, Marsh Harrier, Common Kestrel, Water Rail, Moorhen, Purple Swamphen, Common Coot, Oystercatcher, Black-winged Stilt, Avocet, Kentish and Grey Plovers, Sanderling, Dunlin, Black-tailed and Bar-tailed Godwits, Whimbrel, Curlew, Common Redshank; Black-headed, Lesser Black-backed and Yellow-legged Gulls, Hoopoe, Lesser Short-toed and Crested Larks, Zitting Cisticola; Cetti's, Dartford, Spectacled and Sardinian Warblers, Southern Grey Shrike, Azure-winged and Common Magpies, Jackdaw, Raven, Spotless Starling, Corn Bunting.

Breeding season: Little Bittern; Night, Squacco and Purple Herons, Black Kite, Collared Pratincole, Little and Whiskered Terns, Yellow Wagtail, Savi's, Reed, Great Reed and Melodious Warblers, Golden Oriole, Woodchat Shrike.

Winter: Greylag Goose, Wigeon, Teal, Pintail, Shoveler, Tufted Duck, Common Scoter, Red-breasted Merganser, White-headed Duck, Balearic Shearwater, Gannet, Great Cormorant, Great White Egret, Glossy Ibis, Hen Harrier, Stone-curlew, Golden Plover, Purple Sandpiper, Common Snipe, Great Skua, Mediterranean and Audouin's Gulls, Caspian and Sandwich Terns, Razorbill, Sky Lark, Meadow and Water Pipits, Grey and White Wagtails, Blackcap, Common Chiffchaff, Firecrest, Penduline Tit, Common Starling, Reed Bunting.

Passage periods: Cory's and Balearic Shearwaters, Gannet, Knot, Little Stint, Curlew and Common Sandpipers, Turnstone, Arctic Skua; Little, Mediterranean and Audouin's Gulls, Common and Black Terns, Roller, Crag Martin, Tawny and Tree Pipits, Common Redstart, Whinchat, Northern and Black-eared Wheatears; Grasshopper, Sedge, Olivaceous, Subalpine, Orphean, Garden, Bonelli's and Willow Warblers, Whitethroat, Spotted and Pied Flycatchers, Ortolan Bunting.

DOÑANA NATIONAL PARK   H5

Status: Parque Nacional & ZEPA (50,270ha). World Heritage Site and UNESCO Biosphere Reserve. The surrounding areas comprise a buffer zone, Parque Natural del Entorno de Doñana, of a further 56,930ha.

The survival of this area as a natural wilderness owes a great deal to its history as a hunting preserve, or Coto, of the Dukes of Medina Sidonia. One of the Dukes married a Doña Ana, and it is after her that the region is still called the Coto de Doña Ana; Doñana for short.

The Dukes played host to hunting parties of the Kings of Spain for over 300 years. These were grand outings, with the royal parties accompanied by large retinues of courtiers and camp followers. As many as 12,000 people assembled for the visit by Felipe IV in spring 1624, enough, one would imagine, to have scared off the game for miles around! Such hunting spectaculars were rare events and did nothing to detract from the importance of Doñana as a refuge for wildlife. Today that importance has been recognised and Doñana is one of the 11 National Parks of mainland Spain.

The Coto de Doñana National Park is administered by the Environment Agency, the Consejería de Medio Ambiente, of the regional government, La Junta de Andalucía. It is run by a director and staff based at El Acebuche. There is also an independent biological research and bird ringing station at El Palacio, which is not open to the public.

This account will guide you to all the main attractions of the National Park and its peripheral protected areas. However, this wonderful area really deserves a book all to itself – and, fortunately, in 2006 Paco Chiclana and Jorge Garzón produced a superb and beautifully illustrated specialist guide, *Where to Watch*

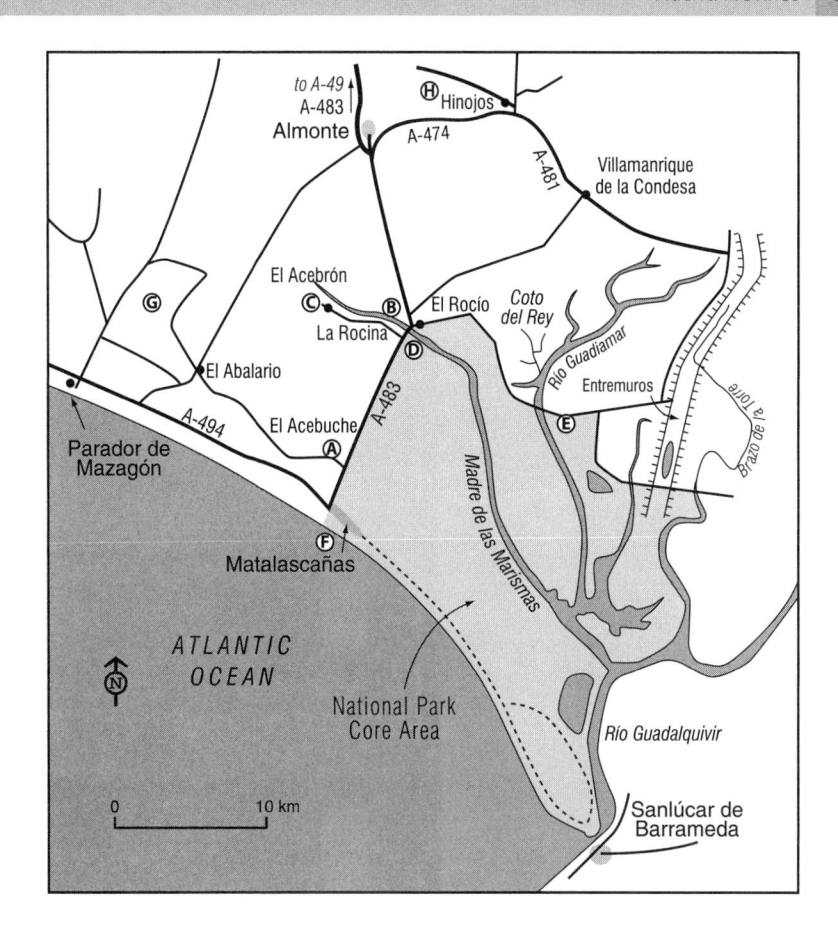

Birds in Doñana, which remains useful and which we heartily commend to anyone on an extended visit and, especially, to regular visitors.

Site description

Doñana is immense. The National Park and its protected margins (Preparques) extend over 1,300km², much of it flat marshlands or marismas. The park comprises the western edge of the Gualdalquivir estuary. Here, the actions of sea and river have combined to build up a large sand bar sheltering an inland sea of shallow lagoons and seasonally flooded salt flats. To seaward lie some 35km of unspoilt sandy beach, its waters paddled not by thousands of holiday makers but by a few dozen fishermen allowed by licence to compete for the cockles with the Scoters and Oystercatchers that flock there. Moving inland there is an extensive system of sand dunes giving on to open woodlands of Stone Pine and extensive stretches of low scrubland, dominated by the Cistus-like Yellow Sunroses. Areas of Cork Oaks support some of the major heronies of the park. Rock roses, lavenders, junipers, tree heaths and a rich variety of other flowering plants provide an aromatic understorey in the woodlands.

Large clearings interrupt the woodlands and here the herds of browsing ungulates suggest a scene from the African veldt. Fallow and Red Deer are scattered around, browsing contentedly and untroubled by the customary presence of human visitors. Parties of Wild Boars, elsewhere shy and retiring, snuffle for roots in the mud in broad daylight, the rangy adults accompanied in spring by groups of striped piglets. By contrast, the mammalian predators remain unobtrusive and mainly nocturnal: you are unlikely to see the Iberian Lynxes which are also characteristic of the park, although you will encounter signs warning of their presence on some roads.

Much of Doñana is a large segment of the Marismas del Guadalquivir, a shimmering expanse of shallow water with islets and reedy lagoons providing cover for a multitude of waterfowl. Doñana is thus, above all, a wetland site and there have been times when it could be considered the number one such reserve in western Europe. Unfortunately, a devastating combination of uncontrolled water abstraction in peripheral farms and drought has meant that some recent years have seen the freshwater habitats dry up for months on end, with obvious consequences for wildlife. Such changes are reversible and Doñana remains an immensely important site, particularly in the wet seasons that still occur. The marismas themselves are saline. They are bounded to the north and west by salt pans and rice paddies and large areas of low scrub dominated by glassworts, fleshy plants capable of standing periodic washing by salt water at high tides. This huge expanse of watery habitats attracts birds in great numbers, especially in winter.

The park has long been at the centre of controversy, and not just from conflicting land use interests that lower the water table and threaten much of the wetland interest of Doñana. There have been damaging incidents of mass deaths of waterfowl attributable partly to outbreaks of botulism but also to the indiscriminate and illegal use of pesticides in adjacent ricefields. The Aznalcóllar mine incident in 1998, where a massive toxic spillage contaminated the drainage basin of the Río Guadiamar, a key feeder stream for the wetlands of the National Park, is now history but similar events could reoccur.

Species

Doñana is one of the major wetlands of Europe and holds internationally important concentrations of wildfowl in winter and during passage periods. Principal among these are the Greylag Geese, some 60,000 of which cross Iberia to spend the winter in the park. They are accompanied by some 250,000 ducks, chiefly Teal, Wigeon, Mallard, Shoveler, Pintail and Pochard, as well as thousands of Common Coots. Large numbers of waders, notably godwits, Ruffs, Black-winged Stilts and Avocets also occur both in winter and on passage.

The park is also of the greatest importance for the species that breed there. There are large heronries, supporting Cattle and Little Egrets; Grey, Night and Purple Herons and the otherwise elusive Squacco Herons, as well as Spoonbills and White Storks. Glossy Ibises have bred successfully in increasing numbers in recent years and are an established feature of the park, with a population of several thousand pairs. The Great Bittern has bred successfully recently but is very scarce. Raptors occur in great numbers: there are hundreds of pairs of Black Kites. Short-toed Eagles also breed commonly as do Marsh Harriers, Booted Eagles, Common Buzzards, Common Kestrels and Peregrines. Several pairs of Spanish Imperial Eagles nest within the boundaries, typically in the crowns of Stone Pines.

A site of this importance cannot fail to attract scarce and elusive species. The Little Buttonquail (Andalusian Hemipode) is now considered extinct in Spain but if it still survives anywhere it will probably be where it was last seen, in 2002 in the north of the park (Coto del Rey), in a protected area not open to the public. There are plans to reintroduce it. Vagrants discovered in Doñana have included such disparate birds as African Marabou, Rüppell's Vulture, Surf Scoter, Long-billed Dowitcher, Slender-billed Curlew, Grey-headed Gull and Pallas's Warbler.

Timing

The park is always of some interest but the combination of low water levels (or no water at all!) and the influx of tourists to the nearby beaches makes summer a time to avoid. Water levels fluctuate with the rains, and there have been some very dry years in each recent decade: 2005, 2012 and 2017 were exceptionally dry. However, spring 2018 was extraordinarily wet and in an average year the park lives up to its wetland status and large numbers of wintering and passage aquatic birds can be seen from October to May. The activity and sound of birds settling to breed and the steady throughflow of migrants makes visits in spring (March–May) particularly pleasant.

For birding purposes at least, it is also a good idea to avoid Doñana at Pentecost (seven weekends after Easter), when up to one million pilgrims converge on El Rocío for the celebration, over several days, of the Fiesta de Nuestra Señora del Rocío (Our Lady of the Dew). This is a statue of the Virgin Mary, venerated locally for over 500 years, which is housed in the large church at El Rocío. The pilgrims include gypsies in caravans and many horsemen with their colourful Señoritas sitting behind them.

Access

The National Park is clearly signposted from the A-49, with access from the north via Almonte on the A-483. The coast road from Huelva (A-494) leads to Matalascañas in the south.

Doñana is run principally with the interests of the fauna and flora in mind. As a result, much of it is closed to the general public. Entry to the core of the National Park is strictly controlled. Basically you can't get in there at all except as part of an organised visit and these need to be booked well in advance at popular times. The visits are run by private companies under licence. The primary concession is held by Cooperativa Marismas del Rocío, Centro de Recepción de El Acebuche, PN de Doñana, A-483, km-37,5, 21730 Almonte, Huelva, Spain (Tel. 959 430 432). You may book in person if you are in El Rocío, especially if you are staying at the Hotel Toruño, which is run by the same company. There are generally two trips a day, morning and afternoon. The standard visit takes four hours, during which you travel in a Land Rover or a 'safari' bus, which generally drives south along the beach from Matalascañas before cutting inland to cross the dunes. The route then visits the pinewoods and the southern fringes of the marismas, before returning towards Matalascañas along the beach. The buses stop from time to time but they aren't an ideal way to see birds, particularly if your party chances to include some of the more raucous members of the general public. Still, the visit is worthwhile if only because it allows you to experience the full range of habitats in the area and there aren't many places in Europe where such large expanses of unspoilt terrain remain. The crossing of the dunes, a veritable sand-sea, is particularly memorable, especially if your bus gets bogged down. Fortunately, if it does there really will be another one along 'in a moment'. It is possible to hire a vehicle with (obligatory) guide for the whole day, for a more leisurely visit, and this would seem a good idea for visiting groups.

Another type of tour, by boat, is available from Sanlúcar de Barrameda (see CA16). Information on tours may also be obtained from the National Park Offices: Centro Administrativo El Acebuche, 21760, Matalascañas, Huelva (Tel. 959 439 629/email *en.donana.cma@juntadeandalucia.es*).

Major tracts of Doñana are relatively inaccessible and there remains scope for improving public access without damaging the environment. However, the accessible sites are first rate. Try them all!

A Centro de Recepción "El Acebuche" A good place to start is the information centre at El Acebuche, which lies 1.5km west of the Almonte/Matalascañas road at km-29. Azure-winged Magpies haunt the car park and adjacent picnic areas during much of the year and a pair or two of photogenic, and much-photographed, White Storks nest on the picturesque roof of the centre itself. The centre offers displays and other information about Doñana as well as a cafeteria and souvenir shops, the latter selling the usual car stickers, T-shirts, books and similar material that may be of interest. Excellent maps are available here too. The centre is adjacent to the large (33ha) lagoon of El Acebuche, which although a kilometre or so long is no wider than 200m and is overlooked in comfort from seven large wooden hides. A telescope will be useful here and elsewhere for top notch views of waterfowl at close quarters. The lagoon is a good site to see Ferruginous Ducks and Purple Swamphens, among other waterfowl. The two easternmost hides overlook an enclosed section of the lagoon, which sometimes houses pinioned exhibits of such species as

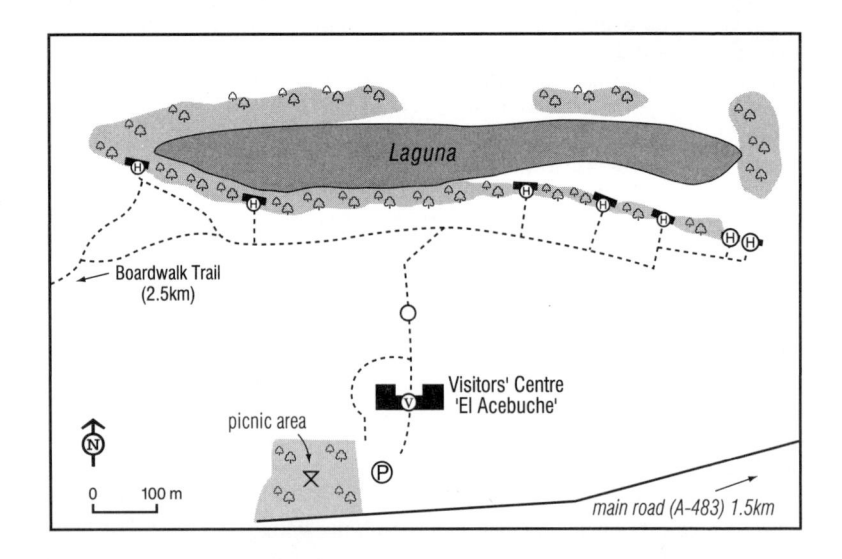

Ruddy Shelduck and White-headed Duck: not tickable except by the most unconscionable of birders! The hides are often surprisingly empty even when the centre itself is busy, many visitors being apparently disinclined to walk any distance.

The surrounding nature trails through the Mediterranean scrub and pinewoods offer a chance to see passerines. There is an extended trail of 2.5km beyond the westernmost hide, passing through pinewoods with many extensive clearings. The entrance is signposted 'Sendero Peatonal Lagunas del Huerto y las Pajas'. The area can be damp at times, and the trail is accordingly on a boardwalk, but it is dry for much of the year and the 'Lagunas' are then nowhere to be found. Both these ephemeral pools are also served by hides. The trail is worth a visit since it is quiet, there are always raptors overhead, and such species as Hoopoes, Wood Larks, Southern Grey and Woodchat Shrikes, Azure-winged Magpies and a range of warblers and finches readily present themselves.

&. Good access to the centre, boardwalks and hides.

B Centro de La Rocina The information centre at La Rocina lies across the road from the village of El Rocío. The centre itself offers general information. The main interest lies in the associated nature trail, 2.5km long, which includes several hides overlooking the marshy Charco de la Boca, a slow stream flowing into the Madre de las Marismas at El Rocío and with numerous boggy inlets and reedbeds as well as small areas of open water. All the hides are worth visiting. This is a reliable site to see Savi's Warblers. Purple, Squacco and Night Herons and Marsh Harriers are often present together with Purple Swamphens and a 'seasonal selection' of other waterfowl. The surrounding woodlands and scrub are worth searching for resident and migrant passerines. The diversity of habitats means that anything can turn up, especially on passage. Southern Water

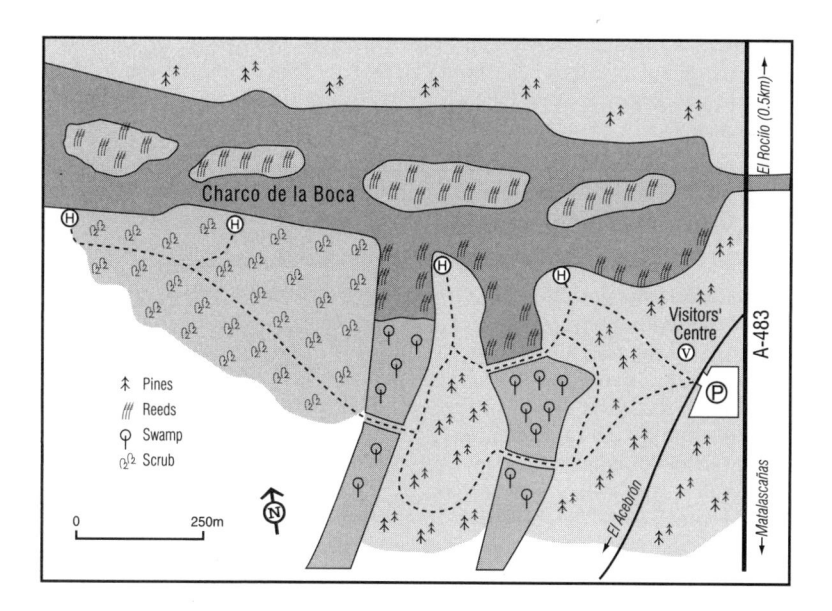

Voles may sometimes be seen at close quarters at the point where the boardwalk descends to the first marshy inlet.

♿ Accessible but assistance may be needed at the steeper boardwalk sections.

C Palacio del Acebrón A 'stately home' in the pinewoods, 7km from La Rocina. A permanent exhibition showing traditional human life and exploitation of the marismas is of interest, especially if it's raining outside. A nature trail, which visits the adjacent woods and the pool of El Charco del Acebrón, is well worth the short walk (1.5km).

♿ Only the Palacio is accessible.

D El Rocío Much of the village of El Rocío, lying just off the Almonte/Matalascañas road (A-483), is preserved as a living relic of the days when Doñana was almost wholly inaccessible except on horseback, although much new building has emerged on the periphery in recent years. The original village still retains its picturesque, if dusty, sand streets and its white houses, these last complete with hitching rails for horses, suggesting the set for a spaghetti western. The only posses here though are the groups of scope-toting birders since El Rocío is adjacent to a large shallow lagoon, a northward extension of the marismas proper, called La Madre de las Marismas (mother of the marshes). Water levels permitting, this is a superb site, the whole area thronged with herons, egrets, waterfowl and waders, notably Black-winged Stilts, Avocets, Ruff and Black-tailed Godwits and the attendant Black Kites, Marsh Harriers and other raptors. Spoonbills and Glossy Ibises are usually present. This site can produce 70 or more species before breakfast, which can be achieved if you spend a night or two in the village.

Most visitors to El Rocío observe the lagoon from the waterfront promenade. It is worth following this eastwards and continuing onwards for about one kilometre along a dirt track leading to the park boundary at Boca del Lobo. The fence-posts and tree-tops here should be inspected for raptors especially. Both Spotted and Lesser Spotted Eagles have been recorded in this general area in some winters. The Sociedad Española de Ornitología (SEO) has an information centre at the southern end of the promenade, from where there are also good views of the lagoon.

♿ Good access to the promenade.

E The Marismas. José Valverde Centre The dirt road leading east from El Rocío through the Stone Pinewoods of the Coto del Rey to the northern boundary of the park and the northern part of the marismas is now closed to private visitors, partly in order to reduce disturbance of Iberian Lynxes. The marismas remain accessible, however, via the approaches to the José Valverde visitors' centre (Centro de Visitantes José Antonio Valverde/Cerro Garrido), named after 'Tono' Valverde, one of the founding fathers of the Doñana reserve. It is well worth the excursion to travel there since the dirt roads serving the centre cross excellent tracts of the northern marismas, offering good views of concentrations of waterfowl and waders, and their attendant predators, except in the dry season. The passerines here must not be overlooked:

Lesser Short-toed Larks and Spectacled Warblers breed commonly for example.

This centre receives relatively few visitors since it is a considerable distance from any tarred roads and the access tracks are of very variable quality. The centre proper is a substantial wooden building, incorporating displays, a shop and a cafeteria. It has picture windows and a short, screened boardwalk overlooking an immediately adjacent permanent lagoon (Lucio de las Gangas), which attracts breeding and wintering waterfowl. Purple Swamphens are prominent here. The chief attraction, however, is the large nesting colony of Glossy Ibises, a relatively recent establishment but which now numbers several hundred pairs in some years. All those Ibises that you may see around Doñana and in the Brazo del Este (SE1) sporting conspicuous rings originate from this colony, where they are marked as nestlings. The Ibises are accompanied by nesting Little and Cattle Egrets, Purple and Squacco Herons and Little Bitterns, all of which may be seen at relatively close quarters in spring and early summer, when the colonies are occupied. This is clearly the best time to visit the centre itself, not least because the roads may not be too muddy then (with luck). Winter visits are still enormously worthwhile, however, since this is the peak season for waterfowl, when the general area attracts great numbers of Greylag Geese – accompanied by occasional rarer species such at White-fronted and Barnacle Geese; ducks, including Marbled Ducks; Cranes, Great White Egrets and many other aquatic birds. It is worth asking at El Acebuche or at the Centro Dehesa Boyal in Villamanrique for advice on the state of the roads if it has recently been wet.

The most reliable access to the centre is available from the east. If you are already in Doñana, for example, in El Rocío, follow signs to Villamanrique de la Condesa. A minor road leads straight there from the eastern side of the A-483 at a point about 2km north of El Rocío. If you approach from this direction, turn right at the Elf petrol station outside Villamanrique and continue for 100m to a roundabout where you take the third exit (left). This brings you to a second roundabout where you take the second exit (right) towards Isla Mayor and the Algodonera (cotton factory). Continue on this road until you reach a ford, where the road rises beyond onto a dyke. Turn left here, reassured by the sign pointing to the Centro de Visitantes. (If the ford is impassable a short detour involving a bridge is available to the right.) The road on the dyke is quite good tarmac. Follow it to the end where it becomes dirt – ignoring other alternative routes to the centre, which are signposted, unless you have 4WD – and continue onwards on that past a conspicuous white pumping station (Casa Bomba). The track then bears right and continues straight on to the centre. The total distance from Villamanrique is 30km but you will probably take a couple of hours over the journey since you will want to stop frequently to scan the marshes and salt flats for birds. The track itself is normally readily passable with care even without a 4WD vehicle. The same track is also accessible from La Puebla del Río and Isla Mayor (see SE2).

A visit to the centre can be followed by a drive further west to view the Caño del Guadiamar watercourse and the flanking marismas, which are excellent for waterfowl in winter especially. Drier tracts here often attract Pin-tailed Sandgrouse and, occasionally, Little Bustards. Turn left out of the centre and continue westwards along the road until you come to a bridge across the Caño del Guadiamar. Turn right and follow the watercourse northwards to view the marismas. It is possible to continue along this route – mud permitting – to intersect your original access road to the pumping station. Alternatively you can return south to the bridge and turn right, continuing for some 6km across further excellent habitat. The road ends at a closed barrier, the Cancela de la Escupidera (Chamberpot Gate; which some wag has adorned with a tin potty), from where you will have to retrace your route.

♿ Good access to the centre. The marismas are readily viewed from the roads.

F The Beach The beach at Matalascañas is a busy resort in summer but at other times it is quiet. Like others on the Costa de la Luz, the beach is broad, sandy and tidal. The portion within the park boundary is accessible on foot from Matalascañas. Seabirds are visible offshore, especially in winter. They include large rafts of Common Scoters in winter, worth scanning carefully since both Velvet and Surf Scoters have turned up among them.

♿ Not readily accessible.

G The Western Pinewoods Huge areas of Stone Pine woodland and intervening tracts of open heathland comprised the western borders of Doñana until 2017 when forest fires destroyed an enormous tract of the pinewoods. This event was seen as an environmental disaster but the Stone Pines were planted in the 1960s and there is now the prospect of a return to a more natural landscape, given regeneration and sensitive replanting. Time will tell. The area has plenty of boggy hollows and temporary lagoons although these are dry in summer and year-round in drought years. Access is from the tarmac road that leads inland opposite the Parador of Mazagón, 22km west of Matalascañas on the A-494. Alternatively, take the turn-off from the A-494 signposted Via Pecuaria Veredal Camino del Loro, on the western boundary of the park at km-35.7, where a good dirt road/cycle track leads inland through the woodlands. Both roads interconnect via a network of sandy dirt roads, offering plenty of opportunities for exploration on foot or by vehicle, including by bicycle on the indicated cycle tracks. It is easy to get somewhat lost here so a compass would be useful to anyone with a poor sense of direction. The area is very quiet and can be something of 'a refuge when El Rocío is choked with visitors. Characteristic species include Spanish Imperial and Short-toed Eagles, Red-necked Nightjars, Bee-eaters, Thekla Larks, Tawny Pipits and Azure-winged Magpies, as well as most of the woodland passerine species of Doñana. Migrants such as Whinchats and Northern Wheatears are often numerous on the open heathlands during passage periods.

♿ Good viewing is possible from or near your car.

H Arroyo del Algarbe and Hinojos pinewoods The vicinity of Hinojos offers good opportunities to see some sought-after but occasionally elusive species: including the Rufous-tailed Scrub-robin, Penduline Tit, Orphean Warbler and Olivaceous Warbler. All of these can be seen in the well-developed riparian woodland along the Arroyo del Algarbe, a stream that approaches Hinojos from the northwest, or in the adjacent farmland and woodland. The area is easily accessed by taking the tarmac road that runs for some 10km alongside the stream, between Hinojos and the A-49. There is no direct access from the A-49 but the southern entrance to the road is immediately north of the bridge across the stream on the A-481, just north of Hinojos. The road passes through areas of olive groves and then alongside Stone Pine woods before crossing the stream and continuing into an area of Cork Oak woodland. You will need to retrace your route but, at a point halfway to/from Hinojos, at the north end of the Stone Pine area, you have the option to cross the stream where a track leads southeast through extensive pinewoods; this track joins the A-482 at km-11.8. All these habitats are worth inspecting at leisure; stop frequently. Other species that may be expected include Black-winged Kite, Quail, Red-necked Nightjar, Golden Oriole, Common Waxbill and Cirl Bunting, among a wide diversity of woodland and open-country birds.

♿ Good viewing is possible from or near your car.

CALENDAR

All year: Red-legged Partridge, Quail, Gadwall, Mallard, Marbled Duck, Red-crested and Common Pochards; Little, Great Crested and Black-necked Grebes; Cattle, Little and Great White Egrets; Squacco, Night and Grey Herons, White Stork, Glossy Ibis, Spoonbill, Greater Flamingo, Red Kite, Griffon and Black Vultures, Marsh Harrier, Common Buzzard, Spanish Imperial and Booted Eagles, Lesser and Common Kestrels, Peregrine Falcon, Water Rail; Spotted, Baillon's and Little Crakes, Moorhen, Purple Swamphen, Common and Red-knobbed Coots, Little Bustard, Oystercatcher, Black-winged Stilt, Avocet, Stone-curlew; Little Ringed, Ringed, Kentish and Grey Plovers, Lapwing, Sanderling, Little Stint, Dunlin, Ruff, Black-tailed and Bar-tailed Godwits, Whimbrel, Curlew, Spotted and Common Redshanks, Greenshank, Green Sandpiper, Arctic and Great Skuas; Black-headed, Slender-billed, Lesser Black-backed and Yellow-legged Gulls, Caspian and Sandwich Terns, Pin-tailed Sandgrouse, Woodpigeon; Barn, Eagle, Little, Tawny and Long-eared Owls, Hoopoe, Iberian Green Woodpecker, Great Spotted Woodpecker; Calandra, Lesser Short-toed, Crested, Thekla and Wood Larks, Barn Swallow, House Martin, Stonechat, Blackbird, Mistle Thrush, Zitting Cisticola; Cetti's, Dartford, Spectacled and Sardinian Warblers, Crested Tit, Short-toed Treecreeper, Southern Grey Shrike, Iberian Azure-winged and Common Magpies, Jackdaw, Raven, Spotless Starling, Spanish and Tree Sparrows, Common Waxbill, Cirl and Corn Buntings.

Breeding season: Garganey, Little Bittern, Purple Heron, Black Kite, Egyptian Vulture, Short-toed Eagle, Montagu's Harrier, Hobby, Collared Pratincole; Gull-billed, Little and Whiskered Terns, Turtle Dove, Great Spotted and Common Cuckoos, Scops Owl, Red-necked Nightjar, Bee-eater, Greater Short-toed Lark, Sand Martin, Red-rumped Swallow, Tawny Pipit, Yellow Wagtail, Rufous-tailed Scrub-robin, Nightingale; Savi's, Reed, Great Reed and Melodious Warblers, Golden Oriole, Woodchat Shrike, Hawfinch.

Winter: Greylag Goose, other goose species, Shelduck, Wigeon, Teal, Pintail, Shoveler, Tufted Duck, Common Scoter, Red-breasted Merganser, Great Northern Diver, Balearic Shearwater, Northern Gannet, Great Cormorant, Black Stork, Black-winged Kite, Hen Harrier, Sparrowhawk, Greater Spotted Eagle (rare), Bonelli's Eagle, Osprey, Merlin, Crane, Golden Plover, Jack and Common Snipes, Woodcock, Marsh and Wood Sandpipers, Razorbill, Kingfisher, Short-eared Owl, Sky Lark; Richard's, Meadow and Water Pipits, Grey and White Wagtails, Dunnock, Robin, Bluethroat, Ring Ouzel, Fieldfare, Song Thrush, Redwing, Blackcap, Common Chiffchaff, Firecrest, Penduline Tit, Common Starling, Siskin, Linnet, Bullfinch, Reed Bunting.

Passage periods: Honey-buzzard, Knot, Temminck's Stint; Curlew, Purple and Common Sandpipers, Turnstone, Little Gull, Common and Black Terns, Short-eared Owl, Alpine Swift, Roller, Crag Martin, Tawny and Tree Pipits, Common Redstart, Whinchat, Northern and Black-eared Wheatears; Grasshopper, Sedge, Olivaceous, Subalpine, Orphean, Garden, Bonelli's and Willow Warblers, Whitethroat, Spotted and Pied Flycatchers, Rock and Ortolan Buntings.

SIERRA PELADA   H6

Status: Paraje Natural & ZEPA 'Sierra Pelada y Rivera del Aserrador' (12,305ha).

Site description
A quiet corner of northwesternmost Andalucía, near the Portuguese border, offering undulating wooded hillsides and areas of more open Cistus scrub. Cork and Holm Oaks are characteristic as well as extensive stands of Stone and Maritime Pines. Pleasant rocky streams cross the site. Some of the area is well preserved notwithstanding the unwelcome encroachment of Eucalyptus plantations, which are an unmissable feature of the western and southern parts of the site. Eucalyptus has fallen from favour in recent years. Some extensive recent reforestation has involved pines and continuing forestry activity causes a lot of disturbance where work is in progress. However, the area is large and unpopulated and quiet locations to explore are always available.

Species
The area boasts one of the largest breeding colonies of Black Vultures in Andalucía, of over 100 pairs. Griffon Vultures also are regularly present, probably commuting from Badajoz province given that Huelva is the only province in our area with no breeding Griffons at all. In general the variety of raptors is high and includes Golden, Short-toed and Booted Eagles. Other interesting breeding species include Black Storks, Eagle Owls and White-rumped Swifts.

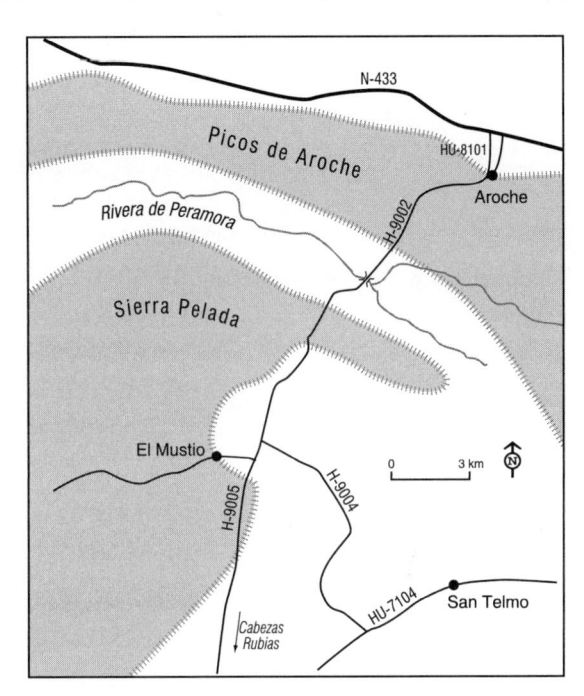

The woodlands and watersides support a good range of passerines, including Rock Sparrows. Otters, Iberian Lynx and Wild Boar are present too but, as ever, elusive.

Timing
Springtime visits are recommended as displaying raptors are obvious and the woodland community is in song and easily located.

Access
The area is crossed by a labyrinth of minor roads and forest tracks, some of which are of very indifferent quality. Straightforward access to some of the better areas is obtained by driving southwest from Aroche. This is a broad dirt road (H-9002), which is in excellent repair. It crosses a small river and several streams before entering the eucalyptus plantations in the vicinity of the largely deserted farm of El Mustio. The raptors, which are the chief attraction of this site, can be seen by stopping and scanning frequently along the road, which has very little traffic.

The northern entry point is found by taking the westernmost (HU-8101) of the two turnings for Aroche off the N-433. The road bypasses Aroche to the west. Once level with the village turn right up a gravel track past a large red-brick building (a school, which accommodates 150+ House Martin nests). This track ascends and then descends to a fork after 2.2km. Take the left fork and keep straight on. This is the main road through the Sierra and crosses all the main habitats. The stone bridge on the Rivera de Peramora, a few kilometres south of Aroche, is a good stopping point with Red-rumped Swallows very much in evidence and a likelihood of waterside species such as Kingfishers and Grey Wagtails. Little Ringed Plovers and White-rumped Swifts also occur. The main ridge across the Sierra is reached 18.7km from Aroche and provides points of vantage to scan for raptors. Black Vultures are often obvious and, if not, soon appear.

The same road may also be reached from the south. The best access here is from Cabezas Rubias, heading northeast for 2km on the tarmacked San Telmo/Cabezas Rubias road, before turning north to follow what becomes a good dirt road (H-9005) that ascends the southern fringes of the Sierra Pelada to join the Aroche road at El Mustio. Stop at the top of the first ascent where there are good views over the whole area. A marked trail leads west for 3km along the top of the ridge here, the open, stony country offering such species as Rock Buntings and Blue Rock Thrushes, with Black Vultures and other raptors overhead. The main ridge is reached by continuing north for another few kilometres.

The Paraje Natural includes the watercourse of the Rivera del Aserrador, a small stream running to the west of El Mustio (not mapped). There is a marked footpath along the streamside on the north bank, signposted from the A-495 at km-72.

♿ Good viewing is possible from or near your car.

> **CALENDAR**
>
> **All year:** Griffon and Black Vultures, Common Buzzard, Golden Eagle, Goshawk, Great Spotted Woodpecker, Grey Wagtail, Blue Rock Thrush, Rock Sparrow, Cirl and Rock Buntings, woodland passerines.
>
> **Breeding season:** Black Stork, Short-toed and Booted Eagles, Little Ringed Plover, Kingfisher, White-rumped Swift, Red-rumped Swallow, woodland passerines.

WESTERN SIERRA MORENA   H7

Status: Largely comprising Parque Natural 'Sierra de Aracena and Picos de Aroche' (186,880ha). ZEPA. UNESCO Biosphere Reserve.

Site description

The western Sierra Morena offers a range of pleasant and tranquil landscapes, easily accessible by road. This is a relatively humid zone: Aracena has an annual rainfall of some 1,000mm. The mountains of the Sierra de Aracena proper, in the south of the area, are accordingly lushly wooded in parts, with extensive forests of Sweet Chestnut being particularly typical. Forests of Cork Oaks, Canarian Oaks and Encinas are also evident as well as extensive stands of Maritime and Stone Pines. Riverine woodland is well developed, for example, along the course of the Río Múrtiga, and again includes stands of oaks and Chestnut as well as White and Black Poplars.

The northern parts of the area include wide tracts of open grazing land and stone-walled pastures interspersed with olive groves and encinares and crossed by stony-bedded rivers. Some of the hillsides are overgrown with dense scrub, predominantly of Cistus. Eucalyptus planting has devastated certain parts of the site, notably north of Aroche, but plans to replant the affected areas with native trees have been implemented in some areas and some large new plantations of Encinas have been established.

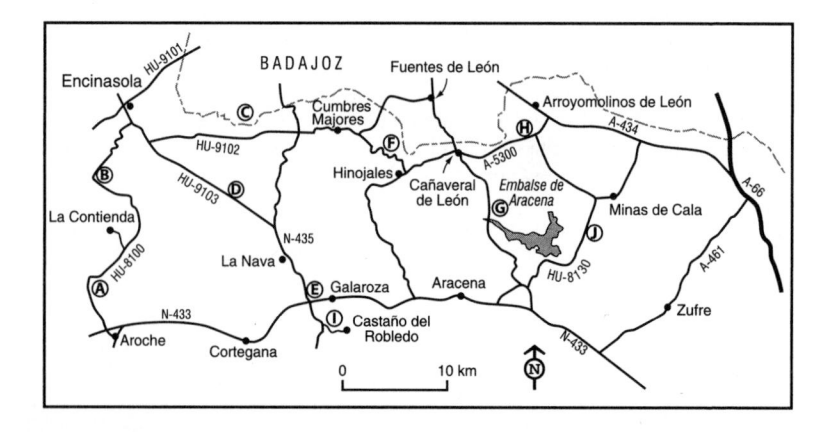

Species
The range of habitats and the ease of access along quiet, tarmac country roads makes this a good place for a leisurely visit in search of a representative range of Mediterranean birds. It offers a good diversity of woodland and mountain species, notably raptors, although none that cannot also be found elsewhere. White and Black Storks breed; the former on every church in every village as usual and the latter along the rocky river valleys. Vultures, including Black Vultures, are frequent overhead and other raptors include Short-toed, Spanish Imperial, Golden and Booted Eagles, Red and Black Kites and numerous Common Buzzards. The woodlands, riverine forest and scrub areas have a rich passerine community, which includes a good variety of warblers. Lesser Spotted Woodpeckers are thinly distributed across the area. Red-rumped Swallows nest under many, probably most, of the bridges and culverts. The latter's presence is undoubtedly an inducement to White-rumped Swifts, which have colonised the region sparsely. Both Bullfinches and Dunnocks are widespread in winter, when Alpine Accentors often appear on Aracena Castle.

Timing
The breeding species are easiest to locate during springtime and early summer visits, when the flowers are also noteworthy. However, interesting species are present all year round.

Access
The area is traversed by the N-433 trunk road, from which numerous minor roads lead to areas of some interest. The region offers plenty of scope for quiet exploration with potential for long walks on woodland tracks or across the uplands, or for just simply driving around and stopping at likely looking places.

&. Good viewing is possible from or near your car in all areas.

Recommended areas include:

A Aroche/La Contienda   Drive north from Aroche on the HU-8100, crossing encinares and a river, where Bee-eaters have a small colony on a sandy bank,

Sardinian Warbler

before ascending through a region of barren hillsides, the remnants of eucalyptus plantations, to reach open meadowland, with scattered mature encinas, new plantations of encinas and large tracts of Cistus heath. Rock Sparrows are common near Aroche and raptors and passerines generally in the upper reaches.

B La Contienda/Encinasola The southern part of this road (HU-8100) follows the contours of undulating country covered in Cistus scrub alive with Dartford Warblers. Vultures and other raptors are common overhead. A short drive up a vertiginous track west of the road, about 1km from its junction with the HU-9101, leads to a mirador, which offers panoramic views and an excellent vantage point to scan for raptors. Along the HU-9101, the Río Múrtiga southwest of Encinasola, is very scenic: a broad stony bed with clumps of oleanders. Black Storks occur here and there are many passerines in the nearby woods of Cork Oaks. The old road bridge, adjacent to the new one, both at km-22, is a good viewpoint and both Crag Martins and Red-rumped Swallows nest on the bridges themselves. White Stork nests on the pylons here also accommodate those of House Sparrows and a few Spanish Sparrows.

C Encinasola/Las Cumbres The winding road (HU-9102) crosses several river valleys and also ascends to traverse open pastureland. The villages have the usual picturesque gatherings of White Stork nests. Raptors are frequent and the passerine community includes Rock Sparrows.

D Río Múrtiga valley The HU-9153 road from Encinasola to the N-435 just north of La Nava follows the river valley, traversing mature woodlands of encina and also excellent areas of riverine woodland and scrub. The road is very quiet and repays frequent stopping and short walks to search for passerines and others. There are many lay-bys and some woodland tracks that facilitate exploration. Wall-to-wall Nightingales are a particular feature in spring but other characteristic species include Black Storks, Little Ringed Plovers, Turtle Doves, Bee-eaters, Hoopoes, Wrynecks, Wood Larks, Red-rumped Swallows, Blackcaps, Rock Sparrows and Serins.

E Río Múrtiga at Galaroza Lush riverine woodland along the N-433, popular with warblers. The road through Galaroza itself suffers from heavy traffic but the short branch to the village of Las Chinas, immediately west of Galaroza, is very quiet and gives good views of the riverine habitats.

F Las Cumbres/Cañaveral de León A picturesque transect across the north of the region, crossing open pastureland, encinares and olive groves, all enclosed by the characteristic stone walls. Rock Sparrows and Cirl Buntings are typical.

G Cañaveral de León/Aracena Woodland habitats, with magnificent encinares in the north. Further south there is some open stony country. The road crosses the western arm of the Embalse de Aracena.

H Cañaveral de León/Arroyomolinos This road traverses encinares near Cañaveral and a pleasant stream (Rivera de Montemayor), with associated riverine woodland, before ascending steeply to a col above Arroyomolinos. Stop south of the col to scan the panorama for raptors. Blue Rock Thrushes are present here and Thekla Larks along the wooded stretches. Cork Oaks on the north side of the col support breeding Common Redstarts.

I Castaño del Robledo Reached by driving south from the N-433 on the N-435, turning east at El Quejigo. Castaño del Robledo is a secluded village, named after the surrounding woods of Sweet Chestnut. The area may be explored on foot or by driving a short way up the tracks that lead east just before entering the village. Drive up the steep concrete ramp (to the local cemetery) and continue upwards to enter the woodlands. Chestnut, Cork Oak and pine woods offer a good variety of passerines. The Natural Park has many signposted walking trails, several of which originate from Castaño del Robledo itself and provide ready access to the woodlands and riversides.

J Ribera de Hierro A very quiet, good tarmac road follows the river valley here for 22km through hilly country, between the eastern end of the Embalse de Aracena and Arroyomolinos de León. The northern entry point is from the A-5300, 2km south of Arroyomolinos. From here the road crosses farmland along the base of the Cistus-blanketed Sierra de la Nava, where Short-toed Eagles and other raptors may be seen quartering the slopes. A large quarry, the Minas de Cala, is something of an eyesore about halfway down the road, but a dam on the river at this point has created a small lake, which serves the quarry workings. The lake has a small reedbed, which may repay scrutiny: park on the old bridge south of the dam and view the lake by walking up the road. The bridge itself has a Crag Martin colony. The road forks south of the dam, where a branch road ascends to the quarry: keep west of the river. The open woodlands and scrub accommodate typical Mediterranean scrub species, including Melodious, Dartford and Subalpine Warblers, and Whitethroats. Both Black and White Storks frequent the river. The southern entry point lies just east of the Embalse de Aracena: itself of little birding interest although it may hold waterfowl in winter. The road (HU-8130) is signposted Puerto Moral at km-79 on the N-433.

CALENDAR

All year: Black-winged and Red Kites, Griffon and Black Vultures, Goshawk, Sparrowhawk, Common Buzzard, Spanish Imperial and Golden Eagles; Eagle, Tawny and Little Owls, Great Spotted Woodpecker, Crested and Thekla Larks, Crag Martin, Blue Rock Thrush, Zitting Cisticola; Cetti's, Dartford and Sardinian Warblers, Blackcap, Nuthatch, Southern Grey Shrike, Azure-winged Magpie, Raven, Spotless Starling, Spanish and Rock Sparrows, Cirl and Rock Buntings.

Breeding season: Black and White Storks, Black Kite, Egyptian Vulture, Short-toed and Booted Eagles, Little Ringed Plover, Turtle Dove, Scops Owl, White-rumped Swift, Bee-eater, Hoopoe, Wryneck, Red-rumped Swallow, Nightingale, Common Redstart, Black-eared Wheatear; Melodious, Subalpine, Orphean and Bonelli's Warblers, Whitethroat, Golden Oriole, Woodchat Shrike.

Winter: Alpine Accentor, Dunnock, Bullfinch.

CÁDIZ PROVINCE

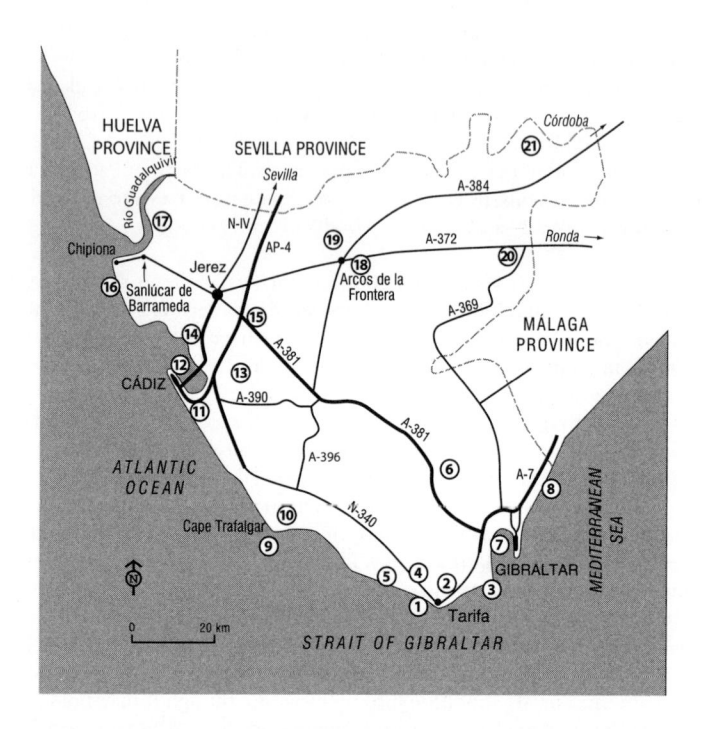

Cádiz province is one of the prime birding regions of southern Spain, offering as it does the great majority of the species of the region. The position in the extreme south of Iberia and the proximity to the African shore across the Strait of Gibraltar make Cádiz the key province in which to observe migration of raptors, storks and many other birds. The Strait itself is also of major significance for seabird migration. The Cádiz coastlands have important sites for both passage and wintering seabirds, waders and other aquatic species. The province offers a range of lagoons attractive to White-headed Ducks, Purple Swamphens and Red-knobbed Coots among a great diversity of wetland birds. The sierras are relatively lush, since Cádiz is one of the wetter Spanish provinces, and they support a full range of raptor and passerine communities. Even steppe species, such as Little Bustards, occur at a limited number of sites. Cádiz is also the only region in Europe to offer five breeding swift species, including both White-rumped and Little Swifts. This is probably the best province for a single-base visit; all its sites are within easy reach of Gibraltar, Jimena or centres in the west of the region.

Sites in Cádiz Province
CA1   Tarifa beach (Playa de Los Lances)
CA2   Central Strait of Gibraltar

CA3	Punta Secreta
CA4	La Janda
CA5	Bolonia, Zahara and Sierra de la Plata
CA6	The corkwoods (Los Alcornocales)
CA7	Palmones estuary
CA8	Guadiaro estuary (Sotogrande)
CA9	Cape Trafalgar
CA10	Barbate pinewoods and estuary
CA11	Sancti Petri marshes
CA12	Cádiz Bay (Bahía de Cádiz)
CA13	Lagunas de Puerto Real
CA14	Lagunas del Puerto de Santa María
CA15	Laguna de Medina
CA16	Chipiona coast
CA17	East Bank of the lower Guadalquivir
CA18	Arcos and district
CA19	Lagunas de Espera
CA20	Sierra de Grazalema
CA21	Peñon de Zaframagón

Getting there
Sevilla (Seville) airport is a convenient gateway to the province, with good motorway links. It is much less busy than Málaga Airport. Gibraltar is an obvious arrival point by air from Britain and Jerez airport provides access mainly from continental Europe. Within the province, rapid travel between the listed sites is possible using the N-340/A-48 trunk road/motorway, which skirts the entire coastline, and both the A-381 (Los Barrios/Jerez) and AP-4 (Cádiz/Sevilla) motorways.

Where to stay
Hotels and campsites are widespread all along the coast and in most of the towns inland, notably Jimena and Grazalema. There are Paradores at Arcos de la Frontera and also in Cádiz city. Gibraltar provides an excellent base for exploring the province: all sites are within day trip distance of the Rock.

TARIFA BEACH CA1
(Playa de Los Lances)

Status: Paraje Natural & Reserva Ornitológica (226ha). Part of the Parque Natural del Estrecho.

Site description
Tarifa beach is the southernmost and one of the best of the many long sandy beaches of the Costa de la Luz. The beach itself is a broad sweep of fine sand some 3km long, fronting on to the Strait of Gibraltar. Several rivers, notably Río de la Jara and Río de la Vega, combine to form a shallow lagoon just

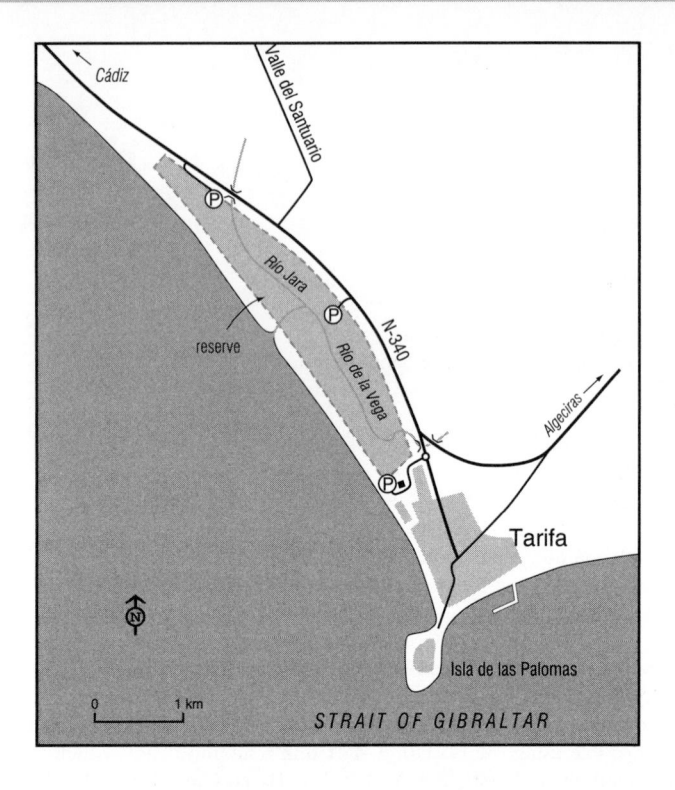

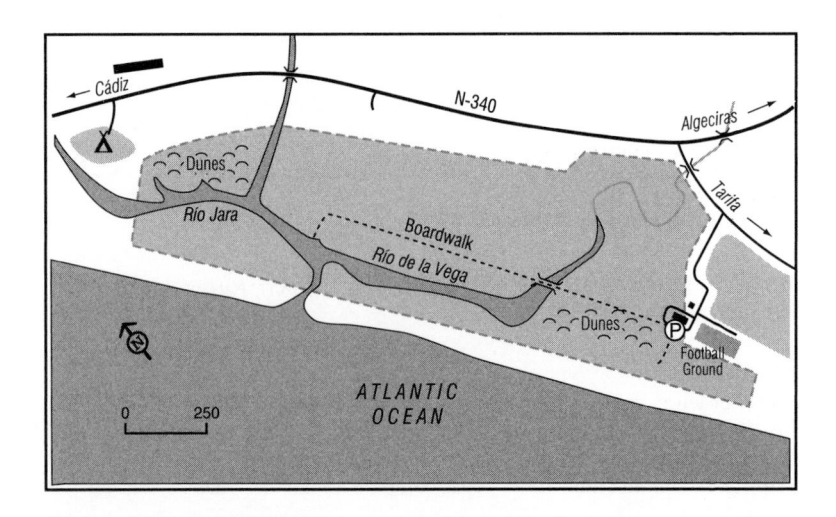

behind the beach proper. The whole area is flooded occasionally by spring tides. The beach is flanked by low sand dunes giving on to rough pasture and Stone Pine plantations. The beach hinterland used to be a favourite and traditional site for hunters, especially finch-trappers in autumn. Happily these were dislodged in the mid-1990s, not without protest, and the area is now a bird reserve.

The beach is popular with tourists, especially the perennial and hardy windsurfers and kitesurfers, and local people. However, its great size allows it to accommodate large numbers of birds even in summer.

Species

The beach is a favourite loafing ground for gulls, including Audouin's Gulls, which are present all year round but are especially numerous from mid-August to October. Tern flocks also form on the lagoon fringes especially and regularly include a few Lesser Crested Terns in May and October/November. Other seabirds are visible offshore. The lagoons attract waders, notably flocks of Sanderlings and Dunlins. Ringed Plovers are usually present and Kentish Plovers breed. The strategic siting means that legions of raptors and other migrants cross the area and many land and rest in the hinterland. Passerines throng the fields and pines at migration times and in winter. The beach is exceptionally well placed to attract scarce or vagrant species, which have included Baillon's Crake, Dotterel, American Golden Plover, Terek Sandpiper, Cream-coloured Courser and Royal Tern.

Timing

Visits are rewarding at all times of year but the largest gatherings of gulls and waders occur in winter. In summer, visits early in the morning will allow you to beat the beach-bums who might later frighten off the birds. At any time of year weekday mornings tend to be the least disturbed periods, although even then the boardwalk is often popular with local people enjoying a brisk stroll. The setting sun inhibits seawatching from late afternoon but is not otherwise a problem. The beach is very exposed and visits on days with strong winds (east or west) are likely to be unproductive and will certainly be most uncomfortable.

Access

Park in the car parks off the N-340 and explore on foot. There are three main points of access: one from the car parks at each end and a rougher central track opposite the CEPSA petrol station at km-82. The northern car park is at km-79.9 or km-80.4. For the southern end take the Tarifa exit at km-83.1 and turn right soon afterwards at a roundabout, to follow the perimeter of the villa development; turn right and then left and continue straight on to the beachside car park. A boardwalk across the inner part of the reserve was inaugurated in 2006, extending from the southern car park, crossing the Río de la Vega as a wooden bridge and continuing to the central area of the beach. The boardwalk is excellent for viewing migrant and wintering passerines on the beachside pastures, and flocks of gulls and waders occur on the fields when the beach is flooded or very disturbed. The boardwalk skirts the main lagoon, where waders and gulls congregate. The northern and southern ends of the beach need to be visited separately since the central river outflow is often too deep and fast flowing to cross.

The southern half is the better choice if time is limited but be sure to walk along the beach itself for views of gulls, terns and waders, as well as to follow the boardwalk trail. Be prepared to wade across the shallow lagoons during spring tides. A telescope is highly desirable.

♿ Only the boardwalk is fully accessible. Other access points are sandy.

CALENDAR

All year: Cory's and Balearic Shearwaters, Northern Gannet, Cattle and Little Egrets, Grey Heron, Ringed and Kentish Plovers, Sanderling, Dunlin, Common Redshank, Arctic and Great Skuas, Audouin's and Yellow-legged Gulls, Thekla Lark, Zitting Cisticola, Spotless Starling, Serin, Greenfinch, Goldfinch, Linnet, Corn Bunting

Breeding season: The above species, the non-passerines mostly represented by non-breeders. Also: Turtle Dove, Common Cuckoo, Hoopoe, Greater Short-toed Lark, Woodchat Shrike.

Winter: Oystercatcher, Golden and Grey Plovers, Lapwing, Knot, Common Snipe, Black-tailed and Bar-tailed Godwits, Whimbrel, Curlew, Common Sandpiper, Turnstone, Grey Phalarope; Mediterranean, Little, Black-headed and Lesser Black-backed Gulls, Kittiwake, Caspian and Sandwich Terns, Razorbill, Puffin, Kingfisher, Calandra Lark, Sky Lark, Crag Martin, Meadow Pipit, White Wagtail, Black Redstart, Common Chiffchaff, Common Starling.

Passage periods: In addition to the above: Purple Heron, Black and White Storks, Spoonbill, Glossy Ibis, Greater Flamingo, raptors, Black-winged Stilt, Avocet, Stone-curlew, Collared Pratincole, Little Ringed Plover, Little Stint, Ruff, Spotted Redshank, Greenshank; Curlew, Green and Wood Sandpipers, Pomarine Skua; Gull-billed, Lesser Crested, Common, Little, Whiskered and Black Terns. All the many species of landbirds crossing the Strait on migration also occur in the area at times: swifts, larks, swallows, pipits, wagtails, chats, warblers and finches are all well represented.

CENTRAL STRAIT OF GIBRALTAR CA2

Status: Parque Natural (in part) & ZEPA: includes the southern portion of PN Los Alcornocales and the eastern part of PN del Estrecho.

Site description

Tarifa is the southernmost point of mainland Europe; the 36th parallel runs through Isla de Las Palomas, the islet just offshore. It lies roughly at the mid-point of the northern shore of the Strait of Gibraltar and only 16km from the nearest point on the African shore (Cape Cires). The hinterland consists of rolling open hillsides, with numerous rocky gullies and outcrops, climbing to the Sierra del Bujeo and Sierra del Cabrito. There are areas of pines and corkwoods, much favoured by resting raptors and other migrants.

Since the 1980s the hilltops north and east of Tarifa have been grotesquely and progressively disfigured by many hundreds of wind turbines. The literal impact of the windmills on birds has been investigated by the Spanish Ornithological Society (SEO/Birdlife) and does not appear to be too significant on a population scale, although some Griffon Vultures and other raptors are killed every year. However, local Egyptian Vulture pairs have been badly affected by fatal collisions of nesting adults with turbines. The original, relatively small turbines, with metal lattice towers, are being replaced by fewer and much larger ones on solid towers, and these are reputedly less perilous for birds since they are more conspicuous and do not provide dangerous perches. However, the greatly increased height and reach of the new turbines may negate these supposed benefits. In any case, the visual impact of turbines, small or large, will always be a problem; in many respects an area of outstanding, indeed world-class, natural beauty has been ruined.

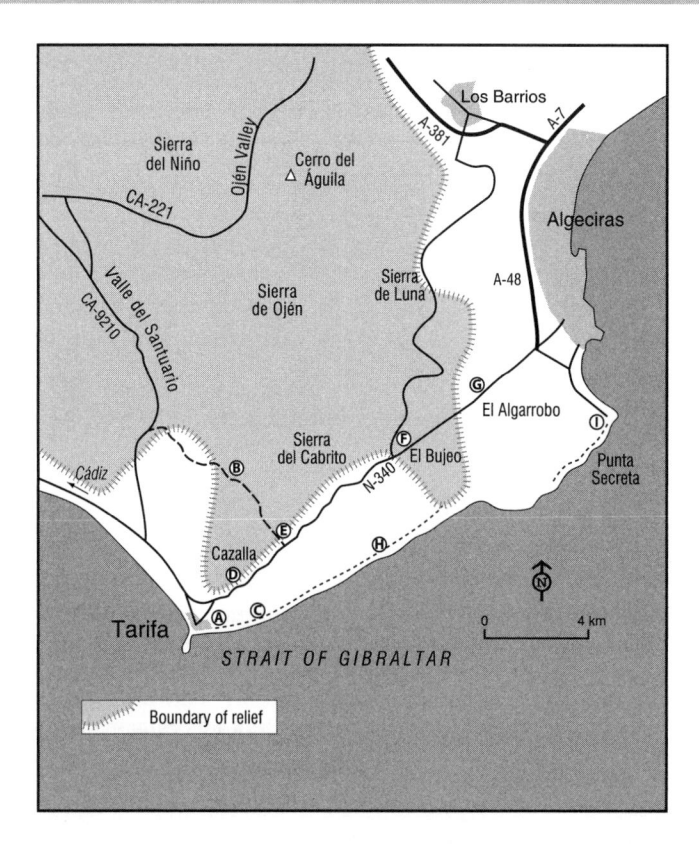

Bird monitoring at the Strait of Gibraltar

The northern shore of the Strait is the area of operations for three highly active organisations which monitor bird activity there: Fundación Migres, Colectivo Ornitológico Cigüeña Negra (COCN) and the Strait of Gibraltar Bird Observatory (SGBO).

The Migres Foundation was established to undertake regular coordinated counts of migrating soaring birds each autumn, from mid-July to mid-October, a project originally funded by the regional government (Junta de Andalucía). Migres is currently supported by the windfarm companies, as part of the mitigation measures for their controversial infrastructure, as well as by local government, Cádiz and Sevilla Universities, and others. The supporting organisations have helped to fund the International Bird Migration Centre (CIMA: Centro Internacional de la Migración de las Aves) – see below. Migres is locally based at CIMA, where in addition to their offices and an observatory (Observatorio del Estrecho), there are library facilities and lodging for volunteers. Volunteer observers are always in demand and should contact the organisers via their bilingual (Spanish/English) website: *www.fundacionmigres.org*, which gives full details. Participants need to commit to help for at least one seven-day period. This is an excellent way to get acquainted with the area and to make a useful contribution to important long-term programmes. Since 2005 Migres has diversified its activities to include monitoring of seabird migration from Tarifa island and regular censusing of waders, gulls and terns at Los Lances (CA1) and Palmones (CA7). They are also involved in a broad range of conservation projects.

The activities of COCN overlap those of Migres and in some ways they may appear to be a rival organisation, although they have collaborated in Migres projects in some years. COCN is conducting a great diversity of specific projects throughout Cádiz province, including bird ringing and the occasional publication of a Cádiz bird report. They maintain observation points around the Strait, notably at Cazalla (see below), and have their own ornithological information centre in Tarifa. See their website for details: http://cocn.tarifainfo.com/spip/.

The SGBO is part of the Gibraltar Ornithological and Natural History Society. See the Gibraltar chapter.

Species

The Tarifa hinterland is rich in the breeding species of rocky, wooded country, including diurnal raptors, White-rumped Swifts and Eagle Owls, and also Iberian Chiffchaffs in the denser riverine woodland. Some very large vulture colonies are here and the Griffons themselves are omnipresent: the strong winds allowing them to fly at any time independently of thermal activity. Other vultures are attracted too: Egyptian Vultures nest and are common on passage, a few Black Vultures occur in some winters and there have been several records of wandering Lammergeiers.

Rüppell's Vultures are a particular birding highlight of the area, especially during the second half of the year, although there are records year-round. A small number arrive every year with Griffon Vultures returning to Iberia, especially in late May and through June. Some linger with the Griffons at and near the Strait, where up to half a dozen individuals may be present at a time. Most of the Rüppell's are immature individuals and most return south annually with the Griffons, in October–November. A few Long-legged Buzzards are another regular feature of the area, which they may be colonising from Morocco, where they are common. Putative hybrids between Long-legged and Common Buzzards, known locally as 'Gibraltar Buzzards', pose perplexing identification challenges in the area.

The chief attraction of the Tarifa region is that it is overflown by the hundreds of thousands, probably millions, of birds using the Strait as a short sea crossing between Europe and Africa. The visible flights of storks and raptors are particularly noteworthy. Important seabird movements also occur through the Strait but these are more easily observed further west, at Punta Secreta (CA3) and Gibraltar.

Timing

The area is of greatest interest during the main migration periods of February–May and August–November. However, it is probably true to say that some migration across the Strait occurs on every single day of the year. Seabird movements through the Strait are also daily events. The breeding species too warrant summer visits and populations of wintering species are also high and of interest.

Access
A number of locations in this area will repay a visit. It is essential to remember that visible migration of raptors and storks is concentrated at the downwind end of the Strait, i.e. towards and east of Tarifa in easterly winds and towards Gibraltar in westerlies. In 'spring' raptors arrive on a broad front but a proportion should be overhead or nearby if you watch from the various points east and west of the mirador mentioned below. The same applies in 'autumn' but then the birds will often be much higher up, having gained altitude for the crossing over the Spanish mainland.

A *Tarifa town* The sea wall in Tarifa town offers a reasonably high vantage point for seawatching; a telescope is highly desirable here. The island (Isla de Las Palomas) is much better but is normally closed to the public, although the local ornithological organisations and those assisting with their survey work have access. Above the harbour, Tarifa castle (Castillo de Guzmán el Bueno) has a small Lesser Kestrel colony.

♿ Not easily accessible.

B *El Mirador/Valle del Santuario* The dirt roads leading inland from the Strait offer outstanding scenery and ample opportunities to stop and search the scrub and woodlands for migrants and breeding/wintering species. A visit is highly recommended. The Cork Oaks here have forsaken the demands of geotropism; the powerful levanter winds have made them all lean decidedly westwards. Spectacular 'falls' of raptors occur during prolonged easterlies in autumn especially: kites and others dot the hills like flocks of small, elongated brown sheep. Bonelli's Warblers are abundant in the cork woods and Cirl Buntings also breed here. The lush vegetation along the streams appears attractive to Iberian Chiffchaffs among others. Traffic is negligible and the roads are in quite good condition but care is needed to avoid potholes and related hazards, such as rockfalls and crumbling verges, along some stretches. The principal road connects with the N-340 at km-91, just west of El Mirador, mentioned below. The dirt road mainly descends northwestwards from El Mirador for 16km, joining the tarmacked road to El Santuario de la Luz at a point 10.8km from the N-340. This tarmac/dirt junction is marked by a house with solar panels and grey chain-link fencing.

The entrance to the Santuario road off the N-340 is on the northern side of the stretch that borders Tarifa beach, just east of Camping Río Jara and conspicuously flanked by tall white masonry brackets. The dry pastures at the southern entrance to the valley and on its northern flanks are excellent for Calandra Larks, which can be quite tame and may be seen at close quarters sitting on the fenceposts in spring. Thekla Larks are very common. Other breeding birds include Montagu's Harriers, Greater Short-toed Larks, Tawny Pipits, Rufous-tailed Scrub-robins and Black-eared Wheatears, alongside a few Little Bustards. Melodious and Cetti's Warblers breed in streamside vegetation. Spanish Sparrows are regular and may breed. The Lesser Kestrels of nearby Tarifa frequently feed in the area. Griffon Vultures, sometimes accompanied by an occasional Black Vulture or Rüppell's Vulture, can often be seen feeding on dead cattle. The area can be visited by driving along any of the roads serving the valley farms. The roads to the wind farms also provide extra avenues of access to the hillsides north of Tarifa.

The Santuario valley itself is a natural conduit from the Strait to the plains of La Janda (CA4) and sees large movements of raptors and, especially, storks. The latter often rest in large numbers at La Janda and fly to or from the Strait via the valley. Hirundine congregations are sometimes spectacular during passage periods.

♿ Most areas are easily viewed from or near your car.

Visible migration watchpoints

C Observatorio del Estrecho. Punta Camorro. The Ornithological Observatory and International Bird Migration Centre run by the Migres Foundation offer a good vantage point during most wind conditions. It is particularly good in spring. The centre (CIMA) is at Punta Camorro, a headland just east of Tarifa. Access is at km-85 on the N-340, along a narrow road that descends to the centre. The watchpoint is a short distance further on.

♿ Easily reached by car.

D Cazalla The Cazalla observatory is on a hilltop just above Tarifa, on the western side of the N-340. It is well placed to receive migrants during light to moderate winds, both easterlies and westerlies. It is particularly useful in autumn since the storks and raptors approaching the Strait from the northwest generally set off southwards from this area. Access is via a short, rough-tarmac track leading off the southbound carriageway of the N-340 at km-87. If you are on the opposite carriageway you should continue to the petrol station between El Mirador (D) and El Bujeo (E) and turn there.

♿ Easily reached by car.

E El Mirador/Alto del Cabrito The public watchpoint over the Strait (Mirador del Estrecho) is actually best avoided since as often as not you will be surrounded by hordes of tourists and the litter they leave sometimes is depressing to see.

The best place to stop, especially if you intend to spend some time here, is the migration watchpoint on the rise of Alto del Cabrito. Take the good dirt road that branches northwards from the N-340 at km-91, just west of El Mirador. Traffic on the main road is often busy and fast and great care is needed here; if you arrive from the south (Tarifa) it may be wiser to overshoot and turn around in the mirador car park instead of cutting across oncoming traffic. Once attained, a signpost at the top of the initial rise indicates a 700m track, which is narrow but easily drivable. Park at the top and view the area from the concrete 'observatory', which is an open, roofed structure, with a table and seats. The site offers commanding views and is only marred by the swishing of the nearby wind turbines. On the positive side, the many access tracks to the turbines offer opportunities for searching the scrub-covered hillsides for grounded migrants and resident passerines.

♿ Easily reached by car.

F Puerto del Bujeo   An excellent and secluded watchpoint for raptor and other migration. It is situated about halfway between Tarifa and the Bay of Gibraltar and as such sees most migration during light winds or on days of moderate westerlies. Access is from the picnic area on the north side of the N-340 exactly at km-95 (marked by a red kilometre post). Take great care when entering/leaving the main road. Once in the picnic area follow the track upwards. The condition of the track is poor but perfectly passable nonetheless and the first 500m are the worst. After this you soon clear the treeline of Cork Oaks and eucalyptus and can enjoy stunning views of Gibraltar, the Bay, the Strait and the Moroccan shore. Migration can be observed from anywhere along here. The intrepid may wish to continue further along this road. It eventually descends through corkwoods to Los Barrios, passing several Griffon Vulture colonies en route. Again, care is urged since the road is in dangerous disrepair in parts. The descent to Los Barrios crosses a few farms; please remember to close all gates.

The woodland of El Bujeo offers a diverse assemblage of breeding passerines. Iberian Chiffchaffs are common and Bonelli's Warblers are abundant. The resident species include the Iberian Green Woodpecker, Dartford Warbler, Firecrest, Crested Tit, Short-toed Treecreeper and Rock Bunting.

&. Accessible by car.

G El Algarrobo   A useful site during westerly winds both in spring and autumn. Access is via a short track from the southbound carriageway of the N-340 at km-99.1, just opposite a collection of radio transmission masts. If you are on the northbound carriageway it is advisable to continue to the bottom of the hill, where you can turn safely. The watchpoint is marked by a concrete shelter on top of a hummock and offers good views along the west side of the Bay of Gibraltar.

&. Accessible by car.

H Tarifa/Punta Guadalmesi coastal path   A footpath inaugurated in 2006 extends east along the coast of the Strait from the sea wall at Tarifa. It follows the route of an old drove road (*via pecuaria*). Look for the green 'VP' marker posts every 50m. The path leads to Punta Guadalmesi, some 10km from Tarifa, offering good opportunities to watch seabirds and other migrants all the way. It is likely to provide good views of spring migration especially. Easy access to the path is available at the Observatorio del Estrecho (see B above).

&. Not accessible.

I Punta Secreta See CA3

CALENDAR

All year: Cattle Egret, White Stork, Griffon and Rüppell's Vultures, Goshawk, Sparrowhawk, Common and Long-legged Buzzards, Bonelli's Eagle, Lesser and Common Kestrels, Lanner and Peregrine Falcons, Little Bustard; Barn, Eagle, Little and Tawny Owls; Calandra, Crested, Thekla and Wood Larks, Crag Martin, Blue Rock Thrush, Zitting Cisticola; Cetti's, Dartford and Sardinian Warblers, Firecrest, Crested Tit, Southern Grey Shrike, Raven, Spotless Starling, Serin, Hawfinch, Cirl and Rock Buntings. (For seabirds, see Gibraltar.)

Breeding season: Quail, Black Kite, Egyptian Vulture, Short-toed and Booted Eagles, Montagu's Harrier, Scops Owl, Red-necked Nightjar; Common, Pallid, Alpine and White-rumped Swifts, Bee-eater, Hoopoe, Greater Short-toed Lark, Red-rumped Swallow, Tawny Pipit, Rufous-tailed Scrub-robin, Black-eared Wheatear; Olivaceous, Melodious, Subalpine, Orphean and Bonelli's Warblers, Whitethroat, Iberian Chiffchaff, Golden Oriole, Woodchat Shrike, Ortolan Bunting.

Winter: Black-winged Kite, Red Kite, Black Vulture, Hen Harrier, Spanish Imperial Eagle, Merlin, Crane, Golden Plover, Lapwing, Woodcock, Alpine Accentor, Fieldfare, Redwing, Common Starling, Brambling, Siskin. (For seabirds, see Gibraltar.)

Passage periods: The potential list is extensive and includes 'frequent rarities'. Regular migrants include: Little Egret, Night, Grey and Purple Herons, White and Black Storks, Spoonbill, Greater Flamingo, Honey-buzzard, Marsh Harrier, Osprey, Hobby, Eleonora's Falcon, Stone-curlew, Collared Pratincole, Stock Dove, Great Spotted Cuckoo, Short-eared Owl, European Nightjar, Kingfisher, Roller, Wryneck, Sand Martin; Tree, Red-throated and Water Pipits, Yellow Wagtail, Bluethroat, Whinchat, Northern Wheatear, Rufous-tailed Rock Thrush, Ring Ouzel; Grasshopper, Savi's, Reed Warbler, Great Reed, Spectacled, Garden, Wood and Willow Warblers, Pied Flycatcher, Spanish and Tree Sparrows, Crossbill, Bullfinch. (For seabirds, see Gibraltar).

PUNTA SECRETA   CA3

Status: Parque Natural & ZEPA. At the eastern end of PN del Estrecho.

Site description
An easily accessible headland on the northern shore of the Strait, just beyond the western entrance to Gibraltar Bay at Punta Carnero. It offers excellent views east to the Rock and southwards to the African shore some 22km away. The hillsides leading down to the shore are largely barren or covered with a thin scrub of Spiny Broom and Fan Palm, with oleanders along the occasional streams, but villa-type developments occupy much of the lower slopes.

Species
An alternative site to Gibraltar for seawatching in the Strait and for observing the northbound arrivals of raptors and storks in spring. The rocky shore attracts resting flocks of gulls, including Audouin's Gulls, as well as waders such as Whimbrels and Turnstones. Single Purple Sandpipers, rare in southern Spain, are seen here with some regularity in winter. A few Shags have been present inshore from time to time in recent years; they quite possibly originate from the small colony in Gibraltar but could be the vanguard of a recolonisation of the Spanish coastline of the Strait. All these birds like to sit on the numerous rocks just offshore. Falls of passerines may occur in spring and, to a lesser extent, in autumn, most frequently during periods of strong, easterly winds.

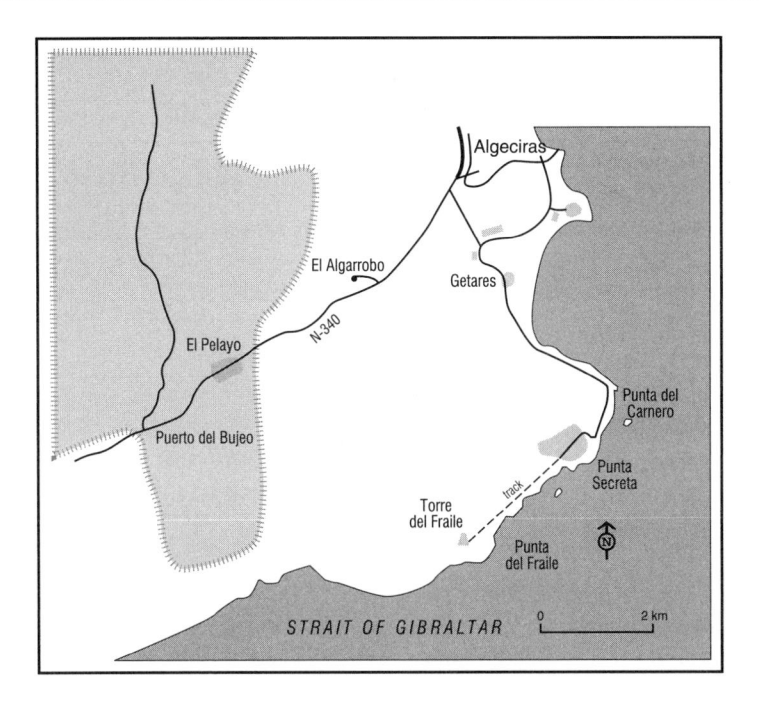

Timing

All year for seabirds. February–June for northward raptor passage: arrivals in this part of the Strait occur chiefly in westerly and calm wind conditions and this is a major watchpoint at such times, when very large movements may occur for long periods. The southward raptor passage passes high over the site and is seen better from Tarifa (CA2) and Gibraltar. February–May and August–November for other migrants.

Access

The coast road from Algeciras to Getares Bay winds on to the lighthouse at Punta Carnero and then round to Punta Secreta. The access road (CA-224/CA-223) is signposted at a palm-adorned roundabout on the N-340 at km-142. Beyond Getares there are frequent small lay-bys giving elevated views of the rocky shore and coastal scrub. This area is always worth searching during passage periods. The roads serving the scattered villas around Punta Secreta provide access to the shore there. Migration watching from Punta Secreta is best done near the small electricity pylon next to the sign Av. Punta Carnero, on the seaward side of the access road. A footpath leads westwards from the seaward side of the coastal villas, following the coastline to Punta Acebuche via Punta del Fraile, a distance of some 3.5km: this provides excellent opportunities to search the scrub and open hillsides for grounded migrants in season.

♿ The migration watchpoint is accessible by car.

CALENDAR

All year: Seabirds: as Gibraltar.

Febuary–May: Storks, raptors and passerines (species as Gibraltar).

August–November: Raptors and passerines. Audouin's Gulls are often numerous in August–September on the rocks.

Winter: Purple Sandpiper.

LA JANDA   CA4

Status: No special protection (20,000ha).

Site description

The plain of La Janda is a remarkable place. A major tract of it, west of Tahivilla almost to Benalup, was formerly an extensive shallow lake, La Laguna de la Janda, which, with its associated reedbeds and marshland, was numbered among the finest of the Spanish wetlands. Among its claims to fame was its status as the last breeding site of the Crane in Spain. Sadly its fate was decided in less conservation-conscious times and it was drained in the early 1960s. In the 1990s the Junta de Andalucía commissioned hydrological studies to see whether the Laguna might be restored, at least in part, but no such developments had materialised by 2018. On the other hand, the southern and western sectors sprouted extensive farms of truly immense wind turbines in 2004–2007, an unhappy development which has quite destroyed the 'endless' open-country vistas, which were a memorable and characteristic feature of La Janda. The wind farms have reduced the area of rough pasture used by open-country birds, including the many storks and other migrants which traditionally use the plain as a staging area. Nevertheless, despite such unwelcome intrusions, La Janda continues to be an exceptionally important and rewarding area to visit.

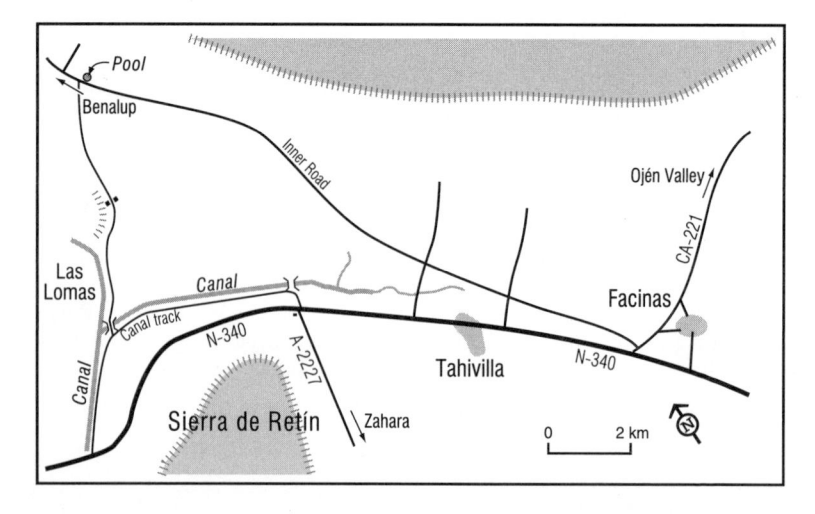

The former lake basin is criss-crossed by deep drainage channels, which retain permanent water and fringing reedbeds. Peripheral reservoirs hold back much of the water that formerly entered the lake. Still, in very wet winters the area floods and La Laguna temporarily reasserts itself. The interest of the site as a wetland was of course vastly diminished once the lake was drained. Nevertheless, the comparatively recent expansion of rice cultivation has helped to sustain the attraction of the region to aquatic birds and other wildlife, although the rice paddies have displaced such dry land species as the Little Bustard.

The remaining plains and pastures are attractive to open-country birds although much habitat has been lost to intensive cultivation in recent years. The area is a magnet for raptors and migrants generally. An extensive and very private estate (Las Lomas), between Vejer and Benalup, includes the large-scale rearing/shooting of pheasants among its sundry activities; that area is thus an unintentional bird-table for raptors, especially in winter. The hillsides along the northern fringe of the plain include a conspicuous triangular escarpment, La Laja del Aciscar, home to a sizable colony of Griffon Vultures.

Species

La Janda offers a taste of the steppes of Extremadura to visitors to southern Andalucía. Little Bustards and Stone-curlews breed although the former have declined and may be elusive. In spring the area resounds to the song of abundant Greater Short-toed and Calandra Larks and Montagu's Harriers quarter the pastures. White Storks breed on most of the farms and occur in very large numbers on migration and on the ricefields in winter. Collared Pratincoles are also characteristic in spring and summer.

The drainage channels have been colonised by Purple Swamphens since the 1990s and these noisy, gawky, conspicuous creatures are readily encountered along the central track. They also inhabit a small reedbed and the adjacent ricefields at the northern entrance to the Facinas-Benalup road (see map). When inundated, the ricefields themselves are popular with migrant waders, including Ruffs and Black-tailed Godwits, and also attract Little Egrets, Squacco Herons, Glossy Ibises, Marsh Harriers and Yellow Wagtails. Black-winged Stilts nest in small numbers. Audouin's Gulls occasionally visit the ricefields in late summer, just as they have learned to do at the Ebro Delta, where many breed, probably looking for crayfish and other invertebrates. Night Herons inhabit the waterside poplars along the road which flanks the western end of the plains; they are elusive in daytime but can be seen flying from their roosts at dusk. Bald Ibises from the nesting colony at Vejer are also present at times.

In winter the calls of Cranes are frequently heard and flocks of up to several hundred are regularly present, mainly from November to February; recent gatherings have exceeded 5,000 birds, attracted apparently by spilled rice on the harvested paddies. Golden Plovers and Lapwings are typical in winter, with other waders, gulls and waterfowl frequenting the ricefields and drainage canals and occurring more widely when the region floods. The Hen Harriers that replace Montagu's Harriers in winter have been joined in recent seasons by one or two Pallid Harriers, which also seem to be overwintering regularly at La Janda. There have also been recent winter records of Greater Spotted Eagles or Greater Spotted x Lesser Spotted Eagle hybrids. Other raptors that occur here outside the breeding season include Black-winged and Red Kites; Bonelli's,

Golden and Spanish Imperial Eagles and Merlins. Black-winged Kites are readily found in La Janda in winter especially, when they may be seen hovering over the pastures and ricefields or sitting on the pylons along the central track. Short-eared Owls are also regular in winter. The western end is a reliable site for finding wintering Booted Eagles, as well as Black-winged Kites, Penduline Tits and Spanish Sparrows.

La Janda has accommodated one or more wintering Long-legged Buzzards, of the North African subspecies *Buteo rufinus cirtensis*, in recent years. Most occur between September and January and there are records of birds arriving across the Strait in late summer, presumably engaged in post-breeding dispersal. Adult Long-legged Buzzards of this form are distinctive, typically almost white-headed, with a dark belly patch, dark flanks and an unbarred orange tail; juveniles are less striking but are also relatively pale headed, darker on the belly, with a brown, lightly barred tail. However, there is evidence that at least some of the Long-legged Buzzards of northernmost Morocco are intergrades between that species and the Common Buzzard, comprising a confusing hybrid form termed the 'Gibraltar Buzzard'. Such apparent intergrades have also nested in the Tarifa hinterland. Pure Common Buzzards are present all year but are more numerous in winter; most are very typical birds and hence much darker than any Long-legged Buzzards. Both species may be found sitting on fence posts and irrigation booms in open country, or hunting over the pastures.

The diversity of habitats and the presence of large numbers of other birds makes La Janda an excellent place to attract, and hence to look for, scarce or vagrant species. Recent records have included African Marabou, Lammergeier, Rüppell's Vulture, Cream-coloured Courser, Sociable Lapwing, Little Swift, Richard's Pipit, Rustic Bunting and Little Bunting.

Timing

The area is rewarding at all times of year. Winter is a particularly interesting season. Visits in high summer are least productive since many birds keep a low profile then. Windy days are also best avoided. Early morning and late evening visits are advisable in hot weather to coincide with peak bird activity and to avoid the heat haze.

Access

La Janda can be accessed from minor roads and farm tracks. The canal track, bordering the main drainage ditch and skirting the ricefields, is recommended. It gives elevated close views of rice paddies, fields and the canal itself, whose reedbeds and muddy margins often repay inspection. The track is usually in good condition although it can become impassable after very wet weather to all but 4WD vehicles. The southern entrance to this track is at km-56 on the east side of the N-340 at the junction with the Zahara road. Access to the northern end is also possible from the N-340 at km-47, via a signposted picnic site.

Further inland, the farm road from Facinas towards Benalup skirts the northeastern edge of the former lake basin for some 20km and gives good views of the pastures and open country frequented by bustards and other 'steppe' species. The road is most easily found by taking the western turning for Facinas off the N-340 and then turning left almost at once on to the wide sandy dirt track marked Camino Agricola Facinas-Benalup, which is the road in question. There is little traffic, allowing frequent stops to be made to scan the fields. Entry to the pastures is not necessary and is downright dangerous wherever cattle are grazing (fighting bulls are reared here). Distances are great and a telescope is advisable. Unfortunately there are several points where the road becomes impassable to non-4WD vehicles in wet weather and you should be prepared to turn back if conditions demand, consoled by the likelihood of seeing more good birds on the return journey.

The northern end of the Facinas-Benalup Road skirts Cork Oak woodland attractive to Common Buzzards and other raptors. There is a small pond and reedbed on the northern side of this road at its junction with the minor road running north/south past Las Lomas. This is worth inspecting, as are the ricefields along the latter road. The 'Las Lomas' road links up with the canal track at its western end, allowing both major sectors of La Janda to be visited in turn; some parts of the road are very poor-quality tarmac, however, demanding caution.

♿ All areas are easily viewed from or near your car.

CALENDAR

All year: Mallard, Cattle and Little Egrets, White Stork, Bald Ibis, Griffon Vulture, Marsh Harrier, Common Buzzard, Little Bustard, Purple Swamphen, Black-winged Stilt, Stone-curlew, Barn and Eagle Owls, Calandra and Thekla Larks, Zitting Cisticola, Common Magpie, Raven, Spotless Starling.

Breeding season: Black Kite, Egyptian Vulture, Short-toed and Booted Eagles, Montagu's Harrier, Lesser Kestrel, Quail, Collared Pratincole, Scops Owl, Red-necked Nightjar, Bee-eater, Hoopoe, Greater Short-toed Lark, Red-rumped Swallow, Tawny Pipit, Rufous-tailed Scrub-robin, Great Reed Warbler, Whitethroat, Woodchat Shrike, Ortolan Bunting.

Winter: Greylag Goose, Wigeon, Teal, Pintail, Shoveler, Little Grebe, Grey Heron, Glossy Ibis, Black-winged and Red Kites, Black Vulture, Hen and Pallid Harriers,

Long-legged Buzzard; 'Spotted', Spanish Imperial, Golden, Booted and Bonelli's Eagles, Merlin, Crane, Avocet; Kentish, Golden and Grey Plovers, Lapwing, Dunlin, Ruff, Jack and Common Snipes, Whimbrel, Curlew, Common Redshank, Greenshank, Common Sandpiper, Great Spotted Cuckoo, Short-eared Owl, Wood Lark, Red-throated and Water Pipits, Penduline Tit, Southern Grey Shrike, Common Starling, Spanish and Tree Sparrows, Siskin, Cirl Bunting.

Passage periods: The main raptor and passerine routes leading to the Strait cross the area and almost anything can turn up here. In addition to all the species mentioned above, the regulars include: Garganey, Squacco and Purple Herons, Black Stork, Honey-buzzard, Sparrowhawk, Osprey, Wood and Green Sandpipers, Roller, Wryneck, Lesser Short-toed Lark, Tree Pipit, Yellow Wagtail, Bluethroat, Common Redstart, Northern and Black-eared Wheatears, warblers and finches.

BOLONIA, ZAHARA AND SIERRA DE LA PLATA   CA5

Status: Parque Natural del Estrecho (in part).

Site description
A sector of the Costa de la Luz, extending some 20–30km northwest of Tarifa. The fine sandy beaches and brilliant light provide a cheerful setting to several interesting sites. The hinterland consists of barren hillsides and rough pastures, strikingly beautiful in spring when the rains bring out masses of wild flowers. These are the famous 'painted fields' and they are worth a visit in their own

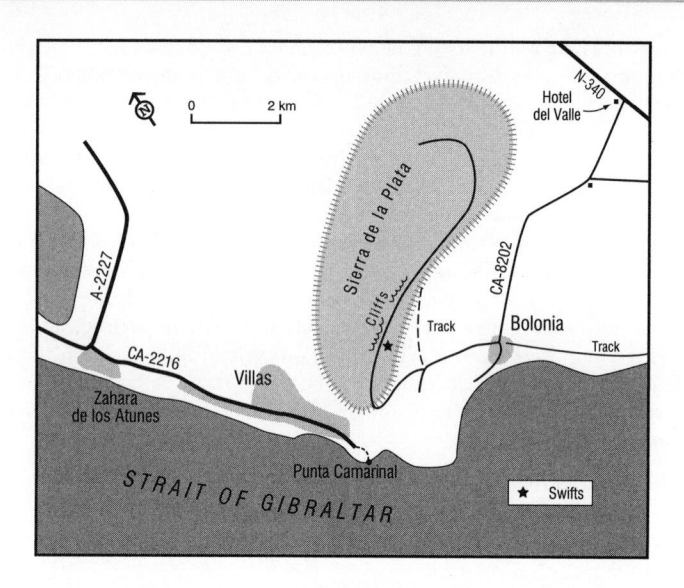

right. The sierras shelter extensive patches of coastal scrub and there are large Stone Pine woods at Bolonia and along the Sierra de la Plata. Watercourses are overgrown with oleanders, which bloom pink in summer. Rocky escarpments on the hillsides support vulture colonies. Zahara offers seawatching and a small estuary.

Species

The area's chief claim to fame is as the first known breeding site in Europe of the White-rumped Swift, a species not previously known to occur closer than tropical Africa until its discovery here in 1964. The swifts use the old nests of the Red-rumped Swallows, which build commonly under overhangs along the slopes of the sierra. Both swallows and swifts are still here, the swifts in particular being present from May to at least September.

The small White-rumped Swift population (fewer than five pairs) may mean a patient wait before you find one but watching in the early mornings and evenings from the southern flanks of the Sierra de la Plata is generally successful. Little Swifts have also colonised the site, much more recently. Sightings of Little Swifts have been annual since 1995, with up to five individuals seen together. Both species could be seen visiting caves on the Sierra de la Plata and Little Swifts were first proved to breed there in 2000, when they were found using a House Martin nest. This was the first recorded instance of breeding Little Swifts in Spain but they have since colonised a few sites elsewhere in Cádiz and Sevilla provinces. The square tails and large, wrap-around white rumps of Little Swifts make them easily distinguishable from the original colonists, with their slender forked tails and small, crescent-shaped white rumps. Since Pallid and Common Swifts are generally present over the area, and Alpine Swifts often occur on passage, this is the only place in Europe where there is any real chance of seeing five swift species at once.

The Sierra de la Plata also has a colony of Griffon Vultures (on the Bolonia side) and Egyptian Vultures have traditionally nested on the facing hillside.

The watercourses at Bolonia are a reliable site for nesting Rufous-tailed Scrub-robins.

Zahara offers seawatching: the species are those that occur generally in the Strait (see Gibraltar). The shallow estuary here holds waders and egrets in small numbers, as well as Garganey and other transient waterfowl on migration.

The whole area is below the flightpaths of storks, raptors and other migrants using the Strait. It is particularly busy during periods of easterly winds: raptors may often be seen arriving at Bolonia Bay and following the Sierra de la Plata during such conditions in spring. Passerine migrants occur abundantly in season.

Timing
May to September for White-rumped and Little Swifts. All times of year will produce interesting birds.

Access
Main roads lead from the N-340 to Bolonia and Zahara, providing simple access to the main sites. White-rumped and Little Swifts may be seen anywhere in the area but the best option is to drive to and past Bolonia and up the eastern side of the Sierra de la Plata to the point where the road levels out as a cliff comes into view straight ahead. Park on the right here, opposite the caves where the swifts nest. The military camp and pinewoods to the west of Bolonia beach are out of bounds and, strictly speaking, it is also not permitted to drive all the way up the tarmac road onto the Sierra, but nobody seems to mind. A migration watchpoint at the col on the CA-2216 Bolonia road, which is sometimes manned during the southward migration, is an excellent site to see massive movements of White Storks in July and August, particularly during easterly winds.

The coastal sites at Zahara are reached by driving southeastwards through the village. The best point for seawatching is from the old, concrete gun emplacements on the rocky headland below the lighthouse at Punta Camarinal. Park in the car park at the foot of the access road and walk up to the headland, which gives elevated views of the coastline. The pinewoods and scrub around the beach and the cove south of the headland are attractive to passerine migrants and worth inspecting during passage periods.

♿ The Sierra de la Plata is easily viewed from or near your car. Punta Camarinal has good (uphill) access but the surrounding sites are not easily reached.

CALENDAR

All year: Cory's and Balearic Shearwaters, Northern Gannet, Cattle and Little Egrets, White Stork, Griffon Vulture, Common Buzzard, Peregrine, Kentish Plover, Arctic and Great Skuas, Blue Rock Thrush, Zitting Cisticola, Dartford and Sardinian Warblers, Raven.

Breeding season: Quail, Egyptian Vulture, Short-toed and Booted Eagles, Montagu's Harrier, Lesser Kestrel, Hobby, Scops Owl, Red-necked Nightjar, White-rumped and Little Swifts, Bee-eater, Hoopoe, Greater Short-toed Lark, Red-rumped Swallow, Tawny Pipit, Rufous-tailed Scrub-robin, Black-eared Wheatear, Melodious Warbler, Whitethroat, Ortolan Bunting.

Winter: Common Scoter, Red-breasted Merganser, Leach's Storm-petrel, Red Kite, Hen Harrier, Osprey, Oystercatcher; Ringed, Golden and Grey Plovers, Lapwing, Sanderling, Dunlin, Whimbrel, Common Redshank, Turnstone, Grey Phalarope; Mediterranean, Little and Audouin's Gulls, Kittiwake, Caspian and Sandwich Terns, Razorbill, Puffin, Crag Martin.

Passage periods: Waterfowl: including Garganey, herons, Black Stork, Spoonbill, Greater Flamingo, raptors including Eleonora's Falcon, waders, Pomarine Skua, Audouin's Gull; Gull-billed, Lesser Crested, Roseate, Common, Arctic, Little and Black Terns, Great Spotted Cuckoo, Short-eared Owl, swifts, Kingfisher, Roller, Wryneck, passerines.

THE CORKWOODS (Los Alcornocales) CA6

Status: Parque Natural & ZEPA (170,025 ha).

Site description

A very extensive attractive region of mainly low, heavily wooded sierras. Cliffs and rocky outcrops are frequent, and numerous streams and small rivers descend the slopes. All or part of seven reservoirs fall within the park boundaries and some of these are of interest. Most of the area is well preserved, being largely devoted to Cork Oak forestry. It is indeed the 'Alcornoques' or Cork Oaks which give the park its name. The wilder forest tracts also include Lusitanian, Pyrenean and Canarian Oaks, White Poplars and Smooth-leaved Elms, with an understorey of rhododendrons. *Rhododendron ponticum* spp. *baeticum*, which causes such problems as an invasive in Britain, is native to the region. However, it is a relic of cooler, wetter periods and can only reproduce vegetatively in its few remaining populations in Iberia (in stark contrast to Britain where the climate is more favourable for seed germination and rhododendrons flourish). Forests of Maritime Pines occur on the higher slopes. Open rocky areas are interspersed by expanses of scrub, including tree heaths and the white-flowered Gum Cistus, and there are large tracts of asphodel steppe in the more heavily grazed regions. Locally there are extensive olive groves, as well as citrus plantations in the lower valleys.

Species

An excellent area for breeding raptors, supporting a key population of Bonelli's Eagles and numbers of Booted and Short-toed Eagles. Both Golden and Spanish Imperial Eagles have a foothold breeding presence. The Griffon Vulture is a flagship species of the park, where there are several large breeding colonies: at least a few always seem to be visible overhead in all areas. The smaller towns, notably Los Barrios and Alcalá de los Gazules, have colonies of Lesser Kestrels. Eagle Owls also have a substantial resident population. The woodlands have a rich passerine community, including a breeding population of Iberian Chiffchaffs, these last in lusher riverine tracts especially, such as those in the central portion of the Ojén valley. Red-rumped Swallows are locally common and have their attendant White-rumped Swifts: the Jimena and Castellar areas are reliable for the latter but they are widespread.

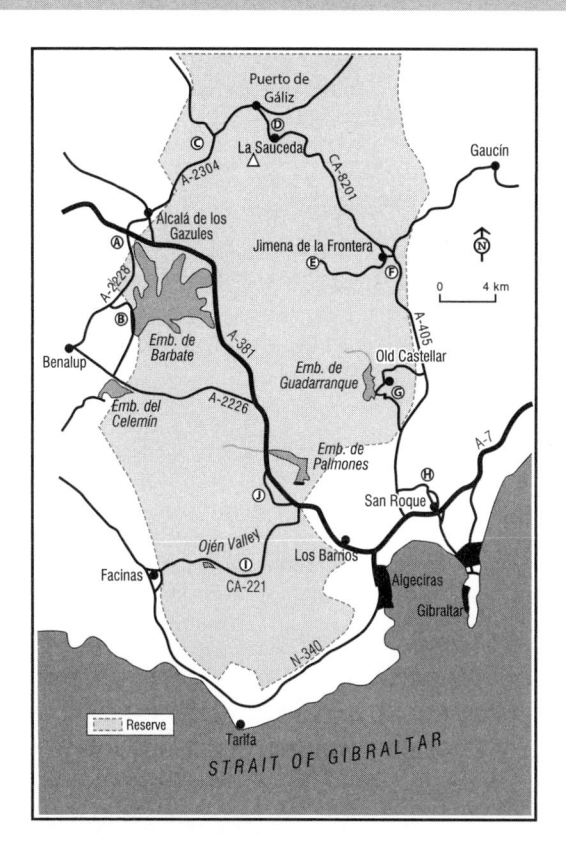

A project to establish nesting Ospreys at the Embalse de Barbate, below Alcalá, has been successful. As usual it involved releasing young birds translocated to the reservoir from elsewhere in Europe (Scotland and Germany), in the hope that they would return to breed there. The site was chosen since Ospreys winter there regularly. A surprising development was that the first Ospreys to nest there, in 2005, were a pair of adults that had not formed part of the project at all: they were winterers that stayed on, becoming the first nesting Ospreys in mainland Spain for over 20 years.

The region is adjacent to the Strait and so receives many passage migrants in transit. The park is crossed by the flyways of raptors and storks and large flocks may be seen throughout the area. The corkwoods themselves are of great importance to migrant raptors, which roost there in droves. At dusk in early May and early September especially, it rains Honey-buzzards here.

In addition to birds, the region has good numbers of Red and Roe Deer, Wild Boar, Genet and Mongoose. The Hozgarganta and other rivers have Otters (rarely seen) and the ubiquitous terrapins.

Timing

The area is rewarding throughout the year. Raptors are always obvious, perhaps especially in spring when the breeding species are soaring and calling over their territories. Visible migration is most evident from late February to early May and again from August to November and involves large numbers of storks, raptors, swifts, Bee-eaters and passerines.

White-rumped Swifts are generally present from May to September. Iberian Chiffchaffs are summer visitors, best detected by song in April and May; they depart in August and are replaced in winter by Common Chiffchaffs, which generally favour more open habitats. There is always considerable passerine activity, even in winter when large numbers of migrants such as Song Thrushes, Robins and Blackcaps arrive from the north.

Access

Much of the area comprises large private estates, including a number of fighting bull ranches. Minor roads provide good access to most parts, however, and numerous forest tracks and well-marked footpaths permit exploration on foot. It is best to drive along slowly, with frequent stops to scan the sky and vegetation (or whenever something interesting shows itself). Areas of corkwoods should be visited to see the full range of passerines. The following sites are especially recommended.

A El Aljibe Visitors Centre An excellent starting point, particularly if you plan to spend some time in the area. This is the visitors' centre for the park and it is located 0.6km south of the A-381 motorway below Alcalá de los Gazules on the Benalup/Alcalá road (CA-2228): leave the motorway at the km-42 exit and follow the brown signs to the Centro de Visitantes. The centre opened in April 2005 and occupies a splendid, airy building featuring displays and presentations of the flora, fauna and traditional lifestyles of the park. There is also a shop selling books, maps and guides. The grounds have been planted with representative native trees and shrubs and are resplendent with wild flowers in spring.

♿ Excellent access.

B Barbate and Celemín reservoirs The Embalse de Barbate is accessible from the A-2228, which skirts the western shore at some distance. A footpath leads down to the reservoir itself and offers opportunities to scan for the Ospreys, which are its chief attraction. Very good views are available from the dam itself: turn off the A-2228 at km-23.1 and drive 2.3km to the Presa (dam) de Barbate. The road crosses the dam and there are parking places and viewpoints at both ends, from where it is also possible to walk down to the margins of the lake. Ospreys may be seen fishing, or sitting on the drowned trees which project from the shallows. Fundación Migres (see CA2) has a field centre at the south end of the dam. It is worth continuing south beyond the dam to join the A-2226 4km east of Benalup. The road is very quiet and skirts woodland and rocky hillsides as well as farmland, all with their characteristic species.

The much smaller Embalse del Celemín is further south and may be inspected from the A-2226. The road is occasionally busy and it is advisable to find suitable spots to pull off the carriageway when stopping. Both reservoirs

Honey-buzzards

attract small numbers of waterfowl in winter, mainly Cormorants and Mallard, but there is always a chance of something more unusual. The large mammals often seen wallowing in both reservoirs look exciting but, unfortunately, are only cows.

♿ The Embalse de Barbate is easily viewed from or near your car.

C Alcalá de los Gazules and El Picacho Alcalá is a reliable site for the colony of Lesser Kestrels, which inhabit the church above the town, chiefly between late February and September. The A-2304 works its way up from Alcalá to the upper reaches of the park at Puerto de Gáliz. A hefty sandstone outcrop, El Picacho, is unmissable to the east of the road 12km from Alcalá. A footpath winds its way towards this crag from a point opposite the field centre (Aula de la Naturaleza) here, providing access to the woodlands and hillsides, with their passerines and flora. The trail ascends for 4km to the very summit (882m) and its panoramic views. The going is heavy and the whole excursion is unlikely to take less than four hours. However, it makes for a very interesting winter or spring walk. A shorter (1.2km) circular trail loops left from the main trail after some 300m and follows the river northwards: it is signposted Garganta de Puerto Oscuro. The less energetic will prefer to stop at the picnic site (Zona Recreativa) on the main road a little north of the field centre, where woodland passerines may be found with comparative ease (except at weekends when it is much disturbed). A further 1.6km north the road crosses open country and there is a watchpoint (mirador) at a col (Puerto de las Palomas) offering views of the whole area, including the curiously bald Sierra de Las Cabras (goat mountain, which may explain the lack of vegetation). This is a good spot to scan for raptors and for open-country passerines such as Cirl Buntings. It also sees significant movements of hirundines, finches and others during passage periods. The whole area offers a strong chance of sightings of Bonelli's Eagles and Peregrines, both of which nest nearby, together with the ever-present Griffon Vultures.

♿ Only the col at Puerto de las Palomas is easily accessible.

D La Sauceda and Puerto de Gáliz La Sauceda is a campsite and picnic area on the CA-3331 and hence often noisy at weekends or on public holidays. It is pleasant at other times and is the starting point of various marked trails through the woodlands. Puerto de Gáliz is a strategic crossroads about 4km north of La Sauceda and reputedly one of the last haunts of the bandits and outlaws who used to hide out in these hills a century and more ago. It is better known now as a useful coffee stop. Both these sites provide panoramic views over the woodlands and are fringed with open pastures. They accordingly provide chances to scan for raptors, which may include the more elusive species such as Bonelli's Eagles and Goshawks as well as the much more obvious Booted and Short-toed Eagles and Common Buzzards. Migrant raptors ascending the Hozgarganta valley are also sometimes numerous in spring. White-rumped Swifts are also present and should be especially looked for around bridges or culverts where Red-rumped Swallows are present, whose nests the swifts usurp. The woodlands contain Wood Larks, Crested Tits, Firecrests and Bonelli's Warblers among their characteristic passerines. A careful inspection of the clearings below La Sauceda in the early mornings and in the evenings often reveals groups of Red Deer as well as the occasional Roe Deer and, with luck, Wild Boar.

♿ Most areas may be viewed from or near your car.

E Jimena television repeater station The road to the TV repeater station west of Jimena provides a peaceful vantage point to watch raptor migration in both seasons, although strong easterly winds will be least useful for this. The same road passes through good cork and pine woods, rock formations, Spiny Broom scrub and open heathland. The views south to Gibraltar and the Strait, and indeed all around, are superb on a clear day. Access to the road is via Jimena. Enter Jimena from the A-405 Gaucín road, take the first significant left turning down to the bridge across the Río Hozgarganta and continue upwards on this road. The road itself is a degraded tarmac track, drivable with due care, which climbs fairly steadily for 12km to the repeater station.

♿ Easily viewed from or near your car.

F Jimena and the Río Hozgarganta The ruined castle tower at the top of Jimena village is a useful perch from which to watch the ubiquitous local Griffons, Crag Martins and Red-rumped Swallows, the latter often accompanied by their presumably unwelcome White-rumped Swift guests. Descend on foot or by car to the river and then follow the path along the north bank; the waterside scrub and trees contain warblers and other passerines. Little Ringed Plovers are present in spring and summer and probably breed. The boulder-strewn river bed shelters a good population of Otters (seldom seen) and terrapins (unmissable) among other wildlife.

♿ Inaccessible.

G Castellar de la Frontera (Old Castellar) The corkwoods between Jimena and Estación de San Roque are known as 'La Almoraima', a place formerly much beloved of picnickers from Gibraltar (including EG during his formative years). The once-abandoned hamlet of Old Castellar overlooks the area from a rocky hilltop, with fine views all round. The original inhabitants of the village were rehoused in the late 1960s in the much more accessible location of New Castellar but their deserted homes came to attract a semi-resident population of 'New Wave' persons and accordingly is not always as tranquil a spot as it might be but the itinerants seem to have declined recently, or perhaps have just become older and less obvious. Castellar also provides a watchpoint for raptor migration and the nearby corkwoods have their passerines. Hirundine and swift flocks should be examined for White-rumped Swifts. Access is along the road to Castillo de Castellar, which branches west from the A-405 at La Almoraima railway station. It is sometimes preferable to park below the village, near the dam, and explore the area from there. The reservoir (Embalse de Guadarranque) is not usually of interest although migrant raptors and others may be seen drinking there at times.

♿ Some good stopping points, including the hilltop itself, allow viewing from or near your car.

H El Pinar del Rey A large area of Stone Pine woods just outside the park to the north of San Roque. The woodlands are often disturbed by picnickers and trail-bikers but are quieter after dusk when the local Red-necked Nightjars may be seen and heard (April–September). Crested Tits are particularly common and Bonelli's Warbler is another characteristic species here. Bee-eaters are numerous in summer. Access is north from San Roque town or east from the A-405 at km-86. The latter road crosses the railway line and after 3km curves south; the entrance gate to El Pinar is here.

The A-405 is notable for the large numbers of White Stork nests occupying pylons between Estación de San Roque and Castellar. They testify to the increase of this species in Spain: I recall that in the 1950s there were only two pairs here (and no pylons). Another change is that the nests are occupied as early as November, many of the birds no longer migrating to Africa. The body of some nests is often tenanted by colonies of House Sparrows.

♿ Accessible by car but the sandy terrain may restrict movement in the Pinar del Rey.

I Ojén valley The scenic road (CA-221) from Los Barrios to Facinas (Ojén valley) is particularly good, crossing 24km of corkwoods, riverine forest, open scrub and rocky hillsides. Raptors and storks pass directly overhead on migration and most of the breeding and wintering species of the region can be found here. Iberian Chiffchaffs are best located by song and are present in the lush riverside woodland where the road reaches the valley bottom; they are summer visitors and present only from April to August. A good site to look for them is at km-8.5 where a 2km footpath (sendero) signposted 'Arroyo de San Carlos del Tiradero' gives access to the riparian woodlands. The mirador at Puerto de Ojén (km-11.1) is a good place to stop and scan the ridgetops and hillsides: Long-legged Buzzard has been recorded here. The small reservoir (Embalse de Almodóvar) at km-16.8 rarely has anything of interest but may be inspected – just in case – from the northern embankment, which is crossed by the road.

The road itself has often been in an appalling state but has generally been good since the pothole-ridden tarmac was stripped off in 2000 and the surface graded. Unfortunately this was done primarily with the interests of competitive cyclists in mind and of late it has been closed to anyone other than pedestrians, cyclists and local farm traffic. This restriction may change but while it is in place you risk a fine if try to drive through. It is still worth walking (or cycling!) for some distance from the Los Barrios end, where you soon reach an area of rocky outcrops and panoramic views. Access to this end of the road requires leaving the A-381 motorway at km-77. If westbound, cross over the motorway. Ascend the short rise and turn left at the green sign (Valle de Ojén). At the other, southern, end the road bypasses Facinas to the north and you should continue straight on at the junction with the CAP-2213 (Valle del Santuario). You can drive as far as the reservoir.

&. Most areas, other than the steep footpaths, are easily reached by car.

J The Devil's Eye   This is a striking rock arch, known locally (and on sounder authority) as La Montera del Torero, or bullfighter's hat, which it certainly resembles. The rock itself is mainly of interest to the Crag Martins that nest within the arch but the location is in picturesque rocky country, with Tree Heath scrub and open cork woodland. The opening of the A-381 motorway has diverted most traffic away from the area and it is a pleasant spot to visit to see such characteristic birds as Wood Larks, Dartford Warblers, Firecrests, Crested Tits, Cirl Buntings and Hawfinches. Common Buzzards, Bonelli's Eagles and Eagle Owls nest in the vicinity. In fact raptors are usually evident, with Griffon Vultures predominating. However, the site is overflown by flight lines to and from the Strait and a large number and variety of soaring birds may be seen from here during migration periods if the wind is right: light or moderate easterly winds are often best. The site is reached by leaving the A-381 motorway at km-77 and turning right on to the camino de servicio (service road), which was the former main road. A picnic area (Area recreativa Montera del Torero) is on the left after 3.6km and is a good stopping site. Another excellent spot is on the left after 5.4km at the entrance to a dirt road, which forms part of a long-range (92km) footpath (Corredor Verde Dos Bahias) linking Cádiz Bay and Palmones; park here and scan the area for raptors and walk along the trail in either direction to find woodland passerines. The service road rejoins the motorway after 6.5km at km-73, where you can also enter the area from the opposite direction.

&. The stopping points have good access. Footpaths are not easily accessible.

CALENDAR

All year: Cattle Egret, White Stork, Red Kite, Griffon Vulture, Goshawk, Sparrowhawk, Common Buzzard; Spanish Imperial, Golden and Bonelli's Eagles, Osprey, Common Kestrel, Peregrine Falcon; Eagle, Little and Tawny Owls, Great Spotted Woodpecker, Thekla and Wood Larks, Crag Martin, Grey Wagtail, Robin, Stonechat, Blue Rock Thrush, Zitting Cisticola; Cetti's, Dartford and Sardinian Warblers, Blackcap, Firecrest, Crested Tit, Short-toed Treecreeper, Raven, Spotless Starling, Serin, Hawfinch; Cirl, Rock and Corn Buntings.

Breeding season: Black Kite, Egyptian Vulture, Short-toed and Booted Eagles, Lesser Kestrel, Little Ringed Plover, Common Cuckoo, Red-necked Nightjar; Common, Pallid, Alpine and White-rumped Swifts, Bee-eater, Hoopoe, Red-rumped Swallow, Tawny Pipit, Rufous-tailed Scrub-robin, Nightingale, Black-eared Wheatear; Melodious, Subalpine, Orphean and Bonelli's Warblers,

Iberian Chiffchaff, Golden Oriole, Woodchat Shrike, Ortolan Bunting.

Winter: Great Cormorant; raptors: including increased numbers of Red Kites and Common Buzzards, occasional Black-winged Kites and Black Vultures, Great Spotted Cuckoo, Sky Lark, large roosts of Crag Martins, Meadow Pipit, White Wagtail, Black Redstart, Song Thrush, Common Chiffchaff, Southern Grey Shrike, Common Starling.

Passage periods: Migrant raptors and storks overhead, passage of Bee-eaters, hirundines and finches. Migrant passerines of a wide range of species can turn up anywhere and falls of such migrants happen in wet, overcast conditions especially. There is always a chance of encountering vagrant species as well as scarce and irregular migrants such as Eleonora's Falcon, Red-throated Pipit, Rufous-tailed Rock Thrush and Fieldfare.

PALMONES ESTUARY CA7

Status: Paraje Natural & ZEPA 'Marismas del Río Palmones' (58ha).

Site description
The flood plain and estuary of the River Palmones lie just north of Algeciras, on the western shore of the Bay of Gibraltar. The western side of the bay has been ravaged by the construction of heavy industry, most conspicuously the large oil refinery. At Palmones the upper floodplain has suffered encroachment by buildings and the construction of a major motorway intersection. However, the Palmones area continues to provide a sizable expanse of open country that often attracts migrants and other species. The upper reaches consist of grazing marsh with reedy ditches. Small patches of salt marsh border the river as it approaches its shallow, sandy estuary.

Species
The principal interest of Palmones is as a staging point for migrant storks, herons, waders and passerines. Flocks of White Storks, often many hundreds of birds, often linger here. Little and Cattle Egrets are everywhere and flocks of waders frequent the riverside and salt marsh creeks. Wader numbers are seldom very high but a good variety of species use the site. Waders and gulls apart, interesting wintering birds include Bluethroats and Penduline Tits. Several Ospreys winter regularly and have attempted to nest on several occasions, on a pylon and on nesting platforms provided for them.

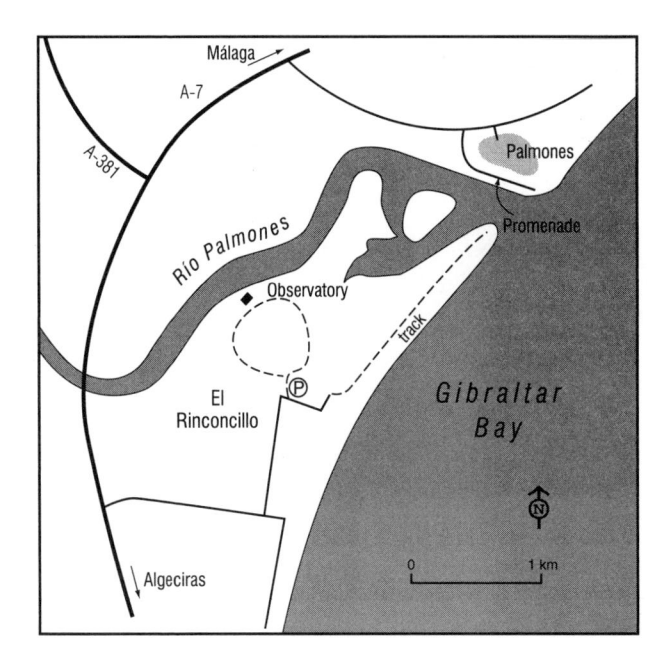

Timing

The site is most rewarding in winter and during the main passage periods.

Access

The estuary and river may be viewed at leisure from Palmones village, where there is an attractive waterfront promenade, conveniently supplied with benches. Palmones is reached from the A-7 motorway; follow the signs from exit (km) 112 southbound or from exit (km) 111 northbound. The marshes are best seen from El Rinconcillo on the south bank. Follow signs to the village from the A-7, south of the river bridge, or from Algeciras waterfront. Turn left on entering the village and follow the road (Calle Camino La Mediana) northwards towards the river; the tarmac gives way to sand before the road curves round to end by the local sewage works, adjacent to the salt marshes. Park by the metal gate, which gives access to a small municipal park, from which there are elevated views across the salt marshes. Explore on foot from here.

Local ornithologists have constructed an observatory on the south bank complete with sun-terrace for open-air viewing of raptor migration overhead. The observatory, managed by COCN (see CA2), and the access tracks give good views across the river and marshes.

♿ The promenade and the observatory access track are accessible.

> **CALENDAR**
>
> **All year:** Cattle and Little Egrets, White Stork, Griffon Vulture, Kentish Plover.
>
> **Breeding season:** Black Kite, Booted Eagle, Kentish Plover, Greater Short-toed Lark, Yellow Wagtail.
>
> **Winter:** Waterfowl, Great Cormorant, Grey Heron, Osprey, waders, Mediterranean and Little Gulls, Sandwich Tern, Bluethroat, Penduline Tit, Reed Bunting.
>
> **Passage periods:** Overhead passage of storks, raptors, hirundines and finches. Glossy Ibis, Spoonbill, Greater Flamingo, wildfowl, waders, terns, chats and warblers.

GUADIARO ESTUARY (SOTOGRANDE) CA8

Status: Paraje Natural & ZEPA 'Estuario del Rio Guadiaro' (36ha).

Site description

Sotogrande is an upmarket residential area and marina built around the estuary of the River Guadiaro, on the Mediterranean coast just north of Gibraltar. Lush gardens, palm groves, golf courses and polo fields appeared 'overnight' in the 1970s and the complex has since expanded very considerably. Nevertheless, the estuary proper is a small reserve protecting an expanse of reeds and tamarisks sheltered from the sea by a sand bar. The area is important as one of the few remaining wetlands on the Costa del Sol and it is popular with waterbirds accordingly.

The area to the north of the estuary includes a small reed-fringed lagoon

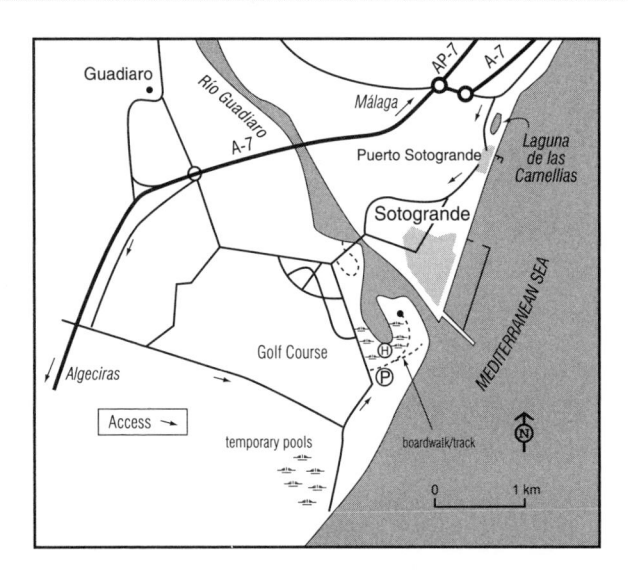

(Las Camelias) and damp scrubby fields, which are also worth visiting, attracting ducks and both migrant and wintering passerines. Temporary pools and tamarisk scrub to the south of the area are of similar interest, especially in winter when flooding occurs.

Species
Gulls and waders are always obvious. The flocks of the former often reward scanning. They usually include Audouin's Gulls, especially in late summer and autumn. Lesser Crested Terns occur with some frequency in autumn especially. A considerable variety of waders occur in small numbers. A range of seabirds are often visible offshore. The freshwater marsh on the south side of the estuary has resident Purple Swamphens. Migrant Ospreys often linger in the area for a few days and occasionally appear in winter. Notable passerines include wintering Bluethroats and Penduline Tits. The area attracts exotic species: a variety of waxbills and other passerines have been recorded here and Monk Parakeets frequent the palm groves.

Timing
Of interest all year round but early morning visits are advisable in summer, when the beach may be busy.

Access
Sotogrande is clearly marked on the A-7 from both directions. The service roads to the complex provide easy access to both banks of the river. The simplest approach is to follow the southern boundary road as far as the sea, where it loops northwards towards the river. Continue along this road until it turns inland again. Park where a large wooden gate, flanked by an information board, gives access to a track leading to the south bank. Explore the beach and sand spit on foot. Here a hide allows the freshwater marsh to be scanned in seclusion and there is also a boardwalk giving access to the central part of the

estuary. The inner estuary may be viewed from a second boardwalk, signposted at the southern end of the east bridge (not the A-7) across the River Guadiaro. Traffic across the bridge is sometimes brisk and you should park off the kerbside between the palm trees by the villas to the south of it.

You can also enter the area from the north. Follow signs from the motorways to Puerto Sotogrande and then bear right for the Guadiaro bridge. Turn left south of the bridge to reach the estuary. The lagoon at Las Camelias is on the beach side of the A-7/Puerto Sotogrande link road. A lay-by adjacent to the south bank of the pool is convenient. From here, a track between a brick wall and a clump of Giant Reeds south of the lay-by provides access to the scrub areas.

♿ The access road to the beach is short and accessible. Other areas are not easily reached.

CALENDAR

All year: Cory's and Balearic Shearwaters, Northern Gannet, Cattle and Little Egrets, White Stork, Yellow-legged Gull, Water Rail, Purple Swamphen, Kentish Plover, Monk Parakeet, Cetti's Warbler, Zitting Cisticola, Common Waxbill (sporadic), Reed Bunting.

Breeding season: Booted Eagle, Little Ringed Plover, Scops Owl, Red-necked Nightjar, Bee-eater, Hoopoe, Greater Short-toed Lark, Red-rumped Swallow, Tawny Pipit, Rufous-tailed Scrub-robin, Nightingale, Reed and Great Reed Warblers, Ortolan Bunting.

Winter: Waterfowl: including Common Shelduck, Common Scoter and Red-breasted Merganser; Little, Black-necked and Great Crested Grebes, Great Cormorant, Marsh Harrier, waders, Arctic and Great Skuas; Mediterranean, Little, Audouin's and Lesser Black-backed Gulls, Sandwich Tern, Razorbill, Kingfisher, Water Pipit, Bluethroat, Black Redstart, Penduline Tit.

Passage periods: Garganey, Little Bittern, Night and Purple Herons, Greater Flamingo, raptors including Osprey, Spotted Crake, waders, Pomarine Skua, Audouin's Gull; Gull-billed, Caspian, Lesser Crested, Common, Little, Whiskered and Black Terns, passerines.

CAPE TRAFALGAR CA9

Status: No special protection.

Site description
Cape Trafalgar, site of the famous sea battle of 21 October 1805, is the rocky, lighthouse-capped tip of a sandy spit west of Barbate. Sandy beaches flank the Cape, the western beach being particularly splendid and unspoilt, sweeping for some 15km towards Conil de la Frontera. A series of low dunes separate this beach from the rough pastures inland. The dune vegetation, and that on the Cape itself, is attractive in spring and is enlivened in summer by the showy, multiple white trumpets of the Sea Daffodil.

Species
The Cape has obvious potential as a seawatching site, during passage periods and in winter. Visible passage of Northern Gannets and terns is often evident and Cory's Shearwaters visit tidal upwellings just offshore. It also marks the westernmost fringe of the raptor flyways across the Strait but soaring birds

generally occur here only when strong easterly winds are blowing. Pools of rainwater near the Cape attract gull flocks in winter, which often include a few Audouin's Gulls. Little Swifts have nested recently in Conil.

Timing
Seawatching from the Cape is recommended in winter and during migration periods. Days with onshore (westerly) winds are likely to be most productive. Late afternoons and evenings can be difficult on bright days as you face the setting sun. The beaches get busy in summer but this should not deter seawatchers.

Access
A branch road leads to the Cape (Cabo Trafalgar) from the Los Caños de Meca/Vejer road (A-2233) to the Cape itself (see map for CA10). This is tarmacked and in good repair although drivers run an excellent risk of getting stuck in the sand if they veer or park off it. Fortunately, numerous hefty windsurfers are reliably available to help extricate bogged-down motorists. The road can be busy in summer when the beaches are quite popular but access is easy at other times. Park at the seaward end and walk up past the barrier to the lighthouse and beyond. The Cape is fairly low-lying but a sheltered spot in front of the lighthouse wall at its northernmost end offers a good point of vantage. A telescope is essential.

♿ Easily reached although wheelchair users will need assistance up the slope.

CALENDAR

All year: Cory's and Balearic Shearwaters, Northern Gannet, Kentish Plover, Sanderling, Arctic and Great Skuas, Yellow-legged Gull.

Winter: Common Scoter, Red-breasted Merganser, Leach's Storm-petrel, Great Cormorant, Grey Phalarope; Mediterranean, Black-headed, Audouin's and Lesser Black-backed Gulls, Kittiwake, Caspian and Sandwich Terns, Razorbill.

Passage periods: Great and Sooty Shearwaters, Pomarine Skua, Audouin's Gull; Gull-billed, Common, Little and Black Terns, Razorbill, Puffin, passerines.

BARBATE PINEWOODS AND ESTUARY   CA10

Status: Partly protected by the 'Parque Natural De la Breña y Marismas del Barbate' (4,817ha). ZEPA.

Site description
The sandy hillsides to the west of Barbate are covered by an extensive wood (2,000ha) of Stone Pines. These are a source of edible pine kernels, which are harvested annually. Large tracts of the woods are of very young trees and there is a well-developed understorey of junipers, Fan Palm and Rosemary. Minor tracks through the pines lead southwards to precipitous sandstone cliffs giving

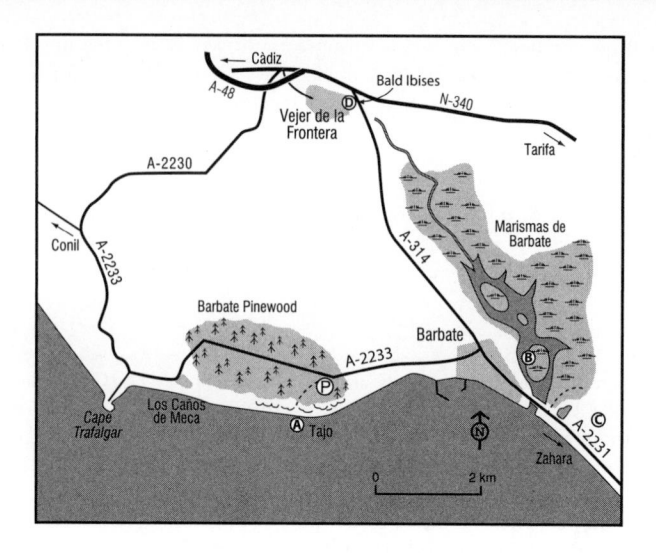

fine views over the western approaches to the Strait and of the Cape Trafalgar promontory (CA9). The Barbate estuary lies immediately east of the town and is flanked by salt marsh and areas of disused salt pans. Pastures and coastal scrub flank the coast road between Barbate and Zahara.

Species

The pinewoods are of most interest during passage periods when they sometimes shelter a wide range of passerine and other migrants. The cliff bases used to house a renowned colony of Cattle Egrets, which here departed from their usual habit of nesting in trees. The nests were on clumps of Oraches on the basal escarpments. The Cattle Egrets have recently abandoned this site, however, after a long period of decline. The cliffs still have a nesting colony of Yellow-legged Gulls as well as Common Kestrels, a pair of Peregrines, Rock Doves, Jackdaws and Ravens. White-tailed Eagles are reputed to have nested here in previous centuries but they are rare vagrants in Iberia nowadays. Ospreys also nested on the cliffs until the mid-20th century and still occur regularly on passage and not infrequently in winter; they do nest at the Embalse de Barbate, a short way inland (CA6-B).

Perhaps the major draw, birdwise, of the area is the nesting colony of Northern Bald Ibises on the cliff face below Vejer. These result from an introduction programme that began in December 2004 at the Sierra de Retín, the rocky ridge just west of Zahara. Some 20 young birds, captive-bred at Jerez zoo, were released from an aviary in the Sierra. Further releases in subsequent years have totalled over 200 birds. Bald Ibises have nested on ledges under overhangs of the low cliff face at La Barca de Vejer since 2012. Here they appear indifferent to the adjacent road and passing traffic, which zooms past just a few metres from the closest nests. The birds forage in open country nearby, including the Barbate estuary, La Janda and the coastal pastures between Zahara and Barbate, sometimes straying as far as Tarifa and the golf courses at Nuevo Sancti Petri. The Barbate cliffs might well attract them too in due course since they use similar escarpments in Morocco.

The estuary and abandoned salt pans immediately east of Barbate attract waders, gulls and terns, especially on passage and in winter, when the flocks are worth scrutinising for rarer species. The estuary is a good site for Caspian Terns, a few of which are usually present in winter.

Timing

Migrants occur in the pinewoods from March to May and from August to November mainly. Nesting activity on the cliff faces, including at the Ibis colony, can be seen from March to May. The Barbate estuary is most interesting in winter and during passage periods.

Access

A Tajo de Barbate (Barbate cliffs) Minor roads and tracks lead off through the pinewoods from the coast road connecting Barbate with the resort of Los Caños de Meca. The cliffs may be approached on foot through the woods. The best starting point is the small car park on the south side of the road, 2.5km above Barbate. Take care when approaching the cliff edges: there is a brisk drop of over 100m to the rocky shore below.

♿ Inaccessible.

B Barbate estuary The estuary may be viewed from the bridge where the Barbate/Zahara road crosses the river. Parking on the bridge itself is not allowed but is possible just before it on the town side. At km-1, just southeast of the town and opposite a cluster of white buildings, a drivable sandy track gives access to an area of disused salt pans, worth exploring for waders and passerines. A roadside muddy pool a short distance south of the entrance to the track is often productive for waders. Further south, some 2.5km from the bridge, there is a broad drivable track along the fringes of the salt marsh north of the road. This can serve to give closer access to the gull flocks in winter too.

♿ Roadside pools and salt pans south of the bridge may be viewed from or near your car.

C The Barbate/Zahara road The A-2231 follows the coast and links Barbate with Zahara. Military signs forbidding access to the shoreline are out of date. Bald Ibises (see above) can occasionally be seen on the coastal pastures here. The coastal scrub and rough grassy patches often hold larks, pipits, wheatears and other passerines, especially during the spring migration. A few sandy tracks allow access. Stopping on the road, except at the few lay-bys, is not recommended since the traffic, though sparse, tends to be fast moving.

♿ Viewable from or near your car at the few possible lay-bys.

D Tajo de Vejer The Bald Ibis colony is on the northern flank of the conspicuous rocky prominence that is crowned by the town of Vejer. The birds are at La Barca de Vejer, where they nest alongside the northern end of the Barbate road (A-314), a short distance from its junction with the N-340. Parking is available on the opposite side of the road a short way past the cliffs. The confiding birds are unmissable when present and offer unsurpassable close views. The total breeding population, some of which nest elsewhere, is only a couple of dozen pairs but numbers are increasing slowly, despite high juvenile mortality.

♿ Viewable from or near your car.

> **CALENDAR**
>
> **All year:** Cattle and Little Egrets, White Stork, Bald Ibis, Greater Flamingo, Griffon Vulture, Peregrine Falcon, Kentish Plover, Rock Dove, Blue Rock Thrush, Raven.
>
> **Breeding season:** Quail, Black Kite, Egyptian Vulture, Short-toed and Booted Eagles, Montagu's Harrier, Lesser Kestrel, Scops Owl, Red-necked Nightjar, Bee-eater, Hoopoe, Greater Short-toed Lark, Red-rumped Swallow, Tawny Pipit, Rufous-tailed Scrub-robin, Black-eared Wheatear; Melodious, Spectacled and Bonelli's Warblers.
>
> **Winter:** Great Cormorant, Hen Harrier, Osprey, Merlin, waders including Grey Phalarope; Mediterranean, Black-headed, Audouin's and Lesser Black-backed Gulls, Kittiwake, Caspian and Sandwich Terns, Crag Martin.
>
> **Passage periods:** waterfowl, herons, raptors, waders, Audouin's Gull; Gull-billed Tern, Common, Little and Black Terns, passerines.

SANCTI PETRI MARSHES   CA11

Status: Paraje Natural & ZEPA (170ha); part of PN de la Bahía de Cádiz.

Site description
A sandbar across the tidal inlet at Sancti Petri shelters an extensive area of salt marsh and saltpans. Just offshore, a ruined castle is picturesquely placed on the islet of Farallón Grande. Sancti Petri itself was an abandoned fishing village until very recently but most of the original buildings have been demolished. There are now three or four bars/restaurants, an active yacht club and (horror of horrors) a shop selling jet-skis. Nevertheless, the waterfront road offers often excellent views of waders on the mudflats, sometimes at very close quarters, providing easy photographic opportunities. Pinewoods and dune scrub south of the town along the splendid Playa de la Barrosa often shelter migrants although most of this area has been heavily developed with golf courses, hotels and chalets (Nuevo Sancti Petri).

Species
Waterfowl, waders and seabirds, especially on passage and in winter. Ospreys are regularly present. Passerine migrants can be abundant at times. Bald Ibises from the Vejer colony occasionally appear on the golf courses here. A particular feature is the enormous abundance on the mudflats of burrowing fiddler crabs, even though the small ones are a particular favourite of predatory Turnstones and Whimbrels. The crab species involved is the West African Fiddler Crab *Uca tangeri*, here near the northern limit of its range.

Timing
Visits may be rewarding at any time of year. However, both the village and the very good beach are very busy in summer and in warm weather generally so visits at other times are advisable.

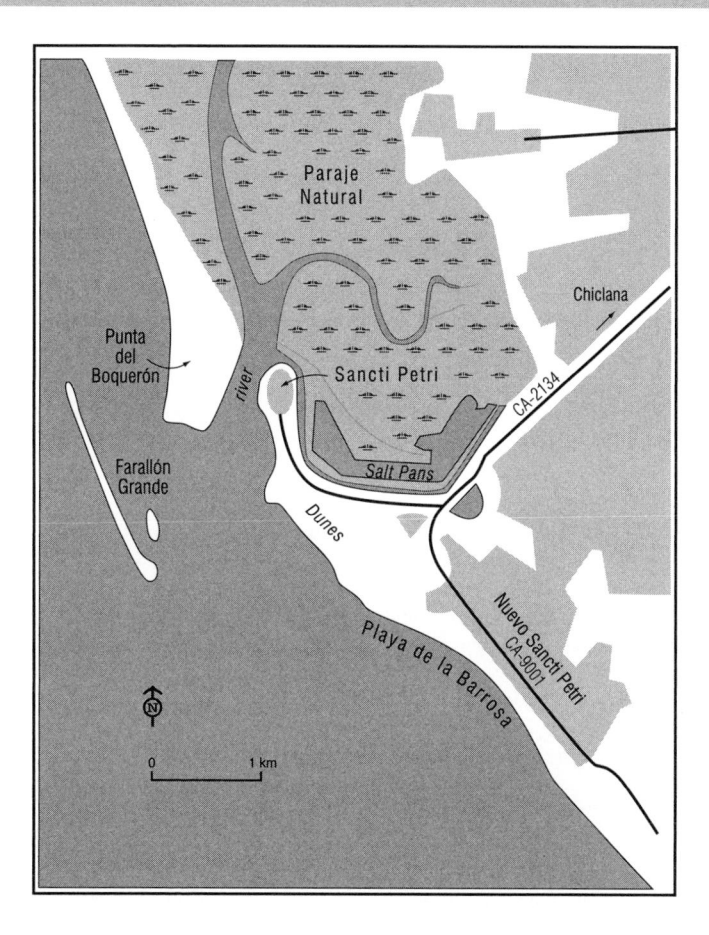

Access

From Chiclana follow signs for Puerto Deportivo Sancti Petri. Straightforward access is available from the A-48: leave the motorway at km-10 and follow signs to Playa Barrosa and then Puerto Deportivo. The river, beach, mudflats and salt marsh are viewed from the village perimeter road, where parking is easy. The salt pans in the south of the area are adjacent to the CA-2134, where there are parking opportunities, and can be viewed through the perimeter fencing. Boat trips – Cádiz Park Safari boats – depart from the jetty in Sancti Petri, where tickets are available from a kiosk. These boats provide an opportunity for closer views of the salt marshes; trips during intermediate states of the tide may be best since the waders then frequent the mudbanks. If migrants are in evidence it is worth following the coast road south from Sancti Petri and visiting the woodlands and dunes there.

♿ Easily viewed from or near your car.

CALENDAR

All year: Cattle and Little Egrets, Grey Heron, White Stork, Spoonbill, Greater Flamingo, Black-winged Stilt, Avocet, Kentish Plover, Sanderling, Dunlin, Common Redshank.

Breeding season: Little Tern.

Winter: Waterfowl: including Greylag Goose, Shelduck, Wigeon, Common Scoter and Red-breasted Merganser; Little, Great Crested and Black-necked Grebes, Northern Gannet, Great Cormorant, Great White Egret, Glossy Ibis,

Hen Harrier, Osprey, Peregrine Falcon, Water Rail, Oystercatcher; Ringed, Golden and Grey Plovers, Lapwing, Knot, Little Stint, Ruff, Jack and Common Snipes, Black-tailed and Bar-tailed Godwits, Whimbrel, Curlew, Spotted Redshank, Greenshank, Turnstone, Grey Phalarope, Arctic and Great Skuas, Mediterranean and Little Gulls, Kittiwake, Caspian and Sandwich Terns, Razorbill.

Passage periods: Seabirds, waders and passerines.

CÁDIZ BAY (Bahía de Cádiz)　　　　CA12

Status: Parque Natural (10,453ha). Ramsar site. ZEPA.

Site description

Cádiz Bay is a natural harbour sheltered behind the rocky spit that bears the provincial capital. The bay is bordered by extensive salt marshes and much of the hinterland comprises salt pans. The Natural Park also takes in large areas of sand dunes and copses of Stone Pines. The Río Guadalete empties into the northwestern portion of the bay. Cádiz Bay is principally a wetland and of great ornithological interest. There is an impressive information centre above the Playa de Camposoto (see B below).

Species

The bay attracts waterfowl and seabirds, including grebes and divers, especially in winter and on passage. At these times too, many thousands of waders frequent the immense expanses of salt pans and salt marshes. There are significant breeding colonies of Yellow-legged Gulls, Black-winged Stilts, Avocets and Little Terns. Spoonbills have also established a small breeding colony. Wintering Ospreys are regular and the Great Northern Diver occurs with some frequency in winter.

Timing

The largest concentrations of waders and wildfowl occur between October and May but the area holds interesting birds all year round.

Access

The A-48 and A-4 motorways intersect in the centre of the bay area and both provide rapid access to the region. Both cross interesting expanses of salt pans but, needless to say, stopping along them is impossible or illegal (but see below). Much of the protected region is inaccessible. The salt pans in particular are privately owned and entry is generally forbidden (although it is always worth asking). Nevertheless, the following sites offer ample opportunities to sample the avifauna of this vast area.

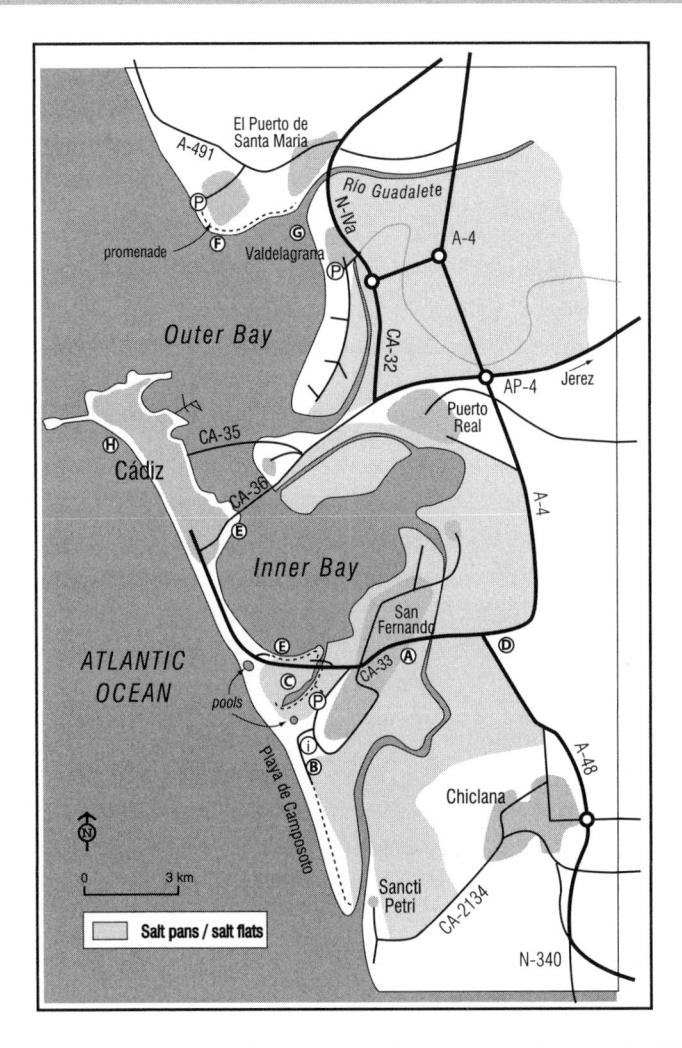

A San Fernando and the southern salt pans (salinas) A bridge on the CA-33 crosses the central creek of the southern saltmarshes, immediately east of San Fernando. Leave the motorway at km-11 and take the San Fernando waterfront road (Avenida San Juan Bosco becoming Avenida Ronda del Estero) south-westwards alongside the creek. Scan the creek and the salt marshes for waders and others; a telescope is all but essential. Waders frequent the muddy fringes of the creek at low tide, affording closer views, and can be viewed from the promenade that follows the creek. Parking is straightforward.

It is especially worth following the Avenida Ronda del Estero to the very end, where signs at a roundabout indicate the Puerto de Gallineras, a marina. A short jetty here provides excellent views, again especially at low tide. Access to the salt pans is also enabled here by a good footpath/cycle track, the Sendero El Carrascón. This heads eastwards through the salt pans and creeksides for up to several kilometres, depending on various options available. It is an obvious way to get closer to waders, flamingos and other birds using the salt pans. The path entrance is off the Puerto de Gallineras roundabout.

♿ Easily viewed from or near your car. The waterfront promenade and Sendero El Carrascón have good access.

B Playa de Camposoto and the Park Visitors' Centre   The Playa de Camposoto is the beach extending southwards from San Fernando to the creek entrance at Sancti Petri (CA11). The beach is the frontage of a long sand bar that encloses an excellent expanse of salt pans. Access is straightforward: drive westwards along the San Fernando waterfront (as above) or skirt the town to the north (see map), to reach a long straight road that separates the beach from the salt marshes and salt pans. The road is a dead end and provides ideal opportunities to view the salt pans along their entire length, about one kilometre. The beach and sand dunes are also worth visiting, especially during passage periods, when migrants may rest on the latter. A footpath leads from the end of the road to Punta del Boquerón, which is the end of the spit opposite Sancti Petri (CA11) and offers additional opportunities to inspect the inlet there.

The visitors' centre for the Parque Natural is on a rise at the northern end of the beach road. It is a sizable two-storey building, offering a very comprehensive range of facilities including a shop, café and panoramic terrace overlooking the salt pans. There is also a short trail through the surrounding area. The centre is signposted 200m east of the roundabout on the road just above the Playa de Camposoto. It is open daily except Mondays during June–September, and except Mondays and Tuesdays during October–May.

♿ The salt pans are easily viewed from or near your car. The information centre and its associated facilities are wheelchair-friendly.

C The northern San Fernando salt pans (salinas)   These lie north of the access road to Playa de Camposoto (B). A car park on the Cádiz side of the road is the starting point of two interlocking footpaths (the Tres Amigos and Río Arillo senderos) giving access to the salt pans, a seasonal lagoon and a small, reed-fringed freshwater lake. The paths extend for some 3.5km and the Río Arillo includes three hides overlooking the salt pans.

♿ Footpaths offer only limited possibilities.

D The central salt pans (salinas)   The A-4 and A-48 motorways cross large tracts of the salt pans but stopping is out of the question. However, an overpass at the southern (Chiclana) end of the A-48, marked 'Cambio de Sentido' (change of direction) gives access to service roads fringing the salinas on both sides of the motorway. It is also obviously possible to transfer between carriageways here. The service roads allow you to stop and view any nearby salt pans that have birds.

The A-4 loops northwards towards Puerto Real from its junction with the A-48. Large numbers of waders are often visible very close to the main road. Again stopping to view is possible from the adjacent service roads. A good entry point is at km-666 on the northbound carriageway of the A-4.

♿ Easily viewed from or near your car.

E The Inner Bay, Salina de Dolores Cádiz Bay is bisected by the two conspicuous road bridges that link the city with Puerto Real and beyond. Good views of the bay are available near the western end of the southern bridge. Follow signs from the CA-33 for Puerto Real via the CA-36 but bear right just before reaching the CA-36 and bridge, where you can see a Ford garage and supermarket. Continue to the waterfront, where you can park to scan the Inner Bay for seabirds and waterfowl.

The southern shore of the bay is also accessible at the Salina de Dolores, near San Fernando. Leave the CA-33 at km-8 and cross over the railway line. This road turns northeast and its central reservation prevents direct access to your destination. Continue a short distance to a roundabout and double back on the opposite carriageway for 200m, where the trail (Sendero) is signposted from a sandy lay-by on the right. You can walk from here or drive in along the access track until you cross the creek (Río Arillo), where you should park. A circular trail (2.5km) from here follows the west bank of the creek to its mouth and then continues along the the bay shoreline westwards, giving views of disused salt pans, salt marsh and the bay itself, before looping back to the start. There are two hides.

♿ Only the northern site is easily viewed from or near your car.

F The Outer Bay Views eastwards over the bay can be had from Cádiz city. The bay is also easily inspected from the eastern side. Access is from the A-491 Rota/El Puerto de Santa María Road at km-24. Follow signs for Costa Oeste just north of El Puerto. Drive straight through the residential area (Urbanización Vistahermosa) to reach the coastal promenades along the bay entrance. The promenade gives elevated views (20m) over the bay and continues right round to the marina (Puerto Sherry). A series of waterfront cafés have the usual outdoor tables so that seawatching can be carried out in unusual comfort. A telescope is essential.

♿ Easily accessed.

G Playa de Valdelagrana This lies just south of El Puerto de Santa María and on the northeastern shore of the Outer Bay. It attracts waders in winter especially. Follow signs to El Puerto via the CA-32 and N-IVa from the A-4 at km-655. Access is via the Avenida del Mar, the first road on the west side of the N-IVa immediately north of the river bridge. Follow signs to the 'Parque Metropolitano'. A footpath leads southwards from Valdelagrana through the Paraje Natural Marismas de los Toruños, a spit comprising salt marshes and dunes at the mouth of the Río San Pedro.

♿ The central track of the marismas is readily accessible.

H Cádiz city The chief attraction here is the sea wall, which offers elevated views and opportunities for seawatching, especially during periods of inshore winds. The city itself has colonies of Monk Parakeets in palms near the Parador and adjacent gardens, which lie at the northwestern extremity of the city peninsula. Pallid Swifts are much in evidence in spring and summer.

♿ Easily accessed.

CALENDAR

All year: Cattle and Little Egrets, Grey Heron, White Stork, Spoonbill, Greater Flamingo, Black-winged Stilt, Avocet, Kentish Plover, Sanderling, Dunlin, Common Redshank, Monk Parakeet, Lesser Short-toed Lark.

Breeding season: Little Tern, Common and Pallid Swifts, Spectacled Warbler.

Winter: Waterfowl: including Greylag Goose, Shelduck, Wigeon, Common Scoter and Red-breasted Merganser, Red-throated and Great Northern Divers; Little, Great Crested and Black-necked Grebes, Northern Gannet, Great Cormorant, Great White Egret, Glossy Ibis, Hen Harrier, Osprey, Peregrine Falcon, Water Rail, Oystercatcher; Ringed, Golden and Grey Plovers, Lapwing, Knot, Little Stint, Ruff, Jack and Common Snipes, Black-tailed and Bar-tailed Godwits, Whimbrel, Curlew, Spotted Redshank, Greenshank, Turnstone, Grey Phalarope, Arctic and Great Skuas, Mediterranean and Little Gulls, Kittiwake, Caspian and Sandwich Terns, Razorbill.

Passage periods: Seabirds and waders. Also passerines.

LAGUNAS DE PUERTO REAL CA13

Status: Reserva Natural & ZEPA (863ha).

Site description

This secluded site comprises three lagoons: Lagunas del Taraje, de San Antonio and del Comisario, surrounded by agricultural land but fringed with reedbeds and Fan Palm scrub. Tamarisks (tarajes) are not particularly prominent. The Laguna de San Antonio, which is largely an excellent reedbed, is fed by water from a nearby waterworks. It drains along a reedy dyke into the Laguna del Taraje, which has extensive open water as well as fringing reedbeds and muddy areas popular with passage waders. Both these lagoons generally enjoy permanent water, thanks to the waterworks, unlike the Laguna del Comisario, which is prone to drying up in some years.

Species

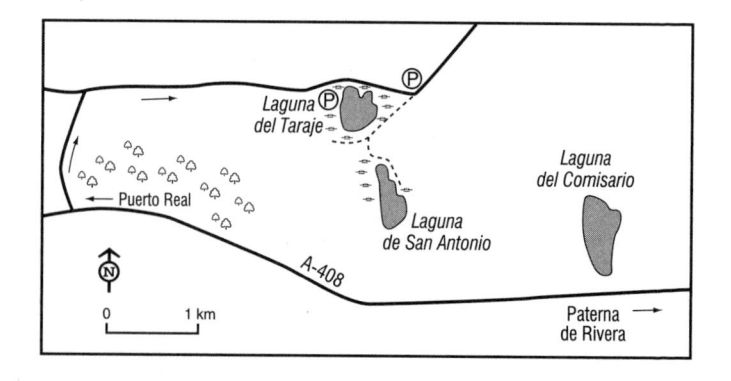

This site is well worth a visit. The reedbeds support nesting Little Bitterns, Purple Herons and Purple Swamphens, as well as Red-knobbed Coots in some years. Raptors are frequent, perhaps attracted by the abundance of rabbits and partridges: they include all three harriers and the occasional Black-winged Kite in winter. Waterfowl and waders occur in winter and on passage especially. Migrant White Storks often descend in great numbers to roost in the vicinity of the Laguna del Comisario in July and August.

Timing
The site is of interest all year round and especially if the Laguna de Medina is dry. Midday visits in hot weather are not advisable because of the haze.

Access
The site lies to the north of the Puerto Real/Paterna de Rivera road (A-408), west of the A-381. Leave the motorway at km-17. A gap in the pines on the right a short distance west of the km-9 marker marks the entrance to a broad dirt road with a green signpost (Parque de Las Cañadas) and a farm sign (La Carrascosa). Follow this road north for 1.5km to an intersection. Turn right and continue for 2.6km to park by the Laguna del Taraje. Explore the lake margin and the surrounding scrub on foot. Good views are available from where the lake first comes into view and also by following the sometimes muddy track which leads from the access road around the western and southern sides of the lagoon. A 500m marked path south of the drainage dyke gives access to the Laguna de San Antonio from the margins of La Laguna del Taraje. La Laguna del Comisario is on private land and has no public access.

♿ Close access by car is only possible to the Laguna Taraje but views are limited.

CALENDAR

All year: Marbled and White-headed Ducks, Little Egret, Marsh Harrier, Purple Swamphen, Common Coot, Calandra Lark.

Breeding season: Little Bittern, Purple Heron, Red-knobbed Coot, Montagu's Harrier, Reed and Great Reed Warblers.

Winter: Waterfowl: including Gadwall, Teal, Mallard, Shoveler, Marbled Duck, Red-crested Pochard and Common Pochard, Grey Heron, Black-winged Kite, Hen Harrier, Osprey.

Passage periods: Waterfowl, waders, Gull-billed and Whiskered Terns, passerines.

LAGUNAS DEL PUERTO DE SANTA MARÍA   CA14

Status: *Reserva Natural & ZEPA (261ha). Ramsar site.*

Site description
One of the most accessible of the several freshwater lagoon complexes in the Cádiz Bay area. There are three lagoons: Laguna Salada, Laguna Chica and Laguna Juncosa. The first of these is the largest and best and the least prone to drying up completely in summer, although it still happens in some years. The three lakes and their surrounding reedbeds and scrub comprise a reserve of 228ha. This is a site of great interest on account of the species that breed and

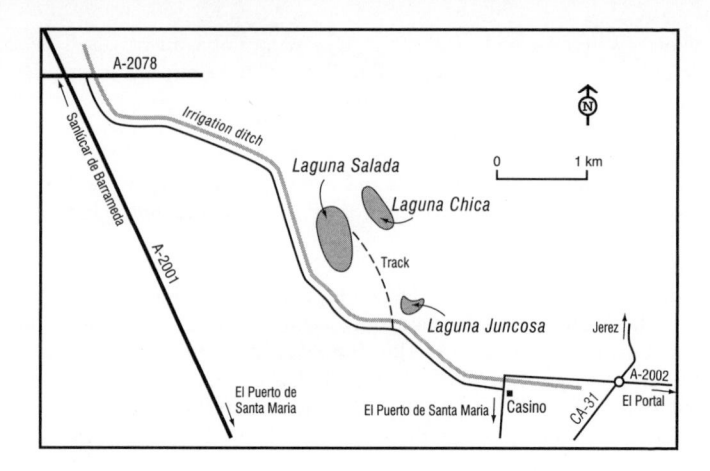

winter here although the general encroachment of villas, vineyards and other developments in the surrounding area has diminished its attraction.

Species
Regular nesters include Little Bitterns, White-headed Ducks and Purple Swamphens. The Red-knobbed Coot has nested in some seasons and Baillon's Crake has been recorded here and may also breed. Hundreds of wildfowl occur in some winters, including Marbled Ducks and Red-crested Pochards as well as White-headed Ducks. The thin bordering scrub of tamarisks can hold a good variety of passerines during migration periods. Waders are also attracted to the muddy lakeshores at most times of the year.

Timing
The site can be rewarding throughout the year except when the lakes dry up completely in dry summers.

Access
A visit here is conveniently combined with one to the Laguna de Medina (CA15). Drive west from the Laguna de Medina and over the A-381 motorway taking the CA-3109 to El Portal. Turn left after the level crossing there and continue on the A-2002 for 6km to reach the CA-31 at a roundabout which is festooned with miniature examples of that local icon, the Osborne bull. Go straight across the roundabout and follow the road round to the Casino Bahía de Cádiz and turn right on to the road that follows an irrigation ditch. After about 3km the ditch is bridged on the right by a short sandy track, which leads through the vineyards to the lakes. Laguna Juncosa is a sedge-filled hollow, which may hold the occasional Common Snipe but not much else. Laguna Chica is also undistinguished. Laguna Salada is accessible on foot: a track leads right around the lake but entry is sometimes prohibited by wardens. In any event, the lake can be inspected from the banks bordering the vineyards nearby.

♿ Not easily accessible.

CALENDAR

All year: (water permitting). White-headed Duck, Little and Black-necked Grebes, Greater Flamingo, Water Rail, Purple Swamphen, Red-knobbed Coot.

Breeding season: Little Bittern, Montagu's Harrier, Baillon's Crake, Black-winged Stilt, Collared Pratincole, Bee-eater, Hoopoe, Reed and Great Reed Warblers.

Winter: Ducks: including Marbled Duck and Red-crested Pochard, Grey Heron, Hen Harrier, Golden Plover, Lapwing,

Common Snipe, Mediterranean Gull, Wryneck, Water Pipit, Bluethroat, Penduline Tit, Southern Grey Shrike.

Passage periods: Purple Heron, storks, waterfowl, raptors, waders: including Little Ringed Plover, Little Stint, Curlew Sandpiper, Ruff, Black-tailed Godwit, Spotted and Common Redshanks, Greenshank; Green, Wood and Common Sandpipers; Common, Little, Whiskered and Black Terns, hirundines and other passerines.

LAGUNA DE MEDINA   CA15

Status: Reserva Natural & ZEPA (355ha). Ramsar site.

Site description
The rolling hillsides of the sherry country near Jerez produce a superb product but they make for a monotonous landscape. The whitish soil is largely covered in vines with extensive areas of cereals and pasture. The Laguna de Medina is a veritable oasis in this agricultural steppe: a shallow lagoon of some 120ha surrounded by an extensive protected zone. The lagoon is fed by a small stream. Water levels are maintained artificially: formerly the lake dried up completely in some summers. The extensive fringing reedbeds and a peripheral scrub of tamarisks and buckthorn hold many birds. There is a panoramic viewpoint near the car park and trails and boardwalks leading to two waterside hides.

Species
The lagoon has long enjoyed a deserved reputation as one of the prime birding sites of Andalucía, on account of the numbers and diversity of waterbirds to be

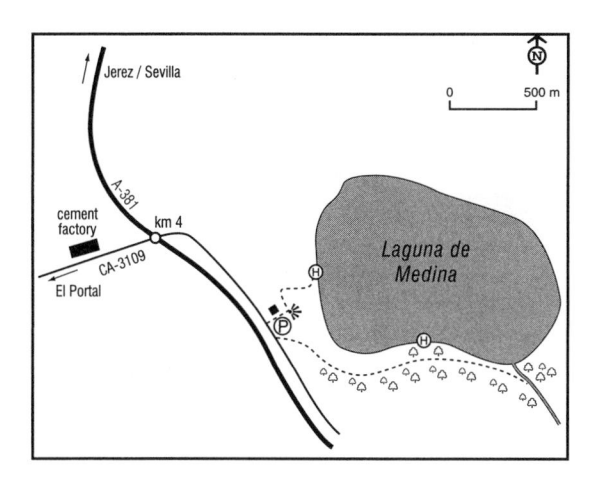

found there, especially in winter. However, in 2005 the lake was reduced to something approaching an ecological desert by the introduction, accidental or otherwise, of carp. These soon increased to great numbers and destroyed most of the submerged vegetation, after which many of the fish themselves died and polluted the water. Waterfowl vanished. Measures to remove the carp and restore the lake were begun in autumn 2006 but the ecosystem was slow to recover. By summer 2018, however, thousands of coots and other waterfowl were once again present there and further improvement may reasonably be expected.

The range of duck species always includes the White-headed Duck, whose small breeding population is greatly increased in winter when several hundred have sometimes been present. Small numbers have reappeared in recent winters. Marbled Ducks may appear in autumn and winter. Garganey occur on passage and Ferruginous Ducks are sometimes present. The ducks share the water with Little, Great Crested and Black-necked Grebes, Greylag Geese and Greater Flamingoes. Whiskered and Black Terns occur on passage and flocks of up to 350 Mediterranean Gulls have occurred occasionally in winter.

Concentrations of Common Coots are a particular feature of the lake. The Coot flocks are largest in late summer and autumn, when many arrive from the Guadalquivir marismas and gatherings of over 25,000 birds form. The lake is also a traditional nesting site of the Red-knobbed Coots, several pairs of which generally nest here. They tend to frequent the shallows and so any coots in the nearby reeds should be inspected carefully. Red-knobbed Coots have longer necks and squarer sterns than Common Coots and their blue beaks are also obvious in good light.

The lake margins are attractive to waders, especially in dry years when water levels drop. Passage waders include Spotted Redshanks, Little Stints, Curlew Sandpipers and occasional rarer species, such as Temminck's Stints and Marsh Sandpipers. The reedbeds have a good population of Purple Swamphens and a few Little Bitterns. Great Reed Warblers and Reed Warblers both breed and Sedge Warblers occur on passage. A spectacular roost of hundreds of thousands of transient swallows and other hirundines develops in the reedbeds in autumn. The peripheral scrub holds a range of maquis species and Quail and Red-legged Partridges frequent the farmland nearby. Wrynecks, Bluethroats and Penduline Tits occur in winter, the last of these being very common.

White-winged Black Tern with Black Terns

The gathering of so many prey species attracts predators. Montagu's Harriers breed nearby and Marsh Harriers are sometimes present. Common Buzzards and Red Kites are frequent, the latter mainly in winter.

Timing

The lake is normally always worth visiting. Winter holds the greatest variety of waterfowl but passage periods are also excellent for waterfowl, waders and passerines. Late spring and early summer are best to see the breeding species, notably Red-knobbed Coots. The largest numbers of waders occur between late July and September in those dry years when a broad muddy margin is exposed as the lake levels fall. As usual, morning and evening visits are essential in hot weather when the heat haze is prevalent.

Access

The lake is clearly visible from the A-381 Jerez/Los Barrios motorway south of Jerez. Leave the motorway at km-4 and follow the signs to the car park. An elevated viewpoint opposite the car park overlooks the whole lake. Closer access is on foot along a boardwalk and track, which follows the southern shore and which gives good views over the reedbeds. A hide halfway down the lake is ideally placed for observing the waterfowl. Follow the track to the end where a feeder stream enters the lake and where muddy stretches between the reeds may enable sightings of Water Rails and, with luck, crakes. A second hide midway along the western shore is reached by a screened trail. This hide is called La Malvasía, after the White-headed Ducks that particularly favour this end of the lake. The other parts of the reserve are out of bounds. Good viewing conditions can be had at any time of day given the many north-facing viewpoints. The lake is large and a telescope is recommended although not essential.

♿ Easily viewed from the boardwalk trail.

CALENDAR

All year: (water permitting). White-headed Duck, Black-necked Grebe, Cattle and Little Egrets, Greater Flamingo, Marsh Harrier, Water Rail, Purple Swamphen, Common and Red-knobbed Coots, Barn Owl, Zitting Cisticola, Cetti's and Dartford Warblers.

Winter: Greylag Goose, Shelduck, Wigeon, Gadwall, Teal, Mallard, Pintail, Shoveler, Marbled Duck, Red-crested and Common Pochards, Tufted Duck, Red Kite, Mediterranean Gull, Wryneck, Bluethroat, Penduline Tit, Southern Grey Shrike.

Spring: Quail, waterfowl: including Greylag Goose (stragglers to early April), Garganey, Marbled Duck, Red-crested Pochard and Ferruginous Duck, Little Bittern, Purple Heron, Red Kite, Montagu's Harrier, Baillon's Crake, Black-winged Stilt, Collared Pratincole, Little Ringed Plover, Whiskered and Black Terns, Bee-eater, Hoopoe, Wryneck, Rufous-tailed Scrub-robin; Sedge, Reed, Great Reed and Orphean Warblers, Penduline Tit, Woodchat and Southern Grey Shrikes.

Breeding season: Quail, Red-crested Pochard, Ferruginous Duck, Little Bittern, Purple Heron, Montagu's Harrier, Baillon's Crake, Black-winged Stilt, Collared Pratincole, Little Ringed and Kentish Plovers, Whiskered Tern, Rufous-tailed Scrub-robin, Black-eared Wheatear; Savi's, Reed and Great Reed Warbler, Woodchat Shrike.

Autumn: Quail, waterfowl including Marbled Duck, Red-crested Pochard and Ferruginous Duck, Little Bittern, Purple Heron, Red Kite, Baillon's Crake, Black-winged Stilt, Avocet, Collared Pratincole, waders including Little Ringed, Ringed and Kentish Plovers, Little Stint, Dunlin, Ruff, Jack and Common Snipes, Black-tailed Godwit, Spotted and Common Redshanks, Greenshank, Curlew, Green, Wood and Common Sandpipers, Whiskered and Black Terns, hirundines: especially Barn Swallow, Penduline Tit, Southern Grey Shrike.

CHIPIONA COAST   CA16

Status: No special protection.

Site description
The town of Chipiona guards the southern approaches to the Guadalquivir estuary: the tall lighthouse (El Faro de Punta del Perro) can be seen for miles around. The rocky foreshore at Chipiona is soon replaced by a sandy coast, running southwards to the Bay of Cádiz at Rota. The beaches here give onto sand dunes, these being better developed at the southern end of the area where dune vegetation and pinewoods often shelter migrants. There are also good stretches of dunes behind the public beaches just south of Chipiona. Nevertheless, much of this coastline has been submerged under holiday homes, golf courses and other touristic developments.

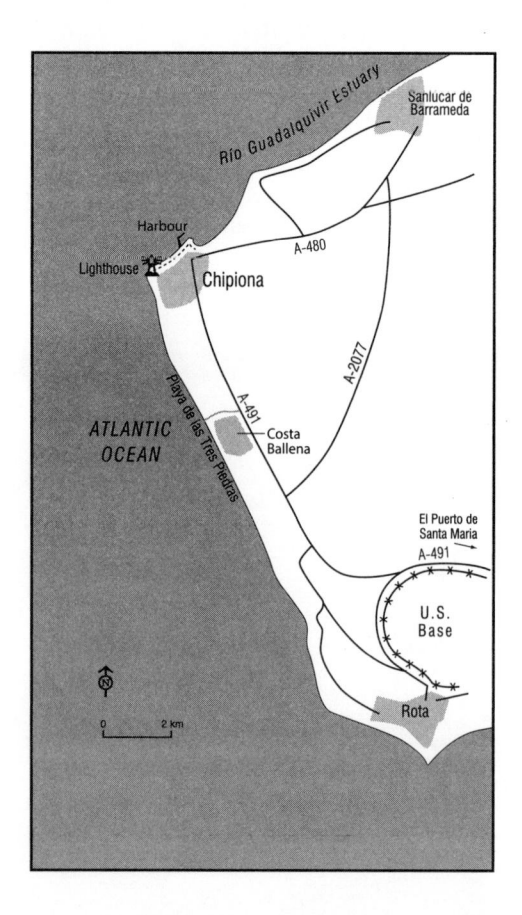

At Chipiona itself a promenade leads northeastwards from the lighthouse, following the southern approaches to the estuary, as far as the harbour. Off the promenade low tide reveals a complex of otherwise submerged rectangular walls, which are the remains of ancient fish traps, reputedly constructed by the Phoenicians.

Species
A major attraction at Chipiona is the small breeding colony of Little Swifts, one of only a handful of such colonies found so far in Spain. The birds nest in the roof of the loading bay of the Port Office at Chipiona harbour, using old House Martin nests that they fill with windblown straw and feathers. They are at least semi-resident, with regular records in mid-winter as well as the breeding season. They are generally easy to see at harbour when nesting.

Seawatching from the lighthouse and along the promenade reveals shearwaters, scoters and other seabirds, notably terns, feeding off the river mouth. Waders and gulls are numerous on the southern shoreline of the estuary, particularly in winter. Gulls, terns and waders often rest on the fish trap walls by the promenade.

Gulls are especially attracted at times to the sandy coastline running south from the lighthouse, especially where a small stream reaches the sea at Playa de las Tres Piedras at the north end of the Costa Ballena development. In winter the gull flocks here have recently included a few individuals of such regionally scarce species as Common, Ring-billed, Herring and Great Black-backed Gulls with some regularity. The dunes often harbour large numbers of migrants during passage periods.

Timing
Summer is best avoided since the beaches are then busy, although seabirds may still be watched for. Seawatching is promising in southwesterlies from autumn through winter to spring. The dunes are most interesting in spring, especially when strong easterlies affect the region.

Access
The lighthouse at Chipiona has a large car park and terrace at its foot, ideal for seawatching: a telescope is advisable. The Little Swift colony is in the white building within the harbour, adjacent to the northern end of the promenade. The harbour and promenade are also easily reached by car. Follow signs to Chipiona town centre (Centro Ciudad) and then those indicating the harbour (Puerto Pesquero/Puerto Deportivo). The promenade also offers convenient bars and restaurants: an excellent place for a birding lunch. The sandy beach running southwards from the lighthouse is readily accessed along the Chipiona/Rota coastal road, from which a number of minor roads lead to the shore. They include the access road to Costa Ballena, which is signposted on the A-491.

At the northern end of the mapped area, the waterfront promenade at Sanlúcar de Barrameda is similarly ideally placed to allow birder-hedonists to consume the local seafood and sherry in some comfort whilst scanning across the Guadalquivir estuary for Spanish Imperial Eagles and suchlike over the Doñana pinewoods opposite.

♿ The lighthouse, riverside promenades, harbour and beach road are readily accessible.

CALENDAR

All year: Seabirds: notably shearwaters, skuas, gulls and terns, Kentish Plover, Little Swift.

Breeding season: Pallid Swift.

Passage periods: Waders, seabirds, Hoopoe, Wryneck and passerines; Woodchat Shrikes and wheatears are often particularly abundant.

Winter: Common Scoter, waders, Gulls, including Ring-billed, Common and Great Black-backed Gulls, Caspian and Sandwich Terns.

EAST BANK OF THE LOWER GUADALQUIVIR   CA17

Status: Within Parque Natural 'Entorno de Doñana'. ZEPA.

Site description

This region forms part of the protected periphery of the Doñana National Park. Its position on the east bank of the river means easier and quicker access from the Costa del Sol and Gibraltar than Doñana proper (the drive from Gibraltar takes about two hours each way) yet most if not all of the species to be seen in Doñana also occur here.

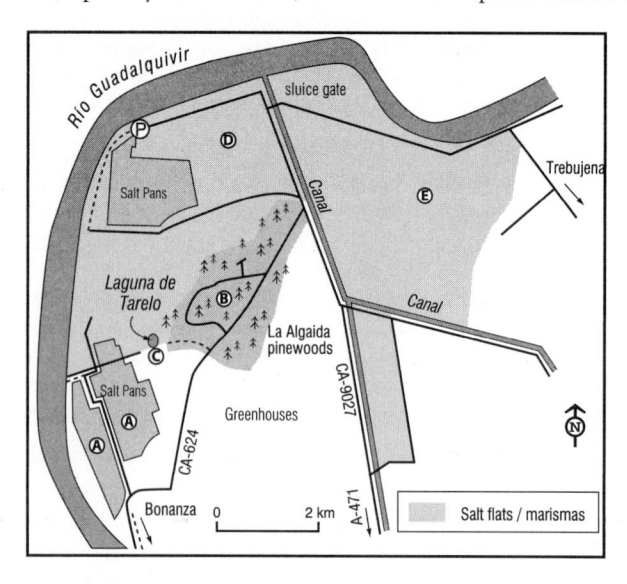

The Bonanza and Algaida Salinas comprise a great expanse of active salt pans and Glasswort flats. The latter continue into the Trebujena marismas, seasonally flooded expanses of salt flats and salt marshes following the river bank northwards from Bonanza. Inland the country becomes gently undulating rough pastureland. Plastic greenhouses are locally an unwelcome intrusion into the salt flats, especially near Bonanza, but their initial expansion seems to have been checked. The town boasts the extensive Stone Pine woods of La Algaida, similar in character to the much larger forests of Doñana just across the river. The freshwater Laguna de Tarelo, adjacent to La Algaida, is always of interest.

Species
The salt pans are of the greatest interest for migrant and wintering waders and waterfowl, which are often very numerous. Scarce species such as Red-necked Phalaropes turn up annually and here a patient observer stands a very good chance of finding vagrant species. Slender-billed Gulls are reliably present for most of the year, especially at Bonanza. Gull-billed Terns breed and Caspian Terns winter. Large flocks of Greater Flamingos are often present. Western Reef Egrets occur occasionally among the many Little Egrets, individuals often staying for many weeks. Great White Egrets and Glossy Ibises are regularly present, especially outside the breeding season. Raptors include Marsh Harriers and great numbers of Black Kites. Spanish Imperial Eagles from the Coto also include the salt pans in their beat occasionally.

The Trebujena area is notable for its views across the river, with its egrets, waders, gulls and terns. Enormous flocks of Lapwings and Golden Plovers occur in winter, when numbers of Stone-curlews and Pin-tailed Sandgrouse are also typical. Lesser Short-toed Larks and Spectacled Warblers breed and are present for all or most of the year. Harriers and other raptors are obvious and include numbers of Red Kites in winter especially.

La Algaida is the only place in Cádiz province which regularly holds Azure-winged Magpies, although they have become very elusive here in recent years. A search through the woods may still result in success. Black Kites breed in numbers and flocks of over 100 form regularly. Red-necked Nightjars also breed locally and may be seen (and heard) readily after dusk from early May and throughout the summer.

The Laguna de Tarelo offers Purple Swamphens and waterfowl. It is particularly noted as a refuge for White-headed Ducks, which breed here and also winter in some numbers. Squacco and Night Herons are often visible on the lake margins. The latter nest on tamarisks on an island in the lake, alongside Cattle and Little Egrets and Little Bitterns.

Timing
All times of year are productive but the largest numbers of birds are present in winter. Wader numbers during migration are also considerable. Heat haze is a problem and midday visits to the salt pans during hot weather will prove frustrating. The Laguna de Tarelo has proved its worth during particularly arid years, when other lakes may dry up in summer.

Access
Take the A-471 Lebrija/Sanlúcar road or the A-480 Jerez/Sanlúcar road to Sanlúcar de Barrameda and continue north on the CA-624 to Bonanza. You can return by the same route or via Trebujena. The CA-9027 that skirts the eastern side of La Algaida pinewoods also leads to/from the A-471 but is somewhat rough in parts.

A Salinas de Bonanza The Bonanza salt pans are reached by driving up a short dirt track from the bend where the CA-624 Sanlúcar/Bonanza road turns inland. They are signposted 'Salinas de Bonanza' and there is a large 'Entorno de Doñana' sign a short way beyond the entrance. The salinas are private but the owners are sympathetic to conservation and allow birdwatchers to enter. Indeed, the site is signposted as a birding site. Waders can be seen at very close range here at times, especially if you use your car as a hide.

♿ Easily viewed from or near your car.

B Pinar de la Algaida Follow the CA-624 northwards to reach the pinewood. Once in the woods the road becomes a generally good, broad sandy track. Numerous trails allow easy exploration of the woods. The Trebujena canal, which runs parallel to the road past the north end of the woods, is a good site for Purple Swamphens.

♿ Easily viewed from or near your car.

C Laguna de Tarelo A permanent freshwater lake situated immediately adjacent to the Algaida pinewoods, at the Bonanza end. A short track leads westwards from just inside the southern entrance to the woods to a watchpoint overlooking the lake. This is not a hide but more of a screen so the birds can see you and will not approach closely. Purple Swamphens are often visible, together with waterfowl such as Black-necked Grebes, both Common and Red-crested Pochards, and White-headed Ducks. A small colony of Cattle Egrets, Little Egrets and Night Herons is established here.

♿ Accessible by car. The 'hide' is 50m away from the car park.

D Salinas de Monte Algaida (Salina de los Portugueses) These salt pans have been easily accessible since February 2000, when the site was inaugurated as a wildlife observation area. This is a wonderful place to visit and strongly recommended. It is far larger than the Bonanza salinas, offering vast open vistas of sky and water and with often very large numbers of a wide range of species. The entrance is a broad sandy track that leads towards the river on the west side of the Trebujena canal sluice gate (see map). Two short stone columns flank the entrance and there is also a sign (Observatorio Salinas de Bonanza). It may be worth noting that the complex of buildings at the sluice gate includes an unmarked bar-restaurant, the only source of food and drink for a long way around.

The access track leads westwards parallel to the river for 2.2km, passing salt flats and shallow, sandy pools. These last often attract wader flocks. Park at the end of the track and explore further on foot along the numerous dyke banks flanking the immense complex of saltpans. Waterfowl and waders frequent these, together with Greater Flamingoes, herons and Spoonbills. The salt flats are covered with glassworts (*Salicornia*) and other halophytic flora. They accommodate both Greater Short-toed and Lesser Short-toed Larks, the former only as summer visitors. Spectacled Warblers and Yellow Wagtails nest here too. The site allows access to the river bank where waders feed along the muddy fringes at low tide and assorted herons lurk in the reeds. Small, scattered trees and bushes should be inspected for passerines, especially during passage periods.

♿ Many interesting areas are easily viewed from or near your car.

E Marismas de Trebujena   The Trebujena area is traversed by long, straight, tarmacked roads giving good views over the river and salt flats. Stop and scan frequently. There are many points where fishermen's paths allow access to the riverbank itself. The drier areas of the salt flats, bordering the hillier agricultural land beyond, are the best place to look for flocks of Pin-tailed Sandgrouse and Stone-curlews in winter. The riverside road that continues northeast beyond the mapped Trebujena turn-off into Sevilla province has long been in poor condition although it is passable. The other roads are in reasonable condition.

♿ Easily viewed from or near your car.

CALENDAR

All the species that occur in Doñana are liable to turn up. The principal species of the region are as follows:

All year: Red-legged Partridge, Quail, Gadwall, Mallard, Marbled Duck, Red-crested and Common Pochards, White-headed Duck; Little, Great Crested and Black-necked Grebes; Cattle, Little and Great White Egrets; Night, Squacco and Grey Herons, White Stork, Glossy Ibis, Spoonbill, Greater Flamingo, Red Kite, Egyptian and Griffon Vultures, Marsh Harrier, Spanish Imperial and Booted Eagles, Lesser and Common Kestrels, Peregrine Falcon, Moorhen, Purple Swamphen, Common Coot, Little Bustard, Oystercatcher, Black-winged Stilt, Avocet, Stone-curlew; Little Ringed, Ringed, Kentish and Grey Plovers, Lapwing, Sanderling, Little Stint, Dunlin, Ruff, Black-tailed and Bar-tailed Godwits, Whimbrel, Curlew, Spotted and Common Redshanks, Greenshank, Marsh and Green Sandpipers; Black-headed, Slender-billed, Lesser Black-backed and Yellow-legged Gulls, Caspian and Sandwich Terns, Pin-tailed Sandgrouse; Barn, Little, Tawny and Long-eared Owls, Hoopoe; Calandra, Lesser Short-toed, Crested, Thekla and Wood Larks, Spectacled Warbler, Short-toed Treecreeper, Southern Grey Shrike, Iberian Azure-winged and Common Magpies, Jackdaw, Raven, Spotless Starling, Tree Sparrow, Cirl Bunting.

Breeding season: Garganey, Little Bittern, Purple Heron, Black Kite, Short-toed Eagle, Montagu's Harrier, Hobby, Collared Pratincole; Gull-billed, Little and Whiskered Terns, Turtle Dove, Great Spotted and Common Cuckoos, Scops Owl, Red-necked Nightjar, Bee-eater, Greater Short-toed Lark, Rufous-tailed Scrub-robin, Nightingale; Savi's, Reed, Great Reed and Melodious Warblers, Spotted Flycatcher.

Winter: Greylag Goose, Shelduck, Wigeon, Teal, Pintail, Shoveler, Tufted Duck, Great Cormorant, Black-winged Kite, Hen Harrier, Sparrowhawk, Bonelli's Eagle, Osprey, Merlin, Crane, Golden Plover, Jack and Common Snipes, Woodcock, Wood Sandpiper, Kingfisher, Sky Lark, Meadow and Water Pipits, Bluethroat, Common Chiffchaff, Penduline Tit, Common Starling, Spanish Sparrow, Siskin, Reed Bunting.

Passage periods: Knot, Temminck's Stint, Curlew and Common Sandpipers, Turnstone, Little Gull, Common Tern, Short-eared Owl, Alpine Swift, Roller, Crag Martin, Tawny and Tree Pipits, Common Redstart, Whinchat, Northern and Black-eared Wheatears; Grasshopper, Sedge, Olivaceous, Subalpine, Orphean, Garden, Bonelli's and Willow Warblers, Whitethroat, Spotted and Pied Flycatchers, Rock Bunting, Ortolan Bunting.

ARCOS AND DISTRICT CA18

Status: Includes two Parajes Naturales/ZEPAs 'Cola del Embalse de Arcos' (120ha) and 'Cola del Embalse de Bornos' (696ha).

Site description

The hilltop town of Arcos de la Frontera is flanked to the south by an imposing sandstone cliff, the Tajo. The Río Guadalete below the Tajo supplies two reservoirs, the Embalses de Arcos and de Bornos, both of which have considerable areas of reedbeds and riparian vegetation attractive to waterbirds. The protected zones of the reservoirs are the northeast sectors in both cases.

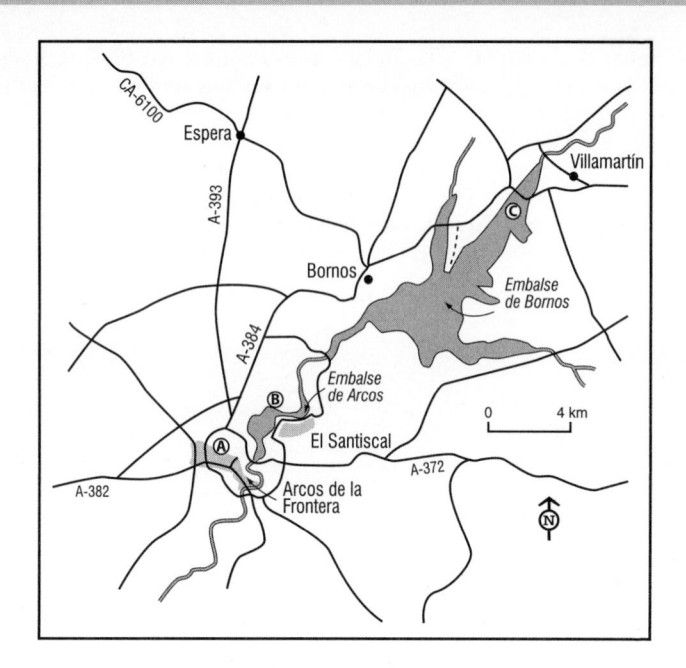

Species

The Tajo at Arcos is home to a colony of Lesser Kestrels, accompanied by Jackdaws and Little Owls. The reservoirs attract numbers of waterfowl, specifically grebes, coots and ducks. Purple Herons and Purple Swamphens nest at both reservoirs and the Bornos reservoir supports a colony of Grey Herons, Night Herons, Little Egrets and Cattle Egrets. The reservoirs are used as staging points by migrating Ospreys, which also occur regularly in winter and have nested at Bornos.

Timing

Interesting species are present all year, although the hot summer months are least productive.

Access

A Arcos de la Frontera   Arcos town lies some 30km east of Jerez on the A-382. The Tajo, which is its chief birding interest, needs to be viewed from above. A mirador is provided for this purpose: follow signs for the Parador, which is adjacent. It is possible to drive all the way to the mirador but it may be preferable to park in the town centre and walk up since the upper streets are dauntingly narrow and there is scant parking available at the mirador. The town has a good population of both Pallid and Common Swifts. The reedbeds on the River Guadalete have Purple Swamphens, which may be seen from the bridge on the A-372 below the town.

♿ Accessible provided you drive to the mirador.

B Embalse de Arcos The site can be viewed from the road that skirts the eastern side of the reservoir and particularly from the village of El Santiscal on the southern bank of the eastern arm. A waterfront road (Avenida Principe de España) gives good elevated views northwards across the western arm from El Santiscal. Reedbeds and sandy islets in the reservoir and open water can be surveyed easily from this position. The best reedbeds are at the eastern end of the reservoir and may be viewed from the road serving the waterfront chalets there. Closer access is provided by a track leading through the waterside vegetation indicated by a Paraje Natural sign.

♿ Only the lakeside is easily accessible.

C Embalse de Bornos This is a much larger reservoir than that at Arcos. The protected area is in the northeast. Access is via a track leading southwards from the A-384 at km-48, west of Villamartín. The track leads to a footpath of some 1.5km, passing through tamarisk woodland to the reservoir.

♿ Not easily accessed.

CALENDAR

All year: Mallard, Great Crested Grebe, Cattle and Little Egrets, Grey Heron, Osprey, Purple Swamphen, Common Coot, Cetti's Warbler.

Breeding season: Night and Purple Herons, Lesser Kestrel, Common and Pallid Swifts, Bee-eater, Nightingale; Savi's, Reed and Great Reed Warblers.

Winter: Waterfowl: including Gadwall, Shoveler, Red-crested Pochard and Common Pochard, Little Grebe.

LAGUNAS DE ESPERA CA19

Status: Reserva Natural & ZEPA 'Complejo Endorreico de Espera' (515ha). Ramsar site.

Site description
A complex of three lagoons: the Lagunas Hondilla, Salada de Zorrilla and Dulce de Zorilla, situated in sheltered, undulating, farming country west of Espera. The lagoons are surrounded by tamarisks and scrub comprising Lentiscs, Olives and Fan Palms. All three have well-developed reedbeds. The site lies near to Arcos (CA18) and a visit to both areas is easily combined. The Espera lagoons are also a worthwhile short detour when travelling between the Jerez district and Grazalema (CA20).

Species
An excellent site for the characteristic waterbirds of the region, with nesting Red-crested Pochards, White-headed Ducks, Purple Swamphens and sometimes Red-knobbed Coots, among others. Marbled Ducks occur in at least some winters and may breed here occasionally. The surrounding area is exceptionally well populated (stocked) with Red-legged Partridges, together with

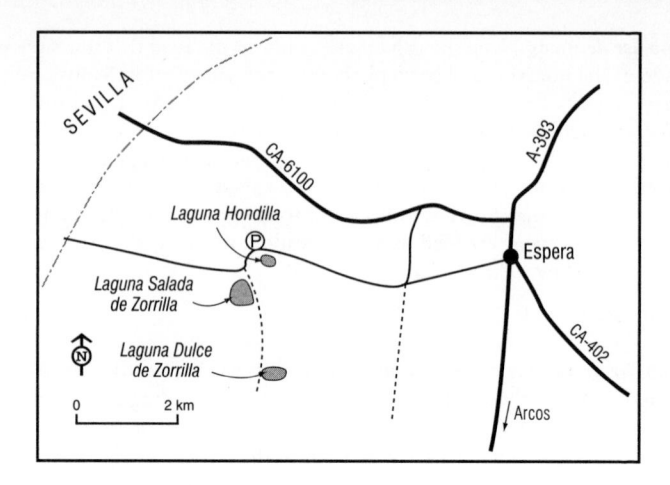

other 'game' such as Rabbits and Iberian Hares *Lepus granatensis*. Stone-curlews and Quail also inhabit the farmland. The area is attractive to raptors, perhaps especially in winter when Hen Harriers are present and a few, chiefly immature, Bonelli's, Spanish Imperial and Golden Eagles occur occasionally.

Timing
The site is usually of interest all year round, the Laguna Dulce usually retaining water even in summer. However, it too dries up completely in drought years. Winter and spring visits are likely to be most rewarding.

Access
Follow signs to the lakes ('Complejo Endorreico de Espera') directly from the southwestern corner of Espera, next to the municipal cemetery, or via the road signposted on the south side of the CA-6100/SE-5207, 2km west of Espera. These are excellent farm roads that lead for some 5km to the Laguna Hondilla, where you should park. The Laguna Hondilla is fenced off but the road is very close to the lake and permits good views. The Laguna Salada is reached by continuing on foot up the road past the Laguna Hondilla and taking the signposted trail that continues south where the road bends to the right. This path leads directly to the Laguna Salada, where a hide offers good views. The trail continues for a further 1.5km to the Laguna Dulce.

♿ Only the Laguna Hondilla is easily accessible by car but it is often well worth visiting.

CALENDAR

All year: Red-legged Partridge, Gadwall, Mallard, Shoveler, Marbled Duck, Red-crested Pochard, White-headed Duck; Little, Great Crested and Black-necked Grebes, Little Egret, White Stork, Greater Flamingo, Marsh Harrier, Common Buzzard, Purple Swamphen, Moorhen, Common and Red-knobbed Coots, Stone-curlew, Barn Owl, Zitting Cisticola, Southern Grey Shrike.

Breeding season: Quail, Short-toed and Booted Eagles, Bee-eater, Reed and Great Reed Warblers.

Winter: Common Teal, Hen Harrier; Spanish Imperial, Golden and Bonelli's Eagles.

SIERRA DE GRAZALEMA CA20

Status: Parque Natural & ZEPA (53,375ha). UNESCO Biosphere Reserve.

Site description
This extensive natural park includes the western portion of the Serranía de Ronda, the limestone mountains of the Betic Cordillera. These are rugged slopes, with bare rocky summits reaching to over 1,600m. The valleys are heavily forested with Cork, Holm and Lusitanian Oaks and stands of Maritime Pine. Locally there are residual woods of the endemic Spanish Fir (Pinsapo). Sheer cliff faces are typical, flanking rocky gorges in which seasonal streams flow briskly and sometimes torrentially: the park boasts the highest rainfall in the entire Iberian Peninsula! The flora is correspondingly lush and exceptionally attractive in spring and early summer.

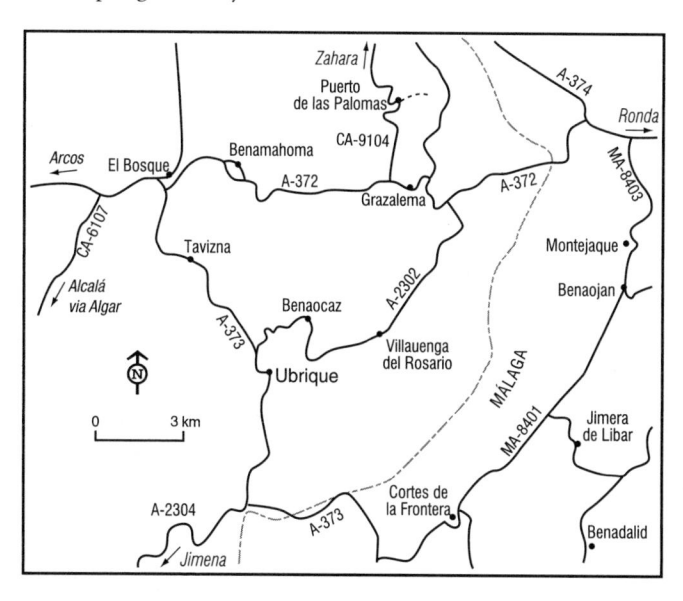

Species
The rocky terrain is attractive to raptors and most sierras of the region have nesting eagles: Short-toed, Booted and either Bonelli's or Golden; the last two are mutually exclusive in their territories. Griffon Vultures nest, commonly accompanied by other cliff-nesting birds, notably Egyptian Vultures, Eagle Owls and Red-billed Choughs. The typical breeding birds also include Northern, Black-eared and Black Wheatears, Rufous-tailed Rock Thrushes, Rock Sparrows and Rock Buntings. Red-rumped Swallows are common and the White-rumped Swifts that they accommodate are also often encountered. Alpine Accentors are regular in winter, when wintering thrushes include Redwings, Ring Ouzels and sometimes Fieldfares. The crags are also shared by herds of Spanish Ibexes. The montane flora is of great interest and includes a number of endemic species.

Timing
The area is worth visiting all year round but is especially rewarding in spring and summer when the breeding species are most evident.

Access
The general area is readily accessible from all points of the compass, notably via Ronda in the east, Arcos in the west, Zahara de La Sierra to the north and Jimena from the south. The climb up from Jimena on the winding road (CA-8201) through Los Alcornocales (see CA6) is scenic but very slow. A quicker alternative is to travel via Alcalá de Los Gazules and Algar, also a scenic and interesting road but much straighter.

The Grazalema area is well served by good (if narrow) roads, which offer opportunities for parking and exploring the hillsides on foot. Grazalema itself is a centre for 'fell' walkers and numerous footpaths ascend the surrounding slopes. Good local maps of the area are available in Grazalema and are a must if you plan to spend some time here and intend to explore widely on foot. It isn't necessary to go far from the main roads to see the typical bird species, however. The roads between Cortes de la Frontera and Benaoján (see MA13), between Ubrique and Grazalema, between El Bosque and Grazalema and between Grazalema and Zahara are potentially rewarding. The Río Majaceite at Benamahoma is a reliable site for Dippers, which are very local in Andalucía. A signposted trail (Sendero del Río Majaceite) follows the river for 5.3km between Benamahoma and El Bosque, with easy access from the car park at the western end of Benamahoma.

The Grazalema/Zahara road rises from Grazalema northwards to a col at Puerto de las Palomas (1,357m). A footpath leads east from the col to the top of the adjacent sierra, offering spectacular views in all directions and a good spot to sit and wait for raptors: Griffon Vultures are common and Bonelli's Eagles also nest nearby and usually put in an appearance if you are patient. Spanish Ibexes are often visible on the opposite peak. The col is also used by migrant raptors in both spring and autumn. In autumn flocks of Black Kites (in August) and Honey-buzzards (in September) tend to be high overhead, gliding southwest towards the Strait. The birds may be more obvious in spring, when many ascend to the pass from the valleys of Los Alcornocales. Migrant raptors here include the full range of species: I have seen both Osprey and Black-winged Kite in transit here. Black Wheatears are also regularly present here as well Rufous-tailed Rock Thrushes, Ring Ouzels and Alpine Accentors in season.

The gorge on the A-2302 below Benaocaz is also a good site for Bonelli's Eagles and for passerines including Black Wheatears and Red-billed Choughs. It is hard to park within the gorge itself but there is a handy stopping place at the southern end on the west side, where there is a mirador. Stop here and scan the area and then walk up the road and across the stony pastures along the base of the gorge.

♿ Good roadside viewpoints are available, including at Puerto de las Palomas.

CALENDAR

All year: Griffon Vulture, Goshawk, Golden and Bonelli's Eagles, Peregrine Falcon, Eagle Owl, Iberian Green Woodpecker, Thekla and Wood Larks, Crag Martin, Dipper, Black Redstart, Black Wheatear, Blue Rock Thrush, Crested and Coal Tits, Nuthatch, Short-toed Treecreeper, Southern Grey Shrike, Red-billed Chough, Raven, Rock Sparrow, Hawfinch, Cirl and Rock Buntings.

Breeding season: Black Kite, Egyptian Vulture, Short-toed and Booted Eagles, Lesser Kestrel, Hobby, Scops Owl, European Nightjar, Alpine and White-rumped Swifts, Bee-eater, Hoopoe, Red-rumped Swallow, Northern and Black-eared Wheatears, Rufous-tailed Rock Thrush; Melodious, Subalpine, Orphean and Bonelli's Warblers, Iberian Chiffchaff, Golden Oriole.

Passage periods: Raptors. Passerines.

Winter: Alpine Accentor, Ring Ouzel, Fieldfare, Redwing.

PEÑON DE ZAFRAMAGÓN   CA21

Status: Reserva Natural & ZEPA (323ha).

Site description
The Peñon de Zaframagón is a conspicuous limestone outcrop, fringed with sheer limestone cliffs, which rises to the north of the Sierra de Lijar and the town of Algodonales. The surrounding pastures on the lower slopes are interrupted by small olive groves and patchy woodland and scrub: of encinas, Hawthorn, Myrtle and Strawberry Tree.

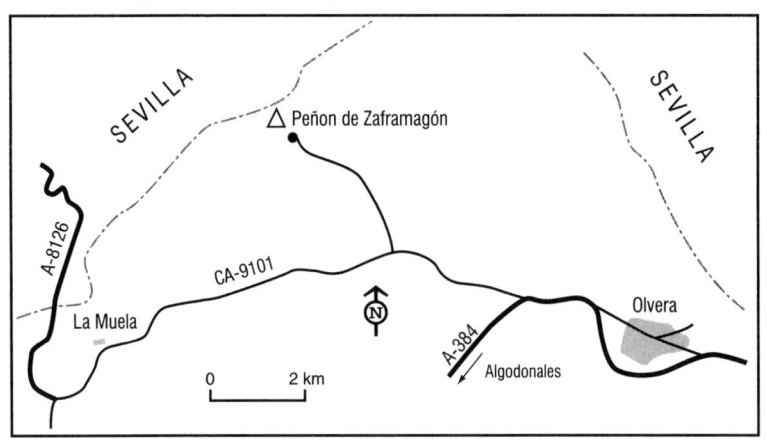

Species
This is the site of a noteworthy colony of Griffon Vultures (c.100 pairs), on the western scarps, with attendant Egyptian Vultures and other raptors and cliff-nesting species.

Timing
Springtime coincides with periods of maximum activity in the vulture colony and other raptors are then at their most visible.

Access

El Peñon lies to the north of the minor road (CA-9101) linking Olvera and the village of La Muela, very near the Cádiz/Sevilla provincial boundary. Turn off this road at km-8, which is about 10km from the junction of the road and the A-8126 to the west. The turning coincides with a cluster of white chalets to the south of the road. The entrance is opposite these: pass under the stone arch beneath the disused railway line, ignoring the alternative right turn. The line itself has been converted into a Via Verde, a hiking and cycling trail, linking Puerto Serrano and Olvera.

The dirt road heads north for 4.1km to the farm (Cortijo de Zaframagón) at the foot of El Peñon. The going is very bumpy but is otherwise good. Drive through the farm and park immediately to the west of it. El Peñon may be inspected from here. Alternatively, follow the obvious footpaths up the olive groves to the base of the cliffs, from where tracks skirt the fringes of the rock. The gritty soil makes for a devastating climb, not recommended in hot weather. The compensation is close views of the vultures and other inhabitants of the cliffs, and fine vistas across the barren hills to the town of Olvera, with its commanding Arab castle on a distant summit.

♿ Inaccessible.

CALENDAR

All year: Griffon Vulture, Bonelli's Eagle, Eagle and Little Owls, Red-billed Chough.

Breeding season: Egyptian Vulture, Lesser Kestrel, Alpine Swift, warblers and other passerines.

SEVILLA PROVINCE

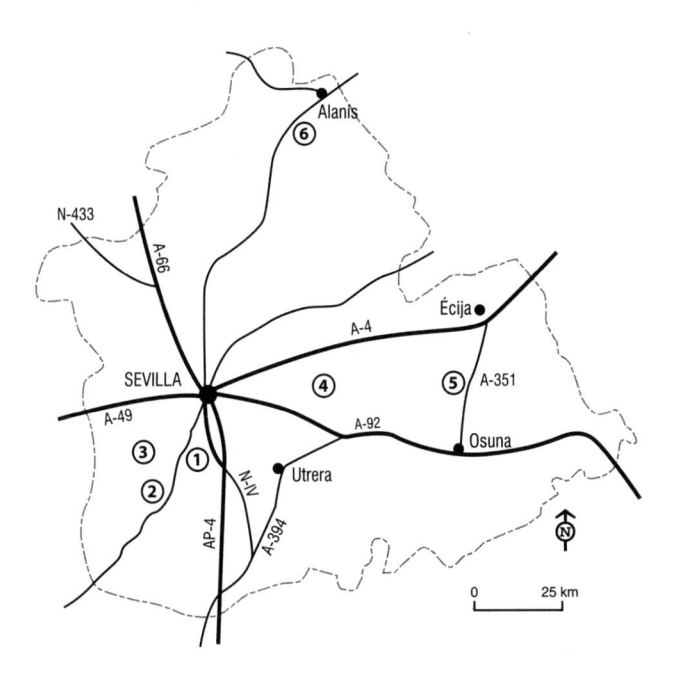

Sevilla province is dominated by the Guadalquivir valley, most of which is intensively cultivated. The greatest birding interest is the southwestern portion, which includes the eastern sectors of Doñana National Park. Sites SE1 and SE2, which fall within this region, are without doubt two of the very best places in Spain for finding and watching birds and you are strongly urged to include them in any itinerary. (The major part of Doñana National Park is in Huelva province, site H5.) The smaller wetlands of the province (e.g. SE3, SE5) often hold a good variety of waterfowl, especially in winter. The agricultural regions of the Guadalquivir valley (SE4) are noteworthy for their thriving populations of Rufous-tailed Scrub-robins and Olivaceous Warblers, which are often elusive elsewhere, and also support small populations of steppe birds, including Great Bustards. The northern sierras (SE6), a section of the Sierra Morena, are also worth a visit: they offer a good range of raptors and Mediterranean woodland species. The interesting reservoirs in the east, on the Sevilla/Córdoba border, are described in the Córdoba chapter (sites CO6 and CO7).

Sites in Sevilla Province
SE1   Eastern Guadalquivir Delta. Brazo del Este
SE2   Western Guadalquivir Delta
SE3   Dehesa de Abajo
SE4   Western Guadalquivir Valley Farmlands
SE5   Lagunas de La Lantejuela
SE6   Sierra Norte

Getting there
Sevilla itself is highly accessible via its international airport and the high-speed train from Madrid. The city is the hub of a number of motorways connecting it with Extremadura, the rest of Andalucía and Portugal. Access to the sites mentioned here is via the AP-4 Sevilla/Cádiz toll motorway (SE1–SE5) and from the A-66 Sevilla/Extremadura motorway for the north of the province (SE6).

Where to stay
Sevilla city has numerous hotels, including motels along the access roads and the Parador nearby at Carmona. Hotels and/or pensiones are also available in the Sierra Norte, at Cazalla de la Sierra, Constantina and San Nicolás del Puerto. With the exception of SE6, the sites described here are easily accessible from a base in Cádiz province, however, and they are (just) within a return day's journey of Gibraltar.

EASTERN GUADALQUIVIR DELTA. SE1 BRAZO DEL ESTE

Status: Paraje Natural & ZEPA (1,343ha).

Site description
The Guadalquivir delta includes a number of subsidiary channels to the main river which delimit several large 'islands'. The Isla Menor, on the east bank, is crossed by a former branch of the river, which in parts now comprises a shallow winding lagoon with extensive fringing reedbeds, surrounded by farmland, ricefields and a veritable labyrinth of drainage ditches. Dirt roads criss-cross the entire area and skirt the main marshes.

The Brazo del Este is a superb site that cannot be recommended too strongly. Indeed, for many species of aquatic birds it is often better than Doñana proper and it is certainly much more peaceful than the National Park itself, since traffic and tourists are absent.

Species
As might be expected, the avifauna resembles that of the marismas in Doñana nearby. However, the numbers of certain species are noteworthy and make visits here so memorable. This is one of those places where 100 species or more may be encountered in a single visit within a fairly limited area.

Great Bitterns, as well as Little Bitterns, are seen here fairly frequently and may breed in some years. Squacco Herons breed and so do large numbers of Purple Herons. Night Herons are common, especially along the river, and appear in numbers at dusk when they leave their daytime roosts. Spoonbills and egrets are often prominent: the latter include a few Great White Egrets, which are regular winter visitors here. Glossy Ibises are locally abundant, especially in winter when several thousand are present.

Waterfowl occur in numbers on passage and in winter especially, with

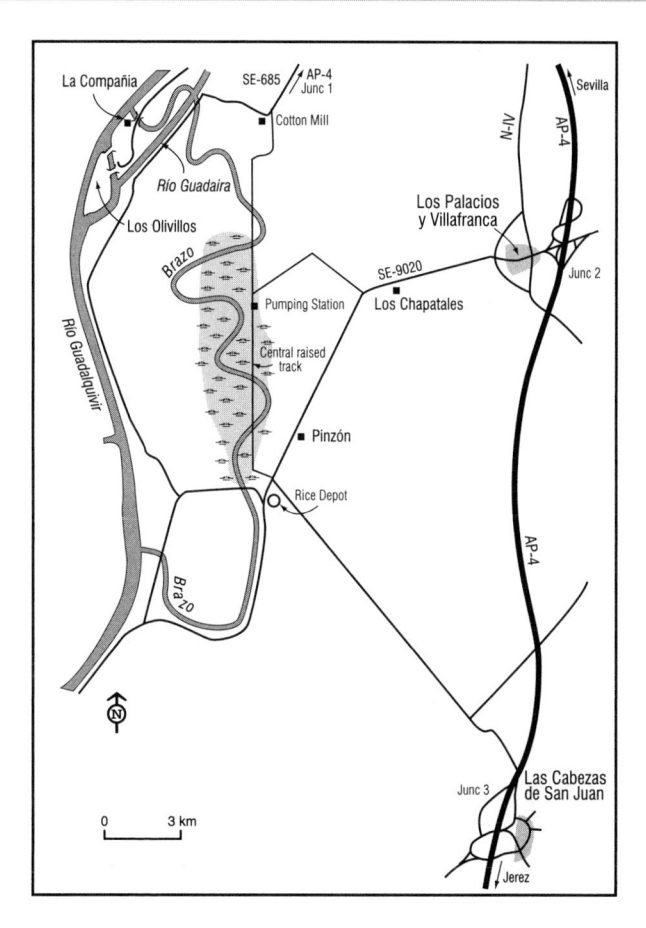

Marbled Ducks being regularly found here and breeding in some years at least. This is a prime site for Purple Swamphens; much of the time they are simply unmissable and records of 30 or more in view at once are quite normal, especially in spring, when territorial squabbles are being noisily resolved with much splashing and chasing, and when the birds are feeding in the adjacent ricefields. Occasionally they abandon the area completely, apparently when the ricefields are not at a suitable stage for them, which seems to happen in early autumn particularly. They are unpopular with farmers because of their depredations of the rice crops, which may explain why they are less abundant than was once the case, although they are a protected species.

It is worth inspecting the marsh margins carefully as Spotted Crakes, Little Crakes and Baillon's Crakes occur, as well as Water Rails: all these may breed locally — although Little Crakes have not been proved to do so. Collared Pratincoles breed in numbers and waders are common, again especially on passage and in winter. The latter regularly include otherwise elusive species such as Temminck's Stint and Marsh Sandpiper.

Passerines include breeding Savi's Warblers, a very local species in Spain. Moustached Warblers have also been recorded here on rare occasions. Water

Pipits are common in winter. A particular feature is the considerable variety of African finches that have been recorded, some of which have established breeding populations. In recent years these have included Black-headed Weavers, Yellow-crowned Bishops, and Common Waxbills with (Asian) Red Avadavats nearby. All of these populations were founded by escaped birds but the Guadalquivir wetlands clearly provide them with an excellent approximation to their native habitats.

Timing

This is an excellent site to visit all year round, although it is important to avoid the hottest part of the day in summer especially. Visits in the mornings and evenings are most productive. Springtime visits will coincide with maximum activity. Passage periods and winter also produce a large variety of interesting species.

Access

The area may be approached from the north (Sevilla), from the south (Lebrija/Trebujena; see CA-17) or from the east, leaving the Sevilla/Cádiz motorway (AP-4) at exits 1 (Dos Hermanas), 2 (Los Palacios) or 3 (Lebrija-Las Cabezas) – see map. The simplest access to the core of the site is via motorway exit 2. Head west from here past Los Palacios y Villafranca and take the SE-9020, passing Los Chapatales and then Pinzón on the left. The eucalyptus strip flanking the canal which borders the Pinzón stretch may be inspected for roosting Night Herons. About 1.5km past Pinzón you reach a shiny grey and blue rice depot. Cross the canal on the right just before the depot to enter the raised central track along the Brazo del Este.

The same point can be reached by leaving the motorway at exit 3, an approach route that crosses extensive rice paddies, which are attractive to waders especially. On leaving the motorway turn left immediately after passing through the toll gate and drive through the town of Las Cabezas de San Juan. Turn left at the roundabout at the foot of the hill and then drive straight on through the rest of the town. The road approaches the motorway again and crosses over it, descending towards a very visible white silo, where there is a T-junction. Turn left here and cross the railway line. From here keep straight on until the tarmac road turns right at a rice depot. Drive straight on here, off the tarmac, and on to either of the two broad, dead-straight, parallel, sandy tracks, which continue northwest for 7.6km. Rice paddies may be inspected on the right all along this drive. Bear right where the dirt tracks rejoin poor-quality tarmac until you pass the grey and blue rice depot mentioned above, where you bear left on to the central track.

The central track is on a raised dyke that crosses the main marshes and gives good views over the pools and reedbeds. A number of subsidiary tracks lead off this road and these are best explored on foot but care should be taken not to disturb nesting birds by approaching too closely. Road traffic along the main track is extremely light and you can stop anywhere. Indeed, waders and crakes are most readily seen by employing the car as a hide. You can return to the SE-9020 at Los Chapatales by bearing right just after the pumping station near the north end of the central track. The whole route may of course be followed in the opposite direction. The network of dirt roads in the area is a total maze so we suggest that you do not wander too far from our recommended routes.

The village of Pinzón, just east of the site, is strategically placed: there are a couple of bar-restaurants and a supermarket here.

It is often worth driving alongside the River Guadalquivir with frequent stops to scan for herons, waders, gulls and terns. The riverside road is good apart from the section to the south of the site, linking it with the Trebujena area. This part is pothole-ridden tarmac and demands careful driving. The river here is extensively obscured by a fringe of Eucalyptus and so is much less accessible. The riverside, and the sector where the Brazo del Este adjoins the main channel, are best accessed from the north. To get here leave the motorway at Junction 1 (Dos Hermanas) in the direction of Coria del Río. Shortly (1km) after crossing the A-4 turn left on to the SE-685 and continue until you see a large building, a cotton mill (Mediterráneo Algodón). You can turn right here to approach the Brazo del Este. Before or after exploring this area you should drive west past the cotton mill to reach the embanked channel of the River Guadaira. Follow the road south (left) along the channel to its confluence with the Guadalquivir and investigate the riverside areas. A productive track follows the southern margin of the Brazo del Este, which you cross shortly after reaching the River Guadaira.

A near-island at the confluence, known as Los Olivillos, is renowned for its mixed heronry, which includes numbers of Squacco Herons. This is best viewed by returning to the SE-685 and heading north to its junction with Coria/Villamarta road. Follow the latter to cross the Río Guadaira and head south (left) for some 11km to reach Los Olivillos. The whole of this road offers excellent opportunities to see aquatic species and riparian passerines.

Most areas are easily viewed from or near your car.

CALENDAR

Most of the species of Doñana (H5) are liable to occur but the following are especially typical.

All year: Cattle and Little Egrets; Night, Squacco and Grey Herons, White Stork, Spoonbill, Marsh Harrier, Water Rail, Spotted Crake, Purple Swamphen, Black-winged Stilt, Tree Sparrow, exotics.

Breeding season: Marbled Duck, Little Bittern, Purple Heron, Black Kite, Montagu's Harrier, Baillon's Crake, Collared Pratincole; Gull-billed, Little and Whiskered Terns, Lesser Short-toed Lark, Yellow Wagtail; Savi's, Reed and Great Reed Warblers.

Passage periods: Garganey, waders: including Ruff, Spotted Redshank, Common Redshank, Little and Temminck's Stints; Marsh, Green and Wood Sandpipers; Gull-billed, Whiskered and Black Terns, Water Pipit, Bluethroat, Sedge Warbler, rarities.

Winter: Waterfowl: including Greylag Goose and Marbled Duck, Great Bittern, Great White Egret, Black Stork, Glossy Ibis, Red Kite, Hen Harrier, Common Buzzard, Little Crake, waders: including Ruff, Common Snipe, Spotted Redshank, Common Redshank, Little and Temminck's Stints; Marsh, Green and Wood Sandpipers, Water Pipit, Bluethroat, rarities.

WESTERN GUADALQUIVIR DELTA   SE2

Status: Partly within the Parque Natural de Doñana.

Site description
The western section of that part of the Guadalquivir delta which lies in Sevilla province retains its natural wetland status on account of the extensive cultivation of rice, the river itself providing a ready source of irrigation even in the driest years. The whole region is sandwiched between the river proper and a subsidiary channel further west, the Brazo de la Torre. A labyrinth of smaller channels serving the ricefields crosses the whole region, encompassing two of the principal delta islands, the Isla Mínima to the north and the Isla Major in the south.

The western part of the site, fringing Doñana National Park, is of the greatest interest. Here the various drainage ditches and accompanying dykes enclose small pools and larger areas of marshland and salt flats. The appearance of the whole area depends on water levels; when these are high the place resembles an inland sea. This is the Zona de Entremuros (between-walls zone). The southwestern part includes a shallow lagoon, El Lucio del Cangrejo, which tends to dry up in summer but which is highly attractive to birds at other times. The whole area offers the characteristic habitats of the marismas of the Guadalquivir and, as with the Brazo del Este (SE1), is in many respects more rewarding to visit for birding purposes than is most of Doñana National Park itself.

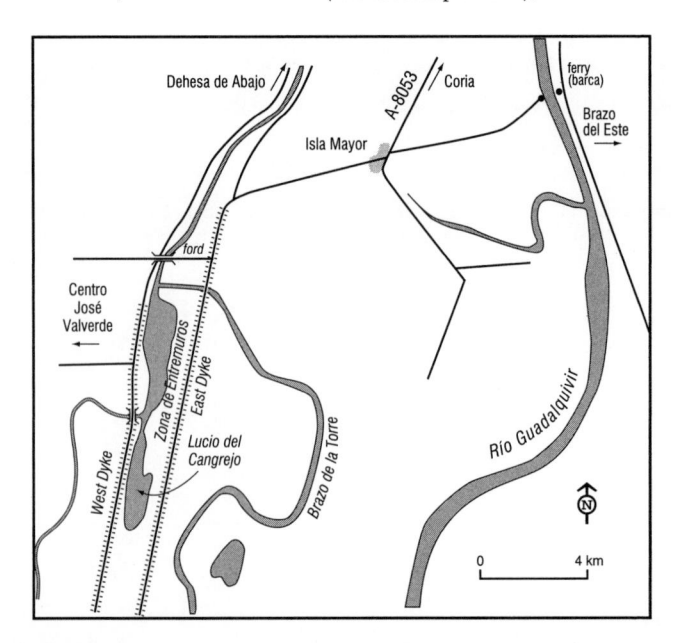

The Isla Mayor is home to a major fish farming enterprise and private wetland reserve, la Veta de la Palma (www.vetalapalma.es), which is carefully managed for birds and other wildlife and accordingly supports numbers of all the characteristic wetland species of the region. There is permanent water here and the reserve acts as a regional refuge to many waterbirds during drought periods. Access by birders is only really possible as part of a conducted tour, a circumstance which will put off some birders. However, it is an excellent site and merits consideration.

Species

All or most of the characteristic waterfowl and waders of Doñana occur here. In particular, El Lucio del Cangrejo and the adjacent wetlands attract Spoonbills, Greater Flamingos, dabbling ducks, including Marbled Ducks and Red-crested Pochards, and numerous waders, notably Ruffs and Black-tailed Godwits. Marsh Harriers are locally numerous. Gull-billed, Whiskered and Black Terns are abundant on passage especially. Lesser Short-toed Larks are typical and may be seen alongside Greater Short-toed Larks; both breed here. Very large roosts of Sand Martins, up to 40,000, have been recorded here, and other hirundines occur during the autumn migration especially. Yellow Wagtails are abundant. Spectacled Warblers breed. The ricefields attract herons and other waterfowl as well as passage waders and waterside species, such as Yellow Wagtails. The period following the rice harvest in autumn sees enormous gatherings of White Storks and Lesser Black-backed Gulls when the gleanings also attract ducks, Glossy Ibises, small numbers of Black Storks and often sizable flocks of Cranes, all of which remain into the winter.

Timing

Visits in springtime are particularly productive. Winter and passage periods generally are also excellent. Much of the natural wetland dries up completely in summer, making the flooded ricefields especially attractive to waterbirds then.

Access

The Isla Mínima is reached by driving south from Sevilla to La Puebla del Río (A-8058), via Coria del Río and continuing south to Isla Mayor town (formerly Villafranco). The approaches to Isla Mayor town from all directions cross enormous expanses of rice cultivation, which may be observed with care from any suitable stopping points such as the entrances to farm tracks. Local traffic is not heavy but it is often fast so stopping on the road itself is inadvisable.

The Isla Mayor proper is the broad sweep of territory south of 'Villafranco' between the Guadalquivir and the Brazo de la Torre. Direct access to (and from) Doñana via El Rocío, crossing the northern marismas, is no longer permitted although it is possible to travel as far as the Centro José Valverde (See H5-E). The key habitats on the western part of the Isla Mínima are easily viewed from the good roads along the parallel fringing dykes of the Zona de Entremuros, which is reached from Isla Mayor. Drive through the main street (A-8053) and turn right (west) on to the Avenida de Villamanrique, which runs alongside a narrow canal. The road is indicated by an enormous red arrow labelled Ruta Turística. Drive 3.5km from this turn-off to reach the dykes. You can drive south along the top of the east dyke to view the enclosed wetland. You should also cross to the west dyke on one of the two concrete fords linking them: these may be impassable when water levels are very high (relatively infrequently), when a long detour to the north is necessary to reach the opposite side.

A gate across the western dyke south of the Lucio del Cangrejo marks the boundary of the National Park. Access to Doñana through here is not permitted but the right turn shortly before the gate leads on to the Centro José Valverde (although 4WD may be necessary to complete the journey in all but the driest conditions). A useful return route is to follow the road along the west dyke northwards. This road also crosses excellent habitat, along the course of the Brazo de la Torre, offering further opportunities to scan the wetlands. It continues to the Dehesa de Abajo (SE3) and beyond to La Puebla del Río.

A visit to this area can usefully be combined with one to the Brazo del Este (SE1) on the east bank. The Guadalquivir can be crossed by ferry at Coria. This is likely to be replaced by the motorway tunnel just north of Coria when the outer Sevilla ring-road (SE-40) is completed. The former ferry service near Isla Mayor no longer operates but the crossing site (see map) is still worth visiting since it provides good views of the river, where Gull-billed Terns, herons and other waterbirds are often present.

The Reserve of Veta La Palma on the Isla Mayor is west of the town of Isla Mayor. Visits can only be made by prior arrangement (email: info@vetalapalma.es).

Most areas are easily viewed from or near your car.

CALENDAR

Many of the species of Doñana (H5) are liable to occur but the following are especially typical.

All year: Cattle and Little Egrets, Night and Grey Herons, White Stork, Spoonbill, Greater Flamingo, Glossy Ibis, Purple Swamphen, Marsh Harrier, Black-winged Stilt, Lesser Short-toed and Crested Larks, House and Tree Sparrows.

Breeding season: Little Bittern, Squacco and Purple Herons, Black Kite, Montagu's Harrier, Collared Pratincole; Gull-billed, Little and Whiskered Terns, Greater Short-toed Lark, Yellow Wagtail; Savi's, Reed, Great Reed and Spectacled Warblers.

Passage periods: Garganey, waders: including Little and Temminck's Stints, Curlew Sandpiper, Dunlin, Ruff, Spotted and Common Redshanks; Green, Wood and Common Sandpipers; Gull-billed, Whiskered and Black Terns, hirundines.

Winter: Great White Egret, Black Stork, Glossy Ibis, Greater Flamingo, waterfowl; including Greylag Goose, Marbled Duck and Red-crested Pochard, Red Kite, Hen Harrier, Common Buzzard, Crane, waders; including Lapwing, Little Stint, Dunlin, Ruff, Common Snipe, Spotted and Common Redshanks; Green, Wood and Common Sandpipers, Lesser Black-backed Gull, hirundines.

DEHESA DE ABAJO SE3

Status: Reserva Natural (617ha).

Site description

A low hill, largely covered in open, stony grassland, dominates the reserve, giving panoramic views across the ricefields of the Isla Mayor to the south and over woodlands (the dehesa) to the north and east. These last are largely of mature wild olives, themselves a significant element of the biological importance of the site, but there are also stands of Stone Pines. A shallow lagoon, the Lucio de la Rianzuela, occupies the valley to the west. A series of trails provide general access to the representative habitats, including the lagoon, which is overlooked by two hides. The information centre, at the top of the hill, is open daily and includes a restaurant among its facilities.

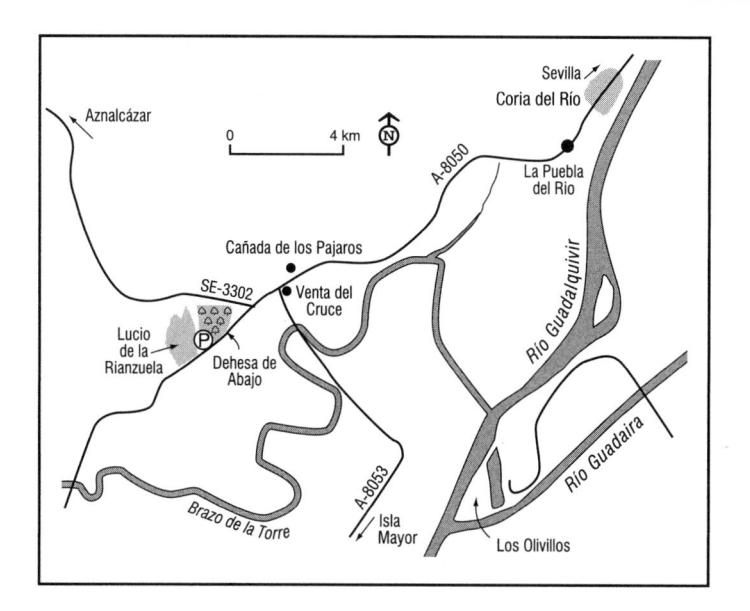

Species

The Dehesa de Abajo was designated a reserve in 2000 partly on account of the presence of an exceptionally large colony of White Storks, censused at over 300 pairs in 2004: the largest single tree-nesting colony of this species anywhere. The nests are on olive trees and also on pylons and buildings nearby. It also supports a variety of breeding raptors, including Black Kites, Common Buzzards, Booted Eagles and Short-toed Eagles. A range of other raptor species visit the area, particularly in winter, including Black-winged and Red Kites, and Marsh and Hen Harriers.

The lagoon supports breeding Purple Swamphens and a diversity of other waterfowl and waders. It is also a very good place to see early hirundines, such as Barn Swallows and Sand Martins, which often congregate here from late December onwards. Red-knobbed Coots sometimes nest here, these very likely released from the captive breeding centre for this species at the Cañada de los Pájaros nearby (see below). Marbled and Ferruginous Ducks, again often originating from the Cañada de los Pájaros, are also regularly present and may breed in some years.

Timing

Visits at any time of year will prove interesting but spring is the obvious time for seeing maximum activity at the stork colony. The lagoon may dry up in summer and is most worth visiting in winter and spring. The Doñana Birdfair is sometimes held here in March.

Access

The reserve is some 12km southwest of La Puebla del Río. If coming from this direction (Sevilla), take the A-8050 from La Puebla del Río towards Isla Mayor. Head straight on (2nd exit) at the roundabout (Venta del Cruce) where the A-3114 continues towards Isla Mayor (Villafranco). A second roundabout is soon reached and here take the first exit and continue for a further 3km where you will see the reserve entrance archway and car park on your right. There is a modest parking charge here. The SE-3305/3302 from Aznalcázar provides an alternative approach from the northwest.

The Cañada de los Pájaros (*www.canadadelospajaros.com*) is a small, privately managed reserve centred around some flooded former gravel pits. It is perhaps comparable with one of the centres of the Wildfowl and Wetlands Trust in Britain, in having a combination of a captive waterfowl collection and a wetland attracting wild birds. The centre is at km-8 on the north side of the A-8050. The centre releases captive-bred Red-knobbed Coots, Marbled Ducks and Ferruginous Ducks in quantity from time to time, although whether these reintroductions culminate in self-sustaining populations has yet to be established.

♿ The general area is readily viewed from or near your car at the hilltop centre. The trails to the lagoon are short but steep in places.

CALENDAR

All year: Little and Black-necked Grebes, Common Buzzard, Purple Swamphen, Red-knobbed Coot, Spanish Sparrow.

Breeding season: White Stork, Black Kite, Booted and Short-toed Eagles, Gull-billed Tern, Bee-eater.

Winter: Greylag Goose, Marbled and Ferruginous Ducks, Glossy Ibis, Spoonbill, Greater Flamingo, Black-winged and Red Kites, Marsh and Hen Harriers, hirundines.

Passage periods: Garganey, Black Tern.

WESTERN GUADALQUIVIR VALLEY FARMLANDS SE4

Status: No special protection.

Site description

The Guadalquivir valley is intensively farmed and the vast tracts of monoculture in some parts are not encouraging on first acquaintance. Nevertheless, there is abundant ornithological interest. The western sector, in the hinterland of the city of Sevilla, has immense, flat wheatfields stretching for miles across the floodplain. Small rivers cross the plain and feed into the Río Guadalquivir. Their courses, and that of the main river, are marked by extensive olive and citrus groves, as well as vineyards. These orchards are also characteristic of the valley fringes and the surroundings of the numerous towns and villages. The river beds have useful stretches of riverine vegetation, including stands of willows and poplars and, most significantly, large clumps of tamarisks.

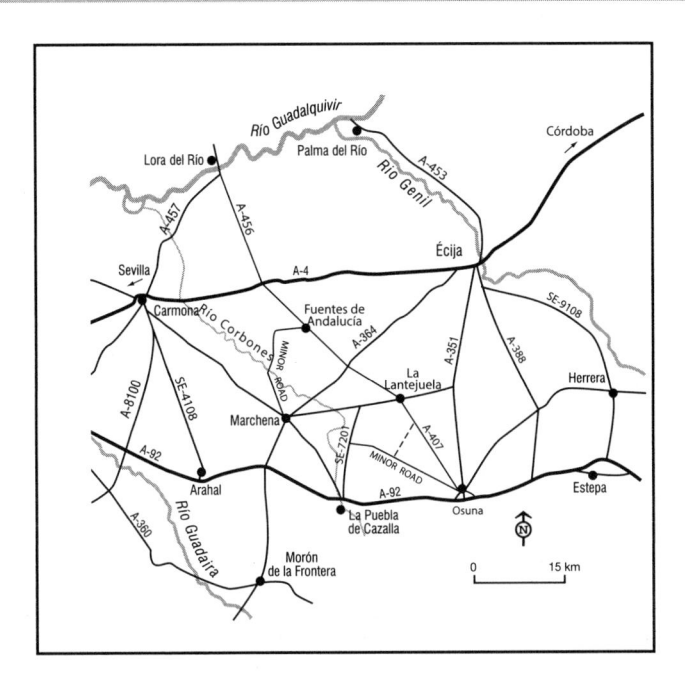

Species
This a regional stronghold of both Rufous-tailed Scrub-robins and Olivaceous (Isabelline) Warblers, two of the most sought-after but often most elusive species of our area. Both occur in numbers in late spring and summer. The Scrub-robins are characteristic of citrus and olive groves whereas the warblers favour tamarisks along river beds and lake margins. They are easily located in season, particularly once you know their songs.

Species of open country may be encountered in the wheatfields. Small numbers of Great and Little Bustards occur but are elusive. Black-bellied Sandgrouse are scarce residents in the area and may be joined by Pin-tailed Sandgrouse in winter. Montagu's Harriers, Quail, Stone-curlews and Greater Short-toed Larks are characteristic and rather more common. Lesser Kestrels nest in some of the towns and villages, including Carmona and Marchena. Flocks of Lapwings, larks, pipits, finches and sparrows, including both House and Spanish Sparrows, occur in winter. Wintering raptors include Red Kites, Common Buzzards, Hen Harriers, Merlins and occasional Bonelli's Eagles.

Timing
Both Scrub-robins and Olivaceous Warblers are summer visitors and both are among the tardiest to arrive in Spain. They begin to turn up in May but some do not appear until early June. They are most easily located from late May through June and into July. By late August most have already returned to Africa. Their stay coincides with the hottest time of year in the hottest part of southern Spain, where shade temperatures (not that there is any shade) regularly exceed 40°C by a good margin. It goes without saying that the mornings and evenings are the most pleasant times in which to locate both species (and most

others) and of course these are the times when both are most vocal. Adequate protection against the ferocious sun and regular fluid-replacement are essential.

Many of the open-country species are resident and they are more easily located in winter and after crops have been cut.

Access

The region is crossed by major trunk roads, notably the A-4 Sevilla/Córdoba and the A-92 Sevilla/Granada motorways. A network of minor road and farm tracks link the well-separated towns and villages and provide good access to the interesting areas, which are spread over a very wide expanse.

Scrub-robins are characteristic of the olive groves around the towns of Arahal, Morón and Puebla de Cazalla. They especially favour older traditional groves that retain a floral ground layer instead of intensively managed orchards that are scraped bare of undergrowth. The many new olive groves unfortunately tend to be of the latter type. The tamarisks along watercourses, including the Río Genil between Herrera and Écija, the Río Guadaira, the Río Corbones and the Guadalquivir itself, all support good populations of Olivaceous Warblers. Stop at suitable bridges and inspect the riverine scrub.

Open-country species may be searched for in the wheatfields between Carmona and Utrera (A-8100), Carmona and Arahal (SE-4108), Écija and Osuna (A-351) and Écija and La Lantejuela. The area between the A-407 and the SE-7201, including the minor road leading from the SE-7201 to Osuna is also often productive, although some open ground has been lost to olive groves in recent years. The open expanses may be scanned with a telescope. It is also worth exploring these regions more closely along the numerous dirt roads and farm tracks that present themselves. Among these, the minor road linking Marchena to Fuentes de Andalucía crosses promising habitat, as well as the Río Corbones.

♿ Most areas are easily viewed from or near your car.

CALENDAR

All year: White Stork, Black-winged Kite, Common Kestrel, Red-legged Partridge, Great and Little Bustards, Stone-curlew, Black-bellied Sandgrouse, Little Owl, Calandra and Crested Larks, Barn Swallow, Zitting Cisticola, Cetti's Warbler, Southern Grey Shrike, Raven, Spanish Sparrow, Corn Bunting.

Breeding season: Short-toed Eagle, Montagu's Harrier, Lesser Kestrel, Quail, Collared Pratincole, Roller, Greater Short-toed Lark, Yellow Wagtail, Rufous-tailed Scrub-robin, Nightingale, Olivaceous Warbler, Golden Oriole.

Winter: Red Kite, Hen Harrier, Common Buzzard, Bonelli's Eagle, Merlin, Crane, Golden Plover, Lapwing, Pin-tailed Sandgrouse, Sky Lark, Meadow Pipit, White Wagtail, Song Thrush, Blackcap.

LAGUNAS DE LA LANTEJUELA   SE5

Status: Reserva Natural & ZEPA 'Lagunas de La Lantejuela y Campiña de Sevilla' (896ha).

Site description
These mainly brackish lagoons are a remnant of the many such wetlands, most of them seasonal, which were formerly scattered across the southern farmlands or Campiña south of the Guadalquivir. Most of them were reclaimed for agriculture in the 1960s. The principal surviving lagoons include La Ballestera (54ha), Calderón Chica (8ha), del Gobierno (20ha) and the recently restored (post-2009) Ruíz Sanchez (358ha), but several others may appear during wet winters. Most of the lagoons tend to dry up in summer but the Laguna del Gobierno has permanent water. Most of the lagoons have sparse fringing vegetation of reeds, sedges and tamarisks but the Laguna del Gobierno is more richly vegetated, with a dense tamarisk fringe.

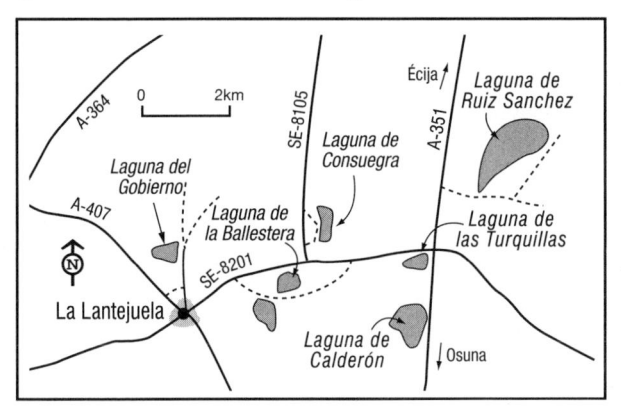

Species
Breeding waterfowl include White-headed Ducks, Black-necked Grebes, Mallard, Water Rails and Purple Swamphens and there is a small colony of Black-headed Gulls. Gull-billed Terns breed in some years. The lakes attract a much greater diversity of aquatic species during passage periods and in winter. Passage waders include Curlew Sandpipers and Greenshanks with occasional rarer species such as Temminck's Stints. Black-winged Stilts and Avocets are among the wintering waders. Flocks of Greater Flamingos occur frequently and may exceed 1,000 birds in winter; a Lesser Flamingo accompanied them in winter 2006/07. Regular waterfowl include large numbers of Mallard and Shoveler and also Greylag Geese, Wigeon, Gadwall, Pintail and Common Pochard, with occasional Shelducks, Marbled Ducks and Red-crested Pochards. Both Ferruginous Ducks and Red-knobbed Coots have been recorded here on a number of occasions. The breeding raptors of the area include Black-winged Kites. Red Kites are quite numerous in winter, when the lakes also often attract one or two of the larger eagles, chiefly Bonelli's Eagles. There is a large winter roost of Lesser Black-backed Gulls, at times attracting over 5,000 birds.

Purple Swamphen

Timing
The site is of greatest interest in winter, and often in spring and autumn, given the seasonal nature of the lagoons and their propensity to dry up in summer and in dry years. However, the permanent Laguna del Gobierno is always worth a visit and is a reliable site for White-headed Ducks.

Access
The main lagoons lie immediately west of the A-351, about halfway between Écija and Osuna, some 3km east of La Lantejuela. View from the roads or follow the tracks leading to the lagoons. The Laguna de La Ballestera is reached along a sandy track signposted from the SE-8201 at km-3.5. The Laguna de Consuegra lies east of the SE-8105, 800m from its junction with the SE-8201: it is reached by a drivable track serving a farm of the same name. The Laguna de Ruíz Sanchez is east of the A-351 3km north of La Lantejuela crossroads: access is via a rough track at the km-14 marker. The Laguna del Gobierno is 900m from La Lantejuela village and is signposted just west of the only roundabout in the village, on the main through road. The entrance to the access road (Av. Vereda de la Huerta) has a yellow pillarbox on the left. You can also reach the access road along a short sandy track on the north side of the A-407 at km-17: turn left at the end of the track. An educational centre and a hide are adjacent to the lagoon. You can see the lake from alongside the hide when this is closed.

♿ Most areas are easily viewed from or near your car.

CALENDAR

All year: (water permitting) Mallard, White-headed Duck, Black-necked Grebe, Water Rail, Moorhen, Purple Swamphen, Common Coot, Black-winged Kite, Marsh Harrier, Stone-curlew, Kentish Plover, Calandra Lark, Cetti's Warbler, Zitting Cisticola, Southern Grey Shrike.

Breeding season: Montagu's Harrier, Black-headed Gull, Gull-billed Tern, Roller, Greater Short-toed Lark, Rufous-tailed Scrub-robin; Reed, Great Reed and Olivaceous Warblers.

Winter: Greylag Goose, Common Shelduck, Wigeon, Gadwall, Teal, Pintail, Shoveler, Marbled Duck, Red-crested and Common Pochards, Ferruginous Duck, Glossy Ibis, Greater Flamingo, Red Kite, Hen Harrier, Common Buzzard, Bonelli's Eagle, Merlin, Peregrine Falcon, Red-knobbed Coot, Crane, Golden Plover, Lapwing, Common Snipe, Lesser Black-backed Gull, Bluethroat, Raven.

Passage periods: Black-winged Stilt, Avocet, Little and Temminck's Stints, Curlew Sandpiper, Dunlin, Black-tailed Godwit, Spotted and Common Redshanks, Green Sandpiper, Greenshank, Whiskered Tern.

SIERRA NORTE   SE6

Status: Parque Natural & ZEPA (177,396ha). UNESCO Biosphere Reserve.

Site description
This region forms the central portion of the Sierra Morena. It lies between the natural parks of the Sierra de Aracena y Picos de Aroche in Huelva province (H7) and the Sierra de Hornachuelos in Córdoba province (CO9). Like those sites, it offers many good birds and lends itself to gentle exploration if your priority is getting away from the tourist-beaten track.

These are smooth rounded hills rather than the jagged sierras so common in Spain. There are extensive woods of Stone Pines, encinas, Cork Oaks and Lusitanian Oaks. Numerous small rivers cross the region, in places fringed by woods of Sweet Chestnuts, Smooth-leaved Elms and poplars. The Río Huéznar is typical and is a renowned trout stream, which attracts Otters as well as fishermen. Extensive meadows and tracts of Gum Cistus scrub are also characteristic.

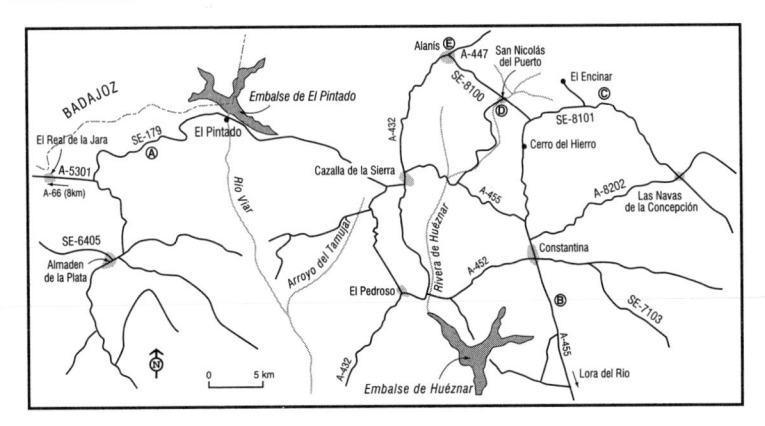

Species
The region offers a good variety of raptors and other species typical of the Mediterranean woodlands. All five eagles breed as do Black and Griffon Vultures and Black and White Storks. Eagle Owls breed despite the general absence of rock crevices and caves, some of them nesting instead on the ground in the woodlands. Characteristic passerines include Thekla Larks, Wood Larks, Red-rumped Swallows, Mistle Thrushes, Azure-winged Magpies and Rock Sparrows. Both White-rumped Swifts and Little Swifts breed locally. This is one of the few places in our area in which Bullfinches occur with some regularity: small numbers are present in winter, usually occurring in small groups along watercourses.

Resident mammals include numbers of Wild Boar, Wild Cats, Red Deer and Fallow Deer. Wolves no longer occur but Lynx are being reintroduced.

Timing
Springtime visits are most productive.

Access

The region is crossed by a network of minor roads and well-marked footpaths, permitting exploration at leisure. The SE-179 road between El Real de la Jara and El Pintado (A) ascends Cistus-covered slopes to cross several high passes, with good views over the region. Very large open pastures are characteristic southwards from Constantina towards Lora del Río (A-455, B) and between San Nicholás and Las Navas (SE-8101, C). Riverside habitats can be visited, for example, from the trails originating from the village of San Nicholás del Puerto, on the Río Huéznar (D).

The A-447 heading northeast from Alanis (E) along the park boundary is especially recommended. Although the road is narrow and winding it is virtually traffic-free and the surface is good tarmac for 26km until the provincial boundary with Córdoba, after which there is a rough stretch beyond the Bembezar bridge. It crosses elevated rocky terrain giving long views over dehesas, deer hunting estates, scrub and pastures, before descending to the River Bembezar. The pools along the river should be scanned for Black Storks and waterbirds and 'genuine' Rock Doves occur here. All along this road there are ample opportunities to stop and scan for raptors and to visit the diverse habitats.

An information centre (Centro de Visitantes 'El Robledo') is near Constantina, at km-1 on the El Pedroso road: it provides information on the various marked trails and other amenities of the park and also features a botanical garden. Another information centre ('Cortijo El Berrocal') is at Almadén de La Plata and is the starting point for three scenic trails passing through representative habitats.

♿ Good viewing is available from or near your car in most areas. The trails are often long and not easily accessible.

CALENDAR

All year: Great Crested Grebe, Black-winged and Red Kites, Griffon and Black Vultures, Common Buzzard, Golden and Bonelli's Eagles, Eagle Owl, Wood Lark, Crag Martin, Blue Rock Thrush, Mistle Thrush, Firecrest, Long-tailed and Crested Tits, Southern Grey Shrike, Nuthatch, Azure-winged Magpie, Raven, Tree and Rock Sparrows, Hawfinch, Cirl Bunting, Rock Bunting.

Breeding season: Black Stork, White Stork, Black Kite, Egyptian Vulture, Short-toed and Booted Eagles, Little Ringed Plover, Turtle Dove; Alpine, White-rumped and Little Swifts, Wryneck, Hoopoe, Red-rumped Swallow, Orphean and Bonelli's Warblers, Iberian Chiffchaff, Golden Oriole, Woodchat Shrike.

Winter: Woodcock, Siskin, Crossbill, Bullfinch.

CÓRDOBA PROVINCE

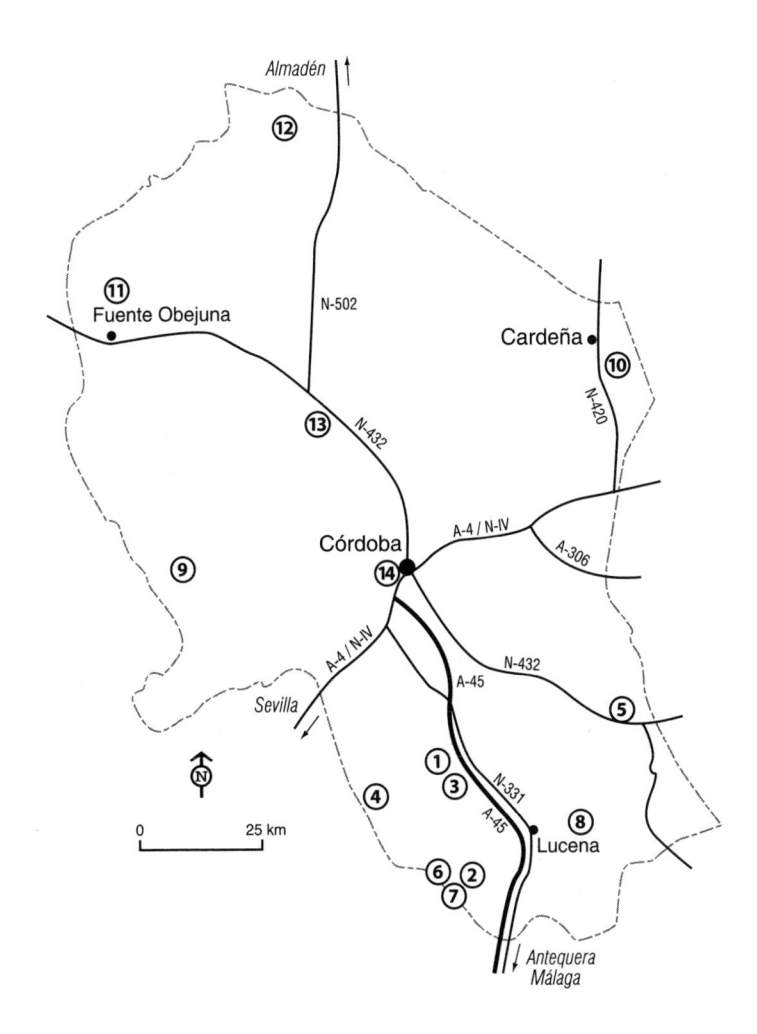

Córdoba province includes the central sector of the Guadalquivir valley, here flanked by the Sierra Morena in the north and outpost mountains of the Betic range in the south.

The habitats of the region are diverse. The Córdoba wetlands are a major attraction of what is one of the hottest and driest regions of Spain in summer. They include a considerable number of lagoons (CO1–5), some of them permanent, and reservoirs (CO6–7), most of which are in the southwest of the province. All are important for waterfowl, in particular, White-headed Ducks and Purple Swamphens. The large Puente Nuevo reservoir (CO13), northwest of Córdoba city, attracts a diversity of waterfowl especially in winter.

In the tranquil north, bordering Badajoz province in Extremadura, the Río Zújar and its tributaries (CO12) offer waterfowl as well as passage waders and riparian passerines. The Río Guadalquivir itself has riparian vegetation that always merits inspecting; the riverside within Córdoba city (CO12) has remarkable bird diversity for an urban location, including mixed heronries, and is well worth a visit.

The Córdoba section of the Sierra Morena largely comprises the Parques Naturales of the Sierra de Hornachuelos (CO9) and Sierra de Cardeña y Montoro (CO10). The avifauna of Mediterranean woodlands is fully represented here. Raptors are much in evidence, with Black and Griffon Vultures being particularly obvious. The increasing population of Spanish Imperial Eagles provides another attraction. In the southeast of the province the Sierras Subbéticas (CO8) also offer a diversity of raptors and a typical Mediterranean woodland passerine community, with a greater presence of birds of rocky terrain.

Large areas are of course occupied by arable land, much of which has been devoted to olive groves, to the detriment of the birds of open country. Nevertheless, good expanses of dry cereal crops and rough pasture survive, the best of these including the steppic habitats of the Alto Guadiato (CO11) in the provincial northwest. Here both bustards, both sandgrouse and a broad range of other farmland species reflect the similar bird communities in the contiguous areas of Extremadura.

Sites in Córdoba Province

CO1 Laguna de Zóñar
CO2 Lagunas Amarga and Dulce
CO3 Lagunas del Rincón and de Santiago
CO4 Laguna de Tíscar
CO5 Laguna del Salobral
CO6 Cordobilla reservoir
CO7 Malpasillo reservoir
CO8 Sierras Subbéticas
CO9 Sierra de Hornachuelos
CO10 Sierra de Cardeña y Montoro
CO11 Alto Guadiato farmlands
CO12 Zújar and Guadamatilla Valleys
CO13 Puente Nuevo reservoir
CO14 Río Guadalquivir at Córdoba

Getting there

Córdoba is linked with other provincial capitals by major trunk roads. Rapid movement within the province is possible using the Autovía de Andalucía (A-4), which follows the Guadalquivir valley and links Córdoba with Sevilla and Madrid. The A-45 links Córdoba and the Costa del Sol at Málaga and intersects the A-92 Granada/Sevilla motorway at Antequera. The N-432 gives access from Córdoba city to the northwest of the province and on to Zafra in Extremadura. By branching off the N-432 on to the N-502 for Almadén you eventually enter western Ciudad Real province.

Where to stay
There are ample tourist facilities throughout the province, especially in Córdoba city itself, where there is a Parador.

LAGUNA DE ZÓÑAR   CO1

Status: Reserva Natural (385ha). ZEPA. Ramsar site.

Site description
The largest of the southern Córdoba lagoons and benefiting from permanent water. The reed-fringed basin is separated by tamarisks from the surrounding low hills, which are covered by vineyards, olive groves and cereal fields. There is a sizable information centre, a public hide and nature trails through mixed Mediterranean woodland.

Species
This lagoon was the last redoubt of the White-headed Duck, whose entire Spanish population reputedly diminished to just 22 birds in 1977. Fortunately, since then, the protection given to this and other sites, plus the captive breeding programme carried out in Doñana, enabled a strong recovery. The national population fluctuates but has numbered around 2,500 individuals post-2000.

As it happens, White-headed Ducks were absent from Zóñar for many years because introduced carp destroyed the subaquatic vegetation. Attempts to

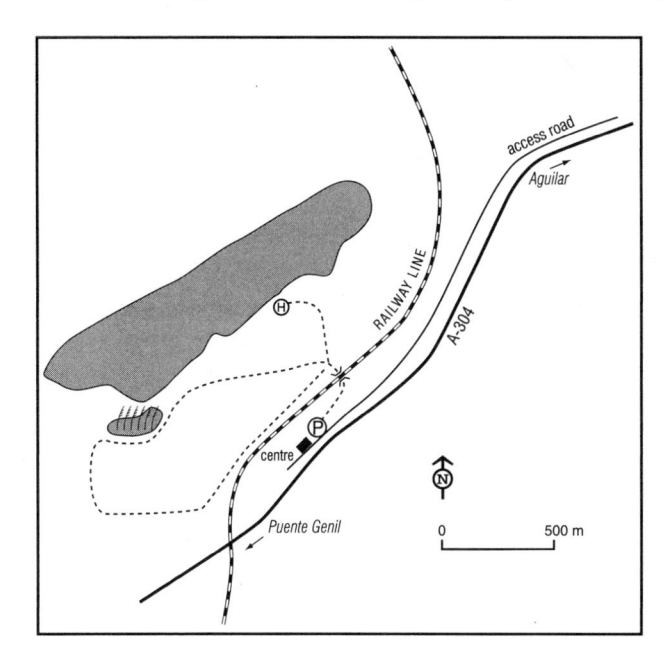

Purple Heron

eliminate the carp eventually succeeded and enabled the ducks to return. The only fish in the lake now is a duck-friendly native species, the Big-scale Sand Smelt *Atherina boyeri*. Several pairs of White-headed Ducks breed and larger concentrations occur in winter. The winter months also see the greatest diversity of waterfowl, including Great Cormorants, Common Shelducks, Wigeon, Gadwall, Teal, Pintail, Shoveler and both Common and Red-crested Pochards as well as occasional Ferruginous, Marbled and Tufted Ducks. Large numbers of Lesser Black-backed Gulls roost here in winter. Garganey move through in March and occasional Black Storks, Ospreys, Gull-billed Terns and Black Terns occur during passage periods. Marsh Harriers and Purple Swamphens are resident in small numbers. Coots are common (literally) but there are occasional reports of Red-knobbed Coots. The lakeside woodlands hold hordes of Nightingales in spring. The reedbeds have colonies of Great Reed and Reed Warblers, with Melodious Warblers on the outermost fringes where there are tamarisks. In winter there are sometimes roosts of several thousand starlings, chiefly Common Starlings, and also large numbers of roosting Cattle Egrets. Wintering passerines include Bluethroats.

Timing
Species diversity is probably highest in winter and during passage periods but there is always something of interest to be seen.

Access
The lagoon is some 4km south of Aguilar alongside the A-304 Aguilar de la Frontera/Puente Genil road but there is no direct access from the A-304. The lagoon is clearly signposted at km-40 on the A-304 just south of Aguilar, the access road being the old Puente Genil road. Follow signs from the roundabout where you leave the A-304. This road ends at the information centre, which overlooks the lagoon. The trails are always open even when the centre itself is closed. A telescope is very useful here.

♿ Adequate in parts but the trails are steep in places.

CALENDAR

All year: Mallard, Red-crested and Common Pochards, White-headed Duck, Little and Great Crested Grebes, Marsh Harrier, Purple Swamphen, Cetti's Warbler, Zitting Cisticola.

Breeding season: Purple Heron, Little Ringed Plover, Hoopoe, swifts, hirundines, Nightingale; Great Reed, Reed and Melodious Warblers.

Winter: Common Shelduck, Wigeon, Gadwall, Teal, Pintail, Shoveler, Ferruginous and Tufted Ducks, Black-necked Grebe, Great Cormorant, Cattle Egret, Grey Heron, Greater Flamingo, Lesser Black-backed Gull, Bluethroat, Common Starling, Reed Bunting.

Passage periods: Garganey, Black Stork, Osprey, Gull-billed and Black Terns, passerines.

LAGUNA AMARGA AND LAGUNA DULCE   CO2

Status: Reserva Natural (250ha). ZEPA. Ramsar site.

Site description

This interesting site, enclosed by olive groves and vineyards, has lakes of contrasting salinity. The freshwater Laguna Dulce, to the left of the entrance track, has relatively little fringing vegetation. The Laguna Amarga (bitter lake), to the right of the track, is surrounded by tamarisks with a few canes, reeds and reedmace. The Laguna Amarga benefits from permanent standing water but the Laguna Dulce is very shallow and often dries up.

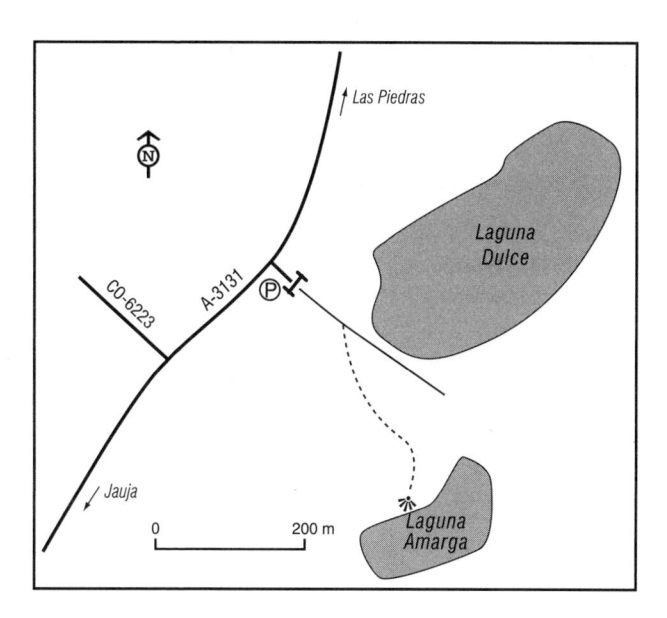

Species
Both White-headed Ducks and Purple Swamphens occur when water and reed conditions are right. Red-crested Pochards have also bred here and Red-knobbed Coots are present at least occasionally. Wintering waterfowl include Gadwall, Wigeon and Tufted Ducks.

Marsh and Montagu's Harriers hunt over the general area and occasional visiting raptors include Bonelli's Eagles. Hoopoes are common along the access track and Nightingales sing loudly at all hours in spring.

Access
The site is just off the A-3131 5km northeast of Jauja, which is southeast of Puente Genil. The reserve is well signposted and has a parking area, from where a closed gate usually denies further vehicular access. Walk along the track to view the Laguna Dulce to the left. The track to the Laguna Amarga rises off on the right, 100m from the car park, to an observation terrace overlooking the lake.

♿ Laguna Dulce is accessible.

CALENDAR

All year: Mallard, Red-crested Pochard, White-headed Duck; Little, Black-necked and Great Crested Grebes, Marsh Harrier, Moorhen, Common and Red-knobbed Coots, Purple Swamphen.

Passage: waders.

Winter: Gadwall, Wigeon, Common Pochard, Tufted Duck, Bonelli's Eagle.

Breeding season: Montagu's Harrier, Little Ringed Plover, Hoopoe, Nightingale.

LAGUNAS DEL RINCÓN AND DE SANTIAGO   CO3

Status: Reserva Natural (137ha). ZEPA. Ramsar site.

Site description
The Laguna del Rincón provides permanent water but levels may be very low in drought years. It is set in a virtually circular depression and has a wide fringe of canes, reeds, tamarisks and reedmace. Vineyards and cereal cultivation surround the lake, which is quite small (9ha). The seasonal Laguna de Santiago nearby was restored in 2006.

Species
A reliable site for White-headed Ducks. Other breeding species include Little, Great Crested and Black-necked Grebes and both Red-crested and Common Pochards. Wintering species include occasional Greylag Geese as well as Teal, Shoveler and Tufted Ducks. There are resident Water Rails and occasional Purple Swamphens. Marsh Harriers hunt over the area from time to time.

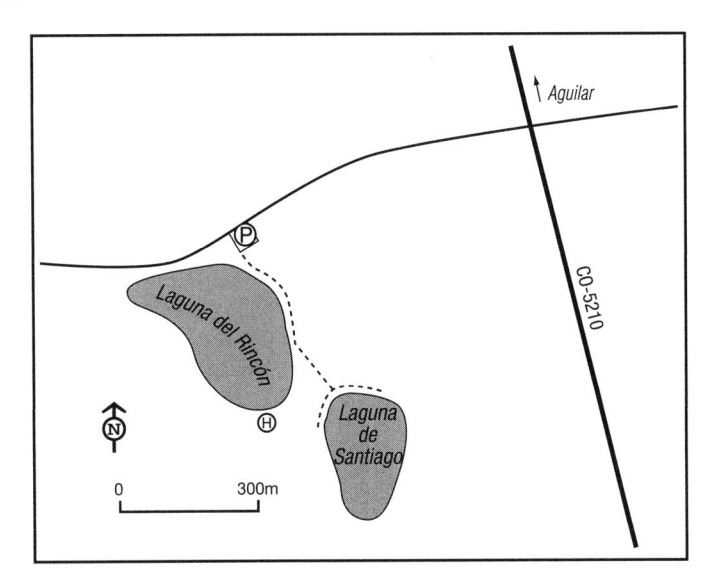

Waders occur occasionally, especially during migration periods. Large numbers of swifts and hirundines feed over the area at times. The reedbeds and surrounding vegetation hold Reed, Great Reed and Melodious Warblers as well as Nightingales. In the vineyards and fields nearby there is a chance of seeing Rufous-tailed Scrub-robins and Greater Short-toed Larks.

Timing
Species diversity is highest in spring. Winter is good for waterfowl.

Access
Access is from the Aguilar de la Frontera/Moriles road (CO-5210), some 6.5km southeast of Aguilar. The entrance to the access track is clearly indicated by a large green Consejería de Medio Ambiente sign featuring a White-headed Duck. The access track is some 750m long and fairly rough. There is a car park by the Laguna del Rincón. The public hide at the far end of this lake is usually closed but there are good views of the the lake from the perimeter track. A short footpath that leads to a firewatch tower also gives access to the shallow Laguna del Santiago.

Limited.

CALENDAR

All year: Mallard, Red-crested and Common Pochards, White-headed Duck, Little and Great Crested Grebes, Water Rail, Moorhen, Purple Swamphen, Common Coot, Corn Bunting.

Breeding season: Marsh Harrier, Black-winged Stilt, Rufous-tailed Scrub-robin, Nightingale; Cetti's, Reed, Great Reed and Melodious Warblers.

Winter: Greylag Goose, Wigeon, Teal, Shoveler, Tufted Duck, Black-necked Grebe, Glossy Ibis, Greater Flamingo.

Passage periods: Waders, swifts and hirundines.

LAGUNA DE TÍSCAR   CO4

Status: Reserva Natural (185ha). ZEPA. Ramsar site.

Site description
This is one of the seasonal lakes and so it lacks the abundant fringe of vegetation of those with permanent water. As compensation it does have more open shore space. The saline nature of the water means that there is some glasswort, as well as reeds and some scattered tamarisks.

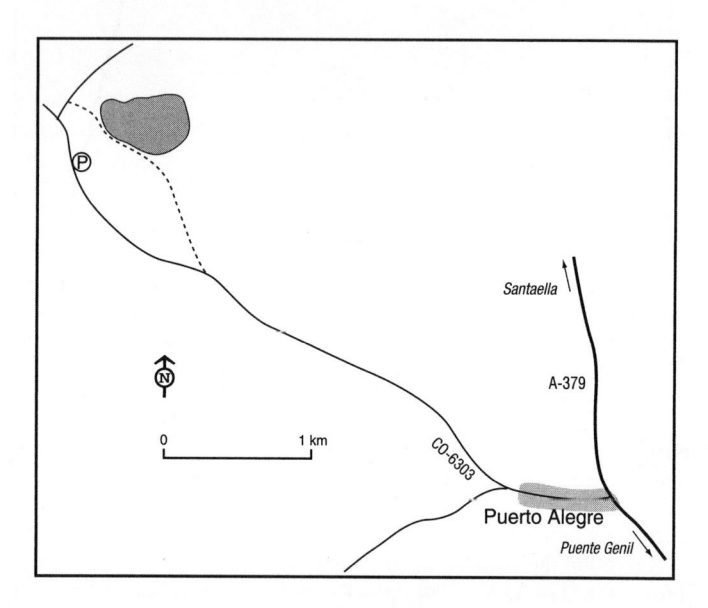

Species
The salinity often attracts small numbers of Greater Flamingos, probably originating from Fuentedepiedra (MA1) as they do not breed here. Breeding ducks include one or two pairs of Red-crested Pochards. Waterfowl diversity is greater in winter and includes Wigeon, Pintail and Shoveler. The open habitat attracts a variety of migrant and breeding waders, the latter including plenty of Black-winged Stilts and a few pairs of Avocets and Little Ringed Plovers. A few non-breeding Gull-billed Terns can often be seen here.

Timing
Spring for migrant waders and winter months for ducks and any wintering waders. The lagoon is generally inundated during December–May.

Access
The Laguna is signposted on the A-379 at the turn-off for the village of Puerto Alegre, 2km west of Puente Genil. Drive through the village on the CO-6303 and bear right where the road forks. Thereafter continue west along the

CO-6303, which soon becomes a track, for 3.5km. The track is normally in good condition. A car park on the left is opposite a signposted footpath leading to a public hide. This overlooks the lake at some distance from an elevated point. A telescope is useful.

♿ Viewable from your vehicle.

CALENDAR

All year: Mallard, Red-crested Pochard, Greater Flamingo, Moorhen, Common Coot.

Winter: Wigeon, Pintail, Shoveler, waders, Lesser Black-backed Gull.

Breeding season: Black-winged Stilt, Avocet, Little Ringed Plover, Gull-billed Tern, Melodious Warbler.

Passage periods: A large variety of waders in small numbers.

LAGUNA DEL SALOBRAL (LAGUNA DEL CONDE) CO5

Status: *Reserva Natural (345ha). ZEPA.*

Site description
A large, shallow, saline lagoon, extending up to 70ha when full. The fringes are only sparsely vegetated with glassworts, tamarisks and halophytic plants. It is prone to drying out by late summer, sooner in dry years.

Species
The lagoon attracts quantities of waterfowl in wet years. Wintering waterfowl include White-headed Ducks and Red-crested Pochards. Greylag Geese and

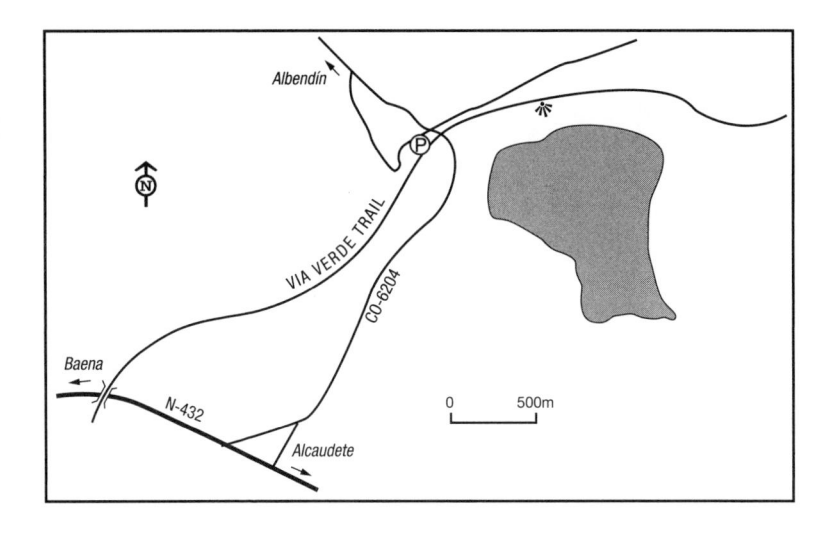

Common Shelducks may also occur in winter. Red-knobbed Coots have been recorded among the numerous Common Coots. Flamingoes are often present. A few migrant waders, such as Common Redshanks and Common Sandpipers occur, especially in spring.

Timing
The winter–spring period is best given adequate rainfall. The lake is apt to be dry at other times.

Access
The lake lies north of the N-432 about halfway between Baena and Alcaudete (Jaén province). It is most easily viewed from the Via Verde, a long-distance cycling/walking trail along a decommissioned railway line that skirts the north shore. There is a viewing platform on this trail. Access the Via Verde by turning north off the N-432 at km-345 onto the CO-6204 (CP-104) local road (signposted Albendín) and bear left until the lake comes into view. You can scan the west arm of the lake from this road and then continue a short distance to a small building at the intersection with the walking route. Park and continue on foot.

♿ Viewable from the CO-6204.

CALENDAR

Winter: Greylag Goose, Shelduck, Wigeon, Shoveler, Mallard, Red-crested Pochard, Common Pochard, White-headed Duck, Greater Flamingo, Common Coot, Black-winged Stilt, waders.

Passage periods: Common Redshank, Common Sandpiper, waders.

CORDOBILLA RESERVOIR   CO6

Status: Paraje Natural (1,460ha).

Site description
The furthest downstream of a series of three reservoirs on the Río Genil. The provincial border with Sevilla runs down the centre of the lake. The reservoir is rather shallow, with significant emergent reedbeds and abundant tamarisks along the inlet streams. Sediment accumulation has affected the reservoir to the point that dredging works were in prospect in 2018, which would have a temporary effect on its wildlife.

Species
At its best this is an attractive wetland that supports a diversity of herons, waterfowl and other aquatic and waterside birds year-round. These include White-headed Ducks and Purple Swamphens. Cranes also winter in the area in small numbers.

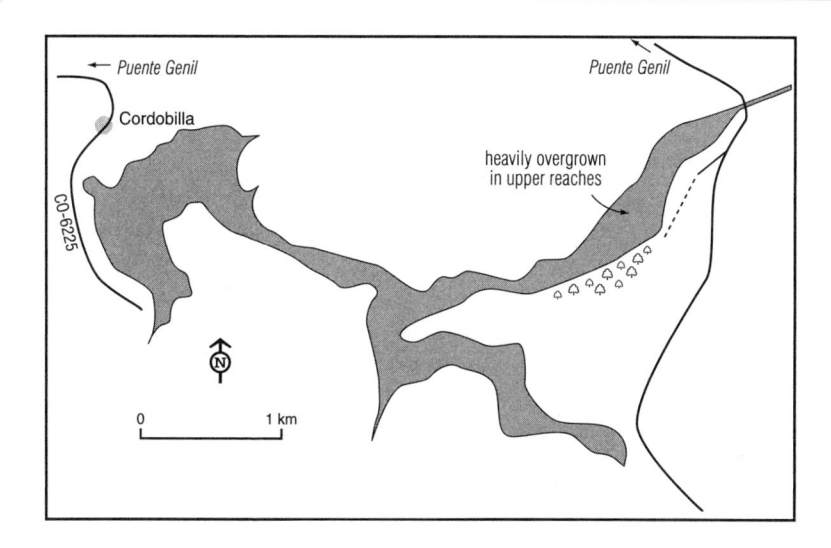

Timing
Winter and spring visits, between November and June, are most productive.

Access
The easiest access to the shoreline is in the northwestern corner, at the village of Cordobilla. Fortunately this is the best sector, with considerable emergent vegetation visible. Take the CV-179 southwest from Puente Genil to Cordobilla village and skirt the south side of the village to reach a bridge over a canal. Continue on foot to the reservoir margin.

&. Not recommended.

CALENDAR

All year: Mallard, White-headed Duck, Cattle Egret, Grey Heron, Marsh Harrier, Purple Swamphen, Cetti's Warbler.

Breeding season: Little Bittern, Night and Purple Herons, White Stork; Reed, Great Reed and Melodious Warblers.

Passage: Osprey, waders.

Winter: Wigeon, Gadwall, Shoveler, Common and Red-crested Pochards, Great Cormorant, Crane, Black-headed and Lesser Black-backed Gulls.

MALPASILLO RESERVOIR   CO7

Status: Paraje Natural (512ha).

Site description
Another reservoir on the Río Genil on the provincial boundary with Sevilla. It is relatively steep-sided but there are flatter areas along its course where there are quite wide reedbeds and occasional stands of willow and tamarisk.

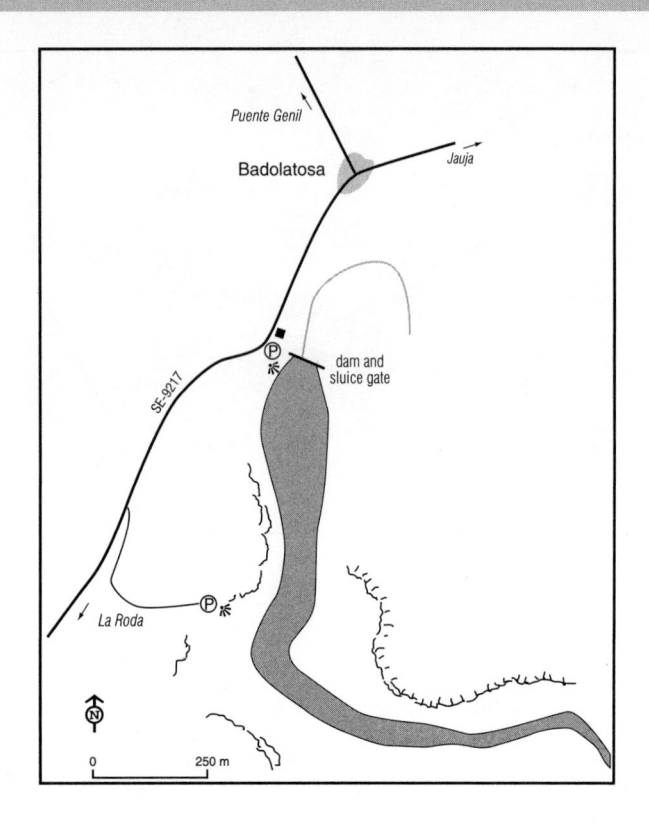

Species
The birdlife is broadly similar to the Embalse de Cordobilla further south. Waterfowl, including White-headed Ducks, are most abundant in winter. Breeding species include the Marsh Harrier and Purple Heron.

Timing
This reservoir may be visited at any time of year although it is most interesting for wintering waterfowl.

Access
This is from the SE-9217, southwest of the village of Badolatosa. There is a lay-by and viewpoint just outside Badolatosa, on the bank above the sluice gates. This part of the reservoir is generally less productive, apart from the Cormorants in winter. A better viewpoint is about 1km further south where a track entrance marked by a Paraje Natural sign turns off the east side of the road, just on a bend. There are two tracks leading from here. Take the right-hand track through the olive groves for 240m parallel to the main road and then turn left. Park just before a hillock on the left. A short walk then brings you to a steep slope with a mirador overlooking reedbeds and a good stretch of the sinuous reservoir.

♿ Somewhat difficult and restricted.

CALENDAR

All year: Mallard, Common Pochard, White-headed Duck, Grey Heron, Common Kestrel, Peregrine Falcon, Purple Swamphen, Common Coot, Cetti's Warbler, Jackdaw, Raven.

Passage: Osprey, waders.

Winter: Wigeon, Teal, Great Cormorant, Greater Flamingo.

Breeding season: Purple Heron, Marsh Harrier, Nightingale.

SIERRAS SUBBÉTICAS   CO8

Status: Parque Natural (31,568ha). ZEPA.

Site description

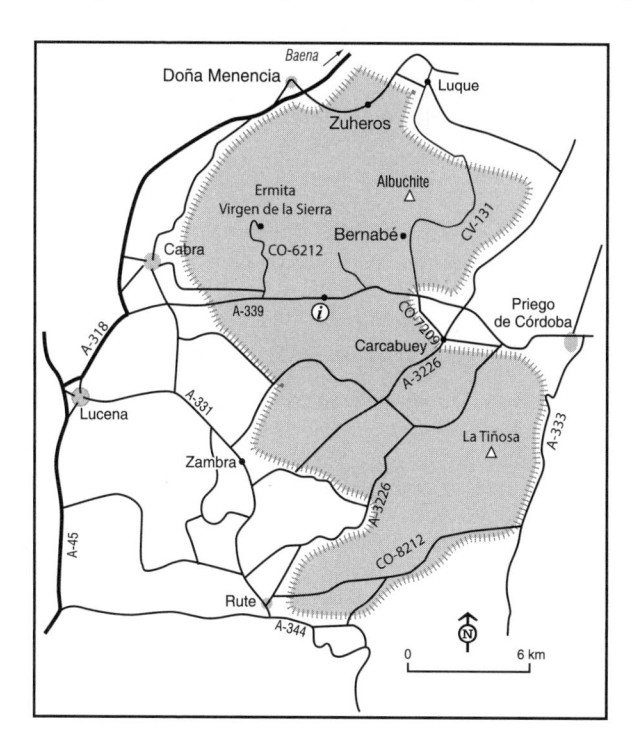

These southern sierras, between Lucena and Priego de Córdoba, comprise a calcareous island massif, an outpost of the Betic range. The geomorphology is limestone with sheer cliffs, rocky outcrops, scree slopes, sharp escarpments and narrow valleys, presenting an impressive rugged landscape. The peaks range between 900m and over 1,500m, Tiñosa (1,570m) being the highest summit in the province. Open terrain, with expanses of montane pasture, and rocky slopes with thin scrub or studded with hawthorn, junipers and other shrubs are characteristic of large areas, with dwarf juniper scrub on the summits. At lower levels there are areas of Mediterranean woodland as well as riparian woodland of willows and White Poplars along the watercourses. Large areas of the lower slopes have been occupied by extensive olive groves and other cultivated land. There is a visitors' centre (Centro de Visitantes de Santa Rita) on the A-339 halfway between Cabra and Priego de Córdoba.

Species

These mountains support a diversity of endemic plant and invertebrate species as well as the typical avifauna of Andalusian woodlands and rocky terrain. The most noteworthy of these last include such raptors as Griffon Vultures, which have a small colony on the cliff face of Albuchite, off the Carcabuey–Luque track. Both Bonelli's and Golden Eagles nest here, joined by Short-toed Eagles in summer. Parties of raptors such as Black Kites and Honey-buzzards move through the area on migration, especially during the southward passage.

Red-necked Nightjars occur at lower levels. More obvious common species include Alpine Swifts, Bee-eaters and Hoopoes. Other characteristic breeding birds include Iberian Green Woodpeckers; Crested, Thekla and Wood Larks, Red-rumped Swallows and Crag Martins. The rocky escarpments have Black Redstarts, Black-eared and Black Wheatears and Blue Rock Thrushes. Rufous-tailed Rock Thrushes breed at the highest levels of the sierras and Alpine Accentors frequent the tops in winter. In spring, the fluting song of Golden Orioles is characteristic of some areas, especially where there are poplars. Both Southern Grey and Woodchat Shrikes occur. The corvids include Red-billed Choughs. Rock Sparrows occur locally within the area. Cirl and Rock Buntings breed.

Migrant passerines are often evident, these including Ring Ouzels and other thrushes in autumn especially, as well as Northern Wheatears, Pied Flycatchers and warblers.

Timing

This site is interesting at any time of year, although it can get very cold, with snow in the winter months. Species diversity is highest in spring.

Access

Access from the west is easy via the A-45 Málaga/Córdoba motorway, leaving at km-72 for Rute on the A-344 or at km-56, bypassing Lucena on the A-318. From the east (Jaén) access is via the A-339 Priego de Córdoba/Lucena road, which bisects the area east/west.

Visitors' Centre (Centro de Visitantes Santa Rita) This is on the south side of the A-339 at km-8. The centre can provide maps and information on walking trails and indeed the Sierras are best explored on foot. The trails include the ascent of La Tiñosa (1,570m), a strenuous 9.5km climb for which permission must be sought two weeks in advance. A simple option giving opportunities to see passerines and overflying raptors is the shortish footpath (2.3 km each way) behind the centre. It ascends through Lusitanian Oak woodland and Mediterranean scrub on the slopes of the Sierra de Cabrera to a mirador. The going is quite steep in places.

Ermita de la Virgen de la Sierra The road up the sierra to this hermitage gives quick access to a scenic viewpoint at the top (1,217m), which is ideally placed to scan for raptors. Swifts and hirundines also forage along the ridge. The hilltop around the hermitage has clumps of pines as well as hawthorn and other shrubs. The Ermita is signposted from the A-339 6km east of Cabra. The access road (CO-6212) zig-zags upwards for 7km through open rocky terrain, where Black Wheatears and Rock Buntings occur. Stop occasionally to scan the slopes and particularly near the top by a disused quarry on the right.

A car park on the right just before the final ascent to the hermitage is the starting point of a walking trail (Sendero del Río Bailén) across a broad mountain meadow (paramo) and following the river. The going is easy for much of the trail, which continues for 11.8km as far as Zuheros village, although a shorter there-and-back walk is of course possible. The trail is an opportunity to find larks, pipits and other open-country species, including migrants, as well as overflying raptors and birds of rocky and bushy habitats.

Touring by car

Most birders will not be lingering long enough in the region to attempt the longer trails. An alternative is to tour representative areas by road, stopping where the habitat looks promising, especially where the ubiquitous olive groves have not replaced native woodland and scrub and where there are views of rocky escarpments and cliffs. A recommended route, giving access to rocky escarpments and a montane plateau, is the Carcabuey/Luque minor road (CV-131) via Bernabé. The southern entrance is at the A-339/CO-7209 crossroads above Carcabuey. The northern entrance in Luque is at the extreme southwestern corner of the village. A longer alternative is the circular route on minor roads from Carcabuey/Rute/Priego de Córdoba (A-3226/CO-8212/A-333). Tracks off these roads allow some wider exploration, as your vehicle and inclination permits.

♿ Many areas may be viewed from your vehicle.

CALENDAR

All year: Griffon Vulture, Egyptian Vulture, Goshawk, Common Buzzard, Golden Eagle, Common Kestrel, Peregrine Falcon, Woodpigeon; Eagle, Little and Tawny Owls, Iberian Green Woodpecker; Crested, Thekla and Wood Larks, Crag Martin, Dunnock, Black Redstart, Black Wheatear, Blue Rock Thrush, Mistle Thrush, Dartford and Sardinian Warblers, Blackcap, Southern Grey Shrike, Jay, Red-billed Chough, Jackdaw, Raven, Rock Sparrow, Crossbill, Cirl Bunting.

Red-rumped Swallow, Tawny Pipit, Grey Wagtail, Nightingale, Black-eared Wheatear, Rufous-tailed Rock Thrush; Melodious, Subalpine, Orphean and Bonelli's Warblers, Spotted Flycatcher, Crested Tit, Golden Oriole, Woodchat Shrike.

Winter: Meadow Pipit, White Wagtail, Alpine Accentor, Ring Ouzel, Song Thrush, Redwing, Fieldfare, Common Chiffchaff, Siskin, Bullfinch, Hawfinch.

Breeding season: Short-toed and Booted Eagles, Turtle Dove, Scops Owl, Red-necked Nightjar; Common, Pallid and Alpine Swifts, Bee-eater, Roller, Hoopoe, Wryneck,

Passage periods: Honey-buzzard, Black Kite, Hobby, European Nightjar, Tree Pipit, Common Redstart, Northern Wheatear, Willow Warbler, Pied Flycatcher.

SIERRA DE HORNACHUELOS — CO9

Status: Parque Natural (67,202ha). ZEPA.

Site description
A particularly beautiful sector of the Sierra Morena, contiguous with the Sierra Norte (SE6) in Sevilla province. The undulating terrain is dissected by river valleys, with some deep ravines and rocky escarpments. Great expanses of encina dehesas include large estates dedicated to Red Deer husbandry and hunting. There are also excellent woodlands of Stone Pines, encinas, Cork Oaks and Lusitanian Oaks. Riparian woodlands, of alders, poplars, willows and ashes, are a particular feature along the streams and rivers, notably along the course of the Río Bembezar, which drains a large part of the park. The upper slopes especially are covered with Mediterranean scrub, with Cistus, Fan Palms, Lentiscs and Terebrinths. Water is plentiful and retained by three reservoirs: the Embalses del Retortillo in the west, del Bembézar in the centre and de la Breña in the southeast.

Species
The typical species of the wooded rocky sierras of southern Spain are all present. Raptors are prominent and include breeding populations of all five eagles.

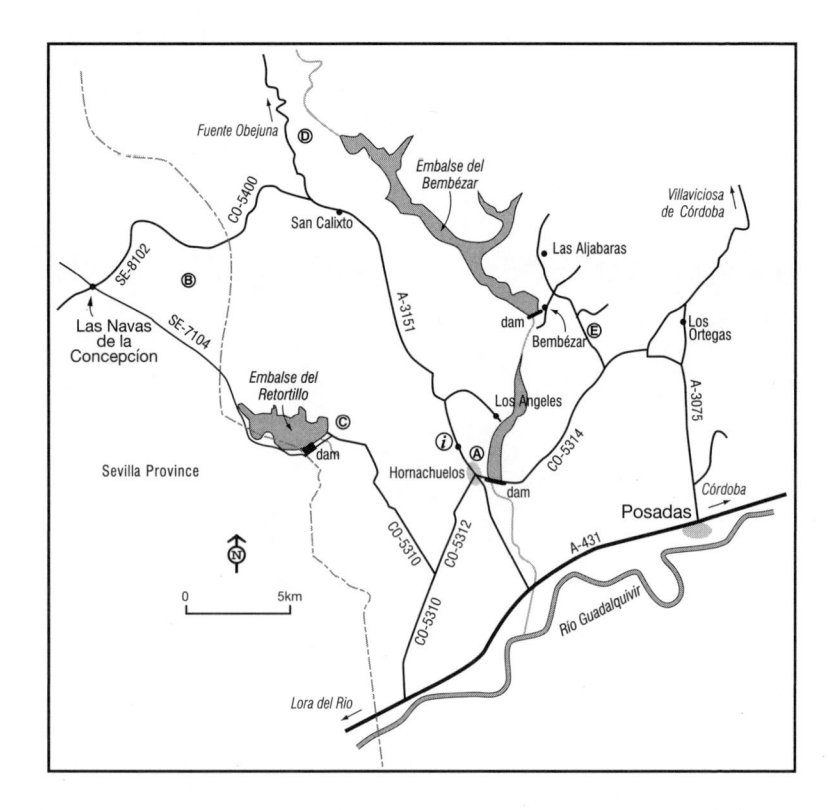

As ever, the Griffon Vulture is the most obvious raptor but the Black Vulture is also numerous here, where it has a sizable breeding popuation. Egyptian Vultures nest in small numbers.

Black Storks are regularly seen fishing in the rivers and several pairs nest locally. They often frequent the Bembézar reservoir. The reservoirs generally have breeding Great Crested Grebes and attract considerable numbers of Great Cormorants and some waterfowl in winter. Other notable breeding species include the Red-necked Nightjar, Alpine Swift, Rufous-tailed Scrub-robin, Black and Black-eared Wheatears, Spectacled and Subalpine Warblers, Southern Grey and Woodchat Shrikes, Red-billed Chough, Crested Tit and Golden Oriole. White-rumped Swifts occur and probably breed, usurping some of the nests of the numerous Red-rumped Swallows. Azure-winged Magpies are abundant, as elsewhere in the Sierra Morena. Rock Sparrows occur locally; a good place to find these is at Mina de la Plata, southeast of Bembézar, where they breed.

Mammals are well represented in the park. Red Deer and Wild Boars are common. The Iberian Lynx occurs but the presence of Wolves is uncertain. Mongooses and Genets occur in small numbers and Otters frequent the reservoirs and rivers.

Timing
The best time is undoubtedly spring and early summer, between April and the end of June or even mid-July. However, the wide variety of species to be seen makes a visit attractive at nearly any time of year.

Access
There is ready access from the north and east but the natural entry point is from the south, from the Guadalquivir to Hornachuelos village. From there you can traverse the park to the northwest, heading for San Calixto and Fuente Obejuna on the A-3151. Alternatively, just south of Hornachuelos, you can turn northeast for Villaviciosa de Córdoba on the CO-5314, which joins the A-3075.

A Hornachuelos village. Huerta del Rey Information Centre   It is worth parking in Hornachuelos village and walking north along the west bank of the river, as far as the Los Angeles monastery. There is a good chance of seeing foraging Black Storks and also both White-rumped Swifts and Red-rumped Swallows. A marked trail along the east bank extends all the way (13km) to the upper Bembézar dam but you may not wish to go that far.

The information centre (Huerta del Rey) is at km-1.5 on the A-3151 Hornachuelos/San Calixto road. Several walking trails of varying lengths originate from here and give access to riparian woodland, which is good for passerines.

B Hornachuelos/San Calixto monastery (17.4km)   The A-3151 is a quiet road that crosses representative areas of the park. There are ample opportunities to stop and view the various habitats. This region, on the western side of the Embalse del Bembézar, is generally productive. A network of farm tracks allows further exploration.

A circular route returning to, or beginning from, Hornachuelos crosses excellent dehesas. For an anticlockwise circuit, take the CO-5400 westwards,

Black-eared Wheatear

3km north of San Calixto, continuing into the northeastern corner of the Sierra Norte (SE4) on the SE-8102 to Las Navas de la Concepción, Sevilla. Head south from here on the SE-7104 to the Embalse del Retortillo (see C below) and back to Hornachuelos. Stop frequently to scan for raptors, especially at bridges and in open country. Black Vultures are frequent.

C Embalse del Retortillo   The CO-5310 southwest of Hornachuelos leads to this reservoir. The road follows the river where there is good riparian woodland. Several parking places allow easy access to the woods. The reservoir itself may be viewed from the dam and at various points on the western shore, where the road enters Sevilla province. Cormorants, Great Crested Grebes and waterfowl occur, the last of these mainly in winter.

D San Calixto monastery towards Fuente Obejuna   The 20km of the A-3151 between San Calixto and the northern park boundary also traverses interesting habitat, including some open country and scenic viewpoints. The road sees little traffic. Red Deer on the hunting estates are an obvious feature as are the Black and Griffon Vultures and other raptors that are often overhead. The park limit is marked by the Río Bembézar. The bridge across this river overlooks the rocky stream bed with deeper pools, attractive to Black Storks. Griffon Vultures nest nearby and the river attracts many passerines to drink.

E Hornachuelos towards Villaviciosa de Córdoba   The CO-5314/A-3075 cross some 30km of representative habitat within the park. A recommended sector here is the minor road 12km northeast of Hornachuelos that leads though dehesas to Poblado del Embalse and Las Aljabaras, at the Embalse del Bembézar.

♿ Most areas may be viewed from your vehicle.

CALENDAR

All year: Great Crested Grebe, Great Cormorant, Egyptian, Black and Griffon Vultures, Goshawk; Bonelli's, Spanish Imperial and Golden Eagles, Common Buzzard, Common Kestrel, Woodpigeon; Barn, Eagle, Little and Tawny Owls, Crag Martin, Black Wheatear, Blue Rock Thrush, Mistle Thrush, Crested and Coal Tits, Nuthatch, Short-toed Treecreeper, Southern Grey Shrike, Jay, Azure-winged Magpie, Common Magpie, Red-billed Chough, Jackdaw, Raven, Spotless Starling, Spanish and Rock Sparrows, Crossbill, Rock Bunting.

Alpine and White-rumped Swifts, Bee-eater, Roller, Hoopoe; Crested, Thekla and Wood Larks, Red-rumped Swallow, Nightingale, Rufous-tailed Scrub-robin, Black-eared Wheatear; Melodious, Spectacled and Subalpine Warblers, Spotted Flycatcher, Golden Oriole, Woodchat Shrike.

Winter: Hen Harrier, Sky Lark, Meadow Pipit, Song Thrush, Redwing, Fieldfare, Common Chiffchaff, Brambling, Siskin. Waterfowl on reservoirs.

Breeding season: Quail, Black Stork, White Stork, Short-toed and Booted Eagles, Turtle Dove, Common Cuckoo, Scops Owl, Red-necked Nightjar; Common, Pallid,

Passage periods: Black Kite, Sparrowhawk, Merlin, Sand Martin, Northern Wheatear, Common Redstart, Whitethroat, Willow Warbler, Pied Flycatcher.

SIERRA DE CARDEÑA Y MONTORO CO10

Status: Parque Natural (41,212ha). ZEPA.

Site description
Another of the protected sectors of the Sierra Morena, contiguous with the Sierra de Andújar (J3). The western part is rather rolling terrain with a considerable amount of cereal farming, not to mention herds of malodorous goats and numerous sheep and cattle. The landscape becomes rougher in the east along the course of the Río de las Yeguas, with high points not exceeding 820m. There are areas of open oak woodlands: dehesas of encinas, Cork Oaks and Lusitanian Oaks, as well as an outpost population of Pyrenean Oaks, the only ones in Córdoba. Plantations of Stone and Maritime Pines are also a feature, as well as expanses of Mediterranean scrub.

Species
Raptors are often evident, especially the Black and Griffon Vultures that visit daily from their nesting areas and roosts nearby. All five eagle species breed, as well as Common Buzzards, Goshawks and Peregrines. Iberian Green Woodpeckers are typical of the open woodlands. Other woodland or scrub birds present include Thekla Larks, Firecrests, Golden Orioles, Southern Grey and Woodchat Shrikes and the inevitable Azure-winged Magpies. Winter sees considerable arrivals of thrushes, including Redwings.

Together with the Sierra de Andújar (J3), the area is important for its mammal populations. The Iberian Lynx occurs at high density and there is a relict population of Wolves nearby. Mongooses, Otters, Genets, Wild Cats and Foxes are relatively numerous. Red and Fallow Deer, and Wild Boars, are numerous. Spanish Ibex, Mouflon and Roe Deer occur locally.

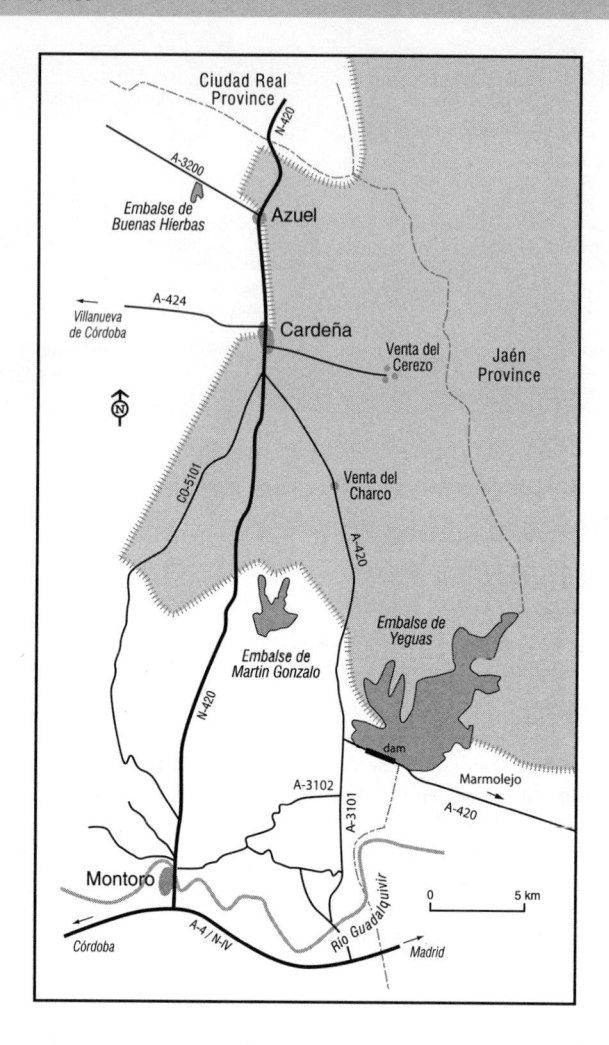

Timing
This area can be visited at any time of year but April–mid-July is best for variety.

Access
Three north–south roads within the park converge on Cardeña. The central road (N-420) that bisects the park is relatively busy but the flanking roads are much quieter and traverse the areas of greatest interest. Ascending one road and descending the other is a good option.

A West road. CO-5101. 34km. The quietest and most scenic of the access routes to Cardeña. Much of it passes through encina dehesas largely devoted to Red Deer hunting. The deer often appear on the road itself. Azure-winged Magpies are obvious and other dehesa birds soon become evident when you stop. They include the Iberian Green Woodpecker, Hoopoe and Crested Tit, among others. Nuthatches and Short-toed Treecreepers are common. Raptors overhead often include Griffon and Black Vultures, as well as Spanish Imperial, Short-toed and Booted Eagles.

The CO-5101 loops west from a point on the N-420 4km north of the Montoro junction. It rejoins the N-420 1km south of Cardeña.

B East road. A-420 via Marmolejo. 47km. This road climbs through olive groves and pine plantations in the south before reaching more productive open areas, between km-30 and km-40 especially, where there are encina dehesas and expanses of grassy pasture. A service track parallel to the main road at km 39–41 is handy for a brief walk through Stone Pine, scrub and encina dehesa to look for passerines and scan for raptors; park by the obvious farm entrances. The road joins the N-420 just south of Cardeña.

Access to the A-420 from the A-4 motorway to the south is possible at the Villa del Río exit (km-351); take the A-3101 north or, further east, at either exit for Marmolejo (km-335 or km-330). The A-420 crosses open country and the dam of the Embalse de Yeguas some 7km west of Marmolejo. Park by the dam, where there are all-round views, and inspect the lake and its margins for waterfowl and passage waders. A short diversion eastwards to the reservoir is necessary if you have taken the A-3102 to the A-420, and is probably worthwhile in winter or during passage periods.

C Cardeña A pleasant small town with the usual central square offering refreshment possibilities. The park information centre, Venta Nueva, is at km-79 of the CO-5101, at its junction with the A-420, 1km south of Cardeña. The centre has details of several walking trails that originate nearby. It is worth taking the (fairly rough) sandy track east from Cardeña to Venta del Cerezo. This track, and others beyond the Venta, offer representative birding through the dehesas.

All areas may be viewed from your vehicle.

CALENDAR

All year: Black and Griffon Vultures, Goshawk, Common Buzzard, Spanish Imperial, Golden and Bonelli's Eagles, Common Kestrel, Peregrine Falcon; Eagle, Little, Tawny and Long-eared Owls, Iberian Green and Great Spotted Woodpeckers; Crested, Thekla and Wood Larks, Black Redstart, Mistle Thrush, Dartford Warbler, Firecrest, Nuthatch, Short-toed Treecreeper, Southern Grey Shrike, Azure-winged Magpie, Magpie, Raven.

Breeding season: Black Stork, Short-toed and Booted Eagles, Roller, Hoopoe, Bee-eater, Scops Owl, Nightingale, Subalpine Warbler, Spotted Flycatcher, Golden Oriole, Woodchat Shrike.

Winter: White Wagtail, Song Thrush, Redwing.

Passage periods: Black Kite, Northern Wheatear, Common Redstart, Pied Flycatcher.

ALTO GUADIATO FARMLANDS CO11

Status: ZEPA.

Site description
A very extensive area of largely open country in the extreme northwest of the province, bordering southeastern Badajoz, Extremadura. Much of it is undulating grassland, grazed by sheep especially, with some large expanses of cereal cultivation. There are also open oak dehesas, some olive groves and areas of low scrub on hillsides.

Species
Birds of open country are the principal interest. The ZEPA Alto Guadiato has one of the two largest Great Bustard populations in Andalucía. However, these do not exceed 150 birds at the most and this species is much more easily found in Extremadura. Other steppic species in the region include Little Bustards and Stone-curlews, Quail, sandgrouse (especially Black-bellied Sandgrouse), Montagu's Harriers and Calandra Larks. There is a population of Black-winged Kites, these being most obvious in winter. Hordes of Spanish and House Sparrows frequent the stubble fields in late summer, when there is a vast sparrow roost in the village of Los Blasquez.

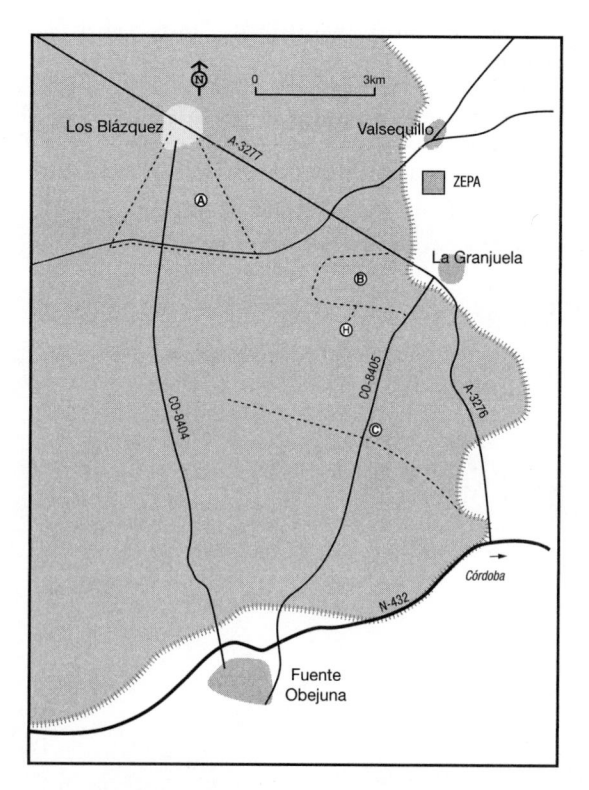

In winter the resident steppic species are joined by large numbers of Lapwings, Golden Plovers, Sky Larks and Meadow Pipits, as well as a few Hen Harriers and Merlins. Several thousand Cranes also winter in the dehesas.

Timing
Winter is good for seeing many of the steppe species, as well as the Cranes, which arrive from November onwards. Black-winged Kites are easiest to find in winter. Otherwise, spring is the best time, between March and early July.

Access
Fuente Obejuna on the N-432 is the natural gateway to the area. The N-432 links Córdoba city to the A-66 near Zafra in Extremadura. Most roads within the area, even quite minor ones, are in good or excellent repair.

The open farmland in the west attracts the largest numbers of bustards and other open-country species. You can explore by car, stopping frequently at farm track entrances to scan the fields and pastures. A useful driving route is the Fuente Obejuna/La Granjuela/Los Blázquez triangle. An information centre in La Granjuela dedicated to steppe birds opened in 2014.

A signposted trail (Ruta Ornitológica Los Blázquez, 10.5km) leads south from that village, from the Avenida de Andalucía at the southwest corner, through representative habitats before returning to the same village. It is an easy walk taking up to three hours. The trail can be driven if dry but walking is recommended. The route is most productive in winter, when Cranes, Black-shouldered Kites and both bustards can be found.

An alternative walking trail (La Piruetanosa), offering similar birds, centred on La Granjuela is also recommended. The trail is 7km but is not circular. You can walk all or part of it there and back or park in La Granjuela and walk to and from that village. Entrances are signposted on the CO-8405 and A-3277, 0.9 and 1.3km respectively from the village.

A crossroads on the CO-8405 at km-7 is an intersection with farm tracks included in a third trail (Itinerario Ornitológico Tierras de Avutardas). The full trail is long (30km) but parts of it are worth walking or driving to on each side of this crossroads, where you can park.

CALENDAR

All year: White Stork, Black-winged Kite; Spanish Imperial, Golden Eagle and Bonelli's Eagles, Common Buzzard, Common Kestrel, Little and Great Bustards, Stone-curlew, Black-bellied and Pin-tailed Sandgrouse, Little Owl, Calandra and Crested Larks, Southern Grey Shrike, Azure-winged Magpie, Raven, Jackdaw, Spanish Sparrow. Dove, Common Cuckoo, Red-necked Nightjar, Bee-eater, Roller, Hoopoe, Greater Short-toed Lark, Black-eared Wheatear, Woodchat Shrike.

Winter: Red Kite, Hen Harrier, Merlin, Crane, Lapwing, Golden Plover, Sky Lark, Meadow Pipit.

Breeding season: Black Stork, Black Kite, Short-toed and Booted Eagles, Montagu's Harrier, Lesser Kestrel, Turtle

Passage periods: Raptors, waders.

ZÚJAR AND GUADAMATILLA VALLEYS   CO12

Status: No special protection.

Site description
The principal attraction of this very quiet sector of northwestern Córdoba is the Río Zújar, which marks the boundary with Extremadura. Here the river and its tributaries, notably the Río Guadamatilla, feed the Embalse de La Serena. The watercourses hold sizable permanent pools even in the driest summers and these attract a diversity of breeding, passage and wintering waterbirds. Reedbeds and riparian scrub are an added attraction. Away from the streams and rivers there are expanses of rocky, sparsely vegetated terrain as well as areas of dehesa and Mediterranean woodland, and farmland in the south of the region.

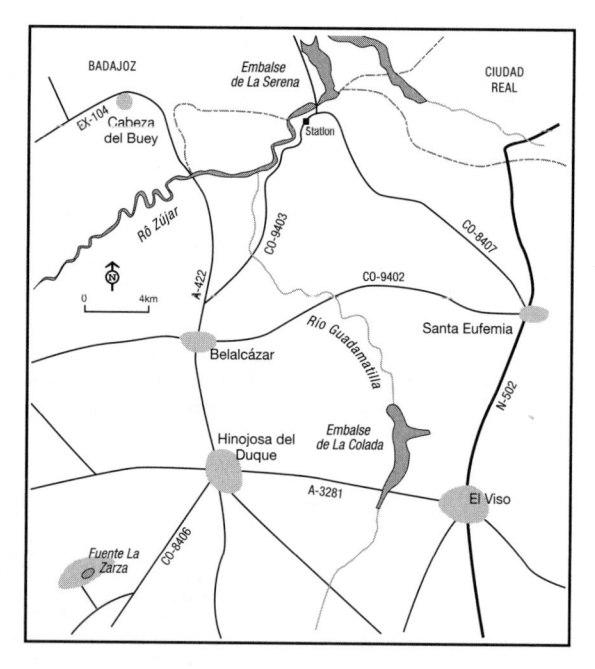

Species
Little Ringed Plovers nest on sandbanks and are joined by small numbers of migrant waders on passage. These typically include Common, Green and Wood Sandpipers, Black-winged Stilts, Greenshanks and Common Redshanks among others. White Storks nest commonly and Black Storks often visit the watercourses. The resident Mallards are joined by other common waterfowl in winter and these have occasionally included White-headed Ducks. Sandgrouse come to drink at the pools in summer. Other typical birds along the rivers include Yellow Wagtails.

Access

A *Belalcázar to the railway station* The excellent road (CO-9403) to the disused Estación de Belalcázar (railway station) is highly recommended. Turn east from the A-422 5km north of Belalcázar. Dry barren terrain, frequented by Southern Grey Shrikes, gives way to encina dehesas, with riparian woodland and reedbeds along the watercourses. The road runs alongside the Guadamatilla to its confluence with the Zújar and continues east along that riverside to the Estación. Close views of the rivers and their many pools, reedbeds and sandbanks are available from the road itself. Traffic is negligible but it is still advisable to pull off when you stop. There is a good stopping place at km-12 (Finca Madoniz), just past a large pool and where a ford across the river gives access to the watercourse. A bridge across the Zújar just past the station leads into Badajoz province but this was closed to traffic in 2017. The bridge gives excellent views of the river and the shallows at the upper extremity of the Embalse de La Serena. The CO-8407 that leads southeast from the bridge to Santa Eufemia is also an interesting drive though dehesas and riverside habitats. However, this road is a rough track and 4WD would be advisable in any but the best conditions. Otherwise it is best to return the way you came, a distance to Belalcázar of some 23km.

B *Río Guadamatilla bridge at El Viso* Bridges are happy hunting grounds anywhere in the region, especially in the drier months, and are often good places to stop. The Guadamatilla bridge on the A-3281 Hinojosa/El Viso road, 6km west of El Viso, crosses the shallow tail of La Colada reservoir. Pull off the busy road at the west end of the bridge, where a track to the dam (Presa de La Colada) is signposted. There are safe elevated views from here and you can also walk along the waterside. Waterfowl gather here and ten or more wader species have been present at times during passage periods.

C *Fuente La Zarza* A public recreational area and picnic area in an area of open dehesas near Hinojosa. The main attraction is in winter, when Cranes are often present in the dehesa. A hide overlooks a small pool that attracts some waders and other waterbirds, as well as an area frequented by Cranes. Overflying raptors and other dehesa birds are also a feature. The site is signposted off the north side of the CO-8406 at km-6.

♿ Most areas may be viewed from your vehicle.

CALENDAR

All year: White Stork, Black-winged Kite; Spanish Imperial, Golden and Bonelli's Eagles, Common Buzzard, Common Kestrel, Stone-curlew, Black-bellied and Pin-tailed Sandgrouse, Little Owl; Calandra, Crested, Thekla and Wood Larks, Black Wheatear, Cetti's Warbler, Southern Grey Shrike, Azure-winged Magpie, Raven, Jackdaw, Spanish Sparrow.

Breeding season: Black Stork, Black Kite, Short-toed and Booted Eagles, Montagu's Harrier, Lesser Kestrel, Little Ringed Plover, Turtle Dove, Common Cuckoo, Red-necked Nightjar, Bee-eater, Roller, Hoopoe, Greater Short-toed Lark, Nightingale, Black-eared Wheatear; Reed, Great Reed, Spectacled and Orphean Warblers, Spotted Flycatcher, Woodchat Shrike.

Winter: White-headed Duck, Red Kite, Hen Harrier, Merlin, Crane, Lapwing, Golden Plover, Common Sandpiper, Sky Lark, Meadow Pipit.

Passage periods: Raptors, waders.

PUENTE NUEVO RESERVOIR　　　　　CO13

Status: No special protection

Site description
A very accessible reservoir on the Río Guadiato, just south of the N-432 trunk road some 30km from Bélmez and a similar distance from Córdoba city. It is worth a detour if you are passing by. This is a sizable (2,032ha), long, narrow reservoir, with over 90km of shoreline. Large areas are devoted to sailing, watersports, bathing and fishing. However, islets, and shallow margins around the southern end especially, attract waterfowl and waders on passage and in winter. A telescope will be useful here.

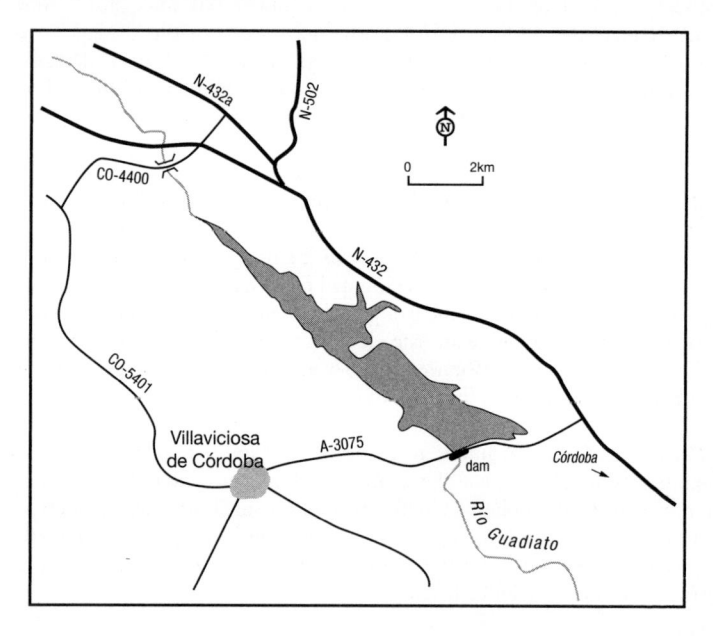

Species
A site with significant potential that has received little attention from birders. Considerable numbers of common waterfowl occur, especially in winter. Great Crested Grebes nest locally. Egyptian Geese are present and rarer occasional waterfowl have included the Ruddy Shelduck. Black Storks visit the lake margin and Ospreys occur on passage. The Bearded Reedling (Tit) has been recorded here: two 'pairs' in March 2001. The surrounding woodlands offer the usual diversity of passerines, including Azure-winged Magpies and Iberian Chiffchaffs.

Access
The southern end can be viewed from the dam and its approaches on the A-3075; follow signs to Villaviciosa de Córdoba from the N-432 at km-238.

There is also vehicular access to the entire eastern side of the reservoir from the A-3075 east of the dam, from a track at the bridge across the river Guadiato on the CO-4400 and at km-231 on the N-432. The CO-5401 is a very quiet road through the western hinterland of the reservoir, crossing hilly terrain with open pinewoods, some Holm Oak dehesa, olive groves and scrub.

♿ Many areas may be viewed from your vehicle.

> **CALENDAR**
>
> **All year:** Great Crested Grebe, Mallard, Egyptian Goose, Little Ringed Plover, Azure-winged Magpie.
>
> **Passage periods and winter:** Common waterfowl including Common and Red-crested Pochards, Great Cormorant, Grey Heron, Little Egret, Black Stork, Common Coot, Osprey, Green and Common Sandpipers.

RÍO GUADALQUIVIR AT CÓRDOBA   CO14

Status: Monumento Natural 'Sotos de la Albolafia' (22ha).

Site description
This site has few equals in the region in terms of species diversity within an easily viewed restricted area, let alone within a provincial capital. The River Guadalquivir is held back by weirs as it flows through the ancient city of Córdoba and is some 250m wide. The section between two major bridges, the pedestrians-only Puente Romano (Roman Bridge) to the east and Puente de

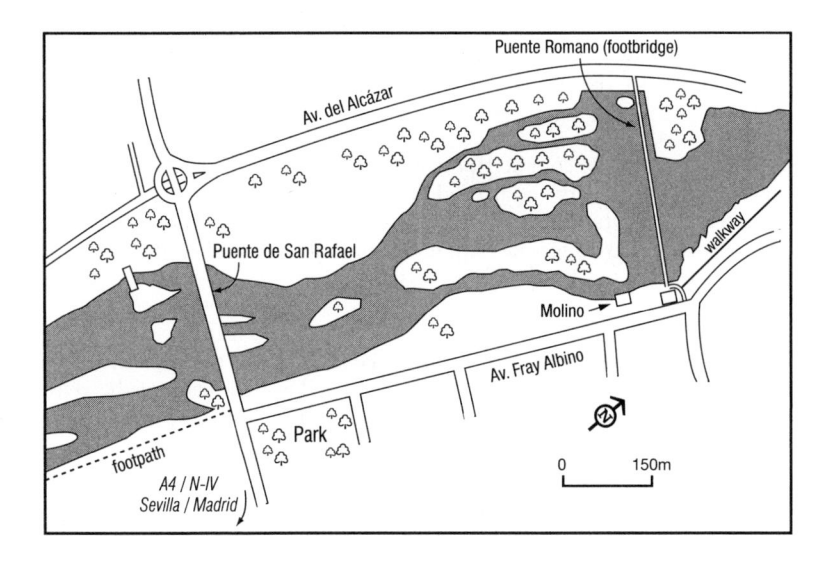

San Rafael to the west, is protected as a Natural Monument. Here there survives a good stretch of mature riparian woodland along both banks and on a number of sizable islands, some with shingle or sandy beaches. Large reedbeds are another well-developed feature. The core site is adjacent to the Alcázar and very near the massive and ancient mosque, with its integrated Catholic cathedral. This magnificent edifice will amply reward inspection by all but the most single-minded of birding visitors.

Species

Well over 100 species, many of them waterbirds, are recorded annually just in the area between the two named bridges. The mixed heronry in the trees near the Puente de San Rafael has nesting Cattle Egrets, Night Herons and Little Egrets. Squacco Herons, Little Bitterns and Purple Swamphens nested formerly but no longer do so, although the first two at least have over-summered recently. Grey and Purple Herons, as well as White Storks, feed along the pool edges and in the vegetation. The same site houses a very large winter roost of Cattle Egrets, over 5,000 birds, with smaller numbers of Little Egrets.

Many birds follow the Guadalquivir during migration. Great Cormorants are increasingly common along the river, especially in winter. Migrant waders regularly include Common Redshanks; Green, Wood and Common Sandpipers, Dunlins and occasionally Ruffs. Common Snipe are frequent in winter. Several thousand Black-headed and Lesser Black-backed Gulls roost in the area in winter. Slender-billed and Audouin's Gulls have also been recorded. Common, Whiskered and Black Terns are seen annually during migration periods.

There are often several thousand swifts and hirundines feeding over the river, especially during migration periods. Large numbers of White Wagtails winter and roost in the area, as do Bluethroats. Small migrants of many species can be looked for in the riverside vegetation during passage periods. Almost any species can turn up here and these include some scarcer migrants such as the Wryneck, Rufous-tailed Scrub-robin and Olivaceous Warbler. Penduline Tits occur in winter and probably nest here. Raptors overflying the area include Lesser Kestrels that nest in the historical buildings, as well as migrants of many species. Black Kites are frequent and Ospreys are annual on passage. Birds apart, Otters are regular visitors.

Timing

This is a particularly accommodating site. The light is nearly always right at any time of day, distances are optimal and the birds are highly visible. A visit is very worthwhile at any time of year, but the best time is between February and July, especially in the evening and early in the morning when the birds are most active.

Access

The A-4 skirts Córdoba city to the south. Follow signs to the city centre and then towards the Mosque (Mezquita). Head for the south bank of the river, where it is usually easy to park for free in the residential streets parallel to the river. Parking on the north bank is harder to find.

A circular walk taking in both bridges is recommended. The best views of the core area are from the promenade on the south side of the river or from either of the bridges. The Molino de San Antonio, on the south bank by the

Puente Romano, has a viewpoint for the heronry. Also on the south bank, an elevated walkway with a wooden balustrade offers good views from an ideal distance of the fringing scrub and woodland upstream of the Puente Romano. A leisurely walk along here, and also along the canopy-level promenade on the north bank between the two bridges, is often rewarding during migration periods. The reedbeds downstream of the Puente de San Rafael are also worth inspecting from the south bank.

♿ Highly suitable for the less mobile.

CALENDAR

All year: Night and Grey Herons, Cattle and Little Egrets, Kingfisher, Common Waxbill.

Breeding season: Little Bittern, Purple Heron, Lesser Kestrel, Little Ringed Plover, Nightingale.

Winter: Great Cormorant, Common Snipe; Black-headed and Lesser Black-backed Gulls, White Wagtail, Bluethroat, Penduline Tit (may breed).

Passage periods: Osprey, raptors, Dunlin, Common Redshank; Green, Wood and Common Sandpipers; Common, Whiskered and Black Terns, Wryneck, passerines.

MÁLAGA PROVINCE

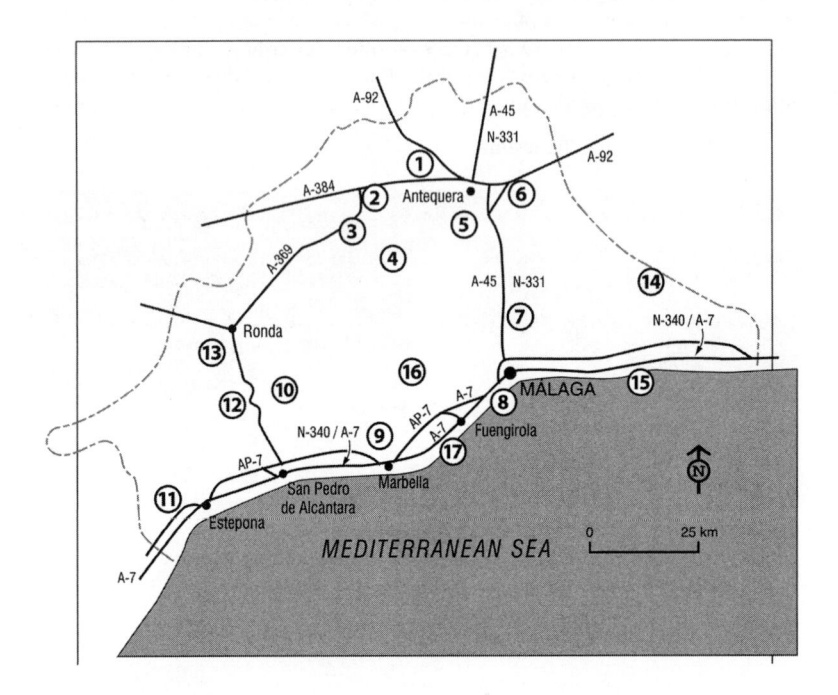

Málaga is the most densely populated of the provinces of the Costa del Sol, with a huge tourist industry employing 60% of the workforce, directly or indirectly. Vast numbers of tourists descend on the coastal zone, especially between July and mid-September. Away from the coastal conurbations there still remain unspoilt sites, of which Fuente de Piedra (MA1) is of international importance on account of its often massive Greater Flamingo colony. The ponds at the mouth of the Río Guadalhorce (MA8), and the small Río Vélez (MA15) at Torre del Mar, are also important given the scarcity of coastal wetlands in eastern Andalucía. Punta Calaburras (MA17), on the coast between Fuengirola and la Cala de Mijas, has proved to be an excellent seawatching site. The province offers a diversity of impressive yet usually peaceful mountain sites, among which the Sierra de Camarolos (MA6), Sierra de las Nieves (MA10) and parts of the Sierra Bermeja and Serranía de Ronda (MA12) are particularly recommended. The Mirador de las Águilas in the Sierra de Alpujata (MA16), near Fuengirola, can be very rewarding for viewing autumn raptor migration. Some sites, notably El Chorro (MA4) and the Montes de Málaga (MA7), suffer considerable tourist pressure but this can be avoided in the summer months by going early in the morning and avoiding Saturdays and Sundays generally.

Sites in Málaga Province
MA1   Laguna de Fuente de Piedra
MA2   Lagunas de Campillos

MA3	Teba: sierra and gorge
MA4	El Chorro gorge and the Abdalajís valley
MA5	El Torcal
MA6	Sierra de Camarolos
MA7	Málaga hills (Montes de Málaga)
MA8	Guadalhorce estuary
MA9	Juanar (Sierra Blanca)
MA10	Sierras de las Nieves National Park
MA11	Sierra Crestellina
MA12	Sierra Bermeja and Central Serranía de Ronda
MA13	Upper Guadiaro valley
MA14	Sierra Tejeda
MA15	Vélez estuary
MA16	Alpujata migration watchpoint (Mirador de las Águilas)
MA17	Punta Calaburras

Getting there

Málaga Airport is an obvious point of arrival in southern Spain. It is adjacent to the A-7/N-340 coastal highway, which provides ready access east/west to other provinces. Virtually all parts of Málaga province are within two hours' drive of the A-7 or the roughly parallel toll motorway, the AP-7. The major trunk roads linking Málaga and other provincial capitals also provide speedy access to all parts of the province. The motorways provide welcome bypasses of the coastal conurbations, through which traffic all too often makes very slow progress.

Where to stay

The tourist industry provides an enormous range of facilities along the coastal strip between Nerja in the east and Estepona in the west, the greatest concentration being between Torremolinos and Marbella. There are nonetheless plenty of places to stay inland. Paradors in the province comprise those at Ronda, Antequera, Málaga Golf, Málaga Gibralfaro and Nerja.

LAGUNA DE FUENTE DE PIEDRA   MA1

Status: Reserva Natural (1,364ha, with surrounding protection area).
Ramsar site. ZEPA.

Site description

This large, natural, saline lake, some 6.8km long by 2.5km wide, lies just off the main Granada/Sevilla road (A-92), close to Fuentedepiedra village. It is set in a basin of surrounding sierras from which small streams provide its only direct water supply. Commercial salt pans operated at the lake until the early 1950s. Cultivation of sunflowers, wheat and olives takes place in the hinterland. The sheltered location makes for near-freezing temperatures in winter and daunting heat in summer, when daytime temperatures can easily exceed 40°C.

It is a seasonal lagoon, only 1.4m deep at its maximum, which dries out in the summer heat, often by June, or July at the latest, except for an area to which water is pumped if the Flamingos have bred and there are still chicks in the crèche. The lake refills from early October onwards. Annual rainfall averages 460mm but is highly variable (range 200–700mm) and this conditions its suitability for waterbirds in any particular season. In poor rainfall years it may be effectively dry as early as March.

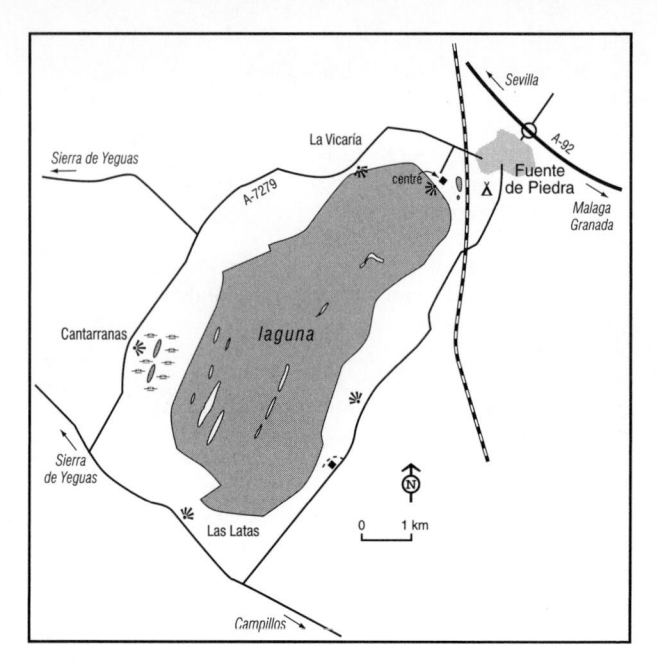

There is a lakeside information centre (Centro de Visitantes José Antonio Valverde) which is open most days (mornings only in August).

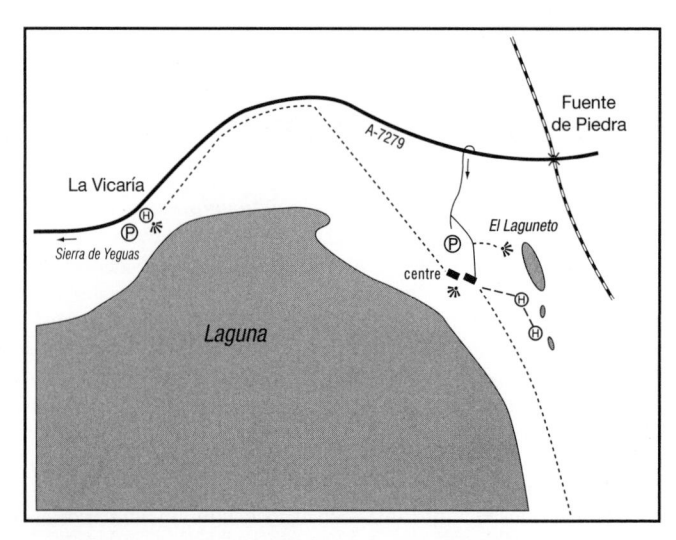

Species

The lake is renowned for its colony of Greater Flamingos. The nesting population averages some 10,000 pairs but has exceeded 20,000 pairs on exceptional occasions. Large numbers of young are produced in good years, the record output being 15,000+ in 1998. Breeding is regular but not annual since it requires a minimum depth of 30cm of water at the end of February. In years when the birds are successful, there is a large-scale ringing session in late July or early August, during which about 10% of the chicks are ringed using standard metal rings and large, easily read, coloured DARVIC rings with a number/letter code. The origins of marked birds can be tracked down, and sightings reported, via the European bird colour-ringing website (*http://www.cr-birding.org/node/787*). In all cases it is important to note the colours, codes and relative positions of all the rings present.

An added attraction is the increasingly regular presence of a few Lesser Flamingos in the colony, where they have nested successfully in some years. Three pairs were present in 2017. A few escaped Lesser Flamingos have been found in Spain. However, the pattern of occurrences suggests arrivals of wild birds from Africa, perhaps accompanying the Greater Flamingos, some of which range as far south as Mauritania.

Breeding populations of ducks and grebes also vary according to water availability but White-headed Ducks and Red-crested Pochards both nest when levels permit. Kentish Plovers, Avocets and Black-winged Stilts breed within the lake boundaries. Gull-billed Terns, usually a few hundred pairs, nest regularly and there is a small Black-headed Gull colony. Common Shelducks, Slender-billed Gulls and Collared Pratincoles have bred occasionally. The lake attracts numbers of waders, especially around the few freshwater inlets, particularly during the spring migration. They include Little Stints, Curlew Sandpipers, Dunlins, Black-tailed Godwits and Ruffs as well as occasional rarer species such as Temminck's Stint and Marsh Sandpiper. Migrant Common, Little, Whiskered and Black Terns occur in variable, usually small, numbers in spring. The small lake (El Laguneto), reedbeds and marshy area alongside the visitors' centre offers such species as the Purple Swamphen (irregular), Squacco Heron and both Reed and Great Reed Warblers.

The surroundings of the lake hold a wide diversity of species. Montagu's Harriers nest in the area, Common Kestrels are widespread, Griffon Vultures occasionally wander over from the colony at Valle de Abdalajís and Peregrines sometimes flash through, causing consternation. Hen Harriers and occasional Merlins occur in winter. A diversity of other raptors occur, especially during passage periods when such species as Short-toed Eagles and Black Kites are frequent. There is always the chance of encountering something more exciting, such as an immature Spanish Imperial Eagle, an occasional winter resident.

Stone-curlews are fairly common in the peripheral habitats and are more evident in winter, when several hundred occur both within the reserve and in the surrounding fields. Little Bustards also breed locally but are more numerous in winter. Other characteristic hinterland birds include Bee-eaters, Hoopoes, Greater Short-toed and Crested Larks, occasionally large numbers of swifts and hirundines, (Iberian) Yellow Wagtails, Black-eared Wheatears, Zitting Cisticolas, Spectacled Warblers, Southern Grey Shrikes and Woodchat Shrikes.

Wintering species on the lake mainly comprise common waterfowl although Marbled Ducks occur occasionally. Several thousand Greater Flamingos are often present by the end of December if the autumn rainfall has been adequate to provide sufficiently high water levels. Wintering waders include Little Stints, which are also quite numerous during migration periods. Up to 400 Cranes winter along the western edge of the lake. A few Hoopoes are often present in winter.

Timing

The best time is normally between December and June or early July, Greater Flamingos occurring throughout this period if there is sufficient water. From late May onwards it is best to visit the lake before 11:00 hrs as heat haze can be a problem later. Wintering species occur from mid-October onwards and spring migration is evident from early March through to the end of May.

The dry period of July to October – or even as late as early November, depending upon the rains – is usually very unproductive and the lake is not really worth visiting then. It usually re-floods from November onwards, after which most wintering species may be easily seen.

Access

The lake is signposted from both carriageways of the A-92 at km-132. Drive through the village and follow the MA-454 across the bridge over the railway line. The entrance to the visitors' centre is signposted 100m beyond on the left. The centre is on a hillock, overlooking the lake. A telescope will prove invaluable here and at other watchpoints around the Laguna. The centre is the starting point of two easy trails. The Sendero del Laguneto (250m) serves the small lagoons and other wet areas nearby. A longer footpath (Sendero de las Albinas) with boardwalks (2.5km) skirts the wet northwestern corner of La Laguna, crossing two inlet streams, as far as a seasonally marshy area (La Vicaría), where there is a mirador. Ducks and waders are often present here. La Vicaría mirador is also accessible from the perimeter road, where you can park.

A drive around the entire perimeter (21km, tarmac) is well worthwhile. There are several signposted viewpoints and the whole route offers opportunities for searching the hinterland habitats. The anticlockwise circuit starting from the visitors' centre along the road to Sierra de Yeguas (A-7279) soon brings you to La Vicaría mirador. Continue southwards for a further 5km on the A-7279 until you reach a fork where you bear left. Two viewpoints are signposted at the fork. The first of these (Mirador de Cantarranas) is reached after 2.2km, where there is a car park. It overlooks seasonal pools that often attract waterfowl and waders and the main lake beyond. The surrounding scrub holds Spectacled Warblers, and Cranes are often on the farmland in winter.

Another viewpoint (Mirador de Las Latas) is on the south side of the lake, a further 3.5km from Cantarranas. Bear left again at the fork on to the A-6213. The small mirador car park is on the left 1.7km from the fork but access here is often restricted. This viewpoint overlooks areas of disused salt pans, whose dykes are used by nesting gulls, terns and flamingos. Return to the A-6213 and continue to a junction, turning left again at the sign marked Ruta Entorno de La Laguna. There are additional viewing opportunities along this road, which follows the eastern side of the lake back to Fuentedepiedra village.

♿ The walkways and hides around the visitors' centre are wheelchair-friendly. The viewpoints around the lake are readily accessible from the car parks.

CALENDAR

All year: Purple Swamphen, Little Bustard, Stone-curlew, Kentish Plover, Black-headed Gull, Hoopoe, Crested Lark, Zitting Cisticola, Cetti's Warbler, Southern Grey Shrike.

Breeding season: Red-crested Pochard, White-headed Duck, Black-necked Grebe, Little Bittern, Squacco Heron, Greater Flamingo, Lesser Flamingo, Montagu's Harrier, Black-winged Stilt, Avocet, Slender-billed Gull (sporadic), Gull-billed Tern, Bee-eater, Greater Short-toed and Crested Larks, Yellow Wagtail; Reed Warbler, Great Reed Melodious and Spectacled Warblers.

Winter: Black-necked and Great Crested Grebes, Grey Heron, Greater Flamingo, Common Shelduck, Wigeon, Teal, Mallard, Pintail, Shoveler, Red-crested and Common Pochards, Marbled and White-headed Ducks, Crane, Black-headed, Lesser Black-backed and Yellow-legged Gulls, Meadow Pipit, Common Chiffchaff.

Passage periods: Spring: Garganey; waders including Little and Temminck's Stints, Dunlin, Ruff, Black-tailed Godwit; Curlew, Green, Wood and Common Sandpipers; Little, Whiskered and Black Terns. Autumn: little to see (lake dry) except for occasional raptors and White Storks.

LAGUNAS DE CAMPILLOS   MA2

Status: Reserva Natural (1,342ha).

Site description
A series of reed-fringed, seasonal lagoons to the east and southeast of Campillos. The principal and most accessible ones are the Lagunas Dulce, Salada, Redonda and Capacete but there are a number of other smaller pools on private land. The water levels in all these lagoons fluctuate considerably with rainfall. The Laguna Salada nearly always has water but the others are often dry in summer or during drought years.

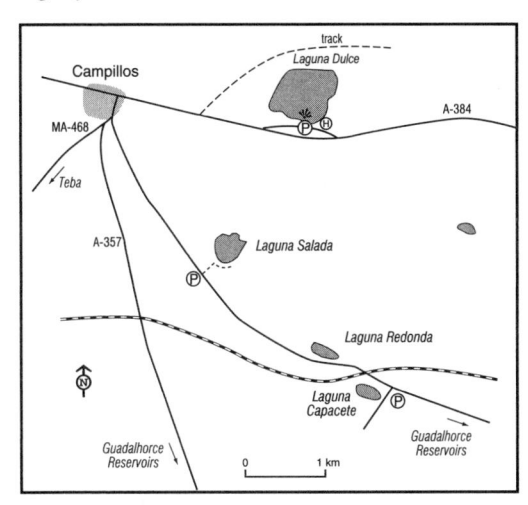

Species
White-headed Ducks are found irregularly at all the lakes, their presence and breeding depending totally on water levels. The lakes also hold varying numbers

of wintering ducks, notably Shovelers. Garganey occur, mainly during spring migration, and there are occasional reports of rarer species such as Ferruginous and Marbled Ducks. Greater Flamingos from Fuente de Piedra nearby also visit in small numbers regularly and indeed there is frequent interchange of waterfowl between Fuente de Piedra and these lakes. The Laguna Dulce in particular may hold an interesting variety of waders when water levels are low, including the ubiquitous Black-winged Stilts, a few Avocets, parties of Black-tailed Godwits; occasional Wood, Green and Curlew Sandpipers, and good numbers of Little Stints.

Timing
All year, but summer and autumn visits are less productive in drought years when the lagoons dry out.

Access
The Laguna Dulce is easily seen just to the north of the A-384, some 3km east of Campillos. This is the most attractive lagoon when there is water. The Lagunas Salada, Redonda and Capacete are also easily accessible from the A-7286, which runs southeast from Campillos. For the Laguna Salada, access from the road is along a dirt track on the north side about 1km from Campillos. Park by the roadside and walk the short distance up the hill to view the lake. The Laguna Redonda is a kilometre or so further east alongside this road; it also has a lay-by and a viewing screen. Further still, beyond the railway bridge, the Laguna Capacete is visible south of the road; to view, park on the track at the eastern end of the lagoon.

♿ The Lagunas Dulce, Redonda and Capacete are easily watched from a vehicle or from the road. The Laguna Salada is much less accessible.

CALENDAR

All year: Mallard, Shoveler, Common Pochard, White-headed Duck, Greater Flamingo, Little Bustard, Lapwing, Crested Lark, Cetti's Warbler, Raven.

Breeding season: Garganey, Reed and Great Reed Warblers.

Winter: Pintail, Red-crested Pochard.

Passage periods: Waders, when water levels are low.

TEBA: SIERRA and GORGE　　　　MA3

Status: No special protection.

Site description
A representative fragment of the Betic range, bisected by a scenic narrow gorge. The sierra, which comprises angled strata with caves and overhangs, lies to the north of the MA-5404 (C-341). It is a long, low barren ridge, higher and steeper at its eastern end where there are small cliff faces. Olive groves occupy the lower flanks. The sierra overlooks an arm of the Guadalteba reservoir. The

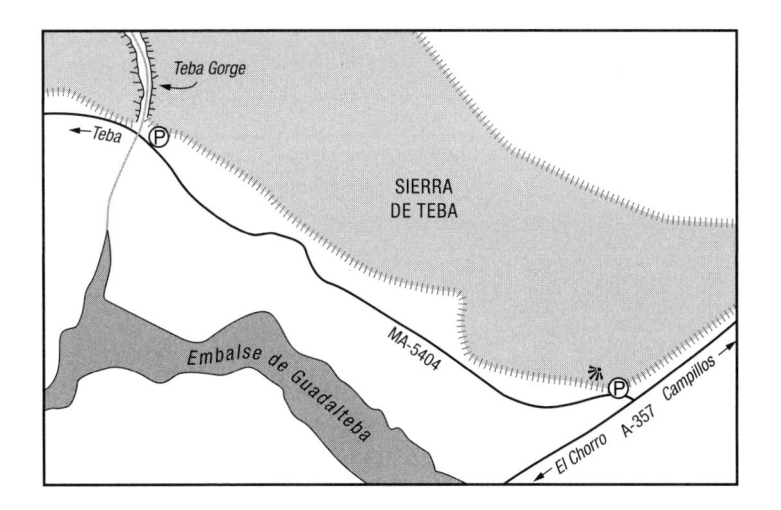

sheer-sided gorge is superbly scenic, although the Río La Venta that flows through is rather unpleasantly contaminated at times.

Species
The sierra and gorge are a traditional site for nesting rupestral (rock-loving) bird species, although disturbance by climbers and trekkers has displaced some from the gorge itself. Bonelli's Eagles, Egyptian Vultures and Eagle Owls nest in the area and diverse other raptors, including Griffon Vultures and Golden Eagles, are often overhead. Alpine Swifts, Ravens, Red-billed Choughs, Blue Rock Thrushes, occasional Rufous-tailed Rock Thrushes, Black Wheatears, Black-eared Wheatears, Melodious Warblers, Rock Sparrows and Rock Buntings all nest in the general area. The gorge also holds nesting Rock Doves, Crag Martins, Jackdaws, Grey Wagtails and Cetti's Warblers. Numerous Nightingales occur along the river downstream of the gorge in spring, when Rufous-tailed Scrub-robins may be sought in the olive groves.

Timing
The area is always worth a look if you are passing by, especially in spring when bird diversity is greatest. A visit here can form part of a trip to Fuentedepiedra and the Campillos lagoons, a short distance further north.

Access
Access is from the MA-5404 Ronda road, between Campillos and El Chorro, east of Teba village. The sierra is easily viewed from the road, where there are parking opportunities. A lay-by on the north side of the road at its western end, where it joins the A-357, is signposted 'Observatorio de Aves' and is a recommended watchpoint for vultures and other raptors. It is possible to walk from the road up to the base of the cliff face here and elsewhere to look for passerines.

The gorge is easily visible where the road crosses the Río La Venta. Parking is available where it is signposted Tajo del Molino, east of the bridge. Entering involves a scramble down to the river bed, water volumes permitting, the easiest access being downstream of the bridge. Once in the gorge, care should be taken as the rocks are somewhat slippery in places and the going becomes more difficult about halfway along.

&. The sierra is easily inspected from the road. The gorge is hard-going and strictly for the fit.

CALENDAR

All year: Griffon Vulture, Common Buzzard, Bonelli's and Golden Eagles, Common Kestrel, Peregrine Falcon, Rock Dove, Eagle Owl, Little Owl, Crested and Thekla Larks, Crag Martin, Grey Wagtail, Wren, Black Redstart, Blue Rock Thrush, Blackcap, Cetti's Warbler, Southern Grey Shrike, Red-billed Chough, Jackdaw, Raven, Rock Sparrow, Rock Bunting.

Breeding season: Egyptian Vulture, Alpine Swift, Bee-eater, Hoopoe, Rufous-tailed Scrub-robin, Nightingale, Black-eared Wheatear, Rufous-tailed Rock Thrush, Melodious Warbler.

Passage periods: Migrant raptors.

EL CHORRO GORGE AND THE ABDALAJÍS VALLEY   MA4

Status: Paraje Natural 'Desfiladero de los Gaitanes' in part (2,016ha).

Site description
An expanse of some 50km^2 to the northwest of Málaga city that includes a diversity of habitats ranging from mountains to chasms: steep slopes wooded with Aleppo Pines, wild olives, junipers and Holm Oaks, Cistus scrub, riverine woodland and reservoirs. The Río Guadalhorce crosses the region and is

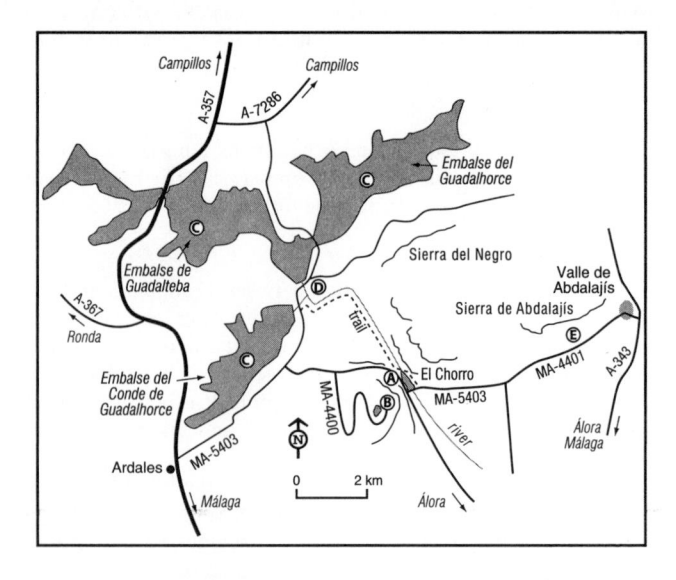

channelled through a spectacular gorge, the Desfiladero de los Gaitanes, better known as La Garganta del Chorro or simply El Chorro. The gorge is some 3km long, with sheer walls rising up to 300m, but is seldom more than 10m across. A vertiginous walkway, part of El Caminito del Rey trail, clings to the wall 100m above the river. This walkway had fallen into dangerous disrepair and was closed until 2015 when it was restored and reopened, becoming a considerable tourist attraction.

Species
The whole area offers a good variety of species, especially those associated with rocky habitats and pinewoods. Griffon Vultures as well as both Bonelli's and Golden Eagles are regularly present and there are occasional Black and Egyptian Vultures. Other attractions include Crested Tits and Crossbills in the pinewoods, Bonelli's Warblers in both pines and deciduous growth, and Alpine Swifts wheeling high overhead.

Timing
The whole area is interesting at virtually any time of year but is undoubtedly best in spring when migrants join the resident breeding species. Weekday visits are recommended. Avoid weekends and holiday periods since the area attracts many visitors, particularly since the cliff walkway was recommissioned, leading to inevitable disturbance and congestion on the narrow roads.

Access
There are several access routes. However, the most straightforward, whether arriving from north or south, is via the A-357, leaving at Ardales to follow signs to El Chorro along the MA-5403. An alternative from the south is to head from Málaga on the A-343 to Valle de Abdalajís. The whole area is served by many small roads, usually narrowish but generally well paved, which allow wider exploration. However, several sites (A–E below) are particularly worth visiting.

♿ Easy watching from a vehicle is possible at sites A, B, C and E. Site D is not easily accessible.

A Desfiladero de Los Gaitanes (El Chorro gorge)   From either direction, the MA-5403 brings you to this spectacular gorge, the centrepiece of the whole area. A bar, El Pilar, is on this road alongside the small reservoir on the Guadalhorce just south of the gorge. This is usually a good site for a first stop, offering views of steep pine-covered cliffs and across to the gorge itself, a narrow chasm that has featured in several films.

From the bar it is possible to watch Griffon Vultures, occasional Egyptian and Black Vultures (the latter usually in winter) and Bonelli's and Golden Eagles, as well as Peregrines and occasional Goshawks. In spring and summer there are often large numbers of Alpine Swifts flying at all altitudes, sometimes whipping through the area at eye-level. Red-billed Choughs are common, as are Jackdaws, although the former usually remain at high levels. The pines beside the bar often provide Crested Tits. Wallcreepers have been recorded very occasionally in winter.

The section of the road just north of the bar area is worth exploring on foot (parking here is difficult). Here the pine-clad slopes rise high south of the road and opposite there is a ravine with cliffs rising up beyond. This area can provide excellent views of Alpine Swifts and Crag Martins. The trees in the ravine often hold Golden Orioles and there are Crested Tits, Bonelli's Warblers, Firecrests and Spotted Flycatchers in the pines. The rock face opposite the road can provide good views of Blue Rock Thrushes, Black Redstarts, and both Black and Black-eared Wheatears.

B Ruinas de Bobastro The Bobastro ruins and a scenic viewpoint (Mirador Tajo Encantada) are signposted from the MA-5403 some 3km for El Chorro, where a narrow winding road (MA-4400) leads upwards for some 5km. There are good opportunities to explore pinewoods and open rocky terrain here, as well as to scan for raptors. Rock Buntings and Red-billed Choughs are among the more obvious species. The view from the top on a clear winter's day, to as far east as the Sierra Nevada, is absolutely fantastic.

C The Reservoirs The reservoirs hold little of interest at any time of year but are always worth a look just in case: one memorable January census produced two Great Northern Divers. More usually they attract small wintering populations of Great Crested Grebes and Great Cormorants as well as roosting gulls. Red-necked Nightjars occur in open areas around the reservoirs.

D Sendero Caminito del Rey This is a long trail (over 7km) that takes in the spectacular aerial walkway of El Chorro gorge. The aerial sector is not of any particular birding interest and is only walkable one way, from north to south: a bus takes you back to your starting point. The initial sector of the trail follows the river and offers good views of riparian woodland, as well as expanses of scrub and conifers. Riverside birds include Scops Owls, Golden Orioles, Nightingales and Bonelli's Warblers in the breeding season. Pinewoods nearby occasionally attract Crossbills, and Crested Tits are regularly present.

There are a number of signposted access points to the Caminito del Rey along the service road to the reservoirs. From the south this road leads northwards at a T-junction with the MA-5403 west of El Chorro. Continue for some 3km to come to a large bar (Bar-Restaurante El Quiosco) on the east side of the road, directly before a mirador overlooking the Conde del Guadalhorce reservoir. You can park here or a little further on, where the entrance to the trail is signposted at the southern end of the dam. Continue on foot as far as you wish and return the same way.

E Valle de Abdalajís The eastern fringe of El Chorro and the home base of the Griffon Vultures and Bonelli's Eagles that overfly the whole area. The interesting features of the site are the cliff face on the northern side of the road and, to a lesser extent, the road through to the Chorro itself. The best time to visit this area is February–June, when cliff birds are nesting.

The valley can be approached from El Chorro or from Abdalajís village. Leave the A-343 to skirt the southern fringe of the village, where El Chorro is signposted. The best site is the cliff face, which is just off the road about 2km west from Abdalajís, and impossible to miss as it looms over you. This is where the Griffon Vultures nest and it is best visited in the early morning, before they begin to disperse from the cliffs for the day's foraging. At the same time it is easy to watch for other large raptors, such as Bonelli's and Golden Eagles, which sail along the cliff face amongst the Jackdaws and Red-billed Choughs, while the Alpine Swifts fly higher still. Peregrines are also frequent in this area. The rocks and fields at the foot of the cliff and the valleyside of the road are good for at least hearing Quail. Black-eared Wheatears, Bee-eaters and Woodchat Shrikes occur commonly all the way to El Chorro in summer.

CALENDAR	
All year: Great Crested Grebe, Mallard, Griffon Vulture, Goshawk, Golden and Bonelli's Eagles, Common Kestrel, Peregrine Falcon, Eagle and Little Owls, Kingfisher, Crested and Wood Larks, Crag Martin, Grey Wagtail, Wren, Black Redstart, Black Wheatear, Blue Rock Thrush, Zitting Cisticola, Dartford and Sardinian Warblers, Southern Grey Shrike, Jay, Red-billed Chough, Jackdaw, Raven, Crested and Coal Tits, Chaffinch, Serin, Linnet, Rock Bunting.	Swallows, Yellow Wagtail, Nightingale, Black-eared Wheatear, Melodious and Bonelli's Warblers, Spotted Flycatcher, Golden Oriole, Woodchat Shrike. **Winter:** Great Cormorant, Grey Heron, Black Vulture, Common Shelduck, Shoveler, White Wagtail, Alpine Accentor, Robin, Common Chiffchaff, Wallcreeper (rare), Southern Grey Shrike, Brambling, Redwing, Siskin, Crossbill.
Breeding season: Quail, Egyptian Vulture, Lesser Kestrel, Scops Owl, Red-necked Nightjar; Common, Pallid and Alpine Swifts, Bee-eater, Hoopoe, Barn and Red-rumped	**Passage periods:** White Stork, raptors: including Osprey.

EL TORCAL MA5

Status: Paraje Natural (1,171ha). ZEPA. UNESCO World Heritage Site.

Site description
A strikingly attractive expanse of highly eroded, surreal, karstic rock formations to the southwest of Antequera. The whole area, a Jurassic seabed 200 million years ago, is a tortuous labyrinth of gullies and passages. Sparse shrubby vegetation, comprising twisted and stunted Holm Oaks, Maples, Elders and Hawthorns among others, grow out of the cracks and crevices between rocks. There is a rich herbaceous flora of rupestral plants.

Species
Bird numbers are not great but interesting residents include Bonelli's Eagles, Peregrines, Red-billed Choughs, Black Wheatears, Blue Rock Thrushes, Rock Buntings and Cirl Buntings, along with a remarkably high concentration of Wrens. Migrants and summering species are generally sparse but include Black-eared Wheatears, Subalpine and Orphean Warblers and Short-toed Eagles occur occasionally. There are regular autumn and winter records of Ring Ouzels, Redwings and Fieldfares and also Dunnocks and Alpine Accentors.

Timing
The area receives a large number of visitors in summer and many school parties from April to June. Early morning visits are recommended then. The views are

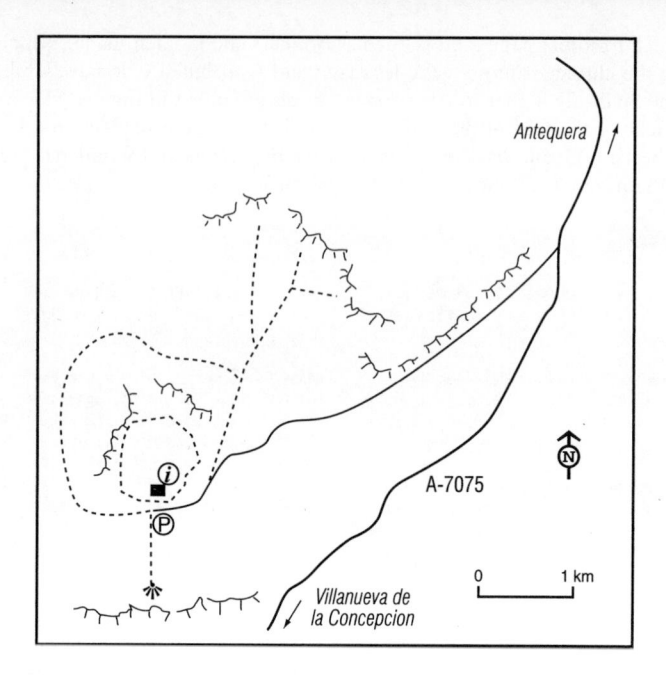

tremendous on clear days, extending all the way down to the coast between Málaga and Torremolinos.

Access

The access road leads southwest from the A-7075 12km south of Antequera. There is a car park with a seasonal restaurant and information centre. Several marked trails lead through the area. Walking uphill to the right just before the car park can also often be productive. (The access road is closed to traffic when El Torcal car park is full and you then need to use the car parks at the A-7075 junction and the shuttle bus provided.)

♿ Only the car park is accessible.

CALENDAR

All year: Sparrowhawk, Bonelli's Eagle, Common Kestrel, Peregrine Falcon, Wren, Black Redstart, Black Wheatear, Blue Rock Thrush, Red-billed Chough, Jackdaw, Raven, Cirl and Rock Buntings.

Breeding season: Short-toed Eagle, Black-eared Wheatear; Melodious, Subalpine and Orphean Warblers, Woodchat Shrike.

Winter: Common Buzzard, Hen Harrier, Meadow Pipit, White Wagtail, Common Chiffchaff, Redwing, Fieldfare, Southern Grey Shrike, Alpine Accentor, Dunnock.

Passage periods: Raptors, Rufous-tailed Rock Thrush.

SIERRA DE CAMAROLOS — MA6

Status: *Special Conservation Area (ZEC).*

Site description
A beautiful place to visit, especially early on sunny, late spring mornings, with the higher crags of the sierra providing a backdrop to the south and a fantastic view away to the northwest. Overall there is much steep, rocky, eroded limestone terrain, with cereal fields and olive groves at lower levels and pastures higher up. The natural vegetation includes Iberian Holm and Portuguese Oaks, with junipers and spiny scrub at higher elevations.

Species
A good site for seeing Golden Eagles and, less frequently, Bonelli's Eagles, as well as Peregrine Falcons. However, the site is most noteworthy for the passerine community. The resident Black Wheatears, Blue Rock Thrushes and Blackbirds are joined for the breeding season by Black-eared Wheatears and Nightingales at the lower levels and Rufous-tailed Rock Thrushes higher up. Northern Wheatears may also breed at the highest elevations. Quite large flocks of migrant Ring Ouzels occur in autumn and a few overwinter in the area. Other wintering thrushes include Redwings and Song Thrushes, and occasional Fieldfares. Bonelli's and Dartford Warblers are common. Both Orphean and Melodious Warblers occur on the lower slopes, and Subalpine Warblers higher up. Iberian Chiffchaffs may also breed locally. Azure-winged Magpies occur in small numbers in the lower areas. This is one of the few areas in the province where seeing Rock Sparrows is virtually guaranteed. Both Rock and Cirl Buntings are very common.

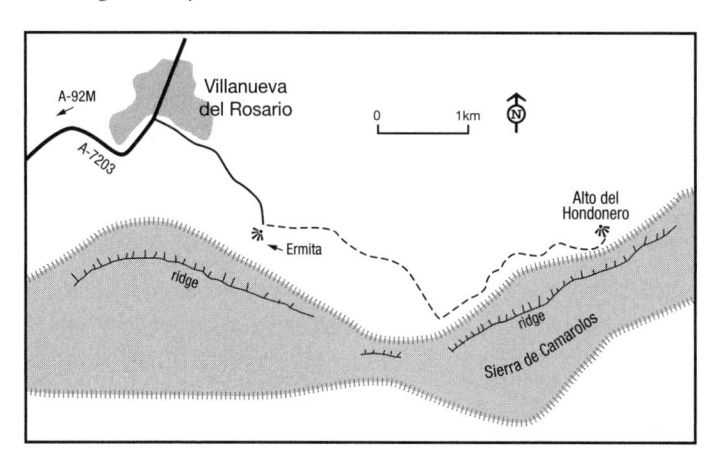

Timing
Any time of year is interesting but spring (April–May) is undoubtedly the most attractive, followed by autumn for migrant raptors and thrushes. Mid-summer can be extremely hot here and mid-winter very cold.

Access
Leave the A-92M at km-20 and take the A-7203 to Villanueva del Rosario. Bear right to skirt the village and look for a sign to the Mirador del Alto del Hondonero, on the right before a bridge. This tarmac road soon ends at a hermitage (Ermita del la Virgen del Rosario), where there is a panoramic view.

Beyond the hermitage a drivable dirt road climbs eastwards to the Mirador del Hondonero, at the foot of the summit of the Sierra, where there is an excellent viewpoint and good access to open rocky terrain. The whole road offers ample opportunities for scrutinising and visiting areas of rocky terrain, pastures, scree, scrub and woodland.

♿ Watching from near the vehicle is possible at the mirador.

CALENDAR

All year: Golden and Bonelli's Eagles, Peregrine Falcon, Thekla Lark; Long-tailed, Crested and Coal Tits, Wren, Blue Rock Thrush, Black Redstart, Black Wheatear, Azure-winged Magpie, Red-billed Chough, Jackdaw, Cirl Bunting, Rock Sparrow, Rock Bunting.

Summer: Alpine Swift, Nightingale, Black-eared Wheatear, Rufous-tailed Rock Thrush; Subalpine, Orphean and Bonelli's Warblers, Iberian Chiffchaff.

Winter: Redwing, Fieldfare, Song Thrush, Ring Ouzel, Bullfinch.

Passage periods: Raptors, Water Pipit, thrushes, Common Redstart, warblers, Ortolan Bunting.

MÁLAGA HILLS (Montes de Málaga)   MA7

Status: Parque Natural (4,956 ha).

Site description
An extensive fragment of the Betic range on the northern outskirts of Málaga city. The steep slopes are well vegetated with scrub, Aleppo Pine plantations and mixed forests and woodlands of Cork and Iberian Holm Oaks. The summits rise to 1,030m.

Species
The passerine community is diverse and includes all the common woodland species. The presence of Azure-winged Magpies is noteworthy. Other species of particular interest include Golden Oriole, Orphean Warbler, Crested Tit and Crossbill. Breeding raptors include Common Buzzards, Short-toed Eagles, Booted Eagles (some of which overwinter) and one or two pairs of Goshawks. Considerable numbers of raptors, notably Black Kites, Honey-buzzards and Short-toed Eagles, overfly and often roost here during the autumn migration. Large flocks of migrant Griffon Vultures are also recorded in October and November.

The Montes are a redoubt of the Mediterranean Chameleon, part of the wide diversity of reptiles and amphibians present. The site is also well known for the large variety of medium-sized mammals it supports. Foxes, Badgers, Mongooses, Weasels, Polecats, Red Squirrels, Beech Martens and Genets all occur, together with numerous smaller mammals, although seeing most of these is pure luck.

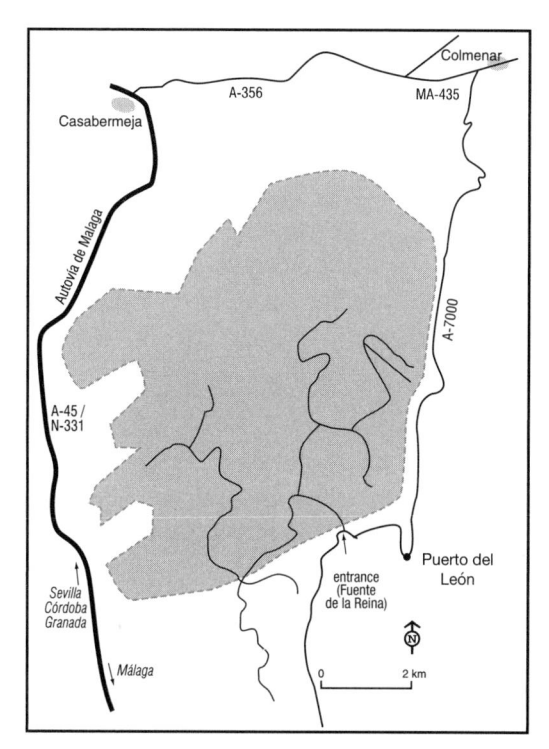

Timing

This very pleasant site is worth visiting at any time of year. However, *Malagueños* in their thousands invade the area, which is on the very doorstep of Málaga city, during weekends and public holidays year-round. Weekday visits are therefore recommended, especially early mornings from March to early June.

Access

The principal access points are on the eastern side, from the A-7000. This road is best reached from the A-45/N-331, leaving at Casabermeja (km-124) to take the A-356 east towards Colmenar. Bear south just before that village, and turn towards Málaga on the A-7000. The park is signposted at the intersection. There are various entry points into the woods on the west side of the road. These are not particularly well marked and are often chained off. The road climbs to the Puerto del León, where there are excellent scanning opportunities, just before the main access point at Fuente de la Reina, some 14km south of Colmenar. The tarmac road into the Montes leads to some of the major viewpoints and other sites of interest. It also gives access to a complex series of tracks and paths, all colour-coded, which can be followed. These are bumpy, and may be somewhat sticky after heavy rains, but perfectly drivable. Select a promising looking spot and explore on foot. An information centre (Ecomuseo Lagar de Torrijos – open Thursdays–Sundays) can provide information on the various walking trails).

A very pleasant three-star hotel, El Humaina, *www.hotelhumaina.es*, benefits from a secluded position within the park. A weekday stay here in the breeding season may prove rewarding. Head for the signposted picnic site 'Area Recreativa El Cerrado', which is close to the hotel.

♿ Suitable in many places.

CALENDAR

All year: Booted Eagle, Common Buzzard, Goshawk, Sparrowhawk, Eagle and Tawny Owls, Jay, Firecrest, Long-tailed Tit, Crested Tit, Coal Tit, Crossbill, Azure-winged Magpie, Rock Bunting.

Breeding season: Short-toed Eagle, Scops Owl, Nightingale, Melodious, Orphean and Bonelli's Warblers, Golden Oriole.

Winter: Song Thrush, Redwing, Common Chiffchaff, tits, Siskin.

Passage periods: Raptors and storks.

GUADALHORCE ESTUARY   MA8

Status: Paraje Natural (130ha).

Site description

A complex of ponds and low scrub between the two arms of the river comprises the Río Guadalhorce estuary. Despite disturbance, this remains one of the best sites in eastern Andalucía, benefiting from its coastal location and the presence of fresh water and mud in a region that has very few wetlands. The ponds on the eastern side of the estuary are the remains of gravel extraction workings around which copious amounts of solid fill have been dumped, much of which is now partially covered by grasses and by tamarisks and other bushes. The area has been partly or totally inundated by the river on several occasions and widening of the river into two arms was carried out to alleviate flooding further upstream and hasten the evacuation of flood waters. The sector nearest the beach is also flooded occasionally by the sea.

There are canebrakes and a certain amount of reed cover along the banks of ponds. The ponds themselves experience great fluctuations in water levels according to season and rainfall. Some are shallow and largely dry out during the summer months. Large pools with islands, providing some open mud attractive to migrant waders, occur along the old eastern arm of the river. There are signposted footpaths and several hides. There is constant human presence and disturbance is a nuisance in spite of wardening by the Environment Agency (CMA). The problem is greatest in summer as the site gives the public access to the beach.

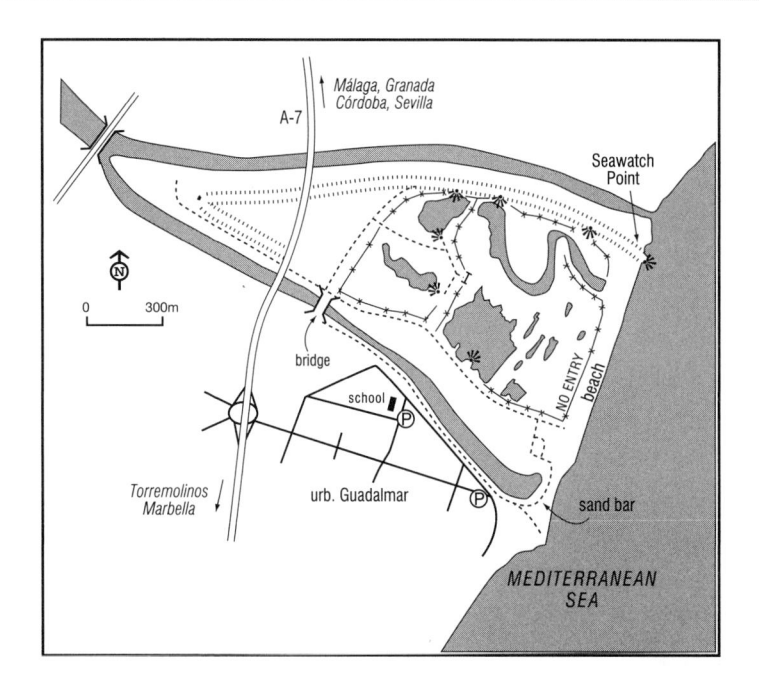

Species

Over 280 species have been recorded here, many of them rarities. These last have included several Nearctic vagrants: Blue-winged Teal, dowitcher sp., Lesser Yellowlegs, Spotted Sandpiper, Laughing Gull, Franklin's Gull and Ring-billed Gull.

Nesting waterbirds include Little Grebes, Mallards, Little Bitterns, Common Coots and a few Purple Swamphens. Others, including Red-crested Pochards and Purple Herons, have bred here sporadically. The numbers and variety of waterbirds increase during autumn and winter, when the above are joined by Great Crested and Black-necked Grebes and a good variety of ducks. Grey Herons and Cattle Egrets are common in winter especially. Night Herons may be seen year-round, although they no longer nest here.

Raptors are most evident in winter, when up to a dozen Booted Eagles remain here together with several Marsh Harriers and one or two Common Buzzards. Peregrines, which breed on the cathedral in central Málaga, are not infrequent and migrant Hobbies occur occasionally.

Breeding waders, which have benefited from site improvements in recent years, include Kentish and Little Ringed Plovers and, especially, Black-winged Stilts. They are joined in winter by Lapwings, Ringed Plovers, Common Redshanks and Common Snipe. The diversity, if not abundance, of migrant waders is noteworthy, with up to 19 species recorded on good days in spring and autumn. Collared Pratincoles are quite common along the shore in spring and Common Sandpipers are everywhere then. There are often small numbers of Wood and Green Sandpipers and Black-tailed and Bar-tailed Godwits during both passage periods. Common Redshanks, Greenshanks, Dunlins and Curlew Sandpipers are frequent on passage and there are occasional records of Spotted

Lesser Crested Tern with Sandwich Terns

Redshanks, Marsh Sandpipers and Temminck's Stints. Terek Sandpipers have occurred at least twice.

Large numbers of Yellow-legged, Lesser Black-backed and Audouin's Gulls congregate on the beach, around the ponds and along the arm of the river near its mouth. Mediterranean Gulls are present in considerable numbers on the western arm of the river during both passage periods. Slender-billed Gulls have been recorded in spring. Sandwich Terns occur year-round and Little, Common, Black and Whiskered Terns are commonly recorded during migration periods, with less frequent records of Caspian, Lesser Crested and, very rarely, White-winged Black Terns. There have been single records of Royal, Sooty and Bridled Terns.

Hoopoes occur all year round. Rollers and Great Spotted Cuckoos are occasionally seen on migration. Little Swifts have been seen here in addition to all the common swifts and hirundines. Zitting Cisticolas and Cetti's Warblers are always present and often highly visible. In winter they are joined by hordes of Meadow Pipits and Common Chiffchaffs. Other wintering species of interest include Bluethroats and occasionally Yellow Wagtails. Both Barn Swallows and House Martins may be seen here in January alongside Crag Martins. In general, anything and everything may turn up and this excellent site often produces surprises.

The site is popular with escaped and naturalised exotic species, notably Monk and Rose-ringed Parakeets. There are small populations of Common Waxbills and Red Avadavats.

Timing
A walk through the area, or even a fleeting visit, is nearly always productive at any time of year, even in winter or mid-summer. A telescope is useful.

Access
Access is from the A-7 motorway on the west side of Málaga at km-3, where it is signposted for San Julián and Guadalmar. Follow signs to Guadalmar from the roundabout under the motorway. Turn left to skirt the housing development and park alongside the river. From here you can either go up the ramp and look

across the river before going across the sand bar or go to the seabird watchpoint (mirador). A bridge provides access further upstream near the motorway.

♿ Not very suitable until planned facilities for disabled access are implemented.

CALENDAR

All year: Mallard, Common Pochard, Shoveler, Little Grebe, Greater Flamingo, Common Coot, Little Egret, Night Heron, Common Kestrel, Purple Swamphen, Kentish Plover, Common Sandpiper, Sandwich Tern, Rose-ringed and Monk Parakeets, Little Owl, Hoopoe, Stonechat, Cetti's and Dartford Warblers, Zitting Cisticola, Common Waxbill, Red Avadavat.

Breeding season: Night Heron, Little Bittern, Squacco Heron, Avocet, Black-winged Stilt, Little Ringed Plover; non-breeding gulls including Mediterranean and Audouin's Gulls, Hoopoe, Pallid and Common Swifts, Barn Swallow, Yellow Wagtail of race *iberiae*, Nightingale, Great Reed and Reed Warbler.

Winter: Wigeon, Gadwall, Teal, Pintail, Shoveler, Marbled Duck (rare), Red-crested Pochard, Common Pochard, Ferruginous Duck (rare), Tufted Duck, Common Scoter, White-headed Duck, Great Crested and Black-necked Grebes, Great Cormorant, Cattle Egret, Grey Heron.

Common Buzzard, Booted Eagle, Lapwing, waders, Arctic Skua, large numbers of gulls, Sandwich Tern, Short-eared Owl, Meadow and Water Pipits, Yellow Wagtail, Bluethroat, Common Chiffchaff, Penduline Tit, Southern Grey Shrike.

Passage periods: Garganey, Purple and Grey Herons, Black Kite, Marsh and Montagu's Harriers, Osprey, Hobby, Peregrine Falcon, Collared Pratincole, Grey Plover, Knot, Common Redshank, Greenshank, Green and Wood Sandpipers, Bar-tailed and Black-tailed Godwits; Pomarine, Arctic and Great Skuas; Mediterranean, Little and Audouin's Gulls; Gull-billed, Caspian, Common, Little, Black and White-winged Black Terns, swifts including White-rumped Swift, Roller, Bee-eater, hirundines, pipits, White Wagtail, Yellow Wagtails of races *iberiae*, *flava* and *flavissima*, Grey Wagtail, Whinchat, Northern and Black-eared Wheatears, warblers including Orphean and Spectacled Warblers, and a wide variety of other passerines.

JUANAR (Sierra Blanca) MA9

Status: No special protection.

Site description
Another section of the southeastern end of the Serranía de Ronda, offering pinewoods, scrub and rocky habitats. It is part of the Sierra Blanca, including La Concha, the rather barren looking mountain that looms conspicuously over Marbella. A mirador at Juanar offers magnificent views of the Costa del Sol, the Strait and Morocco on a clear day. There are marked trails to the summit of La Concha.

Species
Golden, Bonelli's and Booted Eagles all nest in the general area as well as Peregrines. Griffon Vultures often overfly. It is not uncommon to come across small parties of migrant Black Kites and Short-toed Eagles moving along the sierra, especially in autumn. Eagle Owls breed in the area of the mirador.

In spring, the most productive period, Wood Larks are highly audible and somewhat less visible as they circle high in song flight. Blue Rock Thrushes and Rock Buntings occur on rocky outcrops, with Rufous-tailed Rock Thrushes at the highest levels. Vegetated areas offer a diversity of woodland and scrub species, these including Melodious, Orphean and Bonelli's Warblers, Firecrests,

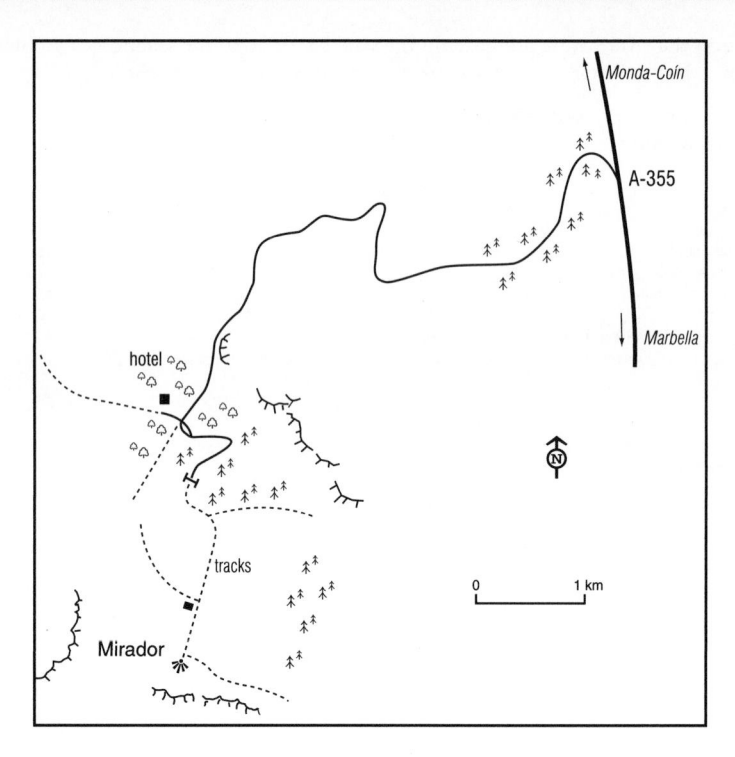

Short-toed Treecreepers, Crested Tits and Crossbills. Both Southern Grey and Woodchat Shrikes are present.

Timing
The Refugio de Juanar area is ideal for a half-day excursion, preferably a weekday morning, at any time of year. Spring visits, between April and June, provide the maximum variety of species. Sunday visits at any time of year are best done early, before the picnickers arrive.

Access
This is from the A-355 Marbella/Coín road. From the south (Marbella) it climbs up past the white village of Ojén and over the pass (Puerto de Ojén). The 5km access road to the Refugio de Juanar hotel is on the left approximately 1km over the top of the pass, just beyond the eyesore of a quarry. From the north (Coín) the turn-off is some 7km past Monda. The access road is narrow and winding and stopping opportunities are limited. However, it is worth looking at the pines immediately after the turn-off, as well as the area around km-4, where there are rocky outcrops on the left above the road. The mirador is signposted at the hotel and is reached on foot along a 1.5km track. This passes through pines initially, then opens out with olive groves on either side. The various tracks around the hotel offer additional opportunities for exploration, and ascent to the summit (one hour each way).

♿ Not recommended. Only the hotel area is easily accessible.

CALENDAR

All year: Griffon Vulture, Common Buzzard, Golden and Bonelli's Eagles, Peregrine Falcon, Red-legged Partridge, Eagle and Little Owls, Iberian Green Woodpecker, Great Spotted Woodpecker, Wood Lark, Dartford Warbler, Crag Martin, Blue Rock Thrush, Crested Tit, Firecrest, Short-toed Treecreeper, Jay, Red-billed Chough, Raven, Crossbill, Rock Bunting.

Breeding season: Booted Eagle, Alpine Swift, Hoopoe, Rufous-tailed Rock Thrush; Melodious, Subalpine, Orphean and Bonelli's Warblers.

Winter: Song Thrush, Hawfinch, Siskin, Serin.

Passage periods: Honey-buzzard, Black Kite, Short-toed Eagle, swifts and hirundines.

SIERRA DE LAS NIEVES NATIONAL PARK

Status: Parque Nacional 'Sierra de las Nieves' (18,530ha). UNESCO Biosphere Reserve. Designated a National Park in 2019.

Site description

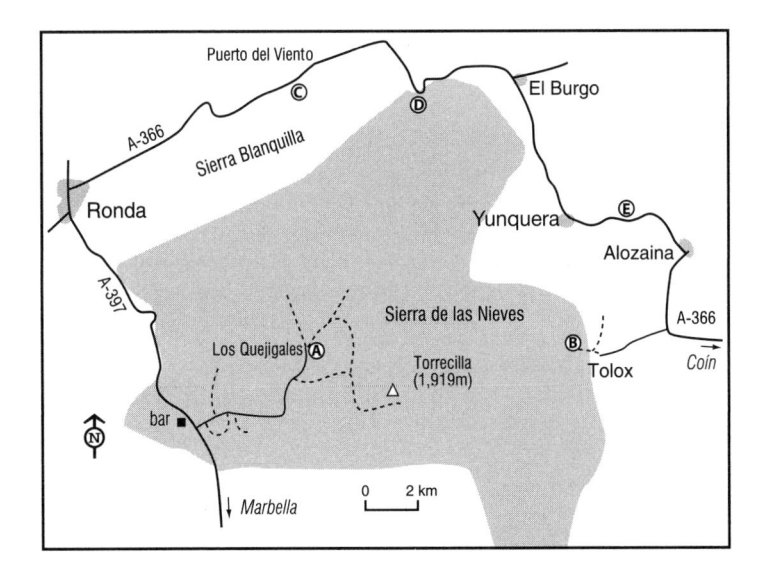

This large area consists of three sierras within the eastern Serranía de Ronda east of the A-376 San Pedro de Alcántara/Ronda road. The Sierra de las Nieves National Park includes the Sierra de Tolox at its eastern end. The Sierra Blanquilla lies north of the Sierra de las Nieves. The terrain is generally high with a mean altitude of 1,100m in the Sierra de las Nieves. The highest point is Torrecilla (1,919m) in the Sierra de Tolox. In large part the ground is very broken with outcrops of limestone and steep slopes and cliffs. The vegetation varies according to altitude. There are extensive mixed oak woodlands and some pine plantations at lower elevations. Dense scrub of junipers and other low shrubs are characteristic at higher elevations, where there are also Yews, Lusitanian Oaks and expanses of Maritime Pines with a scrub understorey. The endemic Spanish Fir (Pinsapo) has its largest stands in the Sierra de las Nieves. The summits are relatively barren above about 1,500m. The whole region is of exceptional botanical interest, with some 1,500 plant taxa recorded.

Species

An excellent area for finding all the typical bird species of the Serranía de Ronda. In particular, Los Quejigales in the Sierra de las Nieves offers high diversity within a reasonably confined and accessible site. Many species are found all year round, although some move lower in winter, when the climate can be relatively harsh.

The sierra tops have breeding populations of the Sky Lark, Tawny Pipit, Northern Wheatear and Rufous-tailed Rock Thrushes, species that are largely confined in southern Spain to high elevations. Other characteristic breeding birds of these sierras include both Black and Black-eared Wheatears, Southern Grey and Woodchat Shrikes, Red-billed Choughs and a large diversity of warblers and other woodland and scrub species. The Common Redstart has an outpost nesting population here. Rock Sparrows occur, most often near the cliff up the track from Los Quejigales (A) and also in the small valley to the east of Ronda in the Sierra de Blanquilla (C) and in the gorge at Jorox (F). Rock Buntings are widespread and common. Alpine Swifts are often overhead and the White-rumped Swift also nests in the region. Alpine Accentors do not nest here but occur in winter. Flocking thrushes, notably Ring Ouzels but also Redwings, Song Thrushes and occasional Fieldfares, are characteristic in autumn and winter.

Diverse raptors nest locally at low density. Griffon and Egyptian Vultures occur and occasional Golden and Bonelli's Eagles may be expected. The sierras are on the flight path of migrating raptors heading to and from the Strait of Gibraltar, especially in autumn when large flocks of Black Kites and Honey-buzzards often pass through the region.

The whole area is outstanding for spring flowers and so, unsurprisingly, for butterflies as well.

Timing

The best time is spring, which at these altitudes means from late April until mid-July. This is not to say that the area is devoid of ornithological interest at other seasons but it can get very hot in summer, over 40°C, and temperatures may fall below -5°C in winter. The name Sierra de las Nieves, 'sierra of the snows', is not without significance. Also, it does get rather crowded anywhere in the region at weekends in spring and autumn, especially in July and August in the Tolox area when the balneario (curative mineral water spa) is open.

Access

♿ None of these sites is suitable for the less mobile except for D and the vicinity of the car park at A.

A Sierra de las Nieves (Los Quejigales) From the south, from San Pedro, the entrance to the access road is clearly signposted by a large green CMA sign soon after the km-14 post on the A-397 San Pedro-Ronda road. From the

Ronda direction, the entrance is on the left exactly 1km past a bar-restaurant, 'El Navasillo'. The access road is tarmac initially but soon becomes good quality dirt, perfectly passable under normal conditions.

A recommended early stop is 2.1km from the entrance, where there is a little stream bed on the left. Park here and take the track off up to the left for about 1.5km. The track passes through dense scrub at first, then widens, before entering a natural rocky basin in which there is an uninhabited house. This site is reliable in season for Black and Black-eared Wheatears, Subalpine Warblers, Dartford Warblers and Rock Buntings, among others.

The access road continues upwards for a further 8km, to the campsite and car park at Los Quejigales, from where a number of signposted walking tracks allow further exploration. You can walk up towards the cliff off to the right, or up through the pines on either side. You can also return to the fork at 9.2km (800m from the Los Quejigales car park) and take a track up to the left, which follows a circuitous course through the woodlands back to the campsite and car park.

The track up to the peak of Torrecilla (6.5km) provides the best chances of seeing the upland species, notably Rufous-tailed Rock Thrushes, Northern Wheatears, Spectacled Warblers and Sky Larks in the breeding season. Song Thrushes, Redwings, Ring Ouzels and Fieldfares may be numerous here in autumn. The Torrecilla track, ascending some 900m, is good going but steep in places, demanding a significant degree of fitness. An option on the way back is to return to Los Quejigales along the streamside of La Cañada del Cuerno, which joins the upper reaches of the Torrecilla track at the Puerto del Oso.

B Sierra de Tolox   Take the A-366 Ronda/Coín road and turn-off for Tolox along the A-7250. Bypass the village at the lower level where the road overlooks the river and follow signs to the Balneario (mineral water spa) at the head of the valley. Park by the spa and follow the riverside track upstream. Raptors often overfly and there are wagtails, pipits and warblers as well as a variety of migrant thrushes in autumn, these last often in considerable numbers.

C Puerto del Viento – Sierra Blanquilla   The A-366 Ronda/El Burgo road, skirting the northern edge of the Sierra Blanquilla, crosses rather barren, often stony, terrain. There are broad expanses of rough pasture and mainly low scrub. Steep rocky outcrops are a feature. The whole region is attractive to the birds of the stony uplands of the Serranía de Ronda, notably Black and Black-eared Wheatears, Red-billed Choughs, Southern Grey and Woodchat Shrikes, Rock Sparrows and Rock Buntings. Nesting raptors include both Bonelli's and Golden Eagles.

A good place to spend some time is the Puerto del Viento (Windy Pass), which usually lives up to its name, on the A-366 at km-11 (1,190m). There is a lay-by and information board. The panoramic view offers opportunities for scanning and it is possible to walk along the hillside at the pass. There are easy stopping places, including open farm tracks along the A-366, allowing wider exploration. Good examples of such tracks are at km-16.5 and km-17.2, east of Puerto del Viento.

D Mirador del Guardia Forestal   A monument to the forest guard corps. The stop and car park are signposted from both directions on the A-366 at km-16.5.

A brief visit is worthwhile, if only for the fantastic views over the Río Burgo and to the Sierra de las Nieves. Crested Tits occur in the pines nearby and Rock Buntings often feed around the monument and on the rock face below.

E Jorox Gorge A small gorge off the A-366 Coín/Ronda road at km-38, between Alozaina and Yunquera. There is parking about 50m east of the bridge. The gorge is difficult going and passable only on foot for little more than 150m. However, in spite of its small accessible area, it holds several interesting and highly visible species such as Crag Martins, Black Wheatears, Blue Rock Thrushes, some Rock Sparrows and Rock Buntings in abundance.

CALENDAR

All year: Griffon Vulture, Goshawk, Golden and Bonelli's Eagles, Peregrine Falcon, Common Kestrel, Woodpigeon; Eagle, Little and Tawny Owls; Thekla, Wood and Sky Larks, Crag Martin, Wren, Dipper, Robin, Black Redstart, Stonechat, Black Wheatear, Blue Rock Thrush; Dartford and Sardinian Warblers, Blackcap, Firecrest, Crested and Coal Tits, Nuthatch, Short-toed Treecreeper, Southern Grey Shrike, Red-billed Chough, Jackdaw, Raven, Hawfinch, Rock Sparrow, Rock Bunting.

Nightingale, Common Redstart, Northern and Black-eared Wheatears, Rufous-tailed Rock Thrush; Melodious, Spectacled, Subalpine and Orphean Warblers, Common Whitethroat, Bonelli's Warbler, Iberian Chiffchaff, Spotted Flycatcher, Golden Oriole, Woodchat Shrike.

Winter: Merlin, Meadow Pipit, Alpine Accentor, Ring Ouzel, Redwing, Song Thrush, Fieldfare, Common Chiffchaff, Siskin.

Breeding season: Quail, Egyptian Vulture, Booted Eagle, Common Cuckoo, Scops Owl, Red-necked Nightjar, Alpine and White-rumped Swifts, Bee-eater, Hoopoe, Tawny Pipit,

Passage periods: Honey-buzzard, Black Kite, Short-toed Eagle, swifts, hirundines, Willow Warbler, Pied Flycatcher.

SIERRA CRESTELLINA   MA11

Status: Paraje Natural (478ha).

Site description

A limestone ridge at the southwestern end of the Sierra Bermeja that runs approximately north–south for some 4km between the picturesque towns of Casares and Gaucín near the Mediterranean coast. The highest point is 926m. The ridge forms the eastern scarp of the Río Genal. Much of the hillsides are open pasture, intensively grazed by goats and cattle, but there are extensive tracts of hawthorn, buckthorn and Cistus scrub, together with stands of Iberian Holm Oaks and Maritime Pines as well as Aleppo Pine plantations. The sierra proper has cliffs and scree slopes. Near Casares there are areas of limestone pavement in what is otherwise a sandstone region. The aspect of the southern approaches to the site is marred by a wind farm, whose giant white windmills are visible from a great distance.

Species

One of the easiest sites to reach from the Costa del Sol and it offers a good range of the mountain species typical of the Serranía de Ronda. The colony of Griffon Vultures is obvious at the northern end of the ridge. This is a traditional site for Bonelli's Eagles, and Golden Eagles visit occasionally. The Casares Castle has nesting Lesser Kestrels. White-rumped Swifts are regularly present. The road between Manilva and the turn-off for Casares is also good for Tawny Pipits. Other characteristic breeding species include Black and Black-eared Wheatears; Dartford, Spectacled and Subalpine Warblers, Red-billed Choughs and Rock Buntings. Rock Sparrows are frequent on the roadside above the Río Genal, at the north end of the site.

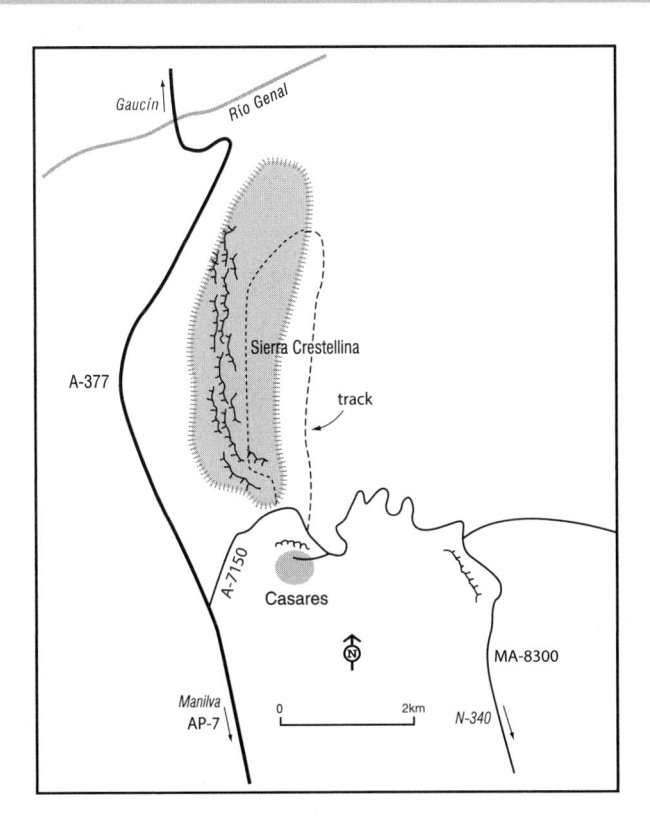

Timing
Morning visits are usually recommended, especially in summer, but afternoon visits allow the sierra to be viewed with the sun behind you. All times of year are rewarding: raptors are usually present flying up and down the ridge. The scrub and woodlands show a great deal of passerine activity and may hold significant numbers of a wide range of species during migration periods.

Access
Straightforward access is via the A-377 Manilva/Gaucín road, leaving the AP-7 motorway at km-142, and following signs for Casares. The sierra is best viewed from the A-377, which closely skirts the western side of the ridge before descending to the Río Genal, although parking spots along this road are few and far between. A few unfenced footpaths and goat tracks allow access to the lower slopes.

Nightingale

A signposted footpath (1.3km) that climbs to a viewpoint at the southern end of the ridge is recommended. Access is from the A-7150 Casares road at km-2. Parking at the entrance is limited and you may need to park in the village. You can return the same way or continue on the trail following the eastern side of the ridge before looping back to Casares village: this longer walk is some 7km. The whole area is of considerable botanical interest in spring especially and there are good opportunities to find passerines and for viewing raptors overhead. It is also worth parking in Casares village and taking the stiffish uphill walk to the castle, where there are good views to be had of up to four swift species and also of the Lesser Kestrels and Jackdaws.

The Genal Valley is a popular and pleasant picnic spot with easy access to the river just north of the bridge. Explore on foot for passerines. The mirador on the A-377 just below Gaucín is another good raptor watchpoint with stupendous views over the valley.

♿ Suitable only for watching from the road. Access in Casares is very steep.

CALENDAR

All year: Red-legged Partridge, Griffon Vulture, Sparrowhawk, Golden and Bonelli's Eagles, Common Kestrel, Peregrine Falcon, Little Owl, Thekla Lark, Crag Martin, Stonechat, Black Wheatear, Blue Rock Thrush, Mistle Thrush, Dartford and Sardinian Warblers, Blackcap, Firecrest, Short-toed Treecreeper, Red-billed Chough, Jackdaw, Raven, Spotless Starling, Chaffinch, Rock Sparrow, Cirl and Rock Buntings.

Breeding season: Black Kite, Egyptian Vulture, Short-toed and Booted Eagles, Turtle Dove, Common Cuckoo; Common, Pallid, Alpine and White-rumped Swifts, Bee-eater, Red-rumped Swallow, Tawny Pipit, Nightingale, Black-eared Wheatear; Melodious, Spectacled and Subalpine Warblers, Woodchat Shrike, Ortolan Bunting.

Winter: Red Kite, Common Buzzard, Meadow Pipit, White Wagtail, Robin, Song Thrush, Common Chiffchaff, Southern Grey Shrike, Common Starling, Siskin.

Passage periods: A wide range of raptors overfly the area, especially on their approach to the Strait during the southward migration. Grounded migrant passerines are sometimes prominent.

SIERRA BERMEJA and CENTRAL SERRANÍA DE RONDA   MA12

Status: Includes Paraje Natural 'Los Reales de Sierra Bermeja' (1,236ha).

Site description
This is a very large region consisting of two mountain ranges. The Sierra Bermeja is to the south, just inland from the coast at Estepona, to which it provides a rugged and largely unspoilt backdrop. The Serranía de Ronda lies further inland. The two are separated by the valley of the Río Genal. The area described here is bounded at its western end by the A-377 Manilva/Gaucín road and the Sierra Crestellina, which is considered separately (MA11). The eastern boundary is the A-376 San Pedro de Alcántara/Ronda road. The eastern sectors of the Serranía de Ronda, including the Sierra de las Nieves, are described above (MA10). The northern boundary is the A-369 Gaucín/Ronda road. Further west the Serranía de Ronda comprises the Sierra de Grazalema (CA20).

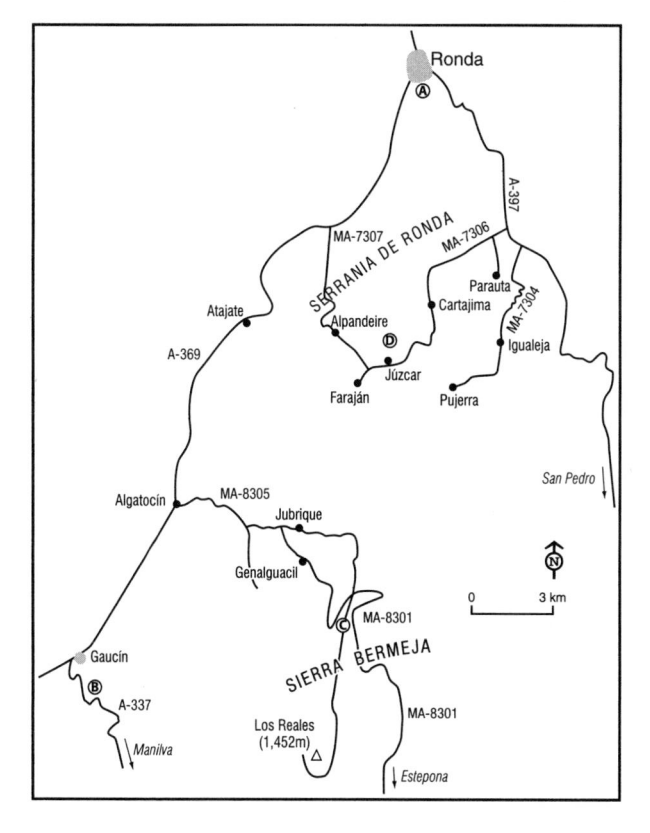

This is a limestone region with a distinctive and interesting flora, boasting a good range of endemic species including the Spanish Fir (Pinsapo), which is a feature of the Sierra Bermeja especially. Areas of open mountain pasture alternate with woods of pine, Cork Oak and encina. Sheer cliffs provide ideal nest sites for raptors and other species. There are also extensive areas of barren, stony uplands around the summits. Fast-flowing streams provide abundant water all year round. It is unhelpful to emphasise any particular site within this very large region, as what might be seen at one spot could be seen at any other, but general pointers to certain areas are offered. The choice is considerable and birding is usually based on the 'stop at a suitable-looking site and watch' method.

Species

The region is of interest for its resident and migrant raptors as well as for its mountain, woodland and scrub passerines. Breeding raptors include Griffon Vultures; Golden, Bonelli's, Booted and Short-toed Eagles and Goshawks. Peregrines are relatively common throughout the area and particularly around Ronda. Large flocks of Black Kites and Honey-buzzards, and the full range of migrant raptors, use the lines of the sierras as flyways during migration, especially in the autumn.

Alpine Swifts are characteristic of the high tops and are easily seen at Ronda. White-rumped Swifts range widely throughout the region and should be looked for among feeding parties of other swifts in late summer especially. Tawny Pipits are present in suitable habitat. Golden Orioles are common along the wooded valleys in the breeding season. Both the pine and broadleaf woodlands hold a wide variety of species. Characteristic species of the open terrain include Black Wheatears and Black Redstarts, with Northern Wheatears and Rufous-tailed Rock Thrushes at the upper levels. Common Redstarts nest locally. Ring Ouzels, Redwings, Fieldfares and Bramblings occur in varying numbers in winter. Rock Sparrows and Rock Buntings are locally common.

Timing

A rewarding area year-round. Spring and early summer will give maximum variety.

General access

The region is readily accessed from the AP-7 or N-340 to the south, and also from the north, along any of the major roads leading to Ronda (see map). The Sierra Bermeja is reached via the tortuous MA-8301 local road, which joins the N-340 at Estepona: it is not directly accessible from the AP-7.

♿ Many of the roads in this region and Ronda (A) provide good observation opportunities.

***A* Ronda** This ancient town has been inhabited since long before the Romans, who settled here, as did the Moors much later. The town bestrides the Tajo, a giant cleft in the rock, which is spanned by a bridge. There are cliffs on the northwestern side of the gorge and a steep slope and fields beyond on the southwest side. The Río Guadalevín runs through the gorge.

The vertiginous views from the bridge and the nearby viewpoint (mirador) are renowned. A variety of species may be seen from either of these. Peregrines often flash through the chasm looking for errant Rock Doves. They pose a lesser threat to the Alpine Swifts, which also whistle through. Common and

Lesser Kestrels are also present in this area. Black Redstarts and Blue Rock Thrushes frequent the walls of the gorge. White and Grey Wagtails are common in the bottom of the Tajo itself and in summer there are Nightingales all the way along the course of the river. If you venture down into the bottom of the gorge, keep an eye open for Melodious Warblers and Rock Sparrows, both of which occur by the river and in the surrounding fields.

Park in the town centre, from where there is ready access to the mirador and bridge. If you are feeling energetic and want to visit the bottom of the gorge, cross the bridge southwards and keep right past the Palacio de Mondragón, until you reach the top of the fairly steep downward path. The descent is a stiffish walk but the return journey is slow going for all but the super-fit.

B Manilva-Gaucín   The most westerly of the entry points into the region. Take the A-377 Manilva/Gaucín road up from the coast, which first passes the Sierra Crestellina (MA11) before joining the A-369 just to the west of Gaucín. The vicinity of the Río Genal, between Gaucín and the Sierra Crestellina, provides ample opportunities to explore the scrub and riverine vegetation for passerines, especially on the north bank of the river.

C Sierra Bermeja   This is the reddish-brown mountain that closely overlooks Estepona. The drive to the top from the coast can be done in an hour but it is worth stopping frequently to view the diverse woodland and rocky habitats along the way. Access is from the MA-8301, the rather narrow and twisting mountain linking Estepona and Algatocín. The ascent of the Sierra Bermeja proper is along a rather rough tarmac road clearly signposted for Los Reales at the crossroads at the Puerto de Peñas Blancas pass, where there is a 'Bienvenidos a la Serranía de Ronda' sign. This road climbs through pinewoods to a parking area, from which a short footpath leads up to a mirador. A signposted footpath also leads to the summit of the mountain (1,452m). The views are absolutely stunning on a clear day and migrant raptors are often evident, especially in autumn.

D Cartajima-Júzcar-Alpandeire   A circular tour (via Ronda) through these villages gives access to a range of habitats as you pass through broadleaf woodland between Cartajima and Alpandeire and also more open areas between Alpandeire and the junction with the Gaucín/Ronda road. A cliff face close to the road about 1.5km north of Cartajima is likely to produce Blue Rock Thrushes and numbers of Rock Sparrows.

CALENDAR

All year: Red-legged Partridge, Griffon Vulture, Goshawk, Sparrowhawk, Golden and Bonelli's Eagles, Common Kestrel, Peregrine Falcon, Little Owl, Thekla Lark, Crag Martin, Dipper, Stonechat, Black Wheatear, Blue Rock Thrush, Mistle Thrush, Dartford and Sardinian Warblers, Blackcap, Firecrest, Short-toed Treecreeper, Red-billed Chough, Jackdaw, Raven, Spotless Starling, Chaffinch, Crossbill, Cirl and Rock Buntings.

Breeding season: Egyptian Vulture, Short-toed and Booted Eagles, Scops Owl, Turtle Dove, Common Cuckoo, Bee-eater, Alpine and White-rumped Swifts, Red-rumped Swallow, Nightingale, Black-eared Wheatear, Melodious Warbler, Whitethroat, Woodchat Shrike, Ortolan Bunting.

Winter: Red Kite, Common Buzzard, Meadow Pipit, White Wagtail, Robin, Song Thrush, Ring Ouzel, Fieldfare, Redwing, Common Chiffchaff, Southern Grey Shrike, Common Starling, Brambling, Siskin.

Passage periods: A wide range of raptor species overfly the area. Migrant passerines are also prominent at times.

UPPER GUADIARO VALLEY MA13

Status: Parque Natural 'Sierra de Grazalema' in part. ZEPA.

Site description
This the area between Montejaque and Jimera de Líbar, west of Ronda, where the upper reaches of the Río Guadiaro run southwest down a beautiful, quite steeply sided and well-wooded valley. The valley is bounded on the western side by the Sierra del Palo and, immediately behind Montejaque and Benaoján, by the Sierra de Líbar. To the east there is a minor sierra of the Serranía de Ronda. The information centre serving the natural parks of Los Alcornocales, Sierra de Grazalema and Sierra de Las Nieves is just south of the area at Cortes de la Frontera.

Species
The Río Guadiaro attracts a number of waterbirds, including Grey Herons, Water Rails, Moorhens and Common Sandpipers. The Dipper, a very scarce

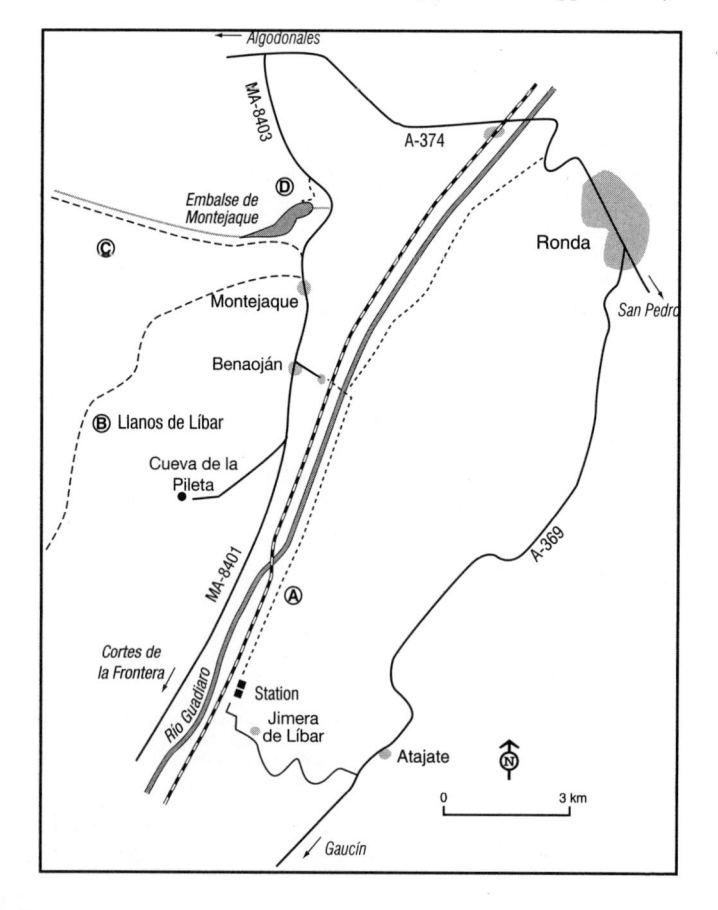

species in southern Spain, nests locally. The sierras have nesting Griffon and Egyptian Vultures and Short-toed, Golden, Booted and Bonelli's Eagles. Common, Pallid, Alpine and White-rumped Swifts occur and this is a reliable area for the last-named species, especially in late summer, when small groups, probably family parties, join feeding flocks of other swift species. Golden Orioles are common along the valley in the breeding season. Rock Sparrows and Rock Buntings are locally common around Montejaque and Benaoján.

Timing
A lovely area at any time of year. It is undoubtedly at its best in spring, especially between late April and the end of May or first week of June. Visit in summer to find White-rumped Swifts.

Access
The valley may be entered from the north, turning off from the A-374 Ronda/Algodonales road for Montejaque and Benaoján. Access from the south is via Jimera de Líbar, off the A-369 Gaucín/Ronda road. This is excellent walking country and a comprehensive range of species may be located within rambling distance of Benaoján.

♿ Observations may be made from the roads in many places.

A *Benaoján-Jimera de Líbar*   A splendid riverside stroll along the path following the east bank of the Río Guadiaro from Benaoján towards Jimera de Líbar (or upstream).

B *Llanos de Líbar*   A longish valley walk, some three hours one way, from Montejaque to the Líbar fountain. The route continues beyond here to Cortes de la Frontera, a total distance of 19km. The entry point is a track by the Bar La Cabaña on the northern fringe of Montejaque (Avenida Europa). The trail crosses pastures that are resplendent with wild flowers in spring. The eroded rocky flanks of the valley have Black Wheatears, Red-billed Choughs and other rupestral species. A shorter walk along the same track is still worthwhile if you don't wish to embark on a major expedition.

C *Valley northwest of Montejaque*   This beautiful valley that potentially offers a very good bird list is well worth visiting. Take the local road that forks northwestwards off the MA-8403 some 3500m north of the MA-8403/8402 junction in Montejaque. There are some 10km of tracks within the valley, most of which are drivable under normal circumstances.

D *Embalse de Montejaque*   A small reservoir to the north of Montejaque. The dam is over a century old but the project failed because the porous limestone basin is unable to retain water for very long, although it floods after prolonged heavy rain. The cliffs and the open rocky hinterland attract the typical birds of the area, including Griffon Vultures. Access is via a short track signposted (for a cave, the Cueva del Hundidero) from the MA-8403 north of the reservoir, 3.5km from the A-374 junction.

CALENDAR

All year: Red-legged Partridge, Grey Heron, Griffon Vulture, Goshawk, Sparrowhawk, Golden and Bonelli's Eagles, Common Kestrel, Peregrine, Water Rail, Moorhen, Common Sandpiper, Little and Eagle Owls, Kingfisher, Iberian Green and Great Spotted Woodpeckers; Crested, Thekla and Wood Larks, Crag Martin, Grey Wagtail, Dipper, Wren, Stonechat, Black Redstart, Black Wheatear, Blue Rock Thrush, Mistle Thrush; Dartford and Sardinian Warblers, Blackcap, Firecrest, Short-toed Treecreeper, Southern Grey Shrike, Jay, Red-billed Chough, Jackdaw, Raven, Spotless Starling, Chaffinch, Crossbill, Rock Sparrow, Cirl and Rock Buntings.

Breeding season: Egyptian Vulture, Short-toed and Booted Eagles, Lesser Kestrel, Turtle Dove, Common Cuckoo, Scops Owl, Red-necked Nightjar; Common, Pallid, Alpine and White-rumped Swifts, Bee-eater, Hoopoe, Wryneck, Sand Martin, Red-rumped Swallow, Rufous-tailed Scrub-robin, Nightingale, Black-eared Wheatear; Melodious Warbler, Subalpine and Bonelli's Warblers, Iberian Chiffchaff, Spotted Flycatcher, Golden Oriole, Woodchat Shrike.

Winter: Common Buzzard, Meadow Pipit, White Wagtail, Robin, Song Thrush, Ring Ouzel, Fieldfare, Redwing, Common Chiffchaff, Long-tailed Tit, Southern Grey Shrike, Common Starling, Siskin.

Passage periods: A wide range of raptor species overfly the area. Passerines are also prominent.

SIERRA TEJEDA MA14

Status: Parque Natural 'Sierras de Tejeda, Almihara y Alhama' (40,662ha).

Site description

A superbly rugged mountain range with scree slopes that straddles Málaga and Granada provinces. Most of it lies within Granada province (see GR5) but the western extremity is in Málaga. Much of the terrain is rocky and barren, with areas of dwarf shrubs. Mixed Mediterranean woodland occurs lower down. The

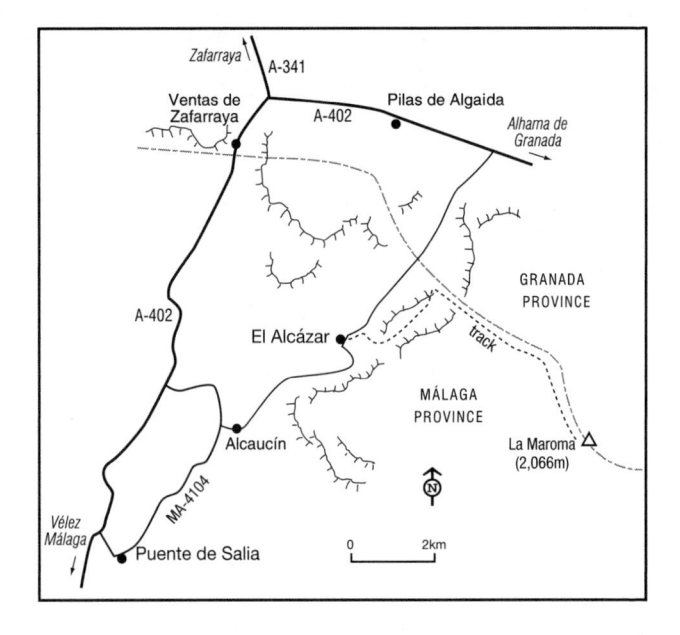

A-402 between Viñuela (inland from Vélez-Málaga) and Granada runs past the western end of the Sierra Tejeda at Venta de Zafarraya. On the western side of the road the land is relatively more open and less rugged generally, while to the east it is strikingly abrupt. The core of the Sierra Tejeda is roadless and largely inaccessible, although there are a number of marked walking trails around the periphery. The very extensive park reaches the coast east of Nerja, where there are opportunities for seawatching from the coastal cliffs. More detailed information is available from the visitors' centre, which is at the western end of the village of Sedella, on the MA-4105 southwest of the park.

Species
The typical rocky sierra species are evident. Nesting raptors include Golden, Bonelli's, Short-toed and Booted Eagles and Goshawks. The resident Black Wheatears are joined in summer by Black-eared Wheatears, Northern Wheatears and Rufous-tailed Rock Thrushes, the last two occurring only at higher levels. Ring Ouzels and Alpine Accentors are often present in winter. The Wallcreeper has been found in winter here and may then be regular but you would be very lucky to find one on a casual visit.

Timing
Interesting species are present all year but spring visits tend to be most productive.

Access
The western Sierra Tejeda is best approached from Alcaucín in Málaga province, just east of the A-402. In Alcaucín head for the upper, eastern side of the village and follow signs to El Alcázar. You can also enter at the road at its northern end, at km-40 on the A-402, where the Natural Park is signposted. This narrow and winding road is fairly rough, especially the northern half, but is passable by ordinary cars except in adverse conditions. The views are fantastic, especially in winter when there is often snow on the tops. A picnic site (Area Recreativa El Alcázar) 5km from Alcaucín is the starting point of a walking trail to the summit of La Maroma (2,068m), the highest peak of the Sierra Tejeda. This walk is well transited at weekends but is a very challenging proposition, the 16km return trip taking over eight hours to complete.

♿ Observations may be carried out from vehicles.

CALENDAR

All year: Goshawk, Sparrowhawk, Common Buzzard, Golden and Bonelli's Eagles, Peregrine, Common Kestrel, Eagle Owl, Crag Martin, Thekla Lark, Stonechat, Blue Rock Thrush, Black Redstart, Black Wheatear, Dipper, Southern Grey Shrike, Raven, Red-billed Chough, Goldfinch, Linnet, Rock Sparrow, Rock Bunting.

Breeding season: Black Kite, Short-toed and Booted Eagles, Scops Owl, Red-necked Nightjar, Alpine Swift, Tawny Pipit, Northern and Black-eared Wheatears, Rufous-tailed Rock Thrush; Subalpine, Spectacled and Bonelli's Warblers, Woodchat Shrike.

Winter: Alpine Accentor, Ring Ouzel, Wallcreeper.

VÉLEZ ESTUARY   MA15

Status: No special protection.

Site description
There are so few wetlands on the Costa del Sol that this estuary is a hotspot for birds, especially during migration periods. As such it is always worth a visit despite its small size. A sand bar usually separates the river from the sea so the water is generally still and comprises a series of shallow, reedy pools, flanked by trees and low scrub. The stretch by the N-340 road bridge and the last 250m nearest the sea are the most interesting. Heavy rains occasionally flush away the sand bar and the attraction of the site is then much reduced, temporarily. The site has been blighted in the past by the traditional presence of persons whose idea of 'a day at the seaside' is far removed from what might be considered mainstream practice. The problem has greatly diminished but still exists so it may be best not to visit the site unaccompanied.

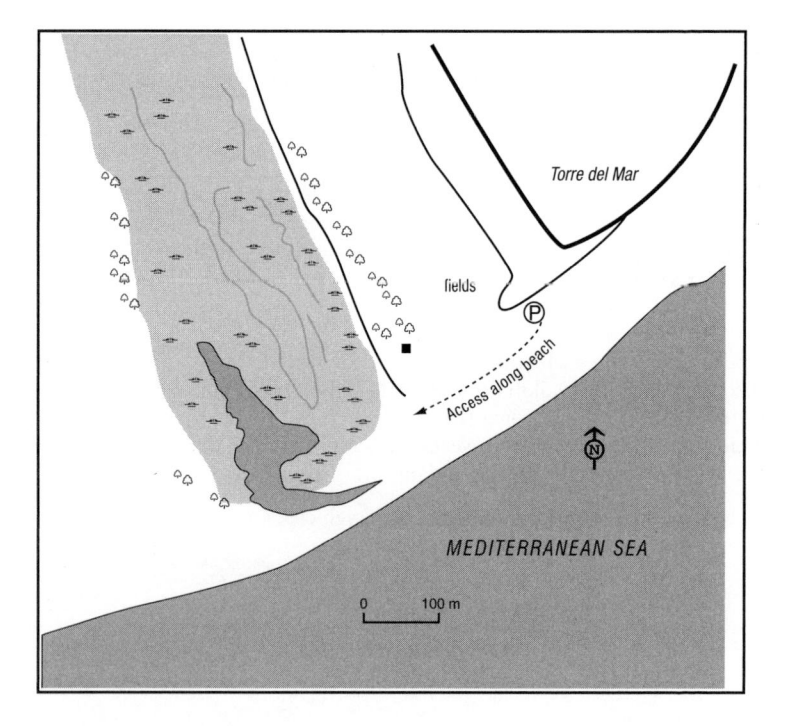

Species
This site is renowned for attracting rallids. Spotted Crakes appear annually on passage and Baillon's Crake is seen occasionally and may have bred here. Moorhens are especially abundant, Purple Swamphens are resident and Water Rails occur in winter. Most heron and egret species, and common waders, occur at times, chiefly on passage. Rare waders that have been found here

include the White-rumped, Pectoral and Broad-billed Sandpipers. Large numbers of gulls often congregate on the beach, among which Audouin's Gulls are numerous on passage and Mediterranean Gulls are especially common in autumn and winter. Both Franklin's and Laughing Gulls have also been reported here. Caspian and Lesser Crested Terns occur occasionally among the commoner species on the beach. Cory's and Balearic Shearwaters, and perhaps Levantine Shearwaters, occur offshore. Among the passerines, almost anything may turn up on passage, with warblers, pipits and wagtails being prominent. Red-throated Pipits have been recorded on several occasions.

Considerable numbers of Common Snipe and the occasional Jack Snipe frequent the marshy edges in winter, when Green Sandpipers are quite frequent along with other wintering waders. Water Pipits, Bluethroats and Reed Buntings are also common in winter, when Wrynecks are also sometimes present. Penduline Tits occur in winter and may nest locally. Small populations of Red Avadavats, Common Waxbills and Yellow-rumped Bishops are present, chiefly in the sugar cane fields, and Monk Parakeets have spread into the area.

Timing
Interesting birds are present year-round.

Access
Leave the N-340a at Torre del Mar but it is easier to enter the town and drive down to the *paseo maritimo* – the coastal promenade. Follow this westwards past a campsite and park. A 200m walk along the beach brings you to the estuary. Both sides of the river are accessible but the east bank, where there is a hide, is recommended. Watching from the sandbar also allows good views in all directions.

♿ May be difficult. Access requires walking along the beach.

CALENDAR

All year: Grey Heron, Purple Swamphen, Moorhen, Common Coot, Monk Parakeet, Tree Sparrow, Zitting Cisticola, Common Waxbill, Red Avadavat.

Breeding season: Little Bittern, Little Ringed and Kentish Plovers, Red-necked Nightjar, Greater Short-toed Lark, Yellow Wagtail, Nightingale, Isabelline and Reed Warblers, Tree Sparrow.

Passage periods: Cory's (Scopoli's) and Balearic Shearwaters, Purple and Squacco Herons, Spotted and Baillon's Crakes; waders including Common, Curlew, Wood and Green Sandpipers, Common Redshank and Greenshank, Audouin's and Mediterranean Gulls, terns including Caspian Tern and rarely Lesser Crested Tern, Yellow Wagtail, hirundines, warblers including Sedge, Reed, Grasshopper and Savi's Warblers, Red-throated Pipit, Tree Pipit.

Winter: Water Rail, Sanderling, Common and Jack Snipes, Common Redshank, Greenshank, Green Sandpiper, Wryneck, Bluethroat, Penduline Tit, Reed Bunting.

ALPUJATA MIGRATION WATCHPOINT (Mirador de las Águilas)   MA16

Status: No special protection.

Site description
A raptor migration watchpoint, named by Paco Ríos of SEO-Málaga, who uncovered its potential. The site has recently (2018) been recognised and developed as a watchpoint (a short distance west of the original location) by the local authority. It is on the south side of the Sierra de Alpujata, between Fuengirola and Coín. The mirador faces east-northeast across a wooded valley enclosed by hills over which migrating raptors and others appear. A major forest fire in 2012 destroyed the original thin woodland cover and there has since been some expansion into the area of the inevitable golf courses and villas.

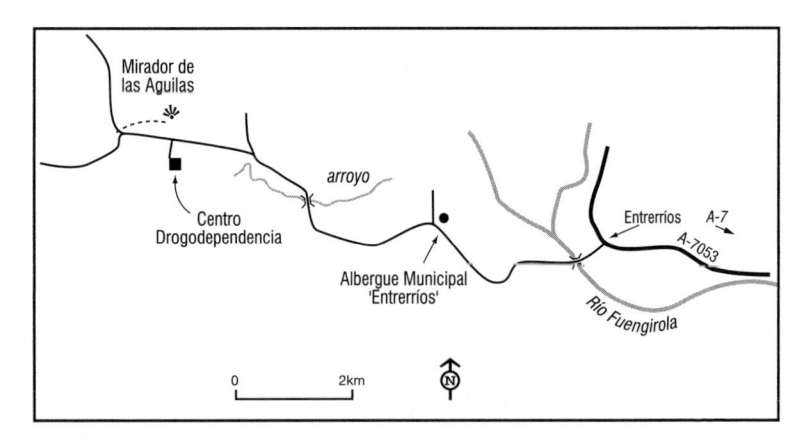

Species
The site comes into its own during the post-breeding passage period, as soaring raptors converge on the Strait of Gibraltar. Black Kites (August) and Honey-buzzards (September) are the most abundant. The other common migrant raptors include Short-toed and Booted Eagles, Marsh and Montagu's Harriers and Griffon Vultures. Ospreys and Egyptian Vultures are recorded occasionally as well as small numbers of Common Buzzards, Red Kites and Hobbies. The timing and species composition of the movements is generally the same as at the Strait (CA2, G).

Non-raptors on the move here in 'autumn' include occasional Black Storks and White Storks. At times considerable numbers of migrant Bee-eaters, European Turtle Doves, swifts and hirundines pour through the area, following the line of the sierra towards the Strait. The swifts include Alpine Swifts and, very occasionally, White-rumped Swifts. In late autumn there are occasional records of flocks of Bramblings and thrushes, which include Redwings.

Resident and summer breeding species in the general area include Bonelli's Eagles, Goshawks, Eagle Owls, Red-necked Nightjars, Iberian Green Woodpeckers and Blue Rock Thrushes.

Timing
The post-breeding raptor migration period extends from late July into November, although the most productive period is from mid-August to the first week in October. See 'Watching raptor and stork migration', pp. 43-48, for more details of the timing and species composition of the movements. Visits are most productive when the winds are from the west or south as easterlies divert the birds inland. Bring a folding chair: it makes watching much more comfortable.

Access
Take the exit for Fuengirola-Coín (208B) from the A-7 Fuengirola bypass on to the A-7053 local road. Follow the A-7053 for about 3.5km under the AP-7 motorway and past Mijas Golf to a roundabout, where you take the left exit signposted 'Entrerríos'. Cross a bridge and then turn left again at another sign for 'Entrerríos'. This takes you back under the bridge you just crossed. Stay on this road, which narrows and is slightly rough in parts. Cross over a small stream (usually dry) and continue until the road forks, where you bear left. The mirador is on the right 1.5km from the fork, where there is space to park. A short track leads up to the observation point.

♿ Not ideal unless you can manage a short uphill slope.

CALENDAR

All year: Goshawk, Sparrowhawk, Golden and Bonelli's Eagles, Peregrine Falcon, Eagle Owl, Tawny Owl, Iberian Green and Great Spotted Woodpeckers, Blue Rock Thrush.

Autumn migration: Black Stork, White Stork, Honey-buzzard, Black and Red Kites, Egyptian and Griffon Vultures, Short-toed and Booted Eagles, Marsh and Montagu's Harriers, Sparrowhawk, Common Buzzard, Osprey, Lesser and Common Kestrels, Hobby, Eleonora's Falcon; swifts including Alpine and White-rumped Swifts; hirundines including Red-rumped Swallow, Whinchat, Common Redstart, warblers, Pied Flycatcher.

PUNTA CALABURRAS　　　　　　　　MA17

Status: No special protection.

Site description
One of the best seawatching sites on the Costa del Sol. The location is hardly prepossessing as it lies just off the main coastal road, the N-340, and the background roar of traffic is constant. Neither has it much height above sea level and it faces south, so that the sun is in the face between mid-morning and mid-afternoon.

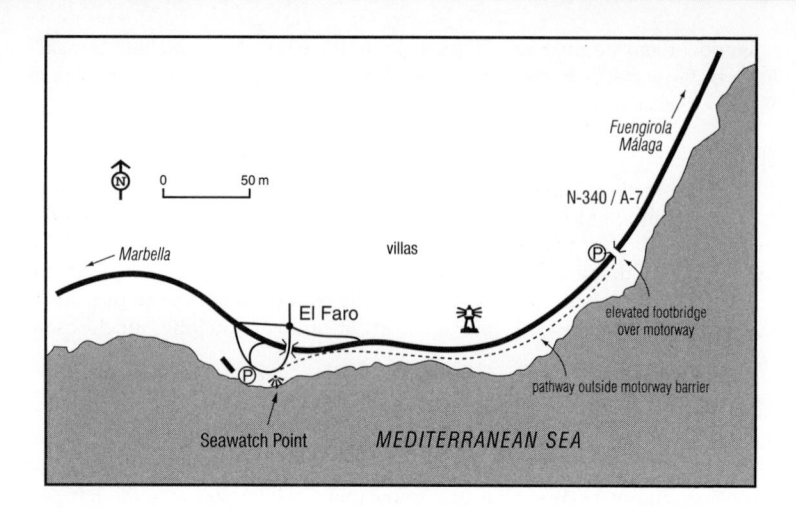

Species
Calaburras is of interest for the presence of Cory's Shearwaters between April and early November. These are often very close inshore during the late afternoon, when they follow fishing vessels heading for Fuengirola. The Mediterranean form of Cory's, Scopoli's Shearwater, also occurs, particularly during its early spring and late autumn migration periods, posing the usual identification challenge. Balearic Shearwaters are present during most of the year but Levantine Shearwaters are very erratic here.

Gannets are often present in relatively large numbers from September to May. There is a year-round presence of Yellow-legged Gulls, accompanied by Black-headed and Lesser Black-backed Gulls in winter. In summer, particularly between July and late September, there may be several hundred Audouin's Gulls migrating westwards towards the Atlantic; some rest on the rocks onshore if undisturbed. Mediterranean Gulls too are common on westward passage between late July and October and again from late January to April, when they may be abundant during periods of inclement weather. Pomarine, Arctic and Great Skuas are commonly recorded during September–May. A few Razorbills feed offshore in winter and are common on passage. During the winter months there is a daily presence of Great Cormorants from the large roost at the Guadalhorce (MA8).

The rocks attract a few species of waders, notably Turnstones, some of which will come scavenging for bread! Two or three Whimbrels and occasional Oystercatchers and Dunlins are present during the winter months. More noteworthy is the presence of up to half a dozen Purple Sandpipers in some winters, with records until late March: this is one of the very few sites in the area where there is a good chance of observing this species.

Timing
The site is worth visiting at any time of the year, with the period late May–late June being the least attractive. In the remaining months of the year, 'anything' might be seen. The sun makes morning and midday watching difficult and sunglasses essential: the afternoon is far easier on the eyes!

Access

Access is off the A-7/N-340 coast road some 3km southwest from Fuengirola at the El Faro interchange. From the Fuengirola direction cross over the road to the coast side. Parking is available by a restaurant. The watchpoint is at the end of a short track leading down to the rocky shore. It is also possible to park in the El Faro residential area, from where a footbridge and a rough path lead to the regular seawatch area.

♿ Access is possible from the entrance by the restaurant, although wheelchairs will need a good push!

CALENDAR

All year: Balearic Shearwater, Northern Gannet, Yellow-legged and Black-headed Gulls, Sandwich Tern.

Passage periods: Common Scoter, migrant ducks; Cory's, Scopoli's, Balearic and Levantine Shearwaters; Pomarine, Arctic and Great Skuas, Audouin's Gull, other gull species; Lesser Crested, Common, Little and Black Terns, Razorbill, passerines.

October–March: Red-breasted Merganser (rare), Sooty Shearwater (rare), European Storm-petrel, Great Cormorant, Turnstone, Whimbrel, Purple Sandpiper, Great Skua, Mediterranean and Lesser Black-backed Gulls, Razorbill.

GRANADA PROVINCE

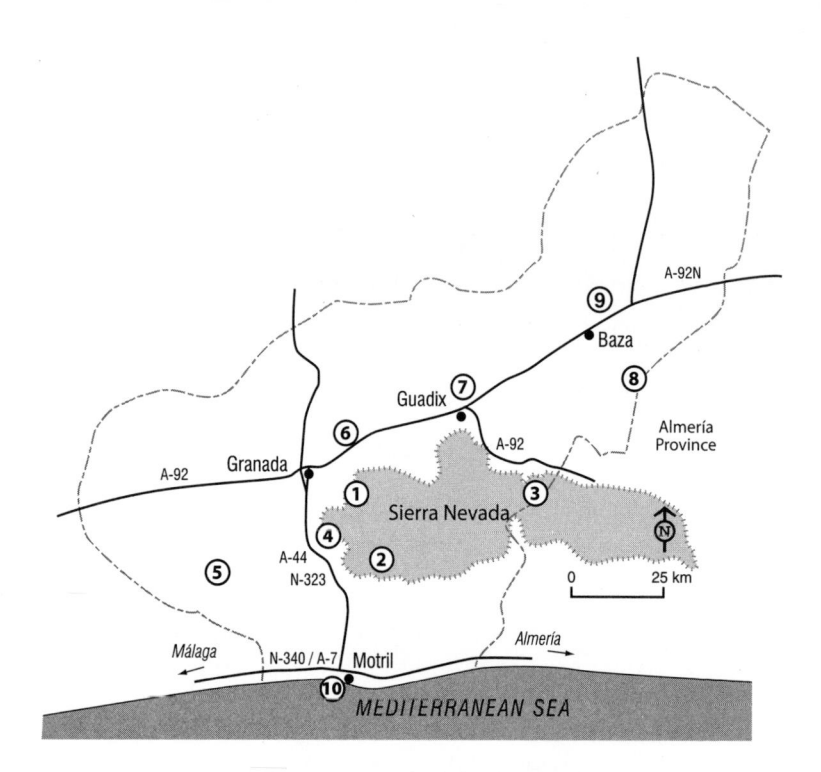

Granada province is dominated by mountains, large tracts of which are inaccessible. The most imposing range is the lofty Sierra Nevada (GR1), notable for its isolated breeding populations of Alpine Accentors and, in the east (GR3), Citril Finches. The remaining selected sites comprise more sierras (GR2, GR6, GR8), two steppe areas (GR7, GR9) and three wetlands (GR4, GR5, GR10), all of which hold some interesting species in small numbers. The species diversity of the steppe areas compares poorly with similar sites in Almería, Córdoba or Extremadura. Nevertheless, such range-restricted species as Dupont's Lark and Trumpeter Finch do occur and the Cream-coloured Courser has nested recently.

Sites in Granada Province
GR1   Sierra Nevada
GR2   The Alpujarras
GR3   La Ragua pass (Puerto de la Ragua)
GR4   Laguna de Padul
GR5   Alhama Reservoir (Pantaneta de Alhama de Granada)
GR6   Sierra de Huétor
GR7   Guadix Basin (Hoya de Guadix)

Cream-coloured Coursers

GR8 Sierra de Baza
GR9 Baza Basin (Hoya de Baza)
GR10 Suárez Wetland (Charca de Suárez)

Getting there

Nearly all the sites are within a three-hour drive from the A-7 coastal motorway, which follows the provincial coast, or from Málaga city. Inland, the A-92 and A92-N enable rapid transit east–west across the centre of the province and to/from Sevilla.

Where to stay

Most sites are within easy range of Granada city, which has a considerable range of hotels, including motels along the approach roads. The tourist accommodation at the Solynieve complex on the top of the Sierra Nevada is geared towards the skiing fraternity and is usually open only between November and the end of April. Accommodation elsewhere is widespread, especially along the coast.

SIERRA NEVADA GR1

Status: Parque Nacional (86,208ha) within a larger Parque Natural (171,646ha). UNESCO Biosphere Reserve. ZEPA.

Site description

The Sierra Nevada is the imposing range to the south and east of the famous city of Granada. It boasts the two highest peaks in Spain: Mulhacén (3,481m) and Pico Veleta (3,392m). The northern slopes are relatively gentle but the southern foothills, the Alpujarras (GR2), are much steeper and deeply dissected by riverine gorges. At its eastern end the range extends into Almería province.

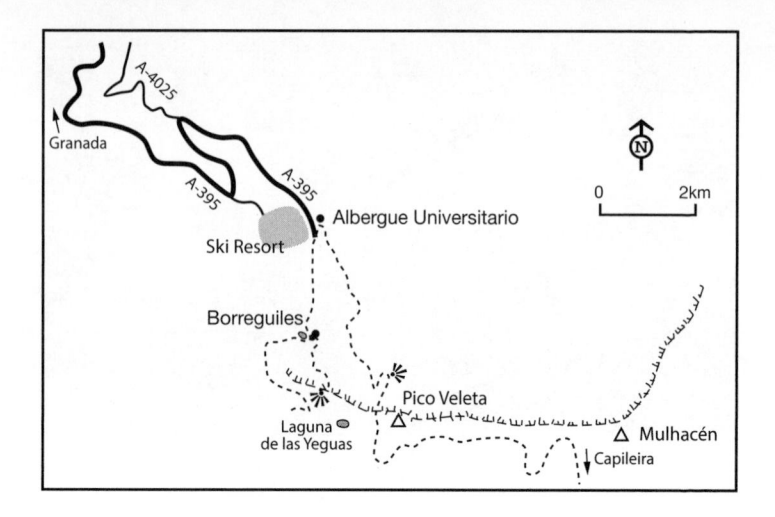

The lower slopes were once extensively wooded with Pyrenean Oak forest, much of which has long been lost to charcoal burning or pine plantations. There is some natural cover of Scots and Black Pines but woodlands generally give way to low spiny 'hedgehog' scrub at 1,700–2,000m, this in turn being replaced by dwarf juniper scrub up to around 2,600m. The sparse vegetation of the higher levels gives the range a striking barren appearance when viewed from afar. The alpine zone above 2,600m has extensive moorland and stony scree.

The Sierra Nevada is of major international importance for its alpine flora and fauna. In fact, it boasts the highest density of endemic species of anywhere in Europe and the Mediterranean basin. The flora is especially distinguished and of the greatest interest: the 200 species of the alpine zone include 40 endemics. The invertebrate community, including butterflies, is also exceptional. The important biodiversity led to designation of the Sierra Nevada as a Biosphere Reserve in 1986 and as a National Park in 1999. Nevertheless, these high levels of protection have not prevented considerable pressures from tourist development, especially for skiing, chiefly around the western summits.

The whole region is of the greatest scenic interest and rewards gentle exploration along the many designated footpaths. Details of these are obtainable from the information centres (see Access). However, relatively brief visits to a few accessible key locations would suffice for birding purposes.

Species

The exceptional biodiversity of the Sierra Nevada does not extend to its avifauna, which is broadly similar to that of other sectors of the Betic range, such as the Serranía de Ronda. It differs in that the highest levels have an isolated nesting population of Alpine Accentors, the most southerly in Iberia. Other interesting breeding species, which do not nest elsewhere in the Betic mountains, are the Citril Finch, Ortolan Bunting and Dunnock. The last of these is a recent colonist of the Sierra Nevada, as is the Nuthatch. The Azure-winged Magpie was first recorded in 2011 and its status here is uncertain. The Rock Sparrow is a common resident of sunny rocky slopes. Sky Larks,

Tawny Pipits, Black Redstarts, Northern Wheatears, Rufous-tailed Rock Thrushes and Rock Buntings also breed on the high tops. A small colony of Pallid Swifts nests within the upper ski station at 2,700m above sea level, the highest colony of this species in Iberia. A small population of Ortolan Buntings nests on north-facing slopes particularly. Another typical montane species, the Water Pipit, sometimes occurs on the alpine meadows but it has never been proved to breed here. Ring Ouzels frequent the juniper scrub on passage and in winter. The Dotterel occurs on passage on the alpine meadows, at least occasionally. In general, there is recent evidence of an upward shift in nesting distribution by such species as the Black Wheatear, with a corresponding range contraction of the birds of the high tops, such as the Northern Wheatear. The Sierra is a good place to see the Iberian Green Woodpecker as well as a range of woodland and scrub species on the lower slopes that includes breeding Common Chiffchaffs, which nest very locally in Andalucía, and wintering Yellowhammers, which are uncommon in the region. Ready access to the wooded slopes is available in the Alpujarras region (GR2).

Raptors are sparse. Common Kestrels are likely to be seen and both Golden Eagles and Peregrines have resident populations. Bonelli's Eagles nest on the limestone escarpments at low elevations. Occasional sightings of Lammergeiers in recent years seem likely to become more frequent given the continuing reintroduction and re-establishment of breeding birds in Andalucía (see J1). The sheep flocks and the plentiful Spanish Ibex in the Sierra guarantee them a good supply of bones.

Las Aves de Sierra Nevada, by Jorge Garzón and Ignacio Henares, is a useful review and analysis of the avifauna. The text is in Spanish but the individual species accounts have English summaries. It is available online on the Environment Agency publications website (www.juntadeandalucia.es/medioambiente/servtc5/ventana/publicaciones) under Sierra Nevada.

Timing

The western end of the range is the habitat of huge numbers of skiers once the winter snows arrive. The winter sports season may extend from early November until late April, and large numbers of tourists tend to occur until late May. Fortunately, fewer people are around in late spring and summer, the best times to visit the higher areas, which become essentially birdless in winter. Other parts of the range attract few visitors at any time but are still most rewarding in spring and summer. Weekday visits are always preferable; avoid Sundays and holiday periods if you can since these are the busiest times year-round.

Warm, waterproof and windproof clothing is advisable when visiting the high tops even in summer and is absolutely indispensable at any other time of year. Snow lingers in some of the corries year-round and it is possible to experience a 30°C temperature drop between the coast and the summits in July and August. A high-factor sun tan cream is also advisable, as well as sunglasses where there is snow.

Access

There are only two routes on to the tops, by road from the north and on foot from the south. The northern approach, from Granada city on the A-395, is drivable all the way to the ski resort (*Estación Invernal*) and is the best option for reaching the alpine zone. The distance from Granada to the hotels complex at the ski resort is approximately 45km, ascending through a variety of habitats. There are limited opportunities for stopping, at small lay-bys and viewpoints, and of course at the visitors' centre at km-26. Follow signs to Pico Veleta and the Albergue Universitario (Granada University Lodge, which offers public hotel facilities year-round). This route reaches an altitude of 2,500m at the resort and lodge. In summer, to go higher you need to walk or take a ski lift, some of which remain open between early June and September. Easy access to the high tops is provided by the regular minibus up to Veleta from the Albergue Universitario. The bus saves you walking about 5km each way and a climb of about 750m. It stops at the Mirador del Veleta (3,100m), from where there are splendid views over a huge area on a clear day: as far as Gibraltar and across the sea to Morocco. Fit walkers may nonetheless prefer to walk up, birding on the way, along the footpath that accompanies the road. The summit of Pico Veleta may be reached on foot from the mirador.

A southern approach is available for determined walkers only. Take the road up from Capileira in the Alpujarras (see GR2 B). Another option is to come out from Granada on the GR-410 signposted for La Zubia and then continue up from there. However, you do not get much above 1,500m.

At the eastern end of the range there are two roads that transverse the Sierra from north to south, as well as a network of mountain tracks (see GR3).

The Sierra Nevada bird guide (see 'Species' above) includes maps and details (in Spanish) of ten recommended birding routes. These are worthwhile if you plan to spend some time in the area but most require some pre-planning, with arrangements perhaps involving a second vehicle to return you to the starting point. The visitors' centres also offer details of walking trails and guided itineraries, useful if you are staying for a while at one of the hotels. The principal visitors' centre is El Dornajo, at km-26 on the A-395, the 'up-road' from Granada. This is open from 10:00 to 17:00 hrs daily except Mondays and Tuesdays. There is another centre in the eastern, Almerian, sector: Laujar de Andarax, at km-1 on the A-348, Laujar/Berja road. You will also encounter Information Posts at viewpoints and other key sites.

♿ The ski area road and the bus to Veleta provide access. Elsewhere difficult or impossible.

CALENDAR

All year: Goshawk, Golden and Bonelli's Eagles, Common Kestrel, Peregrine Falcon, Eagle Owl, Thekla Lark, Grey Wagtail, Dipper, Dunnock, Alpine Accentor, Black Redstart, Stonechat, Black Wheatear, Blue Rock Thrush, Dartford Warbler, Blackcap, Firecrest; Crested, Coal and Great Tits, Nuthatch, Short-toed Treecreeper, Jay, Azure-winged Magpie, Red-billed Chough, Jackdaw, Carrion Crow, Raven, Rock Sparrow, Serin, Citril Finch, Linnet, Crossbill, Cirl and Rock Buntings.

Breeding season: Hobby, Lesser Kestrel, Common Sandpiper, Alpine and Pallid Swifts, Common Cuckoo, Scops Owl, European and Red-necked Nightjars, Bee-eater, Roller, Hoopoe, Sky and Wood Larks, Tawny Pipit, Nightingale, Northern and Black-eared Wheatears, Rufous-tailed Rock Thrush; Melodious, Spectacled, Subalpine and Bonelli's Warblers, Whitethroat, Common Chiffchaff, Spotted Flycatcher, Golden Oriole, Woodchat Shrike.

Winter: Hen Harrier, Merlin, Ring Ouzel, Common Chiffchaff, Wallcreeper (rare), Goldcrest, Brambling, Siskin, Yellowhammer.

Passage periods: raptors, Dotterel, Water Pipit, Whinchat, Common Redstart, Ring Ouzel, Iberian Chiffchaff, Pied Flycatcher.

THE ALPUJARRAS   GR2

Status: The higher levels fall partly within the Parque Nacional of Sierra Nevada. Otherwise no special protection.

General description
The southern flank of the Sierra Nevada, extending eastwards from Lanjarón in Granada province to Ugíjar and on to Padules in Almería province. In the west the numerous small rivers and streams that cascade down from the Sierra Nevada after the autumn rains and during the snow-melt in spring combine as the Río Guadalfeo. A distinctive feature of the region is the network of irrigation channels (acequias), watering orchards and other crops in what is otherwise a relatively dry area.

Sites A, B and C below are representative of the upper Alpujarras. They offer considerable opportunities for exploring interesting habitats in a region of great scenic merit. However, there are plenty of alternatives, especially if you want to combine birding with mountain trekking. The trails above Lanjarón and Cañar are additional options. The Alpujarras tourist offices, in the larger towns, offer the Alpujarras Tourist Guide, which includes maps of all the roads and tracks.

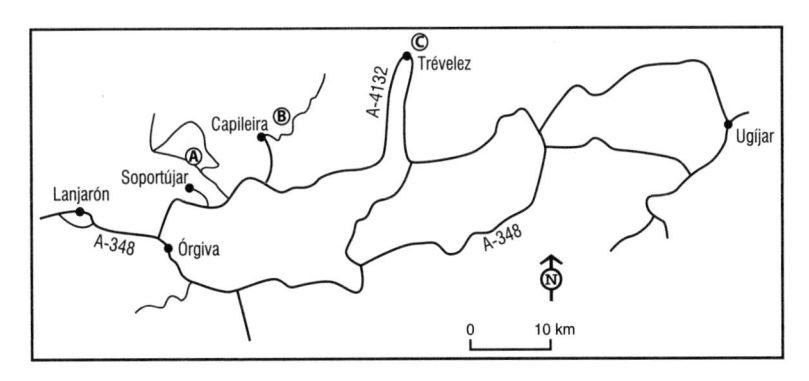

Timing
The best time is undoubtedly spring and early summer, between late March and early July, although some areas are interesting year-round.

General Access
The A-44 and the N-323/N-323a Granada/Motril roads skirt the western edge of the Alpujarras. From Granada take the A-348 for Lanjarón and Órgiva. From Motril take the A-346 for Órgiva. You can also approach from the eastern end by leaving the A-7/N-340 coast road, just east of El Puente del Río, Almería, taking the A-347 for Berja and continuing to the A-348 for Ugíjar.

♿ Some limited access at sites A and C, none at B.

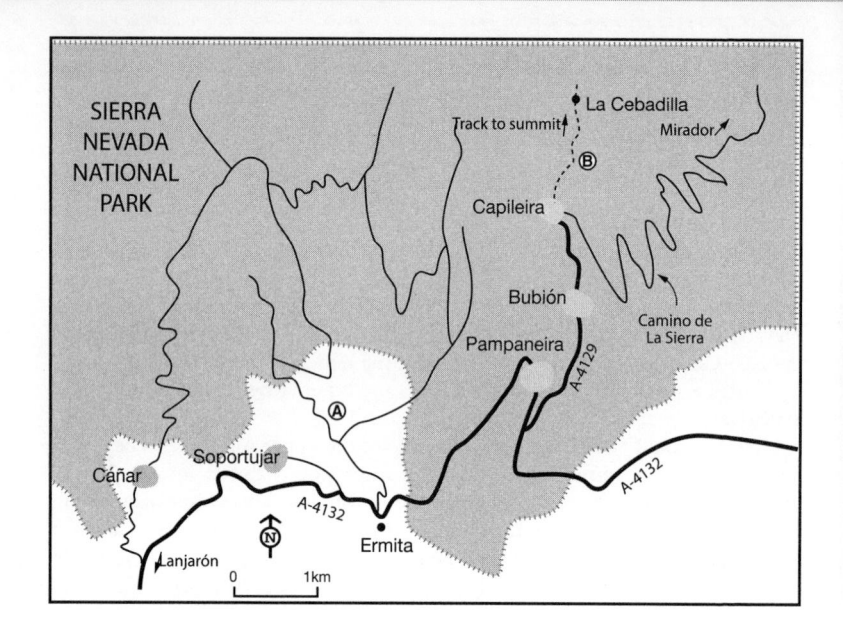

Species

The Alpujarras offer a higher density and diversity of bird species than the high tops of the Sierra Nevada, as would be expected given the greater habitat diversity. Golden and Bonelli's Eagles and other raptors often overfly the area. The open slopes have Hoopoes, Thekla Larks, Blue Rock Thrushes, Black Wheatears, Black Redstarts and Rock Buntings. Dartford and Melodious Warblers occur as well as the ubiquitous Sardinian Warblers. The pinewoods have their typical birds, including Crested Tits and Short-toed Treecreepers. The many small rivers and rushing, clear streams provide a habitat for the Dipper, a very local species in southern Spain, as well as the Grey Wagtail and such species of riparian growth as the Nightingale, Blackcap and Golden Oriole.

A SOPORTÚJAR
Site description

The village of Soportújar basks on the southern slopes of the Sierra Nevada. The general area gives access to a labyrinth of mountain tracks, ascending through open scrubby terrain into the forested flanks of the Sierra. The mapped road, leading into the pine forests above Soportújar, is representative of the region.

Access

Leave the A-348 just west of Órgiva and follow the winding A-4132 upwards. Shortly after passing the turn-off for Soportújar you reach a hermitage (Ermita del Padre Eterno) on the right. A narrow road leads upwards just opposite the hermitage, crossing open hillsides and rocky areas, and offering excellent views and easy opportunities for stopping to look for passerines and other birds. The road enters the pine forest after some 16km.

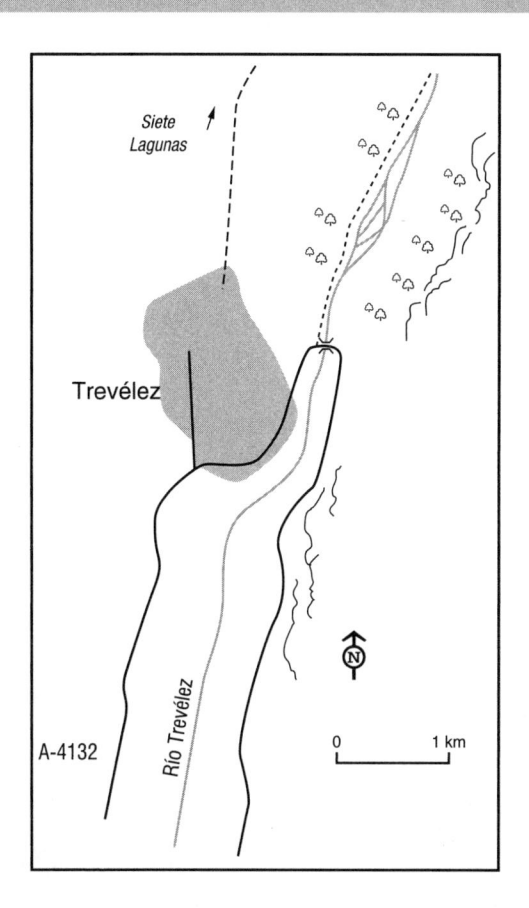

B CAPILEIRA TO SIERRA NEVADA
Site description
Another opportunity to explore the open flanks of the Sierra Nevada on foot is available above the village of Capileira. Indeed, it is possible to walk all the way up to the summit of the range, a distance of some 26km.

Access
Walk through the village of Capileira and continue upwards for some 3km along a track that leads to a power facility, La Cebadilla. In summer, if you are feeling strong enough, are well equipped and provisioned and have support in the form of a pick-up at the far end, you can follow the track to the highest levels of the Sierra Nevada: right over the top, by Mulhacén and Pico Veleta, and down to the Albergue Universitario, and the Estación Invernal (GR1). Most visitors will probably travel a shorter distance to explore the lower slopes.

A guided minibus tour runs occasionally from Capileira village eastwards to the Mirador de Trevélez, high on the slopes to the west of Trevélez village. This service only operates in summer and autumn. Superb views and good walking opportunities are available around the mirador, which is also reachable by car along the Camino de La Sierra, the road that descends east from Capileira before zig-zagging up the slopes from some 13km. The road is narrow tarmac along its first half and in good condition thereafter.

C TREVÉLEZ
Site description
Trevélez, the highest village in Spain (1,560m above sea level), lies at the end of a long, pleasant valley, which produces some of the finest jamón serrano (sierra air-cured ham). The Sierra Nevada, from which the Río Trevélez cascades, rises above the village. The valley sides rise up steeply from the floor, which has open grassy areas and trees. The area above the river bridge is the most interesting.

Access
Access to the village is easy from either west or east. The western approach is more attractive as you look up and across the valley. Park in the village or near the bridge. A footpath along the west bank of the river extends for a considerable distance both upstream and downstream. The upstream walk is recommended although it may be flooded when the river is in spate. Bird trekkers may wish to continue for some 17km far above Trevélez on the signposted footpath to a region known as Siete Lagunas (Seven Lakes) on the southeastern flanks of Mulhacén, where the species of the high tops may be expected. This trail starts by the town hall (Ayuntamiento) at the top of the village.

CALENDAR

All year: Goshawk, Sparrowhawk, Common Buzzard, Golden and Bonelli's Eagles, Common Kestrel, Peregrine Falcon; Eagle, Little, Tawny and Long-eared Owls; Crested, Thekla and Wood Larks, Crag Martin, Dipper, Wren, Robin, Black Redstart, Black Wheatear, Blue Rock Thrush, Mistle Thrush, Dartford and Sardinian Warblers, Blackcap; Long-tailed, Crested, Blue and Great Tits, Short-toed Treecreeper, Jay, Red-billed Chough, Jackdaw, Raven, Chaffinch, Serin, Greenfinch, Linnet, Crossbill, Cirl and Rock Buntings.

Breeding season: Short-toed Eagle, Lesser Kestrel, Common Cuckoo, Scops Owl, Alpine Swift, Bee-eater, Hoopoe, Barn and Red-rumped Swallows, Sky Lark, Grey Wagtail, Tawny Pipit, Northern and Black-eared Wheatears; Melodious, Subalpine and Bonelli's Warblers, Spotted Flycatcher, Golden Oriole, Woodchat Shrike.

Winter: Common Chiffchaff, Siskin, Citril Finch.

Passage periods: Raptors, larks, wagtails, pipits, Common Redstart, warblers, Pied Flycatcher.

LA RAGUA PASS (Puerto de la Ragua) GR3

Status: Within Parque Nacional de la Sierra Nevada.

Site description
A mountain pass where the A-337 crosses the Sierra Nevada at an altitude of over 2,000m, on the boundary between Granada and Almería provinces. The trail network at the pass gives access to open expanses of juniper and spiny scrub as well as open Scots Pine woodlands and Black Pine plantations lower

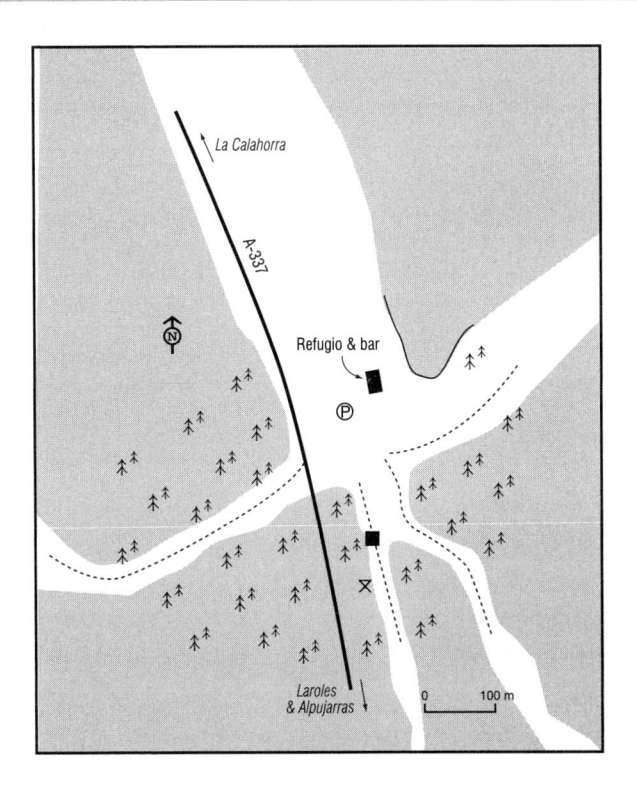

down. There is an information centre at the pass, about 100m south of the roadside mountain refuge and bar.

Species
The Citril Finch, which has a southern outpost population in the Sierra Nevada, is most readily seen at La Ragua, except in mid-winter when the birds descend lower. Anything up to 20 birds may sometimes be found hopping around under the pines near the information centre. Open areas near the pass have nesting Sky Larks, Tawny Pipits and Northern Wheatears. Water Pipits can often be seen here and trips of migrant Dotterels are occasionally found resting in the area.

The pass is a major flyway for migrants confronted with the Sierra Nevada, especially southbound birds in autumn. Hundreds of such raptors as Honey-buzzards and Black Kites have been recorded, together with smaller numbers of a range of other species. The pass is crossed by massive movements of southbound swifts between July and October: these sometimes involve tens of thousands of birds per hour, most of them Common Swifts but with some Pallid and Alpine Swifts as well. All five hirundine species also occur on passage, together with Bee-eaters, larks, pipits and finches.

Timing
The altitude makes for cold and often snowy conditions in winter, making driving potentially hazardous. For birding purposes, it is best to visit during the

more clement weather period, more or less between late April and the end of October, which coincides with the best range of species. Walking the trails can be very pleasant on a fine summer's day but the weather can change very rapidly so be prepared.

Access
From the south, the Alpujarras, take the A-337 north for Calahorra, 6km east of Ugíjar at Cherín. This is the easier road access to the pass. However, taking the A-337 southwards from the A-92 at km-312 (La Calahorra and Puerto De la Ragua) may be more convenient. Once past La Calahorra, with its striking hilltop castle, there are some 13km of unprotected bends before you reach La Ragua. Park at the pass, where various trails are signposted. Very large open expanses open up to either side of the pass, where you can search for upland species.

♿ Two designated footpaths from the parking area are designed for universal use.

CALENDAR

All year: Sparrowhawk, Common Buzzard, Golden Eagle; Sky, Wood and Thekla Larks, Wren, Robin, Firecrest, Coal Tit, Blue Tit, Great Tit, Short-toed Treecreeper, Red-billed Chough, Chaffinch, Citril Finch, Crossbill.

Breeding season: Tawny Pipit, Northern and Black-eared Wheatears, Rufous-tailed Rock Thrush, Common Chiffchaff, Bonelli's Warbler.

Passage periods: Honey-buzzard, Black Kite, Marsh Harrier, Booted Eagle, Dotterel (occasional), Bee-eater; Alpine, Common and Pallid Swifts, Barn and Red-rumped Swallows; Sand, Crag and House Martins, Ring Ouzel, pipits, finches.

Winter: Redwing, Fieldfare, Ring Ouzel, Goldcrest, Brambling.

LAGUNA DE PADUL　　　　　　　　　　　GR4

Status: Within Parque Natural de la Sierra Nevada. ZEPA. Ramsar site. 300ha.

Site description
The largest wetland of Granada province. The lagoon and its associated very extensive reedbeds occupy former peat workings within the Padul depression.

Species
The best site for aquatic species in a province that offers few wetlands. Breeding species include the Little Bittern. A range of wintering and passage waterfowl may be encountered. The site is well known for the relative regularity with which it attracts migrant crakes, particularly Spotted Crakes. The attractiveness of the lagoon to such birds probably reflects the regional scarcity of suitable alternative habitats combined with the regular scrutiny offered by resident and visiting birders, including those that man a local ringing site.

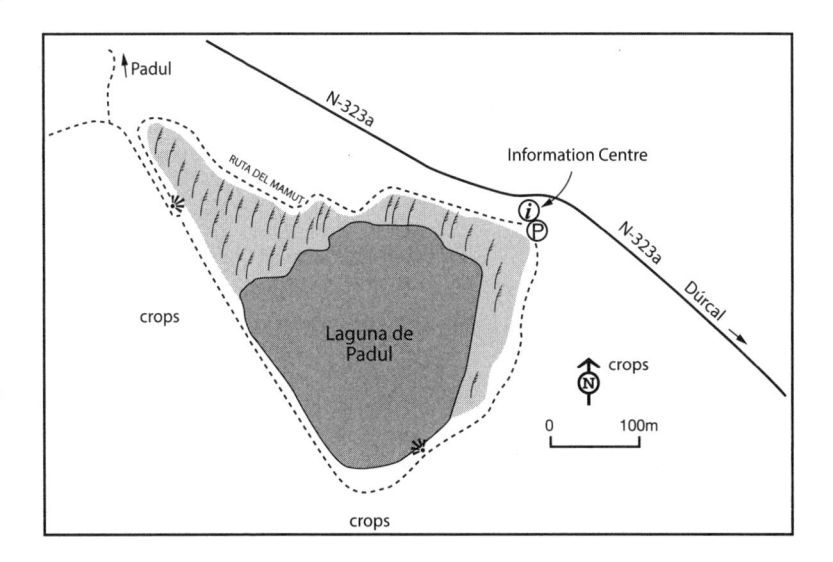

Timing
The lake offers permanent water and so the site is always of interest. The largest numbers of aquatic birds may be expected in winter and during passage periods. Spotted Crakes are most often recorded in spring, particularly in March–April.

Access
The lagoon is some 20km south of Granada city and just south of Padul vilage on the N-322a. Follow signs to the N-323a from the A-44 Granada/Motril motorway. From the north leave the motorway at km-144, signposted Padul. From the south leave at km-157 and continue to Padul via Dúrcal. Stop at 'El Aguadero' information centre, 2km south of Padul (km-153 on the N-323a)

Bonelli's Eagles

and signposted 'Aula Naturaleza'. This is the starting point of trails that give views over the lake and reedbeds. A drivable trail leads south (left) along the lakeside giving views of open water. There is a hide. A longer (2km) 'Ruta del Mamut' loops on boardwalks through reedbeds as well as giving views of the lake and surrounding areas. It is possible to continue further right round the lake – allow at least three hours. The 'mamut' is the mammoth, whose remains have been unearthed locally and which is celebrated here by a life-sized statue, displays of tusks and information boards.

♿ Tracks are level and the Ruta del Mamut is readily accessible.

CALENDAR

All year: Little Grebe, Bonelli's Eagle, Water Rail, Calandra and Crested Larks, Cetti's Warbler, Avadavat.

Breeding season: Quail, Little Bittern; Turtle Dove, Red-necked Nightjar, Bee-eater, Hoopoe; Reed, Great Reed and Melodious Warblers, Woodchat Shrike.

Winter: common waterfowl, Black-winged Kite, Hen and Marsh Harriers, Merlin, Lapwing, Common and Jack Snipes, Sky Lark, Water Pipit, Bluethroat, Common Chiffchaff, Southern Grey Shrike, Penduline Tit, Spanish Sparrow, Reed Bunting.

Passage periods: Night, Squacco and Purple Herons, Marsh Harrier, Spotted Crake (spring); Little Ringed Plover (has bred), Bluethroat; Grasshopper, Savi's and Sedge Warblers.

ALHAMA RESERVOIR GR5
(Pantaneta de Alhama de Granada)

Status: Included in the inventory of Andalucían wetlands.

Site description
A dam on the Río Alhama retains a small reservoir (1ha) in a bowl in the hills. Downstream the river flows through a gorge for about 2km before running through Alhama de Granada. Two hides give good views of the lake, which is fringed by reedbeds. The surrounding vegetation includes eucalyptus, Holm Oaks, tamarisks, the ubiquitous olives, broom, pines and white poplars.

Species
Wintering waterfowl include Wigeon, Teal, Mallard, Pintail and Shoveler, as well as occasional scarce species such as Ferruginous and Tufted Ducks. Grey Herons, Water Rails and Great Cormorants are also regular in winter. Green and Common Sandpipers occur on passage, when other wader species are also possible. A diversity of woodland and scrub species breed around and near the reservoir, including Iberian Green Woodpeckers; Cetti's, Orphean and Melodious Warblers, Iberian Chiffchaffs, Golden Orioles and Southern Grey and Woodchat Shrikes.

Timing
Winter is the best season for waterfowl diversity but some interesting species are present all year round.

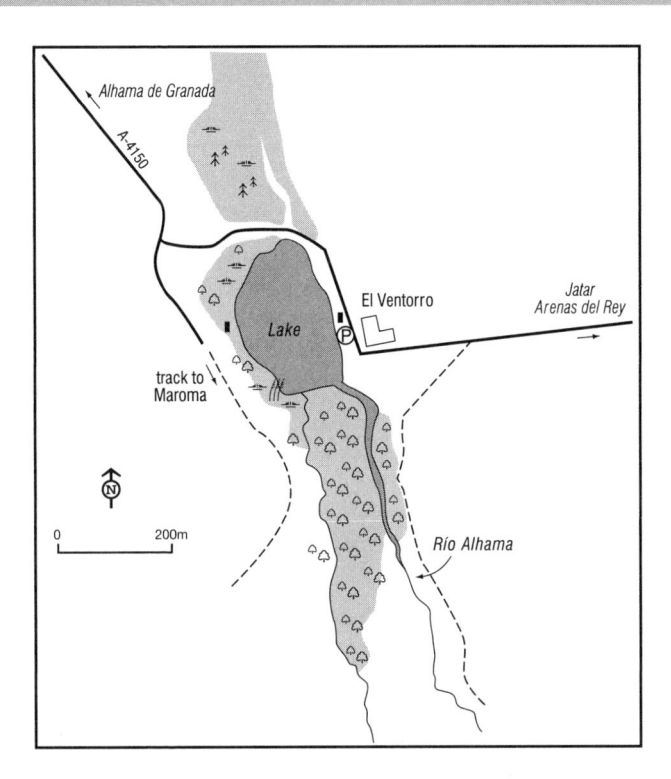

Access

Turn off the A-402 1km south of Alhama de Granada and take the A-4150 for Arenas del Rey. After 3km the road crosses the top of the dam just after passing the sign on the right for Maroma. Park on the right of the road after crossing the River Alhama by the 'Hospedería El Ventorro', a pleasant enough place where it is possible to stay. It is possible to walk right around the lake and there is a drive or walk along an unsurfaced track some 150m on the right after passing 'El Ventorro'. The two hides offer good views of the lake.

The track signposted to Maroma, the highest peak of the Sierra Tejeda (2,065m) is long – over 11km – and hard-going. There is slightly easier access from Málaga province (see MA14).

♿ Accessible in the immediate vicinity of the car park and also along the first sections of the track to Maroma and on the north side parallel to the river.

CALENDAR

All year: Mallard, Little Grebe, Common Kestrel, Moorhen, Common Coot, Woodpigeon, Hoopoe, Crag Martin, Barn Swallow, White Wagtail, Wren, Robin, Blackbird, Cetti's Warbler, Southern Grey Shrike.

Breeding season: Common Swift, Pallid Swift, Turtle Dove, Hoopoe, Nightingale; Melodious, Orphean, Subalpine and Bonelli's Warblers, Iberian Chiffchaff, Spotted Flycatcher, Golden Oriole, Woodchat Shrike, Jay, Common Magpie, Chaffinch, Serin, Greenfinch, Goldfinch.

Passage periods: waders, including Green and Common Sandpipers.

Winter: Common Pochard, Teal, Pintail, Wigeon, Tufted and Ferruginous Ducks, Great Cormorant, Grey Heron.

SIERRA DE HUÉTOR GR6

Status: Parque Natural (12,168ha).

Site description
A large area of medium-sized mountains (1,100–2,070m) a few kilometres northeast of Granada city with views to the Granada depression and the Sierra Nevada. Much of the rock is calcareous, resulting in a rugged landscape. The region is a biotic island surrounded by low-lying terrain and hence the flora includes a large number of endemic species and subspecies. The woodland cover includes areas that were replanted following the devastation of 5,000ha by a forest fire in August 1993. The forest cover includes pinewoods and Iberian Holm Oak woods, with riparian woodland along the rivers. Montane scrub, including hedgehog shrubs and junipers, flanks the bare terrain of the high tops.

Species
Characteristic breeding raptors include the Common Buzzard and Golden Eagle. There is the usual diversity of woodland birds, including Crossbills in the pines and Iberian Green Woodpeckers in the clearings. Subalpine Warblers occur higher up, where the pines thin out, alongside the resident Dartford and Sardinian Warblers, Red-billed Choughs, Ravens and Rock Buntings. Wintering and passage birds include Ring Ouzels.

Birds apart, the Sierra also holds important populations of Red Squirrels, Genets, Badgers, Beech Martens, Wild Cats, Wild Boars and Spanish Ibex.

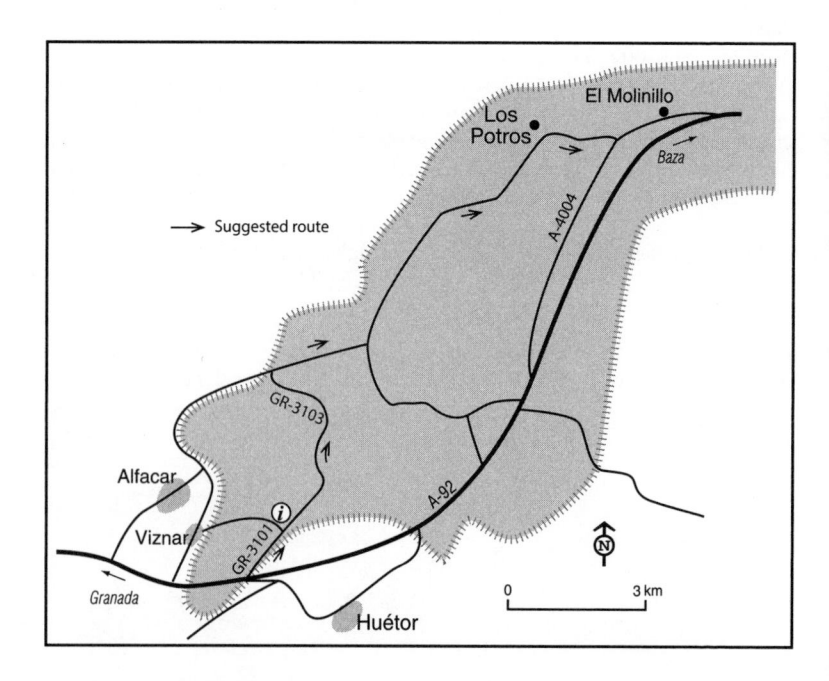

Timing
This area is worth visiting year-round although the highest bird diversity is in spring between March and June. Weekday visits are recommended for peaceful birding since this is a popular camping and picnicking area for visitors from Granada and beyond.

Access
The park is signposted from the A-92 at km-250 (Viznar) but, since the best starting point is the 'Puerto Lobo' visitors' centre, it is simplest to leave the motorway at km-253 (El Fargue, Huétor-Santillán). Head for El Fargue on the A-4002 and turn right to follow the Viznar/Park signs onto the GR-3101, which crosses over the motorway. The visitors' centre is 3km further on. However, it is usually only open on Wednesday–Friday mornings but all day at weekends. The centre provides the usual information, especially on a range of walking trails. These include two easy circular walks (5km and 3km) through a range of habitats that are signposted from the centre itself and are always open.

For exploration by car you can continue north on the GR-103 and bear right to follow the signposted road to Area Recreativa Los Potros. This road passes through deciduous and coniferous woodlands, and gives access to stretches of riverine growth and also some open terrain. There are scenic viewpoints along the way, as well as picnic and camping sites and the starting points of more trails. A trail (Las Mimbres) at Los Potros is of universal access. Thereafter you can head eastwards to the A-404 from which you can rejoin the A-92 either north or south of El Molinillo.

There are frequent stopping places and several trails are of universal access.

CALENDAR
All year: Goshawk, Common Buzzard; Booted, Bonelli's and Golden Eagles, Common Kestrel; Scops, Little, Eagle and Tawny Owls; Crested Tit and Coal Tits, Dartford Warbler, Firecrest, Short-toed Treecreeper, Red-billed Chough, Jackdaw, Raven, Chaffinch, Crossbill.

Breeding season: Subalpine and Bonelli's Warblers, Spotted Flycatcher, Golden Oriole.

Winter: Thrushes, including Ring Ouzel.

GUADIX BASIN (Hoya de Guadix) GR7

Status: ZEPA.

Site description
The Guadix depression is a 500km^2 semi-arid expanse to the north of the Sierra Nevada, at an altitude of around 1,000m. The area has a steppic character and is fairly level, although the chalky ground is much eroded into gullies and ravines in places. There are semi-desert areas and pastures of False Esparto grass, through which flocks of sheep and goats wander with their attendant herders. Cereal crops are also cultivated and there are vineyards and areas of Iberian Holm Oak dehesas, with ever more poplar plantations and some irrigated crops

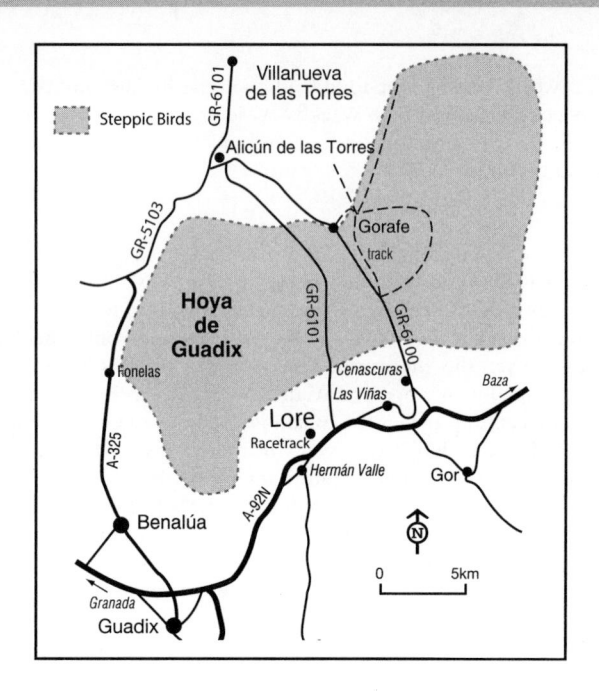

along the watercourses. The recent proliferation of olive groves has led to extensive losses of steppic terrain, unfortunately. There is little human presence in the core of the area except for isolated farms, and along the course of the Río Gor.

Species

The depression is included in the official inventory of steppe bird sites in Andalucía and indeed there are small populations of Little Bustards, Stone-curlews and Black-bellied Sandgrouse. These are thinly spread, however, and may take some finding. Open-country passerines are more easily located. Both Thekla and Crested Larks occur, as well as abundant Calandra, Greater Short-toed and Lesser Short-toed Larks. Tawny Pipits breed in small numbers. Large numbers of Meadow Pipits occur in winter. Black-eared Wheatears are widespread and Black Wheatears are common in the region of Gorafe and in the *ramblas* on the way to Alicún de los Torres. Melodious and Spectacled Warblers breed in the same areas.

Great Spotted Cuckoos have one of their highest densities in Spain here, at about four birds/km^2, reliant as ever upon a good population of Common Magpies, their host species. Red-billed Choughs breed in the surrounding sierras and along the cliffs of the Río Gor, as well as the ubiquitous Ravens and some Carrion Crows. Both Tree and Rock Sparrows are also found in the right habitats and Trumpeter Finches occur locally. Other breeding birds include Rollers, Hoopoes and Bee-eaters. Both Southern Grey and Woodchat Shrikes are widespread.

Montagu's Harriers breed here and small numbers of raptors from surrounding sierras, including both Golden and Bonelli's Eagles, hunt over the area.

Southern Grey Shrikes are widespread, perched ominously in wait for the unwary, and Woodchat Shrikes are also common in season.

Timing
The area is worth visiting year-round although the heat of the summer is best avoided. Early morning and evening visits in spring, between late February and late May, are best for locating the steppe species, when these are most dispersed and vocal.

Access
The area of interest, especially for the steppic species, lies to the north of the A-92N, east of Guadix itself. Turn off the A-92N at km-11 to take the GR-6101 northwards towards Alicún and Villanueva de Las Torres. The road crosses the centre of the depression before rising into rocky, hilly terrain. Occasional farm tracks and other minor roads allow exploration to either side of the GR-6101. The marked track to a motor racetrack (Circuito Permanente), 500m from the A-92N, continues beyond for some distance across typical steppic habitats. Thereafter, it is worth continuing on the GR-6101 as far as Villanueva, through scenic, barren country.

The GR-6100, through Gorafe, and the GR-5103/A-325 south to Fonelas, also cross interesting terrain as well as providing alternative routes to and from the A-92N. Trumpeter Finches may be sought to the east of the GR-6100. Take the track signposted 'Parque Neolítico' (Neolithic Park) at the sharp bend 5km north of Cenascuras. The track beyond the mirador here runs parallel to but above the main road and eventually leads to Gorafe. However, you can continue north beyond Gorafe, where a network of tracks leads into the 'moonscape' of the ramblas 'badlands' at the end of the valley. Trumpeter Finches frequent the ramblas but are elusive: they are easier to find in Almería. It is easy to get lost here so you are advised not to wander far unless you have a guide or a GPS with route markers.

♿ Observation from vehicles is feasible.

CALENDAR

All year: Egyptian and Griffon Vultures, Golden and Bonelli's Eagles, Goshawk, Peregrine, Little Bustard, Stone-curlew; Eagle, Little and Long-eared Owls, Black-bellied Sandgrouse; Crested, Thekla Lark, Calandra and Lesser Short-toed Larks, Black Redstart, Black Wheatear, Dartford and Sardinian Warblers, Southern Grey Shrike, Common Magpie, Red-billed Chough, Jackdaw, Raven, Tree and Rock Sparrows, Trumpeter Finch.

Breeding season: Montagu's Harrier, Hobby, Red-necked Nightjar, Great Spotted Cuckoo, Roller, Hoopoe, Bee-eater, Red-rumped Swallow, Greater Short-toed Lark, Tawny Pipit, Nightingale, Black-eared Wheatear; Melodious, Olivaceous and Spectacled Warblers, Woodchat Shrike.

Winter: Merlin, Sky Lark, Meadow Pipit, White Wagtail.

SIERRA DE BAZA GR8

Status: Parque Natural (53,844ha).

Site description
Another calcareous mountain-island of the Betic range, rising out of the arid lowlands to the north of the Sierra Nevada. The summit rises to over 2,000m so the higher reaches offer a cooler, more humid and verdant environment than the hinterland. The sierra offers deciduous and coniferous woodlands and areas of Mediterranean scrub, with montane grasslands around the high tops, riparian growth along streamsides and some farmland at the lowest levels.

There is a visitors' centre at Narváez, with a nearby mirador giving panoramic views across the park.

Species
The typical avifaunas of woodlands and rocky, open country are present. Black-eared Wheatears are often obvious on dry, sunny hillsides, frequently alongside both Southern Grey and Woodchat Shrikes. Crested Larks are common at low levels in the farmland areas but are replaced higher up by Thekla Larks. Other typical species include the Iberian Green Woodpecker and the Carrion Crow, the latter having a southern outpost presence here. Characteristic raptors include Golden and Short-toed Eagles.

The local mammals include a large Ibex population, Genets, Beech Martens and a reintroduced population of Red Squirrels. The diverse butterflies include a rare subspecies of the Apollo.

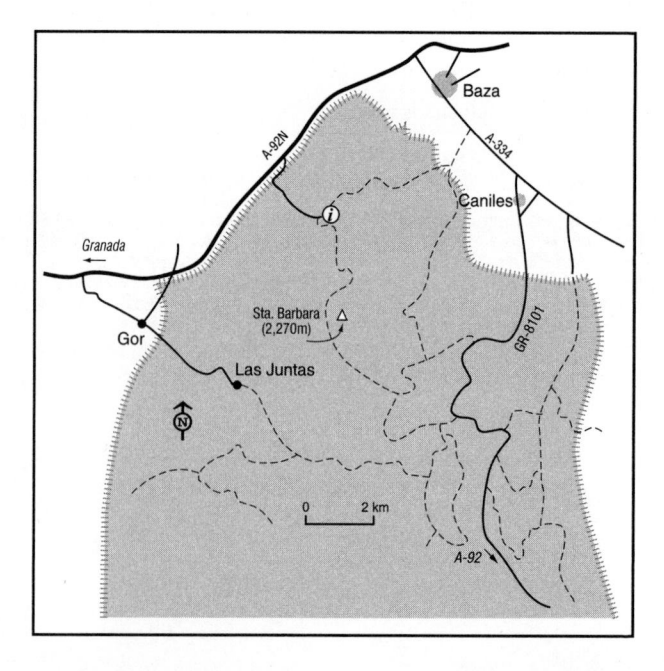

Timing
Worth visiting year-round although the upper levels may become inaccessible in snowy or icy conditions in winter. Bird diversity is highest in spring and early summer.

Access
A useful place to start is the visitors' centre, Cortijo Narváez. This is easily reached from the A-92 at km-28, where the Sierra de Baza is signposted. Follow signs from here to the centre, which is open daily year-round except on Mondays and Tuesdays. The tarmac access road to the centre climbs through pinewoods. An easy circular trail is available just before the centre, winding through aromatic scrub and pinewoods for 4.2km. The centre itself offers details and maps of several other walking trails.

The road forks at the centre and both options for onward travel involve gritty forest roads. These are mostly in good repair but a 4WD vehicle would be useful in winter especially. Both roads give access to a network of forest tracks that allow further exploration. For a representative transect though the area, crossing high open country, you can take the right fork and loop round to Las Juntas. The road from Las Juntas to Gor is good tarmac and offers alternative access to the region via Gor.

Another worthwhile option is the GR-8101, which crosses the east of the region, again through much open, elevated terrain that promises good birding. This is a good tarmac road that runs from between the A-334 near Caniles in the north and the A-92 near Abla (Almería) in the south.

♿ Many areas may be viewed from your vehicle.

CALENDAR
All year: Goshawk, Golden Eagle, Peregrine, Common Kestrel, Red-legged Partridge; Eagle, Little and Tawny Owls, Iberian Green Woodpecker, Crested and Thekla Larks, Wren, Black Redstart, Robin, Stonechat, Dartford Warbler, Firecrest, Southern Grey Shrike, Carrion Crow, Chaffinch, Serin, Greenfinch, Goldfinch, Linnet, Crossbill.

Breeding season: Short-toed Eagle, Common Cuckoo, Turtle Dove, Scops Owl, Hoopoe, Bee-eater, Black-eared Wheatear, Melodious and Bonelli's Warblers, Spotted Flycatcher, Woodchat Shrike.

BAZA BASIN (Hoya de Baza) GR9

Status: No special protection. ZEPA (80,000ha).

Site description
Another very large, flat, semi-arid sector of the lowlands to the north of the Sierra Nevada. Cereal cultivation and sheep pasture are the traditional land uses but the loss of important steppic terrain to planting of olive groves is an ongoing problem. The wilder and drier part, with less agriculture and much more low scrub and tamarisks as well as occasional watercourses, lies to the west of the A-4200.

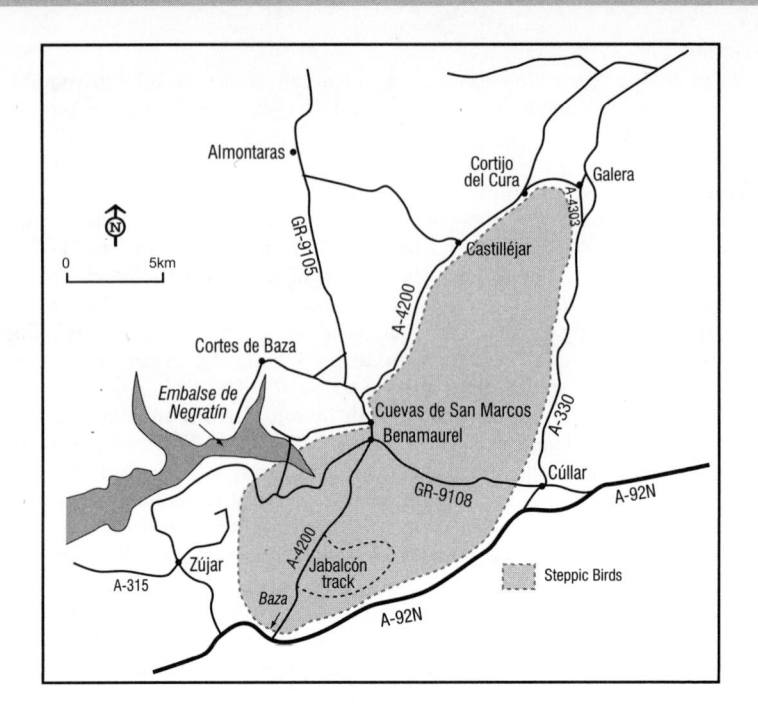

Species
The arid steppic areas hold such typical open-country species as Little Bustards, Black bellied Sandgrouse and Stone-curlews, as well as a few pairs of Montagu's Harriers. Tawny Pipits nest in small numbers. There is a diverse lark community although Dupont's Lark has declined very considerably and is hard to find here. Lesser Short-toed Larks are abundant, however, as are Calandra and Crested Larks. The Trumpeter Finch occurs but is uncommon. The tamarisks and areas with trees, such as along the Río Baza and other watercourses, offer Hoopoes, Rollers, Bee-eaters, Nightingales and Melodious Warblers. Rocky slopes have populations of both Black and Black-eared Wheatears. Common Magpies are generally widespread and attract the Great Spotted Cuckoos that parasitise them.

Timing
The best time is undoubtedly spring, before the end of April, when the cereals are not yet sufficiently high to conceal the bustards, the larks are highly vocal and the heat does not strike so hard before mid-morning. Visit preferably in the first three hours after dawn. Winter visits, when open-country birds are in flocks, may also be rewarding.

Access
The region can be viewed from the main roads, the A-4200 and A-330, which enclose the main steppic area, and from the many minor tracks across it. The A-4200 runs northwards from the A92N at km-43 near Baza; follow signs for Benamaurel. Once on the A-4200 there is a crossroads just before km-5 where

there are signs marking an irrigation channel: 'Canal de Jabalcón – camino de servicio'. It is worth turning east (right) and following the service track alongside the channel. Stop after some 2km where the land rises slightly to the right of the track and where there are some scattered tamarisks. These often hold some species of interest and allow you to scan across the area to the north from a slightly higher vantage point. Further on, the track rises and the land becomes positively arid 'badlands', with only low scrub by way of vegetation. These tracts are worth exploring on foot for some distance.

Further on the road descends to the Río Baza. Cross the river at a ford and bear northwards. The track eventually rejoins the A-4200 at km-11. The A-4200 runs alongside the river and *rambla* from Benamaurel to Castillejar and on towards Galera. There are many olive groves near the road at first but more interesting terrain predominates north of Castillejar. Bear eastwards (right) at the north end of the village of Castillo del Cura to take an unsignposted narrow tarmacked road that leads to Galera, following the west bank of the Río Baza. Cross the river at Galera. From Galera take the A-303 southwards through good, dry farmland and steppic terrain to join the A-336. There is more good habitat to explore all the way to Cúllar, with plenty of stopping opportunities and minor tracks to explore. The A-92 (km-60 junction) is just south of Cúllar.

♿ There are ample opportunities for watching from vehicles.

CALENDAR

All year: Little Bustard, Stone-curlew, Black-bellied Sandgrouse; Dupont's, Calandra, Lesser Short-toed Lark, Crested and Thekla Larks, Black Wheatear, Common Magpie, Raven, Trumpeter Finch, Corn Bunting.

Lark, Red-rumped Swallow, Yellow Wagtail, Tawny Pipit, Nightingale, Black-eared Wheatear, Northern Wheatear.

Winter: Hen Harrier, Merlin, Sky Lark, Meadow Pipit.

Breeding season: Montagu's Harrier, Great Spotted Cuckoo, Roller, Hoopoe, Bee-eater, Greater Short-toed

SUÁREZ WETLAND GR10
(Charca de Suárez)

Status: Local Nature Reserve (14ha).

Site description

A municipal Nature Reserve on the Granada coast, on the waterfront at Motril. There are several freshwater lagoons fringed with abundant reedbeds, canebrakes and dense shrubbery of willows, tamarisks and poplars. There are also sedgebeds, reedmace stands and residual clumps of sugarcane, this last formerly an important crop plant locally. The site is equipped with well-maintained footpaths and there are hides at strategic positions. The coastal location in a region that offers very few wetlands makes it a magnet for waterbirds year-round. There are plans to extend the site to the west, doubling the protected area and adding more lagoons. These plans were in abeyance in 2018 but there are hopes that they will implemented in due course.

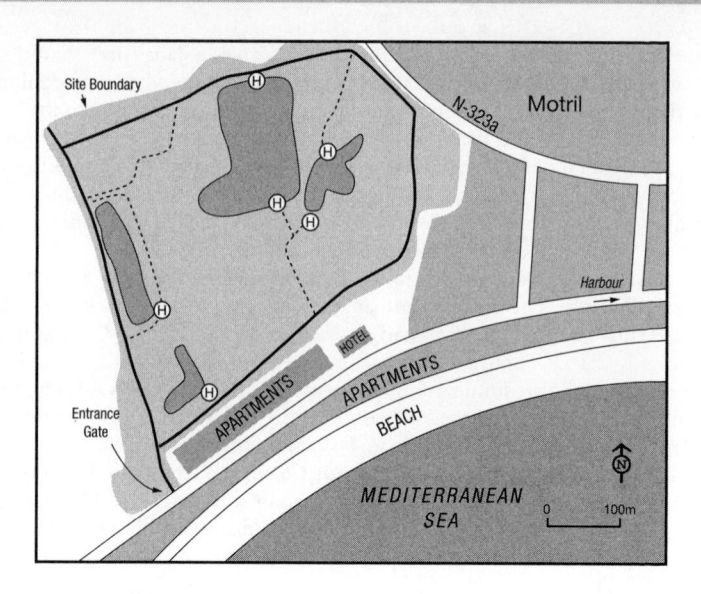

Species
Small numbers of a wide range of aquatic and waterside species are always present. The more notable breeding birds include the Purple Swamphen and Savi's Warbler, as well as a range of exotics. Red-knobbed Coots have been introduced here and nest regularly: four pairs bred in 2018. The site is potentially excellent during passage periods, when all the usual trans-Saharan migrants may be encountered. Such scarce migrants as the Spotted and Baillon's Crakes and even Allen's Gallinule have been recorded and the potential for encountering vagrant species is clearly high. The wintering community includes such typical waterside passerines as the Bluethroat and Penduline Tit, and the Moustached Warbler has been recorded here. The site is well worth visiting if you are passing by on the coast road and would make a rewarding local patch for anyone living close to it.

Access
This is free but is not entirely straightforward since visits during the breeding season (March–May) are only permitted if you are accompanied by a guide. At other periods the site is open at fixed times. Check their website (*http://www.motril.es/index.php?id=1343*) before a proposed visit to ensure that you can get in. Outside the breeding season the site is generally open all morning and for a couple of hours before sunset in the evening. The site gets busy at weekends, which are best avoided.

The Charca is behind the waterfront developments at the western beach (Playa del Oeste) in Motril. Follow signs from the N-340 to Motril harbour (puerto) via the N-323a. The Charca is 900m west of the harbour. Head towards the west beach but turn right to follow the street between the two rows of beachside buildings (Calle del Pelaillo) westwards. You will pass the

Elba Motril Hotel and then five apartment blocks on your right. The entrance gate to the reserve is immediately adjacent to the last of these blocks.

♿ Level and highly accessible.

CALENDAR

All year: Little Grebe, Cattle Egret, Mallard, Common Kestrel, Water Rail, Purple Swamphen, Common and Red-knobbed Coots, Moorhen, Little Ringed Plover, Audouin's Gull, Hoopoe, Monk Parakeet, White Wagtail, Cetti's Warbler, Zitting Cisticola, Tree Sparrow, Red Avadavat, Common Waxbill.

Breeding season: Little Bittern, Sand Martin, Red-rumped Swallow, Yellow Wagtail; Savi's, Reed and Great Reed Warblers.

Winter: Little Egret, Grey Heron, Pintail, Wigeon, Shoveler, Teal, Red-crested Pochard, Common Pochard, Marsh Harrier, Stone-curlew, Black-winged Stilt, Common and Jack Snipes, Kingfisher, Water Pipit, Bluethroat, Moustached Warbler, Common Chiffchaff, Penduline Tit, Reed Bunting.

Passage periods: Garganey, Night Heron, Squacco Heron, Purple Heron, Osprey, Spotted Crake, waders, chats, warblers.

JAÉN PROVINCE

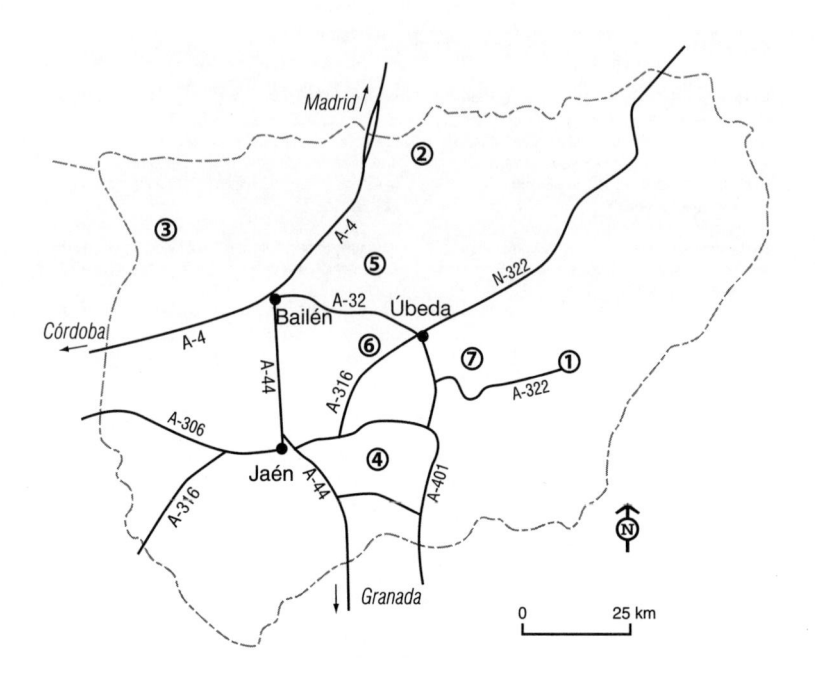

As commented in the Introduction, the wider fame of Jaén province is as the origin of some 40% of the global production of olive oil. There is no denying the enormous visual impact that the horizon-to-horizon olive groves have on the visitor, a negative first impression to be sure if you are looking for wildlife. However, the province is well worth visiting for the birds of its many mountains and their associated forests and woodlands. There are also some remaining areas of cereal and other cultivation that retain their birds of open country. Wetlands are very few but some of the reservoirs can be rewarding to visit. The chief wetland is the Guadalquivir river itself, which has its source at Cazorla and threads its way westwards across the province.

The province includes a large sector of the Sierra Morena, the long, fairly low range of wooded mountains that straddles the boundary between Andalucía in the south and Extremadura and Castilla-La Mancha in the north and which converges in the east with the Betic Mountains at Cazorla. The Sierra Morena (J2, J3) has vast tracts of thinly populated, unspoilt countryside, much of it given over to dehesas of Iberian Holm Oaks that are maintained as estates for 'caza mayor', the hunting of large quarry notably Red Deer and Wild Boar. These estates, together with the natural Mediterranean woodlands, expanses of scrub, riparian growth along rivers and streams and areas of rocky terrain, support very healthy populations of birds and other animals. The most renowned of these is the Iberian Lynx, whose resurgent population has a stronghold in the

Sierra de Andújar (J3). These hills are also excellent for seeing all five eagles on their breeding grounds: in particular, Jaén has by far the largest population of Spanish Imperial Eagles of any Andalucían province – 44 pairs nested in 2017, half of them in Andújar. The full range of the woodland and rock-loving species of Andalucía is available across the province. Highlights include a high density of Azure-winged Magpies, the presence locally of the White-rumped Swift and the successfully re-established Lammergeier population of Cazorla.

The Sierras of Cazorla and Segura (J1) are well worth a visit if only on account of the splendid mountain scenery, some of the grandest available in southern Spain. Lammergeiers apart, they accommodate many other raptor species as well as the communities of pine and mixed woodlands, and of subalpine landscapes on the high tops. Citril Finches are another local speciality there.

A different aspect is provided by the much more sparsely vegetated Sierra Mágina (J4), an island mountain rearing above the olive groves, where birds of scrub and open hillsides are more of a feature. Bonelli's and Golden Eagles are characteristic here, as well as Black Wheatears, Rock Sparrows and other species of rocky terrain. The arid landscapes to the east of the Sierra have some steppic species, including Little Bustards and Black-bellied Sandgrouse.

Waterbirds are best sought elsewhere but they are very far from absent. Two of the reservoirs of northern Jaén (J5) attract interesting birds year-round. Riparian growth along any of the rivers is popular with Nightingales and Golden Orioles, but there are often other less obvious birds to find here. It is always worth stopping at bridges, including those across the Guadalquivir, to see what may be around.

Sites in Jaén Province

J1	Sierras de Cazorla and Segura
J2	Despeñaperros and Aldeaquemada
J3	Sierra de Andújar
J4	Sierra Mágina
J5	Guadalén and Giribaile Reservoirs
J6	Laguna Grande de Baeza
J7	Upper Guadalquivir Reservoirs

Getting there

The A-4, Autovía de Andalucía, connects the province with Madrid to the north and Córdoba, Sevilla and Cádiz to the west. Take the A-44 Autovía de Sierra Nevada exit near Bailén, for Jaén city and to reach the A-92 at Granada. The A-92 provides rapid transit east/west across lower Andalucía. The N-322 gives access from the A-32 east of Bailén to Cazorla and the northeast of the province and thence to Albacete and beyond.

Where to stay

There are reasonably good hotel, lodging and camping facilities, especially centred on the Cazorla region. There much of the accommodation is full between April and September and so should be booked well in advance. The campsites in particular are usually packed over the Easter period and in July and August. There are Paradores at Cazorla itself and also in Jaén city and Úbeda.

There are also small hotels or pensiones in most of the smaller towns, such as Andújar.

SIERRAS DE CAZORLA AND SEGURA J1

Status: Parque natural 'Sierras de Cazorla, Segura y Las Villas' (214,300ha). ZEPA. UNESCO Biosphere Reserve.

Site description
This scenic Natural Park is where two of Spain's larger rivers, the Guadalquivir and Segura, have their sources. The Guadalquivir first flows northeast from the Sierra de Cazorla but is diverted westwards by the Sierra de Segura, to meander through Córdoba and Sevilla provinces to the Atlantic. The nascent Guadalquivir is fed by innumerable rivulets and streams that descend steep-sided, forested valleys and ravines on the flanks of four sierras: Cazorla, Segura, del Pozo and de la Cabrilla. The highest summits exceed 2,000m. The Segura river has its source on the eastern slopes and flows eastwards through southern Albacete and across Murcia to its Mediterranean estuary in southern Alicante.

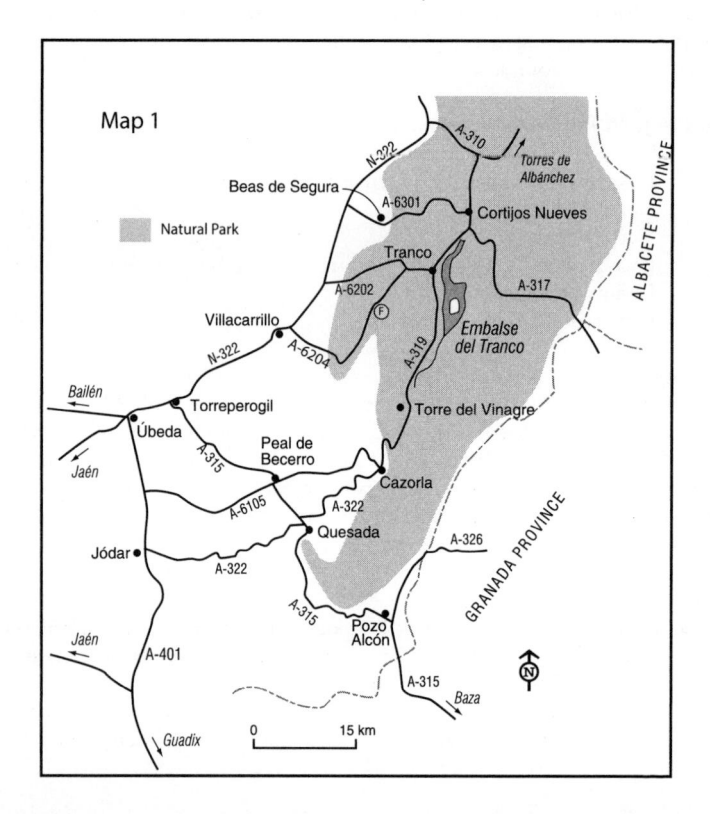

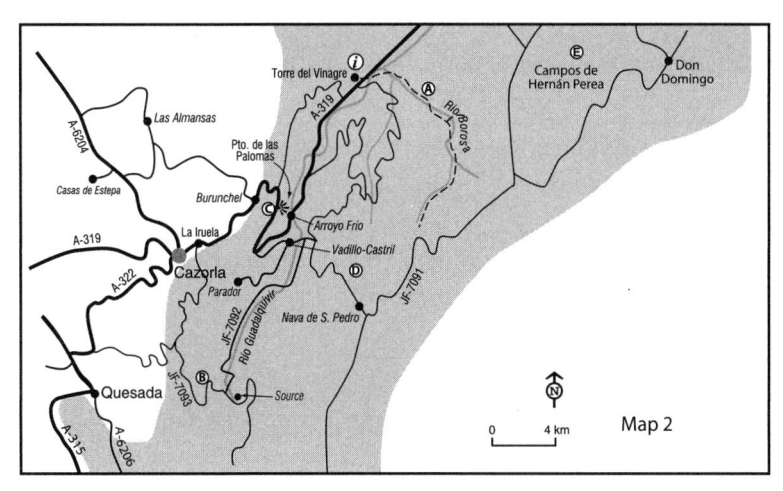

These sierras form a significant sector of the Betic Cordillera, protected by what is the largest Parque Natural in Spain. The limestone terrain includes, as ever, numerous gorges and cliffs. Four small reservoirs lie within the park. The Mediterranean forest – of Strawberry Trees, Lentiscs and diverse shrubby evergreens – that once dominated the lower elevations persists in some areas but elsewhere has given way to pine plantations. Extensive natural forests of Black, Aleppo and Maritime Pines remain as well as mixed woodlands of Holm Oaks and wild olives. Dwarf Juniper scrub is typical of the highest elevations. The varied and diverse flora, over 2,300 species, includes at least 34 endemics, such as the Cazorla Violet. There is also an endemic reptile, Valverde's Lizard *Algyroides marchi*.

The climate is very variable. It rains frequently in spring and into early summer. Winter precipitation is typically very heavy and falls as snow at higher elevations, where roads may be blocked as late as the end of April. Frosts are frequent in winter and even as late as early May. In contrast, it is often extremely hot and dry in summer, the heat sometimes triggering violent storms, with accompanying thunder and lightning, which often cause forest fires.

The Tourist Office in the village of Cazorla (http://www.turismoencazorla.com/) supplies the Guía Práctica, a very useful combined information sheet and road map. There are also two visitors' centres: in the south at 'Torre del Vinagre' at km-18 on the A-319, and in the north at Río Borosa near Torres de Albánchez.

Species
The variety and altitudinal range of habitats make for high biodiversity, birds included, and it is possible to find the great majority of the woodland, forest and rock-loving species of Andalucía here, as well as some open-country species. The star attraction birdwise is the Lammergeier, the famous *Quebrantahuesos* or 'bone-breaker'. Cazorla hosted the last breeding birds in Andalucía until the mid-1980s, when they too ceased to breed. Thereafter, wandering immature birds, presumably from the Pyrenean population, still visited the area occasionally. An ongoing Lammergeier reintroduction project, involving captive breeding, began in 1996. The first three youngsters were released in 2006. The

project has proved successful (see *www.gypaetus.org/* for further news). The first pair of released birds was formed in 2013 and first nested successfully in 2015. In 2017 two pairs each fledged one juvenile.

Other nesting raptors include numerous Griffon and a few Egyptian Vultures, Black and Red Kites; Short-toed, Golden, Booted and Bonelli's Eagles, Goshawks, Sparrowhawks, Peregrines and Common Kestrels, with Lesser Kestrels in some of the villages. Black Vultures are seen occasionally. Spanish Imperial Eagles, mainly immature birds, appear increasingly often and seem likely to colonise the region. Eleonora's Falcons sometimes appear in summer.

The forests and woodlands support Turtle Doves, Orphean and Bonelli's Warblers, Firecrests, Crested Tits, Nuthatches, Short-toed Treecreepers and Crossbills, among others. Dense growth along the many watercourses holds abundant Nightingales, and the riparian woodlands and poplar plantations attract Golden Orioles. Lesser Spotted Woodpeckers also occur with some regularity here. The breeding birds of more open areas and scrub include Wrynecks, Iberian Green Woodpeckers, Melodious Warblers and Spectacled Warblers, and are to be found at the lower levels. The open pastures hold Quail, Red-legged Partridges, Red-billed Choughs, Carrion Crows, Cirl Buntings and Rock Buntings, with Citril Finches (elusive) in the fringing Black Pine forests at high elevations, Northern Wheatears and Rufous-tailed Rock Thrushes also breed on the high tops where Ring Ouzels and other thrushes are attracted to the juniper scrub in autumn. Alpine Accentors occur on passage and in winter.

Waterbirds find little habitat in the park but the Embalse del Tranco supports small numbers of waterfowl, chiefly Mallard, Black-necked Grebes and Little Grebes. Great Crested Grebes also occur occasionally in winter.

Cazorla has high densities of mammals, many of which have become remarkably tame in the absence of hunting and through frequent exposure to visitors. Red Deer and Fallow Deer (introduced) are very common and rocky escarpments have Spanish Ibex. Mouflon (also introduced) are common on the high tops. Wild Boar are abundant but more elusive. Red Squirrels are often obvious in the pinewoods and you may well meet some extremely tame Red Foxes, which approach visitors for food at regular parking places.

Timing

The area attracts huge numbers of visitors at weekends from Easter onwards. In July and (especially) August the park is very busy indeed with humans and their vehicles. Many visitors end up in the central (Guadalquivir) valley at Arroyo Frío resort village, where there are numerous hotels and restaurants. A small but increasing minority range more widely along the walking trails and minor roads. In general, weekday visits between mid-April and June especially, and also from September to early November, are recommended. Winter visits may be worthwhile but snowfall will determine where access is feasible.

Access

The most useful point of entry is through Cazorla village itself, although this is subject to considerable traffic jams at weekends and during holiday periods. The sierras are criss-crossed by a network of minor roads and forest tracks, many of which are accessible to the public. Some of the best are 'dirt' roads but these are kept in excellent repair and normally present no problems to ordinary vehicles, at least in dry weather. Viewpoints (miradors) are widespread and all are worth a visit. The visitors' centre at Torre del Vinagre provides additional details of recommended itineraries, including walking trails. They can also provide details of guided tours, which are a good way of reaching the best and less accessible areas. The following are recommended as an introduction to the park.

♿ Much worthwhile birding is possible from or near your vehicle.

A *Río Borosa valley* Take the minor road southeast from Torre del Vinagre to the nearby fish farm. Continue upriver on foot. A bridge crosses the river after 3.5km. Follow signs to the Cerrada de Eliás and continue for another 3.5km to the hydroelectric station at Salto de los Órganos. The trail passes through a rocky gorge, with further cliff habitat upstream from the hydroelectric station. The whole route offers opportunities to see riparian species, such as Dippers, and cliff species, including Griffon Vultures and Red-billed Choughs, among many others. Return along the same route.

B *Source of the Río Guadalquivir* Take the JF-7092 south from Vadillo-Castril and turn right (west) after 14km on to the JF-7093. The road is unsurfaced after passing a campsite but is easily drivable. The source (Cañada de las Fuentes) is signposted on the left but the main attraction is the road itself, or rather the habitats which it traverses, which offer typical woodland and rocky terrain species. The JF-7093 forks after 12km, where you may bear right for Cazorla village (16km) or left for Iruela (12km). The fork is in rocky terrain above a sheer gorge whose opposite cliff face houses a Griffon Vulture colony. A mirador overlooks the gorge. Other rock-loving species, such as Black Wheatears, also occur here.

C *Puerto de las Palomas (1,290m)* A panoramic viewpoint on the A-319 about 12km north of Cazorla village. Well placed to scan for raptors. The public viewpoint may get crowded so it is often preferable to view from the col a few hundred metres further north, which offers all-round visibility as well as access to rocky terrain and woodland.

D *La Nava de San Pedro* A grassy upland valley surrounded by rocky outcrops. A stream course has riparian woodland and poplar plantations. A broad, level footpath (Sendero Valdetrillo) follows this stream through excellent pinewoods; the path entrance is on the left just beyond the parking area. The few buildings at La Nava include a couple of bar-restaurants. This is also the location of the Lammergeier rearing facility (Centro de cría del Quebrantahuesos), the core site of the reintroduction programme. Guided visits to the breeding centre are possible Thursday to Sunday during June–September, by prior arrangment (email *visitasccq@gypaetus.org*). Access to La Nava is via the JF-7091, which becomes an excellent dirt road after its junction with the JF-7092. La Nava is 6.9km from the fork.

E *Campos de Hernán Perea* An immense open upland area on the Sierra de Segura, extending over some 5,000ha at a mean altitude of 1,600–1,700m. The karstic landscape includes pastures grazed by sheep, both domestic and the introduced Mouflon. The vegetation includes stands of hawthorn and junipers, attractive to Ring Ouzels and other thrushes in autumn. The upland species of the region, including the Tawny Pipit, Alpine Accentor (in winter) and Rufous-tailed Rock Thrush, are best sought here, and Citril Finches occur in the fringing pinewoods. This is perhaps the best site to locate foraging Lammergeiers, as well as other open-country raptors. Eleonora's Falcons are sometimes present in summer, prior to settling at their Mediterranean nesting sites. Access is via the unsurfaced road (JF-7091) that links La Nava de San Pedro (see D) in the south with Don Domingo in the north. Stop frequently along this road or explore more widely along the various tracks.

F *Sierra de Las Villas* **(see Map 1)** A quiet road through the Sierra de Las Villas in the west of the park gives access to rugged terrain attractive to Griffon and Egyptian Vultures, Golden and Bonelli's Eagles and Alpine Swifts. Lammergeiers are a possibility here. The road crosses varied terrain including pinewoods and skirts a small reservoir. Woodland raptors and diverse passerines will also be encountered. Access is via the A-6204 just north of Villacarrillo on the N-322 or from the A-6202 west of Tranco.

CALENDAR

All year: Mallard, Red-legged Partridge, Little and Black-necked Grebes, Lammergeier, Griffon Vulture, Common Buzzard, Golden and Bonelli's Eagles, Goshawk, Sparrowhawk, Peregrine, Common Kestrel, Rock Dove, Woodpigeon; Eagle, Little and Tawny Owls; Iberian Green, Great Spotted and Lesser Spotted Woodpeckers, Crag Martin, Woodlark, Grey Wagtail, Dipper, Wren, Black Redstart, Black Wheatear, Stonechat, Blue Rock Thrush, Mistle Thrush, Dartford and Sardinian Warblers, Blackcap, Firecrest, Crested, Coal, Blue and Great Tits, Nuthatch, Short-toed Treecreeper, Jay, Common Magpie, Red-billed Chough, Jackdaw, Carrion Crow, Raven, Chaffinch, Serin, Citril Finch, Linnet, Crossbill, Cirl and Rock Buntings.

Breeding season: Quail, Red and Black Kites, Egyptian Vulture, Short-toed and Booted Eagles, Turtle Dove, Common Cuckoo, Scops Owl, Pallid and Alpine Swifts, Bee-eater, Hoopoe, Wryneck, Barn and Red-rumped Swallows, Nightingale, Rufous-tailed Rock Thrush; Melodious, Spectacled, Subalpine, Orphean and Bonelli's Warblers, Spotted Flycatcher, Golden Oriole, Woodchat Shrike.

Passage and non-breeding: Honey-buzzard, Black Vulture, Spanish Imperial Eagle, Grey Heron, Common Redstart, Pied Flycatcher.

Winter: Great Crested Grebe, Teal, Alpine Accentor, Fieldfare, Redwing, Common Chiffchaff, Yellowhammer.

DESPEÑAPERROS and ALDEAQUEMADA J2

Status: Parque Natural (7,502ha). ZEPA. Includes the Paraje Natural, Cascada de La Cimbarra (534ha). ZEPA.

Site description
The gorge of the Despeñaperros river offers easy transit between the hills of the Sierra Morena and the plains of Castilla-La Mancha. It accommodates the A-4 Autovía de Andalucía, the motorway linking the south coast to Madrid. The

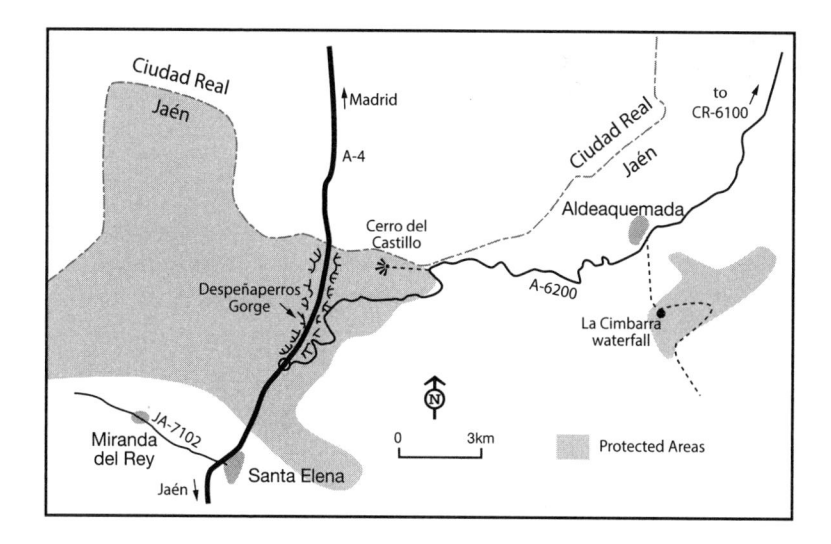

road is partly hidden by tunnels and is usually at least fairly busy. The Natural Park straddles the motorway but there is little scope (or need) for stopping within the gorge itself. Most of the protected area lies west of the gorge but this part is relatively inaccessible. Much of the interest is east of the gorge and includes the hinterland of Aldeaquemada village, outside the park proper. There is an information centre (Puerta de Andalucía) at Santa Elena, off the motorway at km-27, where fuller details of the various walking trails may be obtained.

This sector of the Sierra Morena is comparatively rugged, with cliff faces and rocky outcrops that attract rupestral species such as Griffon Vultures. The vegetation is varied, with mixed woodlands of Iberian Holm, Lusitanian and Cork Oaks, pine plantations and expanses of Mediterranean scrub, with riparian woodland along the streams and rivers. To the east the landscape is quite deeply eroded within the general area of Aldeaquemada, with ravines and waterfalls along the course of the Río Guarrizas, but is rather more gentle elsewhere. The most spectacular waterfall is the Cascada de la Cimbarra, which tumbles some 40m into a deep pool.

Species

Nesting raptors include Griffon Vultures and one or two pairs of Black Vultures, all five eagles, Common Buzzards, Peregrines and Common Kestrels. Other raptors, such as Black Kites, occur on passage. Black Storks nest within the area and there are breeding Eagle, Little and Scops Owls. Alpine Swifts are common around the higher, more rugged areas, as well as both Pallid and Common Swifts. The White-rumped Swift occurs and probably nests locally. Blue Rock Thrushes are characteristic and the more elusive Rufous-tailed Rock Thrushes also nest at the higher levels. The clear waters of the rivers host Dippers, Kingfishers and Grey Wagtails. Nightingales are common along any watercourses in spring.

The diversity of woodland species includes Short-toed Treecreepers and Nuthatches, Azure-winged Magpies and an abundance of tits and warblers, these last including Spectacled and Subalpine Warblers. Southern Grey and

Woodchat Shrikes are not uncommon. Crossbills breed and both Siskins and a few Bullfinches occur in winter. Rock Buntings are common.

The area includes Red Deer hunting estates and the bellows of rutting stags are unmissable in autumn. Wild Boar are also numerous. More elusive mammals include Otters and Iberian Lynxes.

Timing
Species diversity is greatest in spring and early summer, between late April and July, but there are always interesting birds to see. Summer days can be very hot but winters are sometimes cold, with frost or snow on occasion.

Access
The whole area is easily reached from the A-4 motorway. The eastern sector (B and C below) is especially recommended for an initial or single visit.

♿ Much worthwhile birding is possible from or near your vehicle at B and C.

A *West of the Pass*   Leave the A-4 at km-258 for Santa Elena. Take the underpass under the motorway to follow the JA-7102 to Miranda del Rey. Two long circular trails (around 14km) through southwestern sectors of the path are signposted from the village, but it is possible of course to visit just parts of these. The trails are rough and best followed on foot. They pass through woodlands, large expanses of Cistus scrub and other representative habitats. The reserve booklet obtainable at the Santa Elena information centre is recommended for trail details. It may also be downloaded from the Ventana del Visitante website (see p. 23).

B *East of the Pass. Aldeaquemada*   Recommended access is on to the A-6200 for Aldeaquemada, at km-250 just south of the Despeñaperros pass. This excellent quiet but winding road passes through representative habitats, crossing the Despeñaperros river and then climbing through rocky terrain before descending through pinewoods to Aldeaquemada village. Stop frequently to scan for raptors and search the woodlands and rocky escarpments. An excellent vantage point is the Cerro del Castillo, reached by a short uphill walk through Holm Oak woodland from a car park at km-8: there are spectacular views along the Sierra Morena and of the Despeñaperros gorge, where Griffons nest.

C *La Cimbarra waterfall*   A short (3 km) drive south of Aldeaquemada village, where it is clearly signposted (Cascada de La Cimbarra). The sandy but good track follows a stream to a parking place near the waterfall. Two signposted trails both lead from the car park to a rocky platform overlooking the waterfall and its environs, from where it is possible to follow a steep descent to the foot of the waterfall. The track continues left past the car park, descending to the river where there is a picnic site (Arroyo de Martín Pérez), from where a footpath gives access to the riverine and hillside vegetation. You can drive down or walk the 800m from the car park.

Returning from, or access to, Aldeaquemada – avoiding all the bendy roads – is quickest to/from the north, on the A-6200/CR-6100 via Castellar de Santiago, which is linked to the A-4 by the CM-3200.

CALENDAR

All year: Black and Griffon Vultures, Common Buzzard; Spanish Imperial, Golden and Bonelli's Eagles, Goshawk, Sparrowhawk, Peregrine; Eagle, Little and Tawny Owls, Kingfisher, Crag Martin, Iberian Green and Great Spotted Woodpeckers, Thekla and Crested Larks, Grey Wagtail, Dipper, Wren, Dunnock, Robin, Black Redstart, Stonechat, Black Wheatear, Blue Rock Thrush, Mistle Thrush, Dartford and Sardinian Warblers, Blackcap, Firecrest; Long-tailed, Crested, Coal, Blue and Great Tits, Short-toed Treecreeper, Southern Grey Shrike, Jay, Azure-winged Magpie, Common Magpie, Red-billed Chough, Raven, Chaffinch, Serin, Greenfinch, Goldfinch, Linnet, Crossbill, Hawfinch, Rock Bunting.

Breeding season: Short-toed and Booted Eagles, Black Stork, Common Cuckoo, Scops Owl; Alpine, Common, Pallid and White-rumped Swifts, Bee-eater, Roller, Hoopoe, Red-rumped Swallow, Crag and House Martins, Woodlark, Nightingale, Rufous-tailed Rock Thrush; Melodious, Spectacled, Subalpine and Bonelli's Warblers, Golden Oriole, Woodchat Shrike.

Winter: Ring Ouzel, Redwing, Fieldfare, Siskin, Bullfinch.

Passage periods: Raptors, including Black Kite. Spotted and Pied Flycatchers, Common Redstart.

SIERRA DE ANDÚJAR   J3

Status: Parque Natural (74,052ha). ZEPA.

Site description
A 70km stretch of the central Sierra Morena, contiguous with the Parque Natural de Cardeña y Montoro (CO10). The highest point is about 1,290m above sea level. The park offers some of the most heavily wooded areas of the whole of the Sierra Morena, dominated by Cork, Lusitanian and Iberian Holm Oaks, and Stone Pines, as well excellent examples of Mediterranean scrub and stretches of riparian woodland, notably along the Río de las Yeguas. This is in fact one of the best-preserved surviving expanses of Mediterranean forest in Spain. Rocky outcrops and several reservoirs provide further habitat diversity. Large estates devoted to Red Deer hunting cover great expanses.

The visitors' centre (Viñas de Peñallana), on the A-6177 Andújar/Santuario road at km-13, is open all day Friday to Sunday and on Thursday mornings.

Species
This area is of great importance for its diversity of breeding raptors, including all five eagles. It is a particular stronghold of the Spanish Imperial Eagle, which has a thriving and expanding population here: 21 pairs nested in 2017. Other breeding raptors include Black, Griffon and Egyptian Vultures, Common Buzzards and both Red and Black Kites. All these raptors are especially evident when displaying over their territories in spring. There are also Scops, Eagle, Little and Tawny Owls. Other noteworthy species include Black Storks, which are often seen along watercourses and at reservoirs. White-rumped Swifts are present and also nest locally: they should be looked for among aerial feeders.

The species list for the woodlands is very similar to that of the adjacent Sierra de Cardeña y Montoro (CO10). The common chats and thrushes are all present and are joined in winter by considerable numbers of Song Thrushes and the much scarcer Redwings. There is a full range of warblers in the woodlands. Golden Orioles are common along river courses in spring and summer. Both the common shrikes occur. Corvids include large numbers of Azure-winged Magpies.

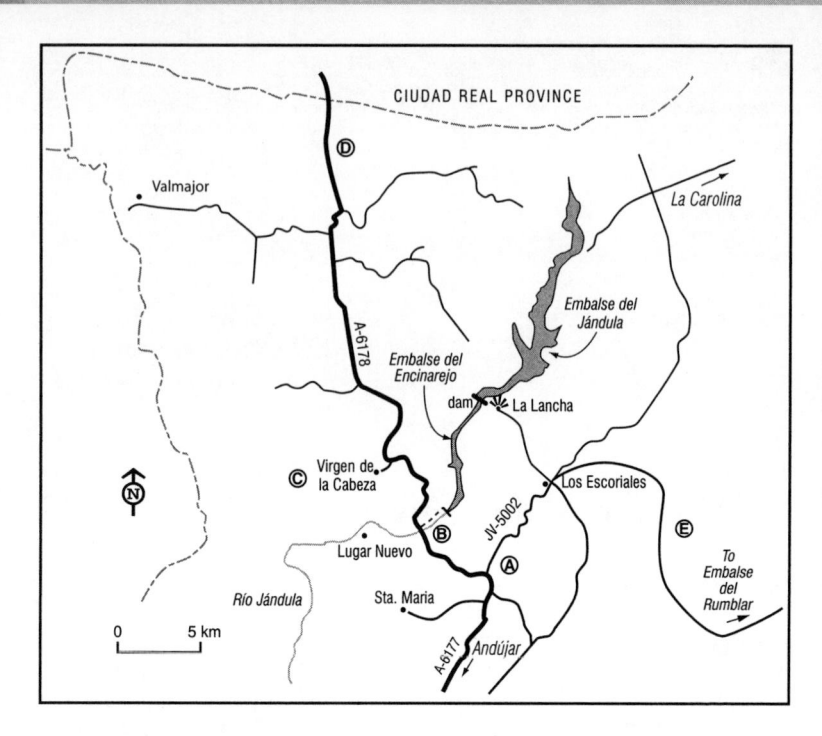

Birds apart, the Sierra de Andújar is best known as a key site for the Iberian Lynx. The national population of this endangered species has been boosted by captive breeding and now numbers over 500 individuals, whose stronghold is in the Sierra Morena. The Sierra has a healthy and increasing population and offers a good chance of encountering a lynx in the wild. Chattering Magpies often provide an indication of lynx presence. Nevertheless, your chances of seeing one will be greatly boosted by employing a local guide: I recommend Iberus (*www.iberusbirdingnature.com*). Red Deer are extremely obvious in the dehesas, where there are also Fallow and Roe Deer and Wild Boar. Other mammals in the region include Mouflon (introduced), Spanish Ibexes, Wild Cats, Mongooses and Otters. Wolves retain a precarious presence in the remote north and there is talk of boosting their numbers in the Sierra Morena.

Timing
Worth visiting year-round but at its best in spring and early summer, between early March and July. Mammals, including lynx, are most easily seen in autumn (September–December), when ground cover is minimal. It can get very cold in winter and considerable snowfalls are not unknown. Sites along the A-6177 are often very busy at weekends, when Sundays especially are best avoided.

Access
The A-6177, which traverses the park north/south, is easiest to access from the A-4 at Andújar in the south. The Santuario of La Virgen de la Cabeza is

signposted from the motorway: follow the brown signs to the Santuario to skirt Andújar town and continue northwards. Recommended areas to visit include:

A *The road to the Embalse del Jándula* Turn eastwards off the A-6177 at km-14.8. The narrow, winding JV-5002 is signposted Pantano del Jándula. Stop at the first col to scan for raptors. After 10km you reach a farm, Los Escoriales, at the end of the tarmac road. Turn left here on to a sandy road with numerous shallow potholes, which descends slowly to the dam of the very isolated Jándula reservoir. The road is signposted Mirador del Embalse del Jándula. The whole of this road is excellent for raptors and birds of the dehesa and rocky habitats generally. The Mirador del Rey, overlooking the reservoir before the final descent to the dam, gives excellent views. Red Deer are often obvious from here and if fortunate you may see Wild Boar or lynx, especially around dawn and dusk.

B *Río Jándula and Embalse del Encinarejo* This well-known site for lynx sightings attracts many visitors looking for the cats but many of the birds of Andújar are also evident here. Access is signposted at km-22 on the A-6177 where, if northbound, a roundabout precedes a metal bridge across the Jándula river. Cross the bridge and turn left onto a broad sandy road that leads to the dam of the Encinarejo reservoir. There are picnic facilities all along this road and it is possible to bathe in the river. Accordingly, this area attracts many people in summer, especially at weekends, when it is best avoided. At other times the road is quiet and allows access to tall riparian woodland and expanses of riverside boulders, with dehesas beyond. An excellent spot to stop is where the road rises and there is a wooden crash barrier: watch for raptors, including Spanish Imperial Eagles, and scan the dehesas for deer and lynx from here. Kingfishers and Otters are also often present.

C *Santuario de la Virgen de la Cabeza* The sanctuary is on a conspicuous rocky prominence some 30km north of Andújar town. Spectacular views are available from here on a clear day. However, avoid the whole area in the week or ten days before and for a few days after the last Sunday in April when there is the *romería* – the religious procession – of the Virgen de la Cabeza, when the place swarms with hundreds of thousands of visitors, making birding and driving quite impossible. The whole A-6177 is also used as an endurance cycling track for competitive events, often on Sundays, when long delays are inevitable.

D *North of the Santuario* The A-6177 narrows after the Santuario and becomes the still excellent A-6188, which continues northwards to the provincial boundary with Ciudad Real and beyond. The road rises and winds through dehesas to reach more open country, with expanses of heath and Cistus scrub. Stop frequently at likely areas, including at two roadside miradors that offer panoramic views over open country. Farm roads and signposted footpaths off this road provide further opportunities for wider exploration. A recommended such diversion is the good dirt road signposted at km-17.5 to the Mirador Valmayor and its adjacent footpath (Sendero El Junquillo); it traverses some 15km of excellent, savanna-like dehesas and open country. Diverse raptors, again including Spanish Imperial Eagles, may be expected as well as birds of woodland, scrub and open country. It is also worth continuing on the A-6178 as far as the provincial boundary, where the road crosses an open valley often teeming with Red Deer. Traffic is very light.

E *Embalse del Rumblar to Los Escoriales* The Rumblar reservoir, southeast of the park proper, is easily reached by leaving the A-4 at the Baños de la Encina exit (km-289). Take the bypass on the left just below the village, which soon reaches a crossroads where the Embalse is signposted. This road (JH-5044) leads through open rocky country to the dam, where you should stop to scan for Black Storks and birds of rocky terrain. Beyond the dam the road becomes rather rough dirt, traversing the dehesas of hunting estates inhabited by numerous Red and Fallow Deer. Birds of the dehesa and open country are also abundant here and the many rabbits and partridges help to account for the considerable lynx population of the area. Visits at dawn and dusk increase the likelihood of lynx sightings. Raptors are common and this area is a local stronghold of both Spanish Imperial and Golden Eagles. Driving as far as Los Escoriales is straightforward and crosses excellent habitat (and allows for a return on tarmac – see A above – but you will probably prefer to retrace your journey). The only vehicles are farm traffic and, increasingly, wildlife watchers and photographers.

Much worthwhile birding is possible from or near your vehicle.

CALENDAR

All year: Black and Griffon Vultures, Red Kite, Goshawk, Sparrowhawk, Common Buzzard; Spanish Imperial, Golden and Bonelli's Eagles, Peregrine, Common Kestrel, Eagle and Tawny Owls, Kingfisher, Iberian Green Woodpecker, Great Spotted Woodpecker; Crested, Thekla and Wood Larks, Grey Wagtail, Crag Martin, Robin, Black Redstart, Stonechat, Blue Rock and Mistle Thrushes, Dartford and Sardinian Warblers, Blackcap, Firecrest, Nuthatch, Short-toed Treecreeper, Southern Grey Shrike, Azure-winged Magpie, Common Magpie, Red-billed Chough, Raven, Hawfinch, Crossbill; Cirl, Rock and Corn Buntings.

Breeding season: Black Stork, Egyptian Vulture, Black Kite, Short-toed and Booted Eagles, Red-necked Nightjar, Alpine and White-rumped Swifts, Bee-eater, Roller, Hoopoe, Wryneck, Red-rumped Swallow, Nightingale, Subalpine Warbler, Spotted Flycatcher, Golden Oriole, Woodchat Shrike.

Winter: Meadow Pipit, White Wagtail, Song Thrush, Redwing, Fieldfare, Brambling, Yellowhammer.

Passage periods: Honey-buzzard, Black Kite, hirundines, other passerines including Common Redstart, Whinchat, Northern Wheatear, Willow Warbler and Pied Flycatcher.

SIERRA MÁGINA J4

Status: Parque Natural (19,957ha). ZEPA.

Site description
One of the typical sierras of southern Jaén, distinguished by the highest peak in the province, Pico Mágina (2,167m). The massif is visible from many miles around, rising from the lowland ocean of olive groves. The rugged limestone terrain includes numerous cliffs. The lower reaches, up to about 1,300m, have woods of encinas and Lusitanian Oaks, with a diverse shrub understorey, and

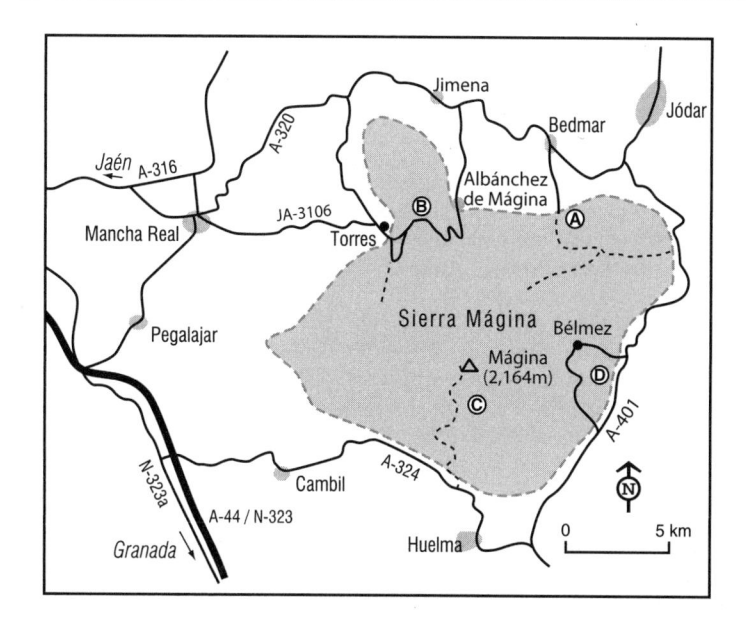

expanses of scrub and pasture on the slopes. Riparian vegetation includes stretches of poplars along the streams. Higher up, there are woodlands of Aleppo and Black Pines, with juniper and broom scrub, montane pastures and rocky outcrops at the uppermost levels. The whole area is of high botanical interest and supports several endemic species, of plants and invertebrates especially.

There is a visitors' centre, 'Castillo de Jódar', in Calle Alhori in the village of Jódar.

Species
The Sierra offers opportunities to see a broad range of Mediterranean bird species, including many of those characteristic of high elevations. Breeding raptors include Bonelli's, Golden and Short-toed Eagles, as well as Griffon Vultures and Peregrines, among others. The resident Eagle, Barn, Tawny and Little Owls are joined by Scops Owls in the breeding season. Other characteristic species, at the lower and intermediate levels, include Bee-eaters and both Southern Grey and Woodchat Shrikes. Golden Orioles occur along the wooded valleys where there are poplars. The ravine and cliff areas support Black Redstarts, Black Wheatears, Blue Rock Thrushes and Red-billed Choughs, with Northern Wheatears and Rufous-tailed Rock Thrushes at the higher elevations. Ring Ouzels occur regularly in winter. Rock Sparrows and Rock Buntings are common. The Spanish Ibex has a good population here.

Timing
Spring visits are recommended but the area can be rewarding at any time of year. There is considerable rainfall at the higher levels particularly, except in summer. Even then, the high tops may be cold and some warm clothing is advisable. Winters may be very cold with snow on the high tops and ice on the roads.

Access

The park is immediately east-southeast of Jaén city. The A-44 motorway skirts the western boundary. The northern sector may be reached by taking the A-316 eastwards from Jaén for 11km before turning right on to the A-320 for Mancha Real and then continuing towards Torres and Albánchez de Mágina on the local roads. The A-324, which also connects with the A-44, gives access to the southern parts of the park. Roads are mainly peripheral but provide useful access to parts of the lower and middle levels. Access to the summit needs to be on foot, along marked trails; C below is the best of these for birding purposes.

A *Bedmar and the Ermita de Cuadros* Bedmar village is off the A-320 to the north of the Sierra Mágina. It lies at the foot of the barren and imposing Serrezuela de Bedmar, which is worth scanning for raptors and choughs. The grassy slopes along the A-320 east of Bedmar occasionally attract Black-bellied Sandgrouse. A road signposted from the bypass below the village leads to and beyond the hermitage, to a picnic site on the Río Cuadros overlooked by a historic watchtower. The road becomes sandy dirt beyond this point. The riparian growth, which includes some impressive figs as well as poplars, elms and oleanders, attracts warblers and other passerines. A footpath follows the river upstream from the picnic site but it is simpler to walk up or drive up the road for some distance. This road, and two other footpaths, allow wider exploration. Watch for raptors overhead and for Red-billed Choughs, Blue Rock Thrushes and other rupicolous species on the escarpments.

B *Albánchez to Torres* This tarmac road cuts across the northwestern corner of the park and climbs to a scenic col (1,200m) above the picturesque village of Albanchez de Mágina. There are far-reaching views from the village (especially if you ascend the 365 steps to the castle) and around the col, where you can scan for raptors, including Golden Eagles, which nest nearby, as well as Ibex. Fruiting hawthorns, wild roses and junipers at the higher levels attract thrushes in autumn and winter. The road passes through encina woodland and expanses of grassy terrain, rocky terraces and low scrub. Follow signs from the A-320 to Albanchez just east of Jimena or to Torres on the JA-3106 from Mancha Real.

C *Pico Mágina trail* This footpath gives access to the summit of the sierra (2,167m). The climb is over 1,000m and is moderately difficult but within the capabilities of persons of reasonable fitness who are properly equipped for mountain walking. Ascent in poor weather or winter is inadvisable. The distance to the top is 11km, returning by the same route: allow at least eight hours there and back. Apart from the magnificent views, the rewards of the climb will include the birds and flora of the sunny hillsides and high tops. Raptors include Short-toed, Bonelli's and Golden Eagles. Typical birds at higher levels include Northern Wheatears and Rufous-tailed Rock Thrushes, with Ring Ouzels regular in autumn especially. Alpine Accentors occur in winter.

The entrance to the trail is on the south side of the Sierra, at km-6.5 on the A-324 between Huelma and Cambil. The entrance is not well marked but a white house with a green roof is a useful indicator. Drive towards the house and soon bear left to follow the broad track for 1km to a car park, where the trail is signposted 'Subida al Pico Mágina y Miramundos'.

D *The eastern flank of Sierra Mágina* This is accessible by driving the A-401, between Jódar/Belmar and Huelma, with a diversion into the Sierra foothills at Bélmez. Stop where conditions permit and the habitat looks interesting: where there are rocky escarpments, natural woodlands and open areas. Olive groves predominate in many areas but the less managed ones, indicated by their vegetated rather than bare understorey, are worth inspecting for characteristic birds, including Turtle Doves, Woodchat Shrikes and Rufous-tailed Bush Robins. The diversion to Bélmez is at km-38, where a tarmac road serves a cluster of hamlets. The road passes through rocky areas and wooded areas, as well as the ubiquitous olive groves and vineyards, again offering possibilities of woodland passerines and rocky terrain birds that include Bonelli's Eagles, Blue Rock Thrushes and Black Wheatears. The road loops back to the A-401.

♿ Much worthwhile birding is possible from or near your vehicle.

CALENDAR

All year: Griffon Vulture, Common Buzzard, Golden and Bonelli's Eagles, Common and Lesser Kestrels, Peregrine, Black-bellied Sandgrouse, Woodpigeon; Barn, Eagle, Little and Tawny Owls, Crag Martin, Crested and Thekla Larks, Dipper, Black Redstart, Black Wheatear, Blue Rock and Mistle Thrushes, Blackcap, Short-toed Treecreeper, Jay, Common Magpie, Red-billed Chough, Jackdaw, Raven, Chaffinch, Serin, Greenfinch, Goldfinch, Linnet, Rock Bunting.

Breeding season: Short-toed and Booted Eagles, Turtle Dove, Scops Owl, Alpine Swift, Bee-eater, Hoopoe, Nightingale, Rufous-tailed Rock Thrush, Subalpine Warbler, Spotted Flycatcher, Golden Oriole, Woodchat Shrike.

Passage: Raptors, including Honey-buzzard and Black Kite, passerines.

Winter: Alpine Accentor, Ring Ouzel, Redwing, Fieldfare.

GUADALÉN AND GIRIBAILE RESERVOIRS J5

Status: No special protection.

Site description
Two large adjacent reservoirs, the Embalse del Guadalén and the Embalse de Giribaile, in the northeast of the province are noteworthy on account not only of the diversity of waterbirds that they attract but also because of the open-country species that frequent their hinterlands in season.

Species
The Giribaile is a relatively recently constructed reservoir, with shallow margins, and the remains of the drowned trees at its eastern end hold a small breeding population of Cormorants. As elsewhere, these are much more abundant in winter at all reservoirs. More emergent dead trees further west on the Giribaile, near the castle on the northern shore, hold a breeding population of Grey Herons and the sparse reed cover there occasionally accommodates breeding Purple Herons, which also breed at the Guadalén. A small mixed heronry at the Guadalén has nesting Grey and Night Herons, and Cattle and Little Egrets. A few pairs of Mallard, Pochard and Gadwall breed and the Red-crested Pochard

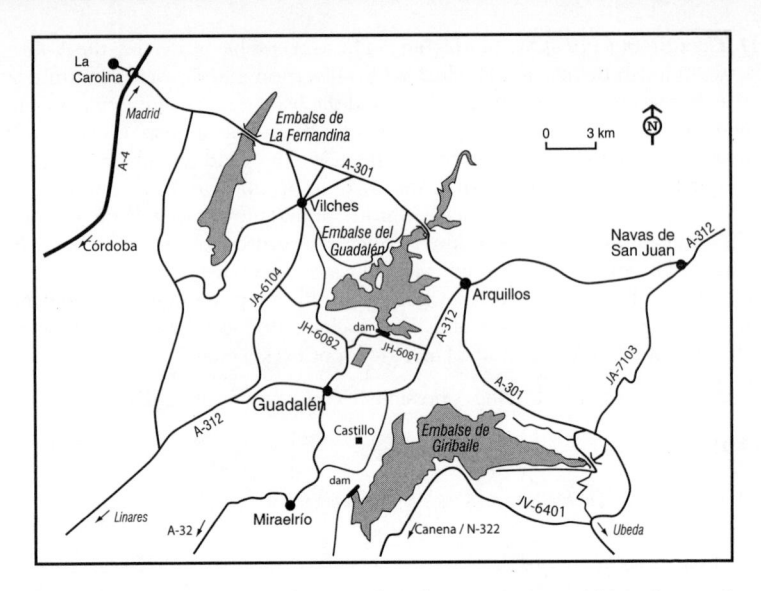

may do so. Waterfowl are much more abundant on passage and in winter, when the greater variety of species includes Wigeon, Tufted Ducks and Pintail. Great Crested Grebes breed at both reservoirs.

Both reservoirs attract large numbers of Black-headed and Lesser Black-backed Gulls, and a few Yellow-legged Gulls, in winter. Black and Whiskered Terns occur occasionally on passage. Greenshanks, Common and Green Sandpipers, Ringed Plovers and other common waders occur sparsely during the migration periods, when occasional rarer waders such as Marsh Sandpipers may appear. Lapwings are present in winter especially.

Short-toed, Golden and Booted Eagles, Marsh Harriers, Common Buzzards and Black Kites breed in the general area. Hobbies occur during passage periods, when Ospreys also visit the reservoirs. Hen Harriers and occasional Merlins hunt the hinterland in winter. Eagle Owls breed in good numbers in the area. The castle at Canena offers excellent views of the small colony of Lesser Kestrels that nest there.

Much of the agricultural land surrounding the reservoirs is devoted to the ornithologically rather barren olive groves, regiments of which march towards the horizon on all sides. Azure-winged Magpies are common here, however, together with Common Magpies and their brood parasites, the Great Spotted Cuckoos. However, the remaining intervening cereal crops support Montagu's Harriers. The flatter land around both the reservoirs houses small numbers of Black-bellied Sandgrouse. Quail call from the long grass in the breeding season. Thekla Larks are common. Several pairs of Black Storks also nest in the area and may be seen southwest of the Guadalén, along the JH-6082 north of Guadalén village.

A few White Storks breed in the area. The church in Guadalén village held a massive nest that at one time was thought to be the largest in Spain. It was over 50 years old when it was demolished by the environmental agency, for (human) safety reasons, in 2010, and was found to weigh some 2,300kg. The storks now occupy a more modest construction.

Timing
Worth visiting at any time of year, although winter and spring are undoubtedly the best periods. Telescope recommended.

Access
The reservoirs are readily accessible from the A-4 at La Carolina to the northwest or from the A-32 in the south. Many of the minor tracks mentioned are rough and may be unsuitable for anything other than all-terrain vehicles after rain.

A *Embalse del Guadalén*   Viewable from tracks along the western and southwestern side and from near the dam in the south. The A-301 from La Carolina to Arquillos skirts the reservoir to the north and there is access to the rather rough tracks from Vilches and also from an entrance at km-15.3 on the A-301. There is also ready access to these tracks via the dam, which is quickly reached from the south via Guadalén village or from the A-312, along a minor road (JH-6081) halfway between Guadalén village and Arquillos. Follow signs to the 'Presa Guadalén' at km-25. The dam itself has a high wall that prevents viewing from there but a picnic site just east of it allows access to the waterside, which can be followed on foot for some distance. The JH-6082, which branches off the JH-6081 west of the dam, approaches the southwestern arm of the reservoir, where tracks allow good views.

B *Embalse de Giribaile*   A track along the northern side of the Giribaile is very good for birding. The easiest access point of several from the A-301 is at km-35.3. A sign at the entrance suggests it is closed but it is perfectly passable unless water levels are high. The tarmac soon gives way to rather rough dirt, which crosses the lake bed at one point and would be blocked when the reservoir is high. Otherwise it gives access to the shallow margins and expanses of rough pasture. The old bridge across the inlet river is in poor repair and has been closed to vehicles but it provides a good vantage point for scanning the area. Access to the southeastern sector is also possible along a track from the A-301 at km-38.1.

A good but narrow tarmac road (JV-6041) follows the southern side of the

Montagu's Harrier

reservoir and gives elevated views of the lake along one stretch. Here it is possible to walk down to the waterside through the olive groves. Access to the dam is signposted from the N-322 west of Canena, although the dam itself has been closed for security reasons recently. However, another worthwhile route is the narrow track leading south off the A-317 3km east of Guadalén village and signposted for the castillo (castle). It leads to the minor road linking Guadalén village, Miraelrío and the N-322, passing through open habitats and offering views from the western side of the lake.

♿ Worthwhile birding is possible from or near your vehicle at both reservoirs.

CALENDAR

All year: Mallard, Common Pochard, Gadwall, Great Crested and Little Grebes, Cormorant, Night and Grey Herons, Marsh Harrier, Golden Eagle, Common Buzzard, Common Kestrel, Eagle Owl, Black-bellied Sandgrouse, Thekla and Crested Larks, Azure-winged Magpie, Common Magpie.

Winter: Tufted Duck, Wigeon, Pintail, Black-winged Kite, Hen Harrier, Merlin, Little Bustard, Lapwing.

Breeding season: Quail, Purple Heron, Black Stork, White Stork, Short-toed Eagle, Black Kite, Montagu's Harrier, Lesser Kestrel, Turtle Dove, Great Spotted Cuckoo.

Passage periods: Raptors including Booted and Short-toed Eagles and Black Kite, Osprey, Hobby, waders, Black Tern, Whiskered Tern, Bee-eater.

LAGUNA GRANDE DE BAEZA   J6

Status. Paraje Natural (206ha). Ramsar Site. ZEPA.

Site description
An artificial irrigation lagoon surrounded by olive groves, with some poplars, ash trees, encinas and tamarisks. There are fringing reedbeds and a hide. The lake itself is the largest lagoon in the province (18ha) and is claimed to have permanent water. However, the water levels vary and the lagoon dried up during the drought year of 2017. A smaller irrigation pool nearby also attracts birds.

Species
Some waterfowl are present year-round, water levels permitting, but are most numerous in winter. Purple Swamphens breed and Penduline Tits nest in the waterside trees. Other breeding species have included Squacco Herons, Black-winged Stilts and Little Ringed Plovers. Azure-winged Magpies are common. Cattle Egrets have a roost here in winter.

Timing
Worth visiting year-round except after very dry summers.

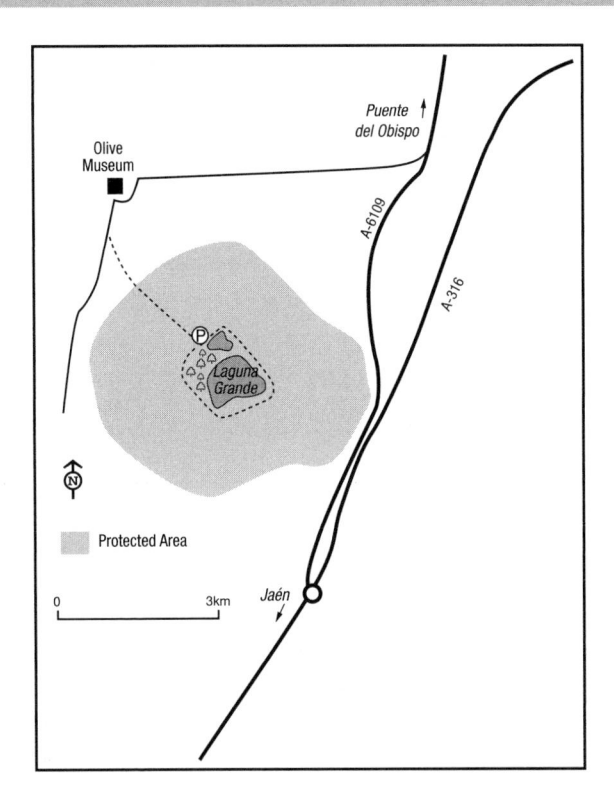

Access

The lagoon is west of the A-316 motorway south of Baeza. Leave the A-316 at km-20 where signposted for the village of Puente de Obispo. The lagoon is accessed via a good dirt track (2.5km) just south of the village, signposted to the 'Museo de la Cultura del Olivo' (Olive Cultivation Museum). Once you reach the museum continue left and shortly left again to reach the site. It is possible to drive round the lake/woodland complex but you will see more birds in the surrounding trees if you park by the information board on the left where you arrive and continue on foot.

♿ Worthwhile birding is possible from or near your vehicle.

CALENDAR

All year: Mallard, Little Grebe, Purple Swamphen, Penduline Tit, Azure-winged Magpie, Spotless Starling.

Breeding season: Night and Purple Herons, Little Bittern, Nightingale; Cetti's, Great Reed and Reed Warblers.

Winter: Gadwall, Teal, Pintail, Shoveler, Common Pochard, Cattle Egret, Grey Heron, Marsh Harrier.

Passage periods: waders, passerines.

UPPER GUADALQUIVIR RESERVOIRS J7

Status: Includes Paraje Natural, Alto Guadalquivir (768ha). ZEPA.

Site description
Three reservoirs, Puente de la Cerrada (A), Doña Aldonza (B) and Pedro Marín (C), are spread out along about 20km of the upper reaches of the Río Guadalquivir. In the past they attracted a range of breeding, passage and wintering waterfowl. The course of the river is marked by riparian woodland. However, the reservoirs themselves have become silted up and choked with tamarisks and have fallen out of use, losing much of their attraction to waterbirds. The possibility remains that they may regain their value in future, especially if there should be an increase in rainfall.

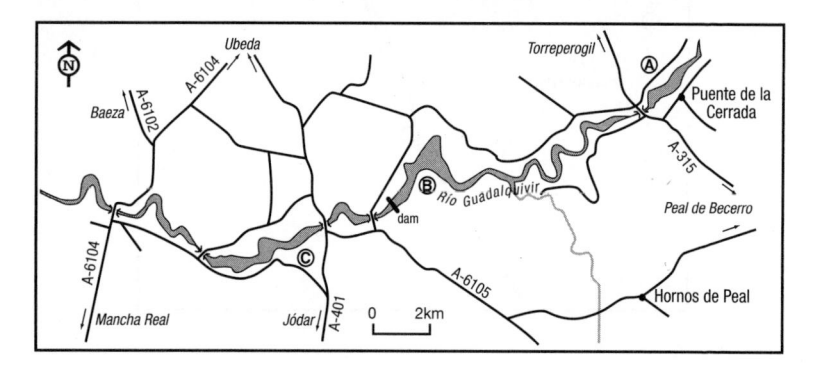

Species
The main attraction of these reservoirs has been the presence of breeding Purple Herons at Pedro Marín and Doña Aldonza, Purple Swamphens at Pedro Marín and Ferruginous Ducks and Marsh Harriers at Pedro Marín and Doña Aldonza. The reedbeds have supported breeding populations of Savi's, Reed and Great Reed Warblers. Penduline Tits occur in the fringing vegetation, especially in winter, and may breed.

Timing
The area is worth checking during and after wet winters, when it is liable to regain some interest.

Access
The three reservoirs are in the centre of the province, between Jódar and Becerro to the south and Úbeda to the north. The A-401 road between Úbeda and Jódar gives access by side roads to both the northern and southern sides of the complex. All three may be viewed from their respective dams or road bridges.

&. Limited observation from or near your vehicle is feasible.

CALENDAR

All year: Grey Heron, Mallard, Ferruginous Duck, Shoveler, Pochard, Little Egret, Common Coot, Moorhen, Purple Swamphen, Penduline Tit.

Breeding season: Little Bittern, Purple Heron, Marsh Harrier, Black-winged Stilt, Savi's Warbler, Reed, Great Reed and Melodious Warblers.

Winter: Gadwall, Teal, Wigeon, Pintail, Red-crested Pochard, Cormorant, Avocet, Lapwing, Dunlin, Common and Jack Snipes, Black-tailed Godwit, Common Redshank, gulls, Bluethroat, Penduline Tit.

Passage periods: Osprey, Black Kite, waders.

ALMERÍA PROVINCE

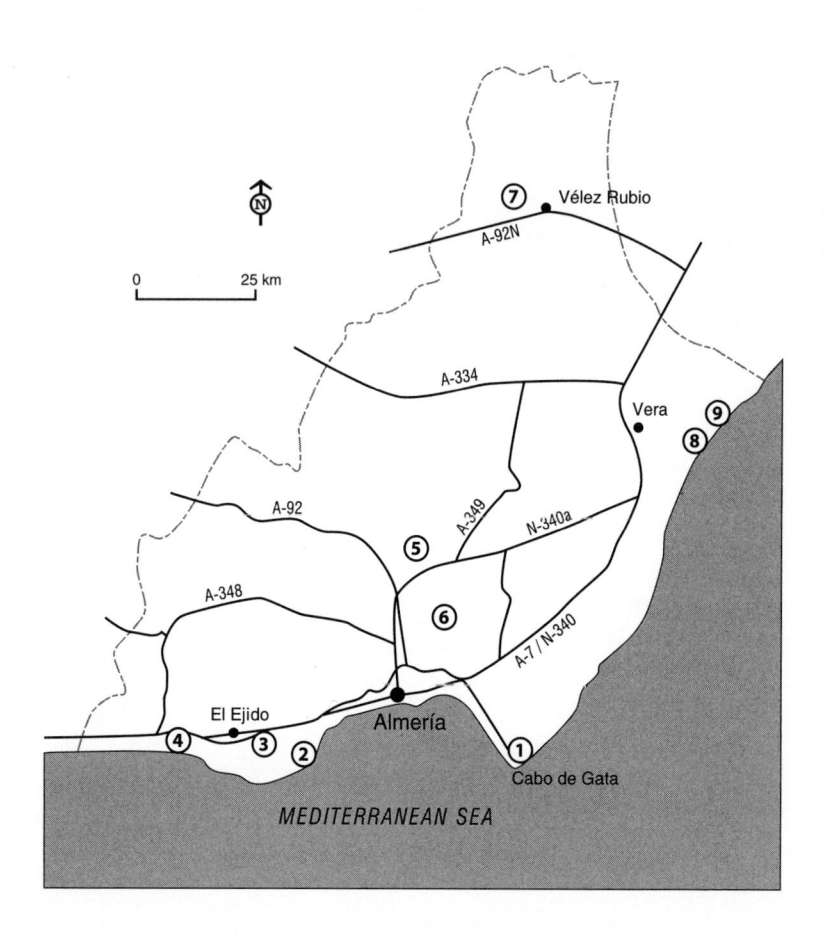

Almería province has a great deal to offer, including sierras (AL1, AL6, AL7), dunes (AL2), some wetlands (AL3, AL4) and the excellent birding at salt pans (AL1, AL2). This is Europe at its most arid: Almería has expanses of true steppe and semi-desert (AL5, AL6), and is the home of the Trumpeter Finch, Dupont's Lark and Lesser Short-toed Lark, among others. Xerophytic scrub once covered most of the sandy coastal plain and foothills, where the dramatic shortage of water made most farming next to impossible. This all changed with the introduction of the ubiquitous and unsightly plastic greenhouses. These now cover thousands of hectares and allow crops to be raised on otherwise unsuitable terrain, but to the huge detriment of the original flora and fauna. This apart, some interesting sites, especially many of the once interesting coastal salt pans at Guardias Viejas and elsewhere, have either long-since disappeared or remain under threat from residential and hotel developments.

A stay in the province can prove very worthwhile although you may well have to work hard to locate some species, especially in the steppe areas.

Sites in Almería Province
AL1   Cabo de Gata Natural Park
AL2   Punta Entinas to Roquetas de Mar
AL3   Las Norias lakes (Cañada de las Norias)
AL4   Adra lakes (Albufera de Adra)
AL5   Tabernas desert
AL6   Sierra Alhamilla
AL7   Sierra de María
AL8   Antas estuary
AL9   Almanzora estuary

Getting there
The A-7 allows excellent transit across the south and east of the province and gives access to the network of minor roads serving the Cabo de Gata region and the inland sites. The A-92N runs across the northern section of the province between Murcia and Granada. Almería airport is a popular destination in southern Spain and the obvious arrival point for visitors to the region.

Where to stay
There are ample tourist facilities at Roquetas de Mar, west of Almería city, which has the excellent salt pans of Salinas Viejas and Salinas de Cerrillo (AL2-B) right on its doorstep. The resort of Retamar, east of Almería city, is well placed for access to the whole of the Cabo de Gata area (AL1) and there are good road links for the Tabernas desert (AL5) and other more northern sites. Retamar is also good for access to the estuaries of the Río Antas (AL8) and Río Almanzora (AL9) and to the east coast in the San José, Mojácar and Carboneras areas, but these resorts themselves are well away from the main sites. Mojácar has a Parador on the coast. There are other places to stay inland. The Roquetas/Almerimar areas, as well as Cabo de Gata, are packed with holidaymakers between July and early September and it can then be very hot.

CABO DE GATA NATURAL PARK   AL1

Status: Parque Natural (land and sea) (49,696ha). ZEPA. UNESCO Biosphere Reserve.

General description
This very large area extends from the western side of Cabo de Gata around to the east coast as far north as Carboneras. The three best sites are the Cabo de Gata headland, the salt pans and the steppe areas, of which Las Amoladeras (D) is recommended as being the most productive ornithologically as well as reasonably accessible. This should not deter you from trying some of the eastern areas of the park, especially between San José and Las Negras, in which such

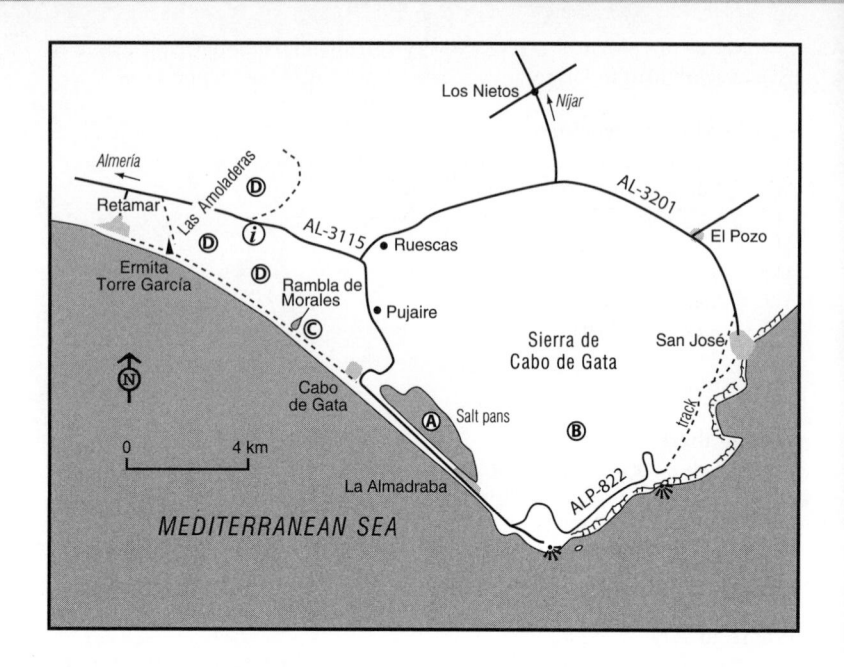

species as the Trumpeter Finch are often encountered. Indeed, any area of natural vegetation may produce surprises. The protected area has some 40km of coastline and also includes the sub-littoral zone to a mile offshore. The extreme aridity of the area is reflected in the diverse community of xerophytic plants, which includes a number of endemics. The visitors' centre for the region is at Las Amoladeras on the AL-3115 Almería/Cabo de Gata road at km-7.

♿ Some degree of access is possible at all sites. There are good opportunities for watching from or near your vehicle.

Timing
The whole area can be visited profitably at any time of year. However, the very high temperatures and crowds of tourists in July and August combine to make these two months the least attractive.

The sierra and steppe areas are best visited in spring and early summer. The salt pans are most productive from March to mid or even late May, and again from late August to October, and also in winter, although they are far from empty in mid-summer. In summer the best birding is to be had in the early mornings and late evenings, especially in the steppes and sierra, since the middle of the day is very unproductive as birds also shelter from the daunting heat. Heat haze must also be taken into account at the salt pans: the shimmering makes optical instruments useless at any distance.

A CABO DE GATA SALT PANS
Site description
The important area of dunes and steppe that extends for some 15km along the eastern side of Almería Bay includes some 350ha of salt pans (salinas). They are

the only surviving commercial salt pans in Almería but also include an ornithological wetland reserve of major importance. The reserve area is, approximately, the sector north from the southernmost hide. The worked salt pans are south of there.

Species
These salt pans are ideally sited to attract and retain passage and wintering birds of a wide variety of species. Virtually anything may occur and scarce and rare birds appear annually. Greater Flamingoes first catch the eye and there are regularly several hundred and occasionally a few thousand present, although they have never nested here successfully. The breeding species include Avocets, Black-winged Stilts, Kentish Plovers and Little Terns. A diversity of herons and egrets occur, especially on passage and in winter. Wintering waterfowl sometimes include up to several hundred Common Shelducks. A full range of the regular waders may be expected, again both on passage and in winter, and close inspection may reveal scarce or vagrant species. Collared Pratincoles are regular on passage and Red-necked Phalaropes appear here in autumn with some frequency. Loafing gull flocks are often a feature and include large numbers of Audouin's Gulls in late summer especially. Slender-billed Gulls occur among the commoner species and have nested occasionally. Marsh terns occur on passage, these including occasional White-winged Black Terns. Southern Grey Shrikes, Black Redstarts and Thekla Larks are among the passerines that may be expected in the general area.

Access
Take the Cabo de Gata road (AL-3115), which is signposted from a roundabout on the coast road (AL-3115/AL-3201) just west of Ruescas. The road skirts the northern edge of the salt pans shortly before Cabo de Gata village, where there is a large lay-by and a hide on the left from which you can view the northern end of the main lagoon. Continue through the village and follow the road southwards alongside the salinas. There are several hides along the western edge of the pans, notably one on the left just before a conspicuous restored church. Stop here and visit the hides on foot. Continue south through the village of La Almadraba to view the last of the pans, just past the end of the village

B SIERRA DE CABO DE GATA
Site description
The sierra, which is mostly of volcanic origin, comes right down to the coast and, birds apart, is scenically most attractive. It is an intensely arid area where rainfall averages just 200mm per year, the lowest in Europe.

Species
The coastal region holds small colonies of Yellow-legged Gulls. Several pairs of Shags, the only ones in the region apart from those at Gibraltar, nest along the coast between the lighthouse and northwards beyond San José. Some European Storm-petrels nest on offshore islets. Seawatching from the Cape often produces Cory's and Balearic Shearwaters, as well as a variety of gulls and terns and occasional skuas during migration periods. A few Common Scoters and Razorbills may be present in winter and migrating Puffins also occur occasionally.

Several pairs of Bonelli's and Booted Eagles nest locally, as well as Lesser

Kestrels and Peregrine Falcons. Eleonora's Falcons occur as scarce migrants, chiefly in autumn. Eagle Owls also breed in the area. The sierra holds a considerable population of Trumpeter Finches. Thekla and Crested Larks, and a few Dupont's Larks, are present in the steppe habitat on the western side.

Access
The Cabo de Gata road (AL-3115) takes you past the salt pans (A) to the lighthouse and the tip of the Cape, where there is a watchpoint and information booth. The ALP-822 branches off the access road just before the lighthouse and is a worthwhile drive over open rocky hillsides covered with low scrub for 3km to a headland dominated by an 18th century watchtower, the Torre de La Vela Blanca, which offers panoramic views of the whole area but is too far from the sea for seawatching purposes. Passerines are the attraction here and include the Black Wheatear, Blue Rock Thrush, Rock Bunting and Trumpeter Finch. It is possible to continue on foot beyond the headland for another 5km, to San José. To visit the more northerly and easterly side of the Sierra de Cabo de Gata by road, follow signs for San José or, further north still, for Las Negras, and explore the various minor roads and tracks throughout the area.

C RAMBLA DE MORALES
Site description
A rambla is a river bed, usually dry except after the infrequent rains in this part of the world. However, the Rambla de Morales is an anomaly in that the last kilometre or so nearest the sea retains water, by infiltration from the sea, even during the driest summers. The last few hundred metres where it widens out into what is in effect a brackish, reed-fringed, coastal lagoon are of great interest.

Species
Waders, especially Kentish Plovers and Dunlins, are often seen pottering around the sandbar at the mouth of the rambla. Wood and Curlew Sandpipers and Little Stints are regular migrants, as are Whimbrel and a few Curlews, which are often to be found resting in the surrounding dunes. White-headed Ducks occur occasionally and have nested here, as do Purple Swamphens. Garganey are regular spring migrants, although they seldom stay long. There have also been fairly frequent records of Marbled Ducks as well as occasional Ferruginous Ducks. Migrant waterbirds often include Collared Pratincoles in spring as well as both Purple and Squacco Herons. Interesting wintering passerines include Bluethroats, Penduline Tits and Water Pipits, these last being annual but scarce. The beach track in winter can produce Trumpeter Finches and Lesser Short-toed Larks, as well as flocks of up to 200 Golden Plovers, which should be scanned for any accompanying Dotterels.

Access
The rambla is most easily reached from the northeast corner of Cabo de Gata village. Here head for the visible *chiringuitos* – the beach bars. You can walk along the beach from here but there is also a parallel, 1.5km, drivable track. It is easy to walk along the southern side of the rambla until it starts to peter out after about 750m.

Access is also possible from the north via Ermita de Torre García (see D) by driving eastwards along the beach track for about 2km but a 4WD vehicle is strongly recommended. This can be rewarding in winter.

D LAS AMOLADERAS
Site description
A sizable residual fragment of the arid dwarf-shrub steppe that formerly covered enormous expanses of coastal Almería. It is contiguous with the large steppic area of the Campo de Níjar. The beach and coastal dunes fringe the largely treeless area of low Mediterranean scrub.

Species
A traditional site for birds of dry steppic regions but numbers of some of the characteristic and most sought-after species, such as Dupont's Lark, Little Bustard and Black-bellied Sandgrouse, have declined markedly in recent years, making these even more elusive than usual. There are nevertheless good numbers of Stone-curlews, Rollers are not uncommon and Bee-eaters and Hoopoes also occur. Other interesting breeding species include the Rufous-tailed Scrub-robin, Spectacled Warbler and Southern Grey Shrike. Trumpeter Finches are mainly winter visitors in moderate numbers, but at least some are present in summer. Another prize species is the Dotterel, a regular migrant and local winterer in small numbers in recent years. Early morning and evening visits are likely to be most successful for all species, particularly in spring when the larks are singing. Any rain pools are likely to attract birds in what is usually a very dry area.

The key target species, Dupont's Lark, is best sought in the northern section of the reserve, in areas of Thyme and False Esparto grass. Both Greater and Lesser Short-toed Larks are reasonably common and both Crested and Thekla Larks also occur, the latter being much more frequent. Sky Larks are winter visitors.

Access
The information centre on the south side of the AL-3115 at km-7 is an obvious starting point, from which there are several signposted walking trails. It is also worth taking the asphalted track on the north side of the AL-3115, a short distance from the entrance to the information centre, which leads to an approach radio beacon for Almería airport, 1.4km from the main road. There are obvious tracks off this road through some of the best steppic habitat, which you can explore on foot. Another option is to drive down to the waterfront road (*paseo maritimo*) in Retamar and head southwards. The road soon becomes a track leading to a small, Byzantine construction, complete with onion dome, the Ermita de Torre García. Several tracks lead northwards across the scrub from here, including one that links the hermitage to the AL-3115 directly.

> **CALENDAR**
>
> **All year:** Shag, Greater Flamingo, Bonelli's Eagle, Common Kestrel, Peregrine Falcon, Red-legged Partridge, Little Bustard, Kentish Plover, Stone-curlew, Black-headed and Yellow-legged Gulls, Sandwich Tern, Black-bellied Sandgrouse, Eagle and Little Owls; Dupont's, Lesser Short-toed, Crested and Thekla Larks, Crag Martin, Black Redstart, Stonechat, Black Wheatear, Blue Rock Thrush; Cetti's, Dartford and Sardinian Warblers, Zitting Cisticola, Southern Grey Shrike, Spanish Sparrow, Linnet, Trumpeter Finch, Rock and Corn Buntings.
>
> **Breeding season:** Cory's Shearwater, European Storm-petrel, Booted Eagle, Lesser Kestrel, Avocet, Black-winged Stilt, Little Tern, Turtle Dove, Red-necked Nightjar; Common, Pallid and Alpine Swifts, Bee-eater, Roller, Hoopoe, Calandra and Greater Short-toed Larks, Red-rumped Swallow, Rufous-tailed Scrub-robin, Black-eared Wheatear, Spectacled Warbler, Spotted Flycatcher, Woodchat Shrike, Cirl Bunting.
>
> **Winter:** Common Shelduck, Pintail; White-headed, Marbled Duck and Ferruginous Ducks, Common Scoter, Little and Cattle Egrets, Grey Heron, Crane, Grey Plover, Lapwing, Curlew, Whimbrel; other waders including Dotterel, Arctic Skua; Mediterranean, Audouin's and Lesser Black-backed Gulls, Razorbill, Puffin, Sky Lark, Meadow Pipit, White Wagtail, Common Chiffchaff.
>
> **Passage periods:** Cory's (Scopoli's) and Balearic Shearwaters, Purple Heron, White Stork, Black Stork; raptors, including Osprey, Marsh Harrier, Booted Eagle and Eleonora's Falcon, Oystercatcher, Collared Pratincole (has bred); waders, including Dotterel, Red-necked Phalarope and Marsh Sandpiper, Great Skua; Mediterranean, Little, Slender-billed and Audouin's Gulls; Gull-billed, Caspian, Common, Whiskered, Black and White-winged Black Terns, swifts and hirundines, Common Redstart, Whinchat, Willow Warbler.

PUNTA ENTINAS to ROQUETAS DE MAR   AL2

Status: Reserva Natural and Paraje Natural (Punta Entinas-Sabinar, 1,949ha). ZEPA. Ramsar Site.

General description

This 15km coastal strip between Punta Sabinar and the salt pans just west of Roquetas de Mar survives as an outpost of natural and semi-natural habitats on the fringes of the worst of the plastic greenhouse badlands. The west of the area comprises expanses of scrub and dunes as well as ponds and areas of salt marsh. The eastern section has scrub, dunes and disused salt pans. This is an excellent and often rewarding site year-round, offering interesting breeding and wintering species and a large variety of migrants during passage periods. However, the former salt pans at Guardias Viejas to the west have long since been reduced to a municipal lake and there are ongoing threats to at least some of the protected area from projects to reclaim the land for residential accommodation. The protected area was reaffirmed as a special conservation area by regional government decree in 2017 so it is hoped that the site will be saved.

General access

Leave the A-7 at km-409 for Almerimar (A-389) and the west of the region, or at km-429 for Roquetas (A-1051) and the eastern end. East/west movement is via the AL-9006 and the AL-3300.

♿ The fringes of the eastern end of the area and the Laguna del Hornillo are visible from the roadside. The tracks are level but sandy and are muddy when wet.

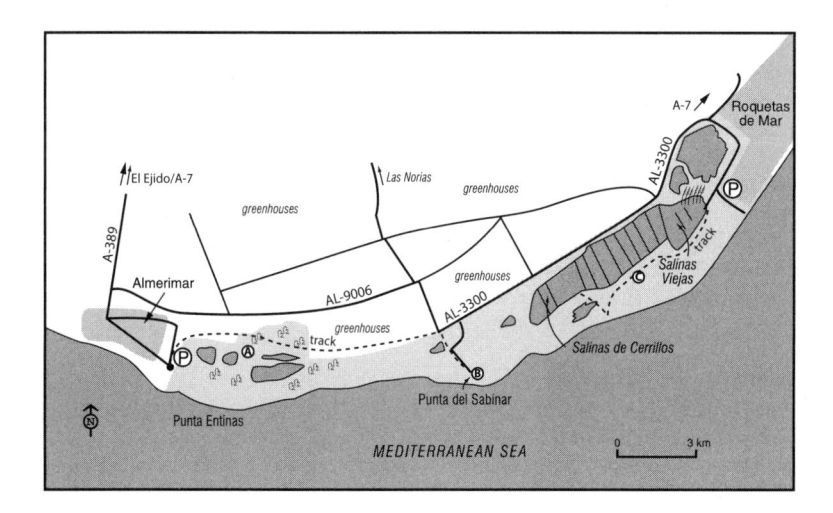

A PUNTA ENTINAS
Site description
Former sand pits now comprise a series of seasonal lagoons within a littoral steppe zone of dunes fixed by tamarisks and dwarf-shrub scrub.

Access
Use the small car park at the extreme western end of Almerimar, from where a 6.5km trail runs westwards through the inland margin of the site. There are several tracks in this area, some leading to the beach, but the relevant track is signposted. The track offers good views in places across the lagoons. The final 1.5km are through scrub habitat. You will need to return along the same route.

B PUNTA DEL SABINAR
Site description
The site features stabilised coastal dunes, with low sandy scrub and the remains of salt pans to the east. The point is marked by a small lighthouse.

Access
Access to Punta del Sabinar is from the AL-3300, which skirts the Salinas de Cerrillos. The access track is a surfaced road that heads south from a point some 650m east of the sharp bend in the AL-3300. Various tracks lead from the lighthouse through the scrub and provide access to the residual saltpans and scrub habitats. A visit here is particularly likely to be worthwhile during passage periods, when there may be interesting grounded passerine migrants.

C SALINAS VIEJAS – SALINAS DE CERRILLO
Site description
This is the eastern part of the coastal strip between Roquetas de Mar in the east and the saline ponds of Punta Entinas to Punta Sabinar near Almerimar in the west. The main attraction is the two sets of disused salt pans (salinas) in different states of disrepair and of varying salinity. There is also an interesting coastal zone of small dunes and ponds between these salt pans and the sea.

Access
On foot along a sandy trail that originates at the western end of Roqueta de Mar. Take the AL-3300 to Roquetas and the first exit (Urb. Playa Serena) at the first roundabout. After 250m the second turning on the right is the Avenida de Cerrillos, which skirts a sizable lagoon (Laguna del Hornillo), which can be viewed from the roadside. Continue for some 500m until the road bends sharp left and park. The entrance to the access trail is at the bend, marked by gates with a flamingo emblem and an information board. The trail leads westwards for some 4.6km, offering excellent views of the salt pans and scrub habitats. The habitat to both sides of the track should be inspected with great care as this is one of the prime areas for birding locally.

Species
A number of regionally scarce or local species nest here although some of them only occur irregularly. In a good season a sizable Black-headed Gull colony (1,000+ pairs) also accommodates 100 or more pairs of Slender-billed Gulls and a few pairs of Mediterranean Gulls. Gull-billed and Little Terns nest in small numbers. Nesting waterfowl include the White-headed Duck occasionally. Red-knobbed Coots were introduced here early this century but seem to have moved away. The Purple Swamphen is usually present, however, and nests locally. Stone-curlews breed in the dunes and sandy areas, especially in the strip between Punta Entinas and Punta del Sabinar, as do Kentish Plovers. Little Ringed Plovers occur and may also breed occasionally. Black-winged Stilts and a few pairs of Avocets breed in suitable areas of the salt pans.

Nesting passerines include Lesser Short-toed Larks, which are common at Punta Entinas and abundant along the track from Roquetas westwards past the salt pans, making this one of the easiest sites to find this species. The Iberian race of the Yellow Wagtail is abundant and Cetti's, Reed and Great Reed Warblers all breed in suitable reedy areas.

The salt pans attract numbers of Greater Flamingos, Little Egrets and Grey Herons, occasional migrant Purple Herons, Glossy Ibises, Common Coots and some migrant and wintering ducks, including Garganey and occasional Red-crested Pochards. There may be 100 or more White-headed Ducks present in winter. Significant numbers of Black-necked Grebes are often offshore in winter, sometimes with a few Common Scoters.

The list of migrant and wintering waders is extensive and scarce or vagrant species are always a possibility. The latter have include Cream-coloured Courser and Terek Sandpiper. Temminck's Stints appear occasionally and Marsh Sandpipers occur annually, alongside Curlew and Wood Sandpipers, Little Stints and many commoner species. Migrant Collared Pratincoles are common in spring.

Occasional skuas, shearwaters and Gannets can be seen offshore in the appropriate seasons. Thousands of gulls use the area on passage and in winter, and many non-breeders remain in summer. Mediterranean Gulls are common migrants and winter visitors. Audouin's Gulls may be seen year-round but especially between April and October; they are most numerous during July–October. Sandwich Terns also occur year-round. Caspian Terns occur occasionally and all three marsh tern species may be seen on passage; the White-winged Black Tern is most likely in autumn after easterlies.

Passerines are especially diverse during falls of migrants, particularly during easterly winds in spring and, less often, in autumn. They include Greater Short-toed Larks and a range of Yellow Wagtail subspecies. All the migrant warbler and chat species may appear and once again there is a real possibility of encountering rarities.

Timing

The whole area is interesting at any time of year but is probably at its best in middle and late spring, from April to mid-May, when there are breeding summer visitors alongside species that are still migrating north. The aftermath of an easterly blow, ideally with some rain and thunder, is likely to be especially productive. Under such conditions the east coast of Spain receives migrants that are very scarce or rare anywhere else in Iberia: the Red-footed Falcon, Wood and Icterine Warblers and the Collared Flycatcher are examples. Autumn wader numbers are linked to water availability but the birds may be helpfully concentrated in wet areas during dry years. In winter there are considerable numbers of ducks, waders, gulls and terns. The area is least attractive between early June and early August.

CALENDAR

All year: Common Shelduck, White-headed Duck, Mallard, Little Egret, Greater Flamingo, Water Rail, Purple Swamphen, Common and Red-knobbed Coots, Avocet, Black-winged Stilt, Stone-curlew, Kentish Plover, Dunlin, Common Redshank; Black-headed, Slender-billed, Audouin's, Lesser Black-backed and Yellow-legged Gulls, Sandwich Tern, Crested Lark, Stonechat, Zitting Cisticola; Cetti's, Sardinian and Dartford Warblers, Southern Grey Shrike.

Breeding season: Black-winged Stilt, Little Ringed Plover, Gull-billed and Little Terns, Bee-eater, Roller, Hoopoe, Lesser Short-toed Lark, Yellow Wagtail, Reed and Great Reed Warblers, Woodchat Shrike.

Winter: Wigeon, Gadwall, Teal, Mallard, Pintail, Shoveler, Red-crested and Common Pochards, Great Cormorant,

Gannet, Cattle Egret, Common Scoter, Marsh Harrier; Ringed, Grey and Golden Plovers, Lapwing, Sanderling, Common Snipe, Whimbrel, Curlew, Turnstone, Marsh Sandpiper, Arctic Skua, Mediterranean and Lesser Black-backed Gulls, Razorbill, Puffin, Meadow Pipit, Penduline Tit, Reed Bunting.

Passage periods: Garganey, Marsh Harrier, Spoonbill, Glossy Ibis, Collared Pratincole, Ringed Plover, Knot, Little and Temminck's Stints, Curlew Sandpiper, Ruff, Black-tailed and Bar-tailed Godwits, Spotted Redshank, Greenshank; Marsh, Green, Wood and Common Sandpipers, Great Skua, Little and Slender-billed Gulls; Kittiwake; Gull-billed, Caspian, Common, Whiskered, Black and White-winged Black Terns, swifts, Greater Short-toed Lark, Yellow Wagtail, hirundines, Northern and Black-eared Wheatears, Pied Flycatcher, warblers, rarities.

LAS NORIAS LAKES   AL3
(Cañada de Las Norias)

Status: Local reserve (140ha).

Site description

Las Norias is a shoo-in contender for the title of most unbecoming birding site in southern Spain but, nevertheless, it is well worth visiting. The White-headed Duck continues to be the star attraction. The lakes were formed by the natural flooding of deep claypits that were excavated in the 1980s. The water is saline and hence of no use to crops in the fringing plastic greenhouses. A veritable ocean of plastic surrounds the lakes but this has not put off the wildlife, notably the waterfowl and other waterbirds that inhabit the water bodies and nest in the fringing vegetation of reeds, canes and tamarisks.

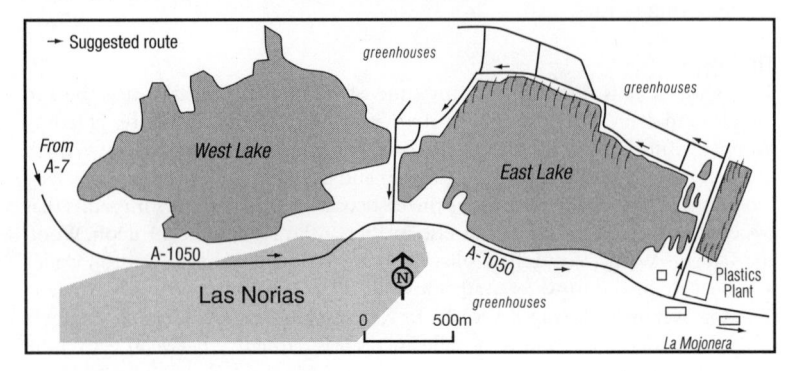

Species
The White-headed Duck may be seen here year-round. There is a small nesting population but numbers increase greatly outside the breeding season. In winter 2010/2011 they exceeded 1,000 birds, the largest known site-count in Iberia. A few Marbled Duck and Red-crested Pochards also nest here in some seasons. Other breeding waterbirds include Purple Swamphens and both Little and Black-necked Grebes. A small heronry is shared by Cattle Egrets and Night Herons. Squacco Herons and Little Bitterns also nest, at least in some seasons. Little Terns have bred and there are also small nesting populations of Collared Pratincoles, Black-winged Stilts, Avocets, Stone-curlews, Little Ringed Plovers and Kentish Plovers. Ducks and waders are commonest on passage. Passage ducks have included Ruddy Shelducks, Garganey, Tufted Ducks and Ferruginous Ducks. The wader variety is greater during the post-breeding migration than in spring. The many wader species recorded have included both Little and Temminck's Stints and both Great and Jack Snipes. The site attracts Gull-billed, Black and Whiskered Terns on passage. A variety of hirundines and swifts also feed over the area during passage periods.

Timing
White-headed Ducks can be found year-round. Spring and early summer are best for nesting species and migrant waders. Autumn offers migrant waders and there are interesting ducks, waders and passerines both then and in winter.

Access
Las Norias village is signposted from the A-7 at km-411. The A-1050 leads to Las Norias and on to La Mojonera, passing along the southern edge of the site and alongside the eastern lake. A recommended route, starting at Las Norias village, is marked on the map. The A-1050 is busy but you can park on the adjacent service road and view the eastern lake from there or cross the road and view from behind the safety barrier. A more comfortable opportunity presents itself at the eastern end of the lake, where a causeway road leads northwards between a clump of eucalyptus trees and a plastic recycling plant. The causeway gives views over water to both left and right, over the eastern lake and also a

Kentish Plovers

tamarisk-fringed pond on the opposite side of the road. Both Purple Swamphens and White-headed Ducks may be expected. Migrant waders and breeding Collared Pratincoles are to be found on the islets to the left of the causeway and waders around any muddy margins.

Turn left at the crossroads at the end of the causeway, follow the track around for 0.9km and park where there is an obvious gap in the greenhouses. From here there is easy viewing of the breeding Cattle Egrets, Squacco Herons and Night Herons amongst the half-submerged trees. Bluethroats and Kingfishers have been seen here in winter, as well as a wide variety of ducks. Continue until you reach an obvious tarmac road to your left. Turn into this road, which becomes the western causeway where there is easy parking and good views of both the western and eastern lakes. At the end of this road you rejoin the A-1050.

The lakes may be viewed from your vehicle.

CALENDAR

All year: Mallard, Red-crested and Common Pochards, Gadwall, White-headed Duck; Great Crested, Black-necked and Little Grebes, Little Egret, Water Rail, Moorhen, Common Coot, Purple Swamphen, Black-winged Stilt, Kentish Plover, Black-headed Gull, Stone-curlew, Lesser Short-toed and Crested Lark, Zitting Cisticola, Cetti's Warbler, Southern Grey Shrike.

Breeding season: Marbled Duck (has bred); Cattle Egret, Night and Squacco Herons, Little Bittern, Collared Pratincole, Black-winged Stilt, Avocet, Little Ringed Plover, Little Tern, Turtle Dove, Bee-eater, Hoopoe; Reed, Great Reed and Melodious Warblers, Woodchat Shrike.

Passage periods: Garganey, Tufted and Ferruginous Ducks, Great White Egret, Purple Heron, White Stork, Osprey, Ringed and Golden Plovers, Knot, Dunlin, Little and Temminck's Stints, Curlew Sandpiper, Ruff, Spotted and Common Redshanks, Greenshank, Green Sandpiper; Gull-billed, Sandwich, Common, Little, Whiskered, Black and White-winged Black Terns, swifts, hirundines, Yellow Wagtail, warblers.

Winter: Grey Heron, Common Shelduck, Wigeon, Teal, Shoveler, Pintail, Great Cormorant, Great White Egret, Common Buzzard, Marsh Harrier, Grey and Golden Plovers, Lapwing, Dunlin, Common and Jack Snipes, Common Sandpiper, Black-tailed Godwit, Turnstone, gulls, Bluethroat, Black Redstart, Penduline Tit, Common Starling.

ADRA LAKES (Albufera de Adra) AL4

Status: *Reserva natural (132ha).*

Site description
An important site in a region that offers few wetlands. It is included for completeness but proper access is difficult, requiring prior arrangement (see below). It consists of two lakes, the Albufera Honda (Deep Lake) and the Albufera Nueva (New Lake). Both are fringed by reeds, sedges, bulrushes and tamarisks and sit surrounded by a sea of agricultural plastic on the south side of the N-340a Málaga/Almería road, close to the sea. The waters are slightly saline because of infiltration.

Species
The White-headed Duck nests here in some numbers, this site at times holding the principal population of eastern Andalucía. Other breeding wetland species include Little, Great Crested and Black-necked Grebes, Little Bitterns and both Red-crested and Common Pochards. Tree Sparrows frequent the area.

The winter sees a notable increase in numbers and variety of waterbirds with the arrival of more Black-necked Grebes, Great Cormorants and a variety of common waterfowl. Over 100 White-headed Ducks winter here. Greater Scaup have been recorded, as have Marbled Ducks, and Tufted Ducks are quite frequent. Bluethroats and Penduline Tits are also present in winter.

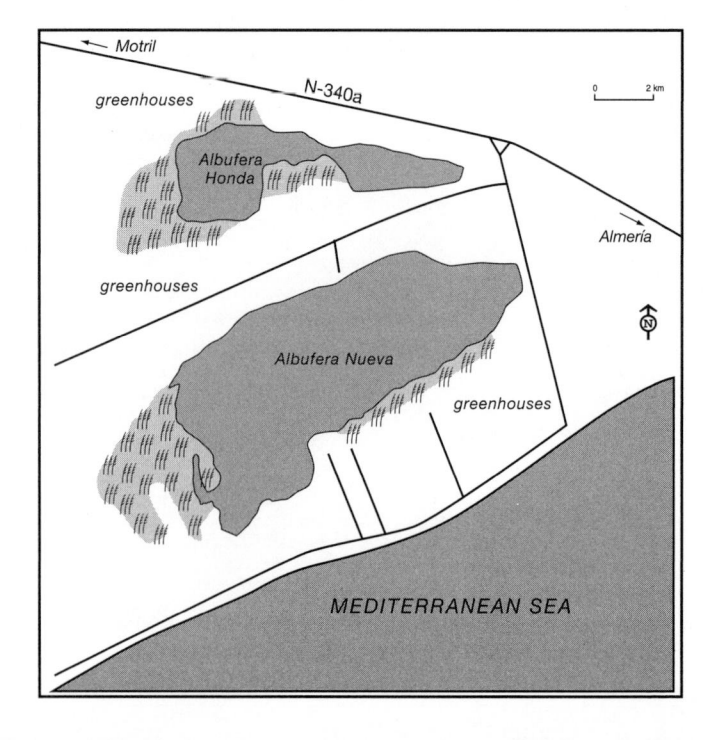

The presence of fresh water attracts a diversity of species during migration periods, including Purple Herons. Marsh Terns are frequent on passage: most are Black Terns but there are occasional Whiskered Terns and, very rarely, White-winged Black Terns. Very large numbers of hirundines feed over the area and roost in the reedbeds during migration periods.

Timing
The area can be visited at any time of year if you want to see Red-crested Pochards and White-headed Ducks, although these are most in evidence in winter and spring. The most productive period is between November and June.

Access
The site is closed to the public except by prior arrangement, so is probably not worth considering for a casual visit, even though the lakes are partly visible from outside the perimeter fencing. A proper visit can be arranged by contacting the Almería city office of the Environment Agency (tel. 950 101 676), Adra Village tourist agency (Tel. 950 400 400) or Adra Museum (tel. 950 403 546); it is probably best to apply in person at any of these unless your Spanish is fluent. Visitors are accompanied by Environment Agency staff and have access to three hides. Arrangements should be made at least a week in advance. Otherwise, the physical difficulties of observation are such that it is not worth stopping unless you are desperate to see White-headed Ducks. There are other sites, such as at Las Norias (AL3) and in other provinces, where this species can be seen far more easily.

Leave the N-340a Málaga/Almería road (*not* the A-7) at km-66, some 6km east of Adra, where the lakes come into view. The turn-off track is marked by a Consejería de Medio Ambiente sign, but is easily missed. Having successfully entered the track, follow the tarmac for about 200m before stopping and trying to see the smaller lake through the fence and reeds. Continue along the rough track for about 200m before stopping near the locked gate and try to look at the larger lake, again through the same screen of fence and tall reeds.

♿ Limited observation from your vehicle is possible.

CALENDAR

All year: Red-crested and Common Pochards, White-headed Duck, Little and Black-necked Grebes, Tree Sparrow.

Breeding season: Little Bittern, Great Reed and Reed Warblers.

Winter: Shoveler, Wigeon, Gadwall, Pintail, Tufted Duck, Great Crested Grebe, Great Cormorant, Marsh Harrier.

Passage periods: Marbled Duck, Purple Heron, Water Rail; Whiskered, Black and White-winged Black Terns (uncommon), hirundines.

TABERNAS DESERT AL5
(Desierto de Tabernas)

Status: Paraje Natural (11,625ha). ZEPA.

Site description
This area is the depression to the northwest of the Sierra Alhamilla (AL6). It is semi-arid, receiving less than 250mm of rain per year, and is the only true sub-desert area in Europe. The landscape is deeply eroded with cuttings and ravines in the bottom of which there are the ramblas, the dry river beds that abound in this part of Andalucía. Vegetation is scarce, low and scrubby, except in the bottom of the ravines where there are tamarisks and oleanders. The northern part of the area is higher, where you enter the southern ramparts of the Sierra de los Filabres. The 'Wild West' landscape has made the Tabernas desert a favourite location for numerous films, especially the so-called spaghetti westerns. Cowboy towns are available for nostalgics.

Species
The region offers a pleasing diversity of the birds of dry, open and steppic landscapes. These include Little Bustards, Stone-curlews and Black-bellied Sandgrouse, as well as Great Spotted Cuckoos, Red-necked Nightjars, Bee-eaters, Rollers and Trumpeter Finches. Thekla Larks are common and there is a small population of Dupont's Larks. Other characteristic breeding species include Eagle and Little Owls, Rufous-tailed Scrub-robins, Black Redstarts, Black-eared and Black Wheatears and Blue Rock Thrushes. Nesting

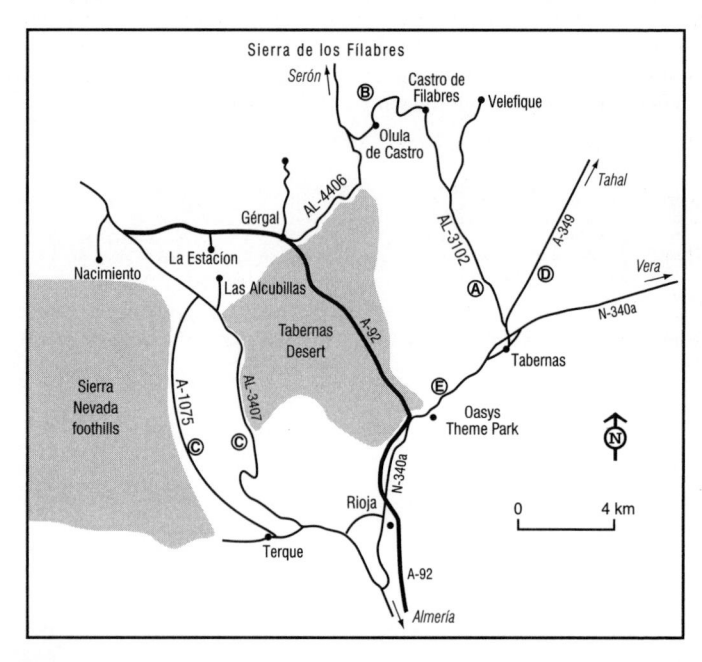

warblers include Spectacled and Subalpine Warblers at higher levels, Orphean Warblers in the semi-wooded areas around the edges, and Olivaceous Warblers in the ramblas. Rock Sparrows occur locally and Rock Buntings are to be found at higher levels. The area is frequented by such raptors as Bonelli's Eagles, Peregrines, Lesser Kestrels and Common Kestrels, some of them visiting from nearby sierras.

The Cream-coloured Courser was found nesting at Tabernas in 2001, for the first known time in Europe, when three pairs bred and some young were fledged. The birds have not been detected at Tabernas since but nesting by at least two pairs was reported not too far away in Albacete (Castilla-La Mancha) in 2012 and by two or three pairs on the plains west of Granada city in 2017. This is an easy bird to overlook and it may be that we are seeing the initial stage of colonisation of Mediterranean Iberia as the climate turns drier.

Timing

The best period is March–May, although the area may be visited profitably at any time of year, even in December and January. Early mornings are best. It can become unbearably hot thereafter, especially in the ramblas, from mid-May onwards and an ample supply of drinking water is indispensable. Given the desert nature of the region, one of the best methods of finding birds is to go down into a rambla, preferably one that still retains some of the precious liquid, and walk along – or wait and sit in the shade and watch.

Access

The A-92 traverses the area. This a fairly busy dual carriageway and you will need to get off it to explore promising areas. Much of the A-92 has parallel service roads alongside the main carriageways and these are recommended for birding purposes. A diversity of rough dirt tracks lead off the service roads and may be worth exploring if conditions and your vehicle permit. Recommended routes for easier access, all of them good tarmac, include:

A The Tabernas/Castro de Filabres road (AL-3102), which crosses expanses of open terrain before climbing into the foothills of the Sierra de los Filabres in the north.

B The AL-4406 between Castro de Filabres and Olula de Castro, which offers a quiet, pleasant if winding drive through hilly terrain, covered with low scrub.

C A circular route along the western fringe of the desert proper and the easternmost foothills of the Sierra Nevada, again through open, hilly, rocky terrain, following the AL-3407 and the A-1075, between Terque in the south and Nacimiento in the north.

D The southern section of the A-349 Tabernas/Tahal road.

E Rambla trails. Any ramblas can reward brief exploration, particularly if they retain some water or where there are stands of tamarisks, a favoured habitat of the Olivaceous Warbler. Sudden storms in the hills can and do produce flash flooding so it would be unwise to linger within a rambla in the event of rain or the sound of distant thunder. There are (poorly) signposted trails through the ramblas in the vicinity of Tabernas village. A convenient access point to the Rambla de Tabernas is from the N-340a near its junction with the A-92 at km-376. Park at the Tourist Office (Oficina de Turismo), which is next to the entrance to the Oasys theme park. There are trails into the rambla on the right to the rear of the Tourist Office and also from the opposite side of the N-340a. It is possible to walk all the way to Tabernas along the rambla from here and there are optional diversions into other ramblas all the way.

&. Many areas may be viewed from your vehicle. The off-road terrain is difficult.

CALENDAR

All Year: Bonelli's Eagle, Peregrine Falcon, Common Kestrel, Little Bustard, Stone-curlew, Black-bellied Sandgrouse, Eagle and Little Owls; Dupont's, Crested and Thekla Larks, Crag Martin, Black Redstart, Black Wheatear, Blue Rock Thrush, Sardinian Warbler, Jackdaw, Raven, Rock Sparrow, Trumpeter Finch, Rock Bunting.

Swallow, Tawny Pipit, Rufous-tailed Scrub-robin, Nightingale, Black-eared Wheatear; Olivaceous, Melodious, Spectacled, Subalpine and Orphean Warblers, Spotted Flycatcher, Woodchat Shrike.

Winter: Meadow Pipit, White Wagtail, Common Chiffchaff.

Breeding season: Lesser Kestrel, Scops Owl, Great Spotted Cuckoo, Bee-eater, Roller, Hoopoe, Red-rumped

SIERRA ALHAMILLA AL6

Status: Paraje Natural (8,392ha). ZEPA.

Site description
A small sierra and steppe site, about 25km long, with some flattish sections below the mountain proper, which rises to a peak at Calativí (1,387m). It is extremely dry, bordered as it is by the Tabernas desert region (AL5) in the north and the Campo de Níjar in the south, and is cut by deep ravines on its flanks. At higher levels most of the vegetation is low scrub but the habitats include abandoned cultivation and pine plantations, the latter largely within a relict Iberian Holm Oak forest. Parts of the southern side, running up from the Campo de Níjar, are deeply eroded.

Species
An important site for Trumpeter Finches. Dupont's Lark has been recorded and could nest here but any population will be very small so this species is best sought elsewhere. Other steppe or semi-arid country species present include a few Little Bustards, Stone-curlews and Black-bellied Sandgrouse, as well as Thekla Larks, Black-eared Wheatears and Southern Grey Shrikes. Black Wheatears, Rock Sparrows and Rock Buntings are present in the rocky terrain. Great Spotted Cuckoos occur in the olive groves, orchards and sparsely wooded areas. Raptors include Booted, Bonelli's and Golden Eagles, Lesser Kestrels and Eagle Owls.

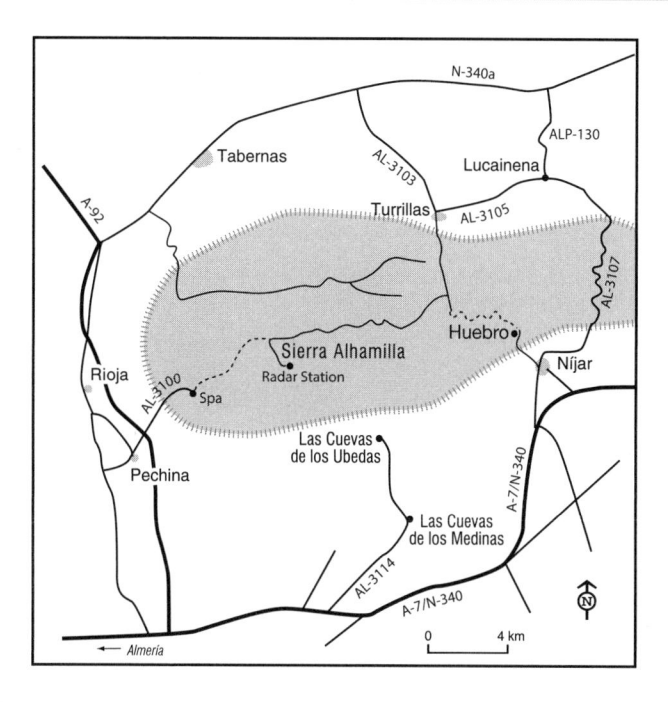

Timing

Visits can be profitable at any season but, as in other dry areas, the best time is early in the day between March and early June, when the summer visitors are present and breeding. It gets extremely hot in summer when temperatures may exceed 40°C.

Access

Access to the core of the area is available from the A-7 (km-479) in the south, taking the Níjar road northwards. Alternatively, from the north, take the AL-3103 southwards from the N-340a some 8.5km east of Tabernas, signposted for Turrillas. From Níjar you can continue on the AL-3107 to Lucainena and on to Turrillas. A still quieter alternative is to bear west from Níjar to Huebro. This latter road is good tarmac winding through the foothills. Beyond Huebro, however, the road is good gravel for a few kilometres before joining the tarmac road to Turrillas on the tops of the sierra. The uplands proper are served by generally good gravel tracks that run west–east more or less across the top. There is though a tarmac track heading south from Turrillas and leading eventually to a radar station that crosses excellent typical habitat of the higher elevations. Access to the foothills along the western flank of the sierra is possible by taking the AL-3100 from Pechina up to a spa (Baños de Sierra Alhamilla); beyond here the road is a gravel track.

All these tracks and roads offer good opportunities to see the species of rocky mountain terrain. The open-country, steppic, species are best sought in the southern approaches to the Sierra, for example, along the road to the village of Cuevas de los Ubedas. Turn north at the roundabout on the N-344, 500m west of Retamar, on to the AL-3114 following signs for Cuevas de los Medinas and

Cuevas de los Ubedas. Open country before Cuevas de los Medinas offers some small chance of locating Dupont's Larks. Black-eared Wheatears occur and are much more readily encountered. Trumpeter Finches are also present, especially further up, beyond Cuevas de los Medinas, but these too are often elusive. The 'badlands' along this road may reward exploration.

♿ Observation from your vehicle is possible. The off-road terrain is difficult.

CALENDAR

All year: Bonelli's and Golden Eagles, Common Buzzard, Peregrine Falcon, Goshawk, Little Bustard, Stone-curlew, Little and Eagle Owls, Black-bellied Sandgrouse, Thekla and Dupont's Larks, Black Wheatear, Blue Rock Thrush, Dartford and Sardinian Warblers, Blackcap, Southern Grey Shrike, Jackdaw, Raven, Rock Sparrow, Trumpeter Finch, Rock Bunting.

Breeding season: Booted Eagle, Lesser Kestrel, Great Spotted Cuckoo, Bee-eater, Hoopoe, Alpine Swift, Red-rumped Swallow, Black-eared Wheatear, Common Chiffchaff, Golden Oriole, Woodchat Shrike.

SIERRA DE MARÍA AL7

Status: Parque Natural (25,562ha).

Site description

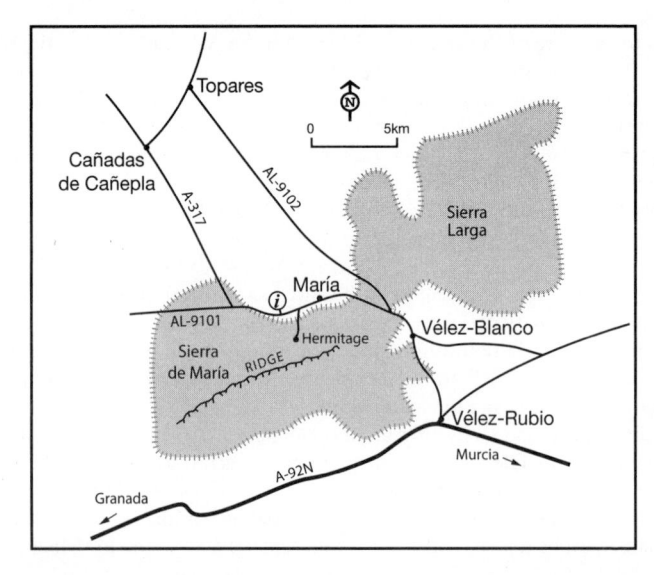

This sierra of dolomitic chalk lies in the relatively humid northernmost part of Almería, where it forms part of the Betic Cordillera. The rock formations include some impressive cliff faces and the region is renowned for its concentration of rupestral art sites. The vegetation is typically Mediterranean, with zones of evergreen scrub, followed by pines and oaks at the higher levels. This is a beautiful and well-protected area, very different from the arid regions that characterise much of Almería. There is considerable precipitation in winter, including significant snowfall in some years.

Species

The park offers a good and representative selection of the species of Mediterranean woodlands and rocky habitats, such as are typical of the Betic range as a whole. These include Short-toed Treecreepers, Crested Tits; Northern, Black and Black-eared Wheatears, Subalpine and Orphean Warblers, Southern Grey and Woodchat Shrikes, and many more. Golden Orioles, Crossbills and Rock Sparrows breed near the hermitage. Winter visitors include Ring Ouzels, Dunnocks and Alpine Accentors. There are Griffon Vulture colonies and other breeding raptors include Golden Eagles, Goshawks and Hobbies. Nearby lowland areas of cereal cultivation and open terrain have nesting Calandra Larks and also attract small numbers of Little Bustards, Stone-curlews and Black-bellied Sandgrouse.

Timing

This sierra is worth visiting at any time of year, even in winter, when the relatively mild climate attracts species that descend from the higher and harsher Sierra Nevada to the west. It can still be very cold here in winter, however.

Access

Proper exploration of the region requires walking some of the many hiking trails, which are the only way to access much of the core of the Natural Park. Details and maps of such trails are available from the visitors' centres, as well as online. There is a visitors' centre and viewpoint (Mirador de La Umbría) west of María at km-2.7 on the A-317 and another (Almacén de Trigo) in the village of Vélez Blanco. For general birding purposes, however, the vicinity of the hermitage (Ermita de la Virgen de la Cabeza) on the northern slopes above María village is convenient. The short access road is signposted on the A-317 1km west of María.

Park at the hermitage and continue on foot for another 500m to the Botanic Garden and visitors' centre there. Disabled parking (only) is available at the garden. The garden itself may be of interest for woodland birds but the circular walking trails beyond are more promising. There are three options of differing lengths and degrees of difficulty, the longest of which covers 2.9km. The trails offer panoramic views and access to scrub and woodland habitats. It is possible to continue beyond the highest point of the marked trail, to reach the rocky summit ridge at the Puerto del Portal Chico; this takes at least four hours, there and back, and demands a high degree of physical fitness.

Birds of open country may be sought in the cereal farming region to the northwest of the Sierra. A circular route using the A-317 and AL-9102, taking in María, Cañadas del Cañepla and Topares, traverses representative areas. Stop and scan frequently.

♿ Observation from your vehicle is practical. The shorter Botanic Garden tracks are conditioned for those of limited mobility.

CALENDAR

All year: Griffon Vulture, Golden and Bonelli's Eagles, Goshawk, Peregrine Falcon, Thekla Lark, Wren, Black Redstart, Stonechat, Black Wheatear, Blue Rock Thrush, Mistle Thrush, Dartford and Sardinian Warblers, Blackcap; Long-tailed, Crested and Coal Tits, Short-toed Treecreeper, Southern Grey Shrike, Jay, Jackdaw, Raven, Carrion Crow, Rock Sparrow, Crossbill; Rock, Cirl and Corn Buntings.

Breeding season: Short-toed and Booted Eagles, Hobby, Scops Owl, Common Cuckoo, Bee-eater, Roller, Hoopoe, Wood Lark, Golden Oriole, Nightingale, Northern and Black-eared Wheatears, Rufous-tailed Rock Thrush; Melodious, Subalpine and Orphean Warblers, Woodchat Shrike.

Winter: Robin, Ring Ouzel, Fieldfare, Redwing, Common Chiffchaff, Dunnock, Alpine Accentor.

Cereal fields and steppic habitats only: Quail, Little Bustard, Stone-curlew, Lapwing (winter), Calandra and Crested Larks, Greater Short-toed Lark (summer).

ANTAS ESTUARY AL8

Status: Special conservation area (50.98ha).

Site description

A tiny estuary hemmed in by the 'villa steppe' of touristic developments. There is abundant reed growth along both margins of the river, as well as a quite ample area of water ringed by reeds that is easily visible from the sand bar that closes off the river mouth. The river generally only flows after the scarce rains when the sand bar breaks and the pent-up water is let out. The breach is usually closed again fairly rapidly by wave action once the rains stop.

Species

Small numbers of Purple Swamphens occur here. The resident Cetti's Warblers are joined by Reed, Great Reed and Spectacled Warblers in summer.

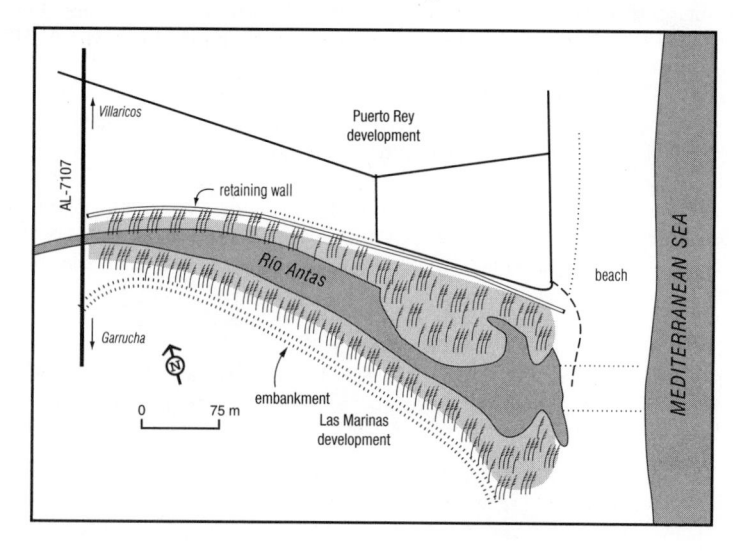

Black-necked Grebes appear in winter. Migrant and wintering seabirds that use the estuary or may be seen from the beach include Balearic Shearwaters and Audouin's Gulls. The principal interest perhaps is the possibility of transient migrants, especially given the scarcity of coastal wetlands in this region. Such species have included White-headed and Ferruginous Duck, Little Bittern and Glossy Ibis. Penduline Tits may winter.

Timing
This site is worth a look at any season if you are passing through the area although it is best in winter and early spring. Visit early in summer and at weekends to avoid human disturbance.

Access
The estuary is at Puerto Rey, between Garrucha and Villaricos. The AL-7107 coastal road crosses the river. Turn shorewards at the roundabout just north of the bridge, and bear left until you reach the beach, where you can park. There is also a track along the southern embankment. Walk down to the sandbar and along the footpaths along both sides of the lagoon. Parking is also available on the waterfront on the south bank.

&. Only limited viewing is possible from your vehicle.

CALENDAR

All year: Little Grebe, Purple Swamphen, Yellow-legged Gull, Sandwich Tern, Hoopoe, Cetti's and Sardinian Warblers, Jackdaw.

Breeding season: Reed, Great Reed and Spectacled Warblers.

Winter: Black-necked Grebe, Great Cormorant, Lesser Black-backed Gull, Razorbill, Common Chiffchaff, Penduline Tit, Black Redstart, Southern Grey Shrike, Jackdaw, Common Magpie.

Passage periods: Balearic Shearwater, Audouin's Gull, Common and Little Terns.

ALMANZORA ESTUARY　　　　　　　　　AL9

Status: No specific protection.

Site description
The mouth of the Río Almanzora is channelled between dykes that enclose dense reedbeds and variable pools of standing water. The river discharges into the Mediterranean only occasionally, after rains. There are low, rocky outcrops just off the beach.

Species
The channel and its reedbeds shelter Little Bitterns, both Grey and Purple Herons and Purple Swamphens. Teal hide in the reeds in winter, when Black-necked Grebes, Great Crested Grebes and Razorbills are often present offshore, as well as occasional Red-breasted Mergansers.

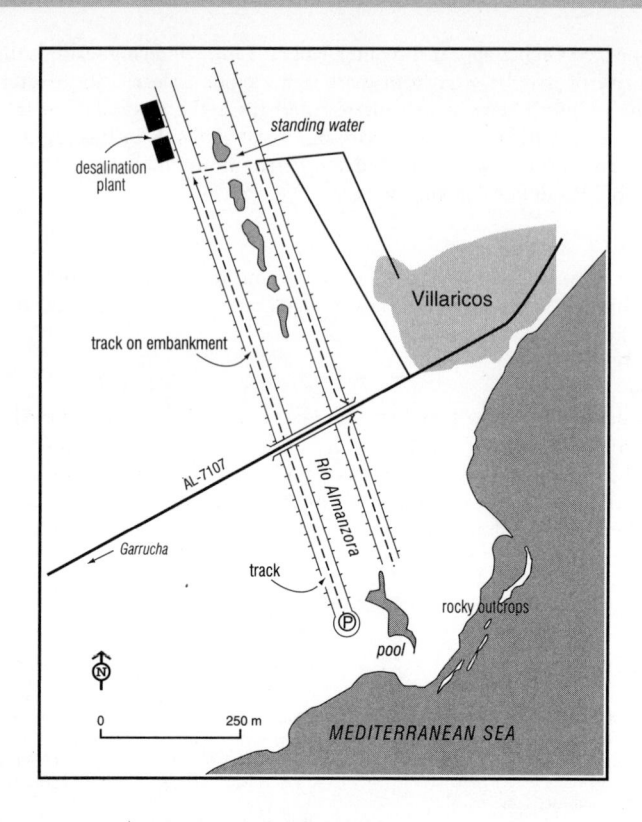

The rocks and shoreline are attractive all year to Kentish Plovers and Sanderlings and a variety of other waders, notably Common Redshanks, Black-tailed Godwits, Whimbrels and occasional Turnstones. Great Cormorants often hang themselves out to dry on the rocks in winter. Flocks of Audouin's Gulls often rest here, especially in late summer. Numbers of Sandwich and Common Terns occur on migration and Caspian Terns have also been recorded.

Timing
Interesting species may occur at any time of year. Early morning visits are necessary in summer and at weekends to avoid the often intense human disturbance later.

Access
The channel is just 1km south of Villaricos on the AL-7107. The river embankments are visible from some distance. Access by car is possible along the top of the western embankment, both downstream and upstream of the bridge. A usually wet area alongside the desalination plant, 400m upstream, attracts Common Snipe and a variety of migrant waders. Alternative viewing is available from the eastern embankment, where there is a pedestrian/cycling track.

♿ Observation from your vehicle is possible all along the embankment.

CALENDAR

All year: Little Grebe, Grey Heron, Purple Swamphen, Kentish Plover, Sanderling, Sandwich Tern, Hoopoe, Kingfisher, Dartford Warbler, Common Magpie, Linnet, Tree Sparrow, Corn Bunting.

Breeding season: Little Bittern, Common and Pallid Swifts, Red-rumped Swallow.

Winter: Black-necked and Great Crested Grebes, Teal, Red-breasted Merganser, Northern Gannet, Great Cormorant, Black-winged Stilt, Common Redshank, Common Snipe, Lesser Black-backed Gull, Razorbill, Crag Martin, Common Chiffchaff, Black Redstart.

Passage periods: Balearic Shearwater, Purple Heron, Osprey; Ringed, Little Ringed and Grey Plovers, Ruff, Whimbrel, Little Stint, Little and Audouin's Gulls; Caspian, Common and Little Terns.

EXTREMADURA

Extremadura covers 41,634km², a little less than half the extent of Andalucía but still making it the fifth largest of Spain's autonomous regions. It comprises only two provinces, Badajoz in the south and Cáceres in the north. Together they offer excellent birds, other wildlife and peaceful, often unspoilt landscapes. Extremadura is widely recognised, both within the country and abroad, as one of the prime and unmissable birding areas of Spain. As a result, increasing numbers of overseas visitors and enthusiasts from other parts of Spain come to Extremadura in search of birds and no doubt this trend will continue. The conservation of this outstanding region is being helped by the growing realisation among the powers that be that the wildlife interest is one of its major assets.

Although official interest in conservation has increased, Extremadura still has a long way to go before it can match the comprehensive inventory of officially protected zones that is such a welcome feature of Andalucía. Most of all, there still remains a particular need for the establishment of large reserves to protect the communities of steppe birds and other open-country fauna and flora, which are arguably the most outstanding of the region's natural assets and among its most endangered.

The inclusion of Extremadura in the first edition of this book in 1994 was intended to increase awareness of the potential for ecotourism in this wonderful area and to encourage local conservation efforts. It is therefore particularly pleasing to note that local and regional authorities in Extremadura are now keen to promote ecotourism. Tourist offices offer a diversity of free publications, including details of the many walking trails that have been established throughout the region, and information on sites of birding and other wildlife interest.

Why Extremadura? I cannot recommend it too strongly. It really is essentially just one big birding site and interesting species can be encountered just about anywhere. Birdlife International has developed an inventory of Important Bird Areas (IBAs) worldwide. Under their criteria, over 75% of the area of Extremadura qualifies as IBAs, hugely more than for any other Spanish region.

This is by far the best part of the area covered by this book in which to find the open-country 'steppe' birds. Many sites offer excellent opportunities, among the best in Spain, to see both Great and Little Bustards, Stone-curlews, both Pin-tailed and Black-bellied Sandgrouse and a full range of the other species of open terrain. The birds occur in important numbers too. It is not at all unusual to find flocks of over 100 Great Bustards, for example.

The plains of Extremadura and their birds are a major draw but the defining habitat of the region is the grazing woodland or dehesa, which covers thousands of hectares. The typical dehesa consists of more or less widely spaced Iberian Holm Oaks, known locally as encinas, separated by grassland and sometimes cereal crops, which are planted as part of a rotation. Dehesa of Cork Oaks is also widespread. Both formations are typically used as foraging areas for livestock, for which branches are cut from the trees periodically to provide browsing opportunities. Herds of pigs gorge on the acorns in autumn. The same habitats are also used by Cranes, which also eat acorns as well as spilt grain. Extremadura

Little Bustards

is a key wintering area for Cranes and flocks of hundreds or even thousands occur here then.

Plains birds apart, the many forested and rocky sierras support a comprehensive raptor population, with significant numbers of Black-winged Kites, Black Vultures and Spanish Imperial Eagles as well as a full range of commoner species. Black Storks occur in good and increasing numbers: the breeding population of at least 200 pairs represents some 40% of the Iberian population. A comprehensive range of smaller species includes good numbers of Great Spotted Cuckoos, Rollers, Azure-winged Magpies, Spanish Sparrows and Rock Sparrows. All these species can be found relatively easily.

The region is not renowned for its natural wetlands but human activities have created important breeding, staging and wintering sites for very large numbers of waterfowl and waders. The many reservoirs are especially attractive to wintering ducks, which occur in thousands at favoured localities. The development of rice farming in the Guadiana valley has also created great expanses of artificial marshland, popular with flocks of storks and Cranes and attractive to waders for much of the year.

When to come

Extremadura is an excellent choice for a touring holiday of a week or more. It can also be combined with visits to Andalucía or to regions further north, such as the Sierra de Gredos. It is important to realise that there is plenty on offer all year round. However, local guide John Muddeman tells me that over 50% of all birding visitors to Extremadura come in April. This is undoubtedly a good time to visit, especially if you are able to miss the busy Easter holiday period. A springtime expedition will provide most of the sought-after species, including the summer visitors; the countryside is resplendent with flowers, birdsong is everywhere and displaying bustards and raptors are at their most conspicuous. I certainly recommend April and May to first-time visitors but urge you to consider autumn and winter visits as well, to see the spectacle of the wintering Cranes, the hordes of wildfowl on the reservoirs and the steppe birds in their winter flocks as opposed to scattered on their breeding territories. February can be particularly good since the wintering birds are still present and raptors are beginning to occupy their nest sites. Off-season visits are of course particularly quiet, even at sites as popular as Monfragüe. High summer is probably best avoided for obvious reasons but good birding is available even then.

Where to go

Many birding visitors to Extremadura actually only ever see the Cáceres/Trujillo plains and Monfragüe and they return home happy to have found all or most of the steppe birds and raptors for which the region is renowned. These key areas should certainly figure on your itinerary but you really should explore further afield – and many rewarding places are suggested here. You will see some additional species and, more importantly, you will experience some wonderful unspoilt landscapes, representing much of what makes the Spanish countryside such a vital part of Europe's natural heritage.

Where to stay in Extremadura

Visitors often base themselves in one spot and sally out daily to visit the various sites from there. This will be the most practical arrangement if you are only coming for a few days and, in that case, the best base undoubtedly is Trujillo. Longer visits are better arranged around two or more bases since you will otherwise have to drive considerable distances on some days. Long drives are tiring and time consuming but the biggest problem is that you tend to arrive everywhere in the late morning, having missed the productive early morning period. This inconvenience is greatly exacerbated in the hotter months, when bird activity in the middle of the day is much reduced.

Within Badajoz province, Zafra is best placed for the southern and southwestern woodlands (BA1 & BA3) and is also convenient for Hornachos (BA4). Badajoz city is in the far west, adjacent to several interesting wetlands (BA12–14) especially. Mérida is ideal for many of the central Badajoz sites (BA4–7), including Hornachos and the Canchales and Cornalvo reservoirs. Both those cities have Paradores and a range of lesser-grade hotels. To visit the eastern sites of the Sierra de Tiros (BA8) and La Serena (BA9), where early morning/evening visits are often particularly important, it is useful to find accommodation in Castuera or Puebla de Alcocer. At the former, the Hotel Los Naranjos (two-stars, Carretera Benquerencia km-39, 06420 Castuera, Badajoz. Tel. 924 761 054), is inexpensive and strategically placed.

Within Cáceres province, both Cáceres city and Trujillo are well placed for Monfragüe (CC6), the Cáceres–Trujillo plains (CC3) and Cuatro Lugares (CC4). However, Trujillo is much more accessible and it is also handy for the Sierra Brava and Alcollarín reservoirs (CC10, CC12) and the Guadiana Valley ricefields (CC11). Cáceres is, however, better placed for the Sierra de San Pedro (CC1) and the Malpartida steppes (CC2). Both Cáceres and Trujillo have Paradores and a range of other hotels.

To visit Monfragüe, especially if you intend to stake out the Eagle Owls at dusk or dawn, you may find it convenient to stay close by at the village of Torrejón el Rubio, just to the south of the park boundary. There are several hostales and hotels here. The park itself offers a campsite at Villareal de San Carlos. There is also plenty of suitable accommodation nearby to the north, at Malpartida de Plasencia and in Plasencia itself.

Staying in Guadalupe is convenient for visiting the easternmost sites of Extremadura, including the sierras of Las Villuercas (CC4), the eastern Badajoz reservoirs (BA10) and Cíjara (BA11). As well as the Parador, Guadalupe offers a full range of accommodation, including a campsite.

Finally, Plasencia is the best base for northernmost Extremadura, including the nearby Jerte valley (CC8) and also the Sierra de Gata (CC7) and the Campo de Arañuelo (CC9), but it is also within very easy reach of Monfragüe, which lies less than 30 minutes unhurried drive to the south: the park boundary is 17km from Plasencia and the key sites of El Salto del Gitano and La Portilla del Tiétar are both only 33km from the city. Plasencia has a Parador and a wide range of hotels. Jarandilla de la Vera, in the Gredos foothills northeast of Plasencia, also has a Parador.

BADAJOZ PROVINCE

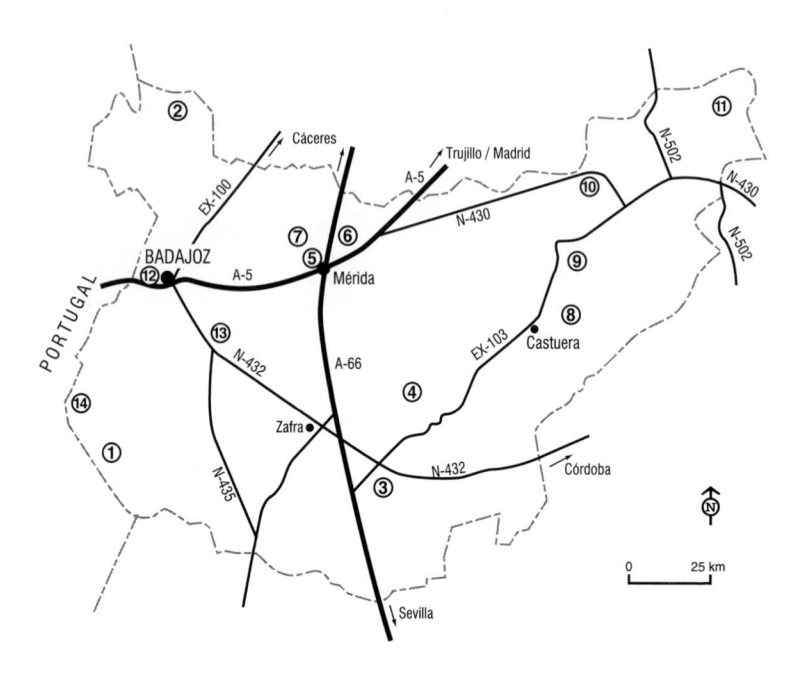

Badajoz province provides a mixture of wooded hilly country and large tracts of open land. Much of the latter is rough pasture and it includes the impressive expanse of La Serena (BA9), one of the best areas of steppe-like habitat in Spain. Bustards and other steppe species are widespread. The River Guadiana flows east to west across the top of the province before turning southwards for the Atlantic. Dams on the river have created a complex of large reservoirs in the north-east of the province (BA10, BA11), as well as along the watercourse at Mérida (BA5), Badajóz city (BA12) and on the Portuguese border (BA14). These provide a guaranteed supply of water in a naturally parched land and many of them are of birding interest, especially in winter when some attract large numbers of waterfowl and roosting Cranes.

A result of the increased availability of water for irrigation has been a rapid expansion of rice cultivation, chiefly east of Mérida and around Madrigalejo (CC11). The rice paddies attract a good variety of waders and other aquatic birds, chiefly in winter and on migration, although some species also breed there. Badajoz provides ready opportunities to see raptors, steppe birds and some wetland species but it also offers many extensive areas of tranquil and enjoyable country that will reward patient exploration.

Sites in Badajoz Province
BA1   Dehesas and reservoirs of Southwestern Badajoz
BA2   The Western Sierras. Sierra de San Pedro
BA3   Southern Badajoz

BA4   Sierra de Hornachos
BA5   River Guadiana at Mérida
BA6   Cornalvo reservoir
BA7   Los Canchales reservoir
BA8   Sierra de Tiros
BA9   Plains of La Serena
BA10   Orellana and other eastern reservoirs
BA11   Cíjara Game Reserve (Reserva Regional de Cíjara)
BA12   River Guadiana at Badajoz City
BA13   Lagoons and farmlands of La Albuera
BA14   Alqueva reservoir inlet

Getting there

The A-66 motorway, linking Gijón on the Biscay coast with Sevilla, crosses Badajoz north/south and provides the obvious link with Andalucía. Visitors arriving in Spain via ferry to Santander or Bilbao will also find it convenient for reaching Extremadura generally. Rapid access is also available via Madrid, using the A-5 Madrid/Badajoz toll-free motorway, which crosses the province diagonally.

DEHESAS AND RESERVOIRS OF SOUTHWESTERN BADAJOZ   BA1

Status: Includes the ZEPAs 'Dehesas de Jerez' (48,000ha) and 'Embalse de Valuengo' (274ha).

Site description

The grazing woodland or dehesa dominates this peaceful corner of Extremadura and vast areas are covered by encinas, the Iberian Holm Oaks. The denser formations are bird-rich if somewhat monotonous to view. However, the southern part of the region has the best woodland, with mature stands of encinas and Cork Oaks from Villanueva to Oliva and, especially, between Higuera de Vargas and Jerez de los Caballeros where there are some particularly fine trees set in very open dehesa, giving excellent opportunities to scan for raptors. The road from Oliva to Valencia del Mombuey also has a more open, park-like aspect with widely spaced trees and excellent views across undulating open countryside.

Wetlands include a number of lagoons, especially southeast of La Albuera (BA13), and several small reservoirs. The latter in particular attract waterfowl, especially in winter, when roosts of Cranes also occur: perhaps the best of them are the Embalse del Aguijón and the Embalse de Cuncos. The Embalse de Valuengo is also worth visiting.

Species

The region offers excellent opportunities to see woodland species of the Extremaduran woodlands, including the many species which winter there, which include several thousand Cranes. The dehesas support an important population of Black Storks of some 30 pairs: here they nest in trees and not on cliffs, as is usual elsewhere in Extremadura. Breeding raptors include Black-winged and Red Kites. Higuera has a significant colony of Lesser Kestrels. The resident passerines of the dehesa include Rock Sparrows and Hawfinches. Orphean Warblers are present in summer. Other attractions include a good population of Otters and Genets, the latter occasionally seen on roads at night.

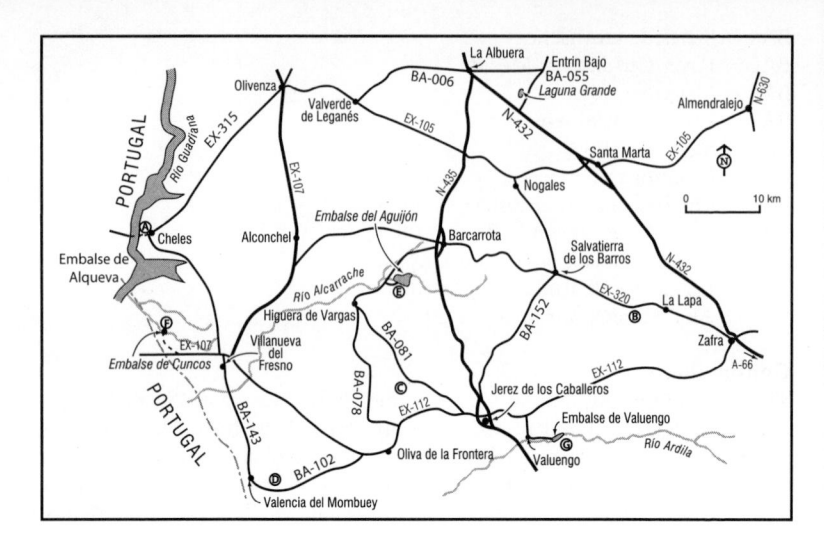

Timing
The dehesas are of interest all year round but are at their finest in spring, when huge numbers of wild flowers are in bloom and the woodlands resound to the songs of Hoopoes, Common Cuckoos, Bee-eaters and many passerines. Raptors are obvious displaying over their territories. They include Black-winged Kites, which are most easily observed at dusk. The heat of summer is best avoided. The reservoirs attract most birds, including roosting Cranes, in winter but visits in late summer may reveal Black Storks and passage waders.

Access
The region can be visited on a circular course starting from Badajoz or from Zafra, to the east. A gentle drive, visiting all the main habitats and with frequent stops to scan for raptors, will take all day since at least 200km will be covered. The minor roads indicated give access to all the main habitats. At Cheles (A) a road descends to the shoreline of the Embalse de Alqueva, on the River Guadiana; worth visiting except in summer when watersports are based here (see instead BA14).

A good route to sample the woodland species is to take the EX-320 (B) from Zafra via La Lapa to Salvatierra de los Barros. From there take the BA-152 to Jerez de los Caballeros. Both these very peaceful roads offer elevated views over the woodlands and there are many stopping places. Another worthwhile drive is to take the BA-081 northwest to Higuera de Vargas from the EX-112 4km west of Jerez (C), returning south via the BA-078, visiting excellent dehesas along both roads. You may then descend on the EX-112 to Oliva de la Frontera, to take the long loop (31km) westwards through more open country to Villanueva del Fresno via Valencia de Mombuey (D): stop at the bridges to look for Black Storks along the watercourses, notably at the Río Alcarrache, where footpaths follow the riverbank both upstream and downstream.

The Embalse del Aguijón (E) is reached from the BA-025, which skirts the reservoir some 8km from Barcarrota. The entrance is signposted 7.1km from Higuera. There is an information centre but this is often closed. Bear right to drive on to the dam for views of the lake and dehesa. A dirt road continues around the lake margin from here, offering good opportunities to walk through the dehesa.

The Embalse de Cuncos (F) is 7km west of Villanueva de Fresnos, north of the EX-107 and near the Portuguese border: take the signposted road at km-70.8, which crosses some good steppe habitat – offering bustards and sandgrouse among others – for 3km to reach the reservoir, where there are several hides.

The Embalse de Valuengo (G) occupies a quiet valley, 3km from Valuengo. Follow the signposts east from the village. Stop before the dam and view from there or walk across and view from the track around the southern side of the lake.

♿ Most areas, including the reservoirs, may be viewed from or near your car.

CALENDAR

All year: Mallard, Gadwall, Great Crested Grebe, Cattle Egret, Black-winged and Red Kites, Griffon and Black Vultures, Common Buzzard, Thekla and Wood Larks, Crested Tit, Nuthatch, Rock Sparrow, Hawfinch.

Breeding season: Night Heron, Little Egret, Black Stork, White Stork, Black Kite, Short-toed and Booted Eagles, Montagu's Harrier, Collared Pratincole, Little Ringed Plover, Scops Owl, Red-rumped Swallow, Orphean Warbler.

Winter: Pintail, Shoveler, Common Pochard, Tufted Duck, Great Cormorant, Greylag Goose, Crane, Lapwing, Golden Plover, Song Thrush, Redwing, Blackcap.

THE WESTERN SIERRAS. SIERRA DE SAN PEDRO BA2

Status: ZEPA (115,032ha).

Site description
This sparsely populated area of low, wooded country to the southwest of the city of Cáceres mainly comprises a large tranche of northwest Badajóz province. Much of the region described here falls south of the N-521 Portugal/Cáceres trunk road where, parallel to the road, a line of sierras appears as a long, low green wall. Cork Oak woodlands, with an excellent understorey of Cistus, broom, lavenders and other shrubs, cover these hills. Further south still, the

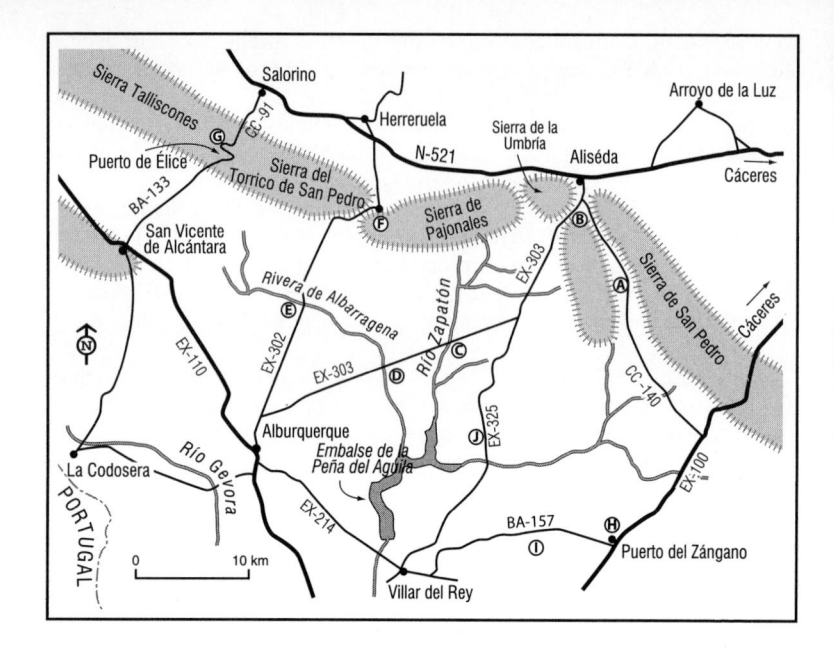

terrain becomes much more open with expanses of cultivation near the towns and villages and large areas of severe, boulder-strewn steppelands, the latter especially north and east of Alburquerque. Very quiet open dehesa is also characteristic of parts of the south.

The long ridge in the east of the region, within Cáceres province, is the Sierra de San Pedro. Further west there are a series of parallel ranges, interspersed with small river valleys. Rocky escarpments along the sierras make them attractive to Black Storks and crag-nesting raptors. Indeed, the area is renowned as an excellently preserved remnant of the original Mediterranean forest cover of the region, with stands of Cork Oaks and encinas and matorral of Cistus and lavender.

The quiet road along the Portuguese border, via La Codosera, is also worth a visit. This is a region of small isolated villages characterised by stone-walled fields, rustic cottages with Hoopoes on the TV aerials, and an ever-thriving population of misshapen little dogs of untraceable pedigree. Raptors and woodland species are much in evidence.

Species

This is among the best areas in Extremadura for seeing Spanish Imperial Eagles. Black Vultures are also particularly common and easily seen overhead: the breeding population of several hundred pairs is among the largest in Spain. The density of breeding raptors is high generally and soaring shapes are omnipresent over the woodlands and ridges. There are good populations of the other four eagles as well as Black Storks, Black-winged Kites and Eagle Owls. Alburqueque houses a sizable Lesser Kestrel colony. The forests and scrub support a rich community of woodland birds, including a truly massive wintering population of Woodpigeons, these last attracted by the abundance of acorns: roosts of up to one million birds are on record. The region is also the best place in Extremadura to see Bullfinches, also in winter.

Timing
Springtime and early summer visits are recommended but interesting species are present all year round.

Access
The region is easily explored by car, starting from the N-521 at Aliseda, from Cáceres via the EX-100 or from the south at Alburquerque. The imposing castle at Alburquerque is a commanding landmark and a good watchpoint for raptors in its own right.

The drive from the N-630 north of Mérida to Alburquerque via the EX-214 through La Nava de Santiago and Villa del Rey passes through orchards and olive groves that are alive with passerines, especially in winter. Once in the region, a zig-zag route may be followed, stopping frequently to scan the slopes and woodlands or simply whenever interesting-looking birds appear, which is very often. The elevated vantage points given below are especially recommended but it is often necessary to spend some time there before such sought-after species as Spanish Imperial Eagles appear, although you may be lucky sooner. Apart for the EX-100, which links the two provincial capitals, most of the roads are quiet and stopping is easy. The Alburquerque/Herreruela stretch can be busy occasionally but even here there are plenty of farm tracks enabling safe parking off the carriageway.

The minor tarmac road (CC-140, A) heading southeastwards from the N-521 at Aliseda, past the Sierra de San Pedro, is of interest. For the first 10km nearest Aliseda the road follows a valley between two low, rocky ridges, which should be scanned for soaring raptors. Frequent stops along this road are recommended, with exploration on foot into the adjacent woodlands. The road crosses mature encinares and corkwoods.

The EX-303 road is often productive, especially in the northern sector (B). An elevated viewpoint and picnic spot at km-19.4 on the EX-303, alongside the Sierra de la Umbría, is well placed for scanning across the woodlands. The picturesque bridges across the Río Zapatón (C) and Río de Albarragena (D)

Spanish Imperial Eagle

are also good stopping points, where Black Storks may sometimes be seen fishing (and frogging) in the stony beds. Similarly, stopping on the EX-302 at the south side of the bridge across the Río Albarragena (E) is worthwhile: the woodlands and river may be scanned here and walks along the banks reveal waterside and woodland passerines. Also on the EX-302, panoramic views across the region are available from the disused railway station of La Herreruela (F); follow the signs up the access track to the open area near the station.

Panoramic views are also to be had from the Salorino/San Vicente de Alcántara road (BA-133), at its high point of the Puerto de Elice (G), which overlooks the woodlands and sierras and provides an excellent scanning point for raptors. A good place to stop is on the north side of the road just south of Puerto de Elice, from where you can explore the woodland and scrub on foot as well as view the southern face of the ridge. In addition, further south still, the entrance to Finca La Galana at km-68.2 is well sited to scan the ridge for raptors: Black Vultures may be seen on their nests here but a telescope is needed since the distance is about 1km. The northern slopes above Salorino should also be visited. Both La Herreruela and Puerto de Elice often provide good sightings of Spanish Imperial Eagles.

Rapid access to the southern part of the area, and the CC-140 alongside the Sierra de San Pedro, is possible from the EX-100. This busy road crosses some good country, including some excellent plains for the 15km nearest Cáceres, but opportunities for safe stopping are limited. Still, there are points of vantage where the road crosses wooded ridges, at km-20, km-26.8 and km-36.8 and at the Puerto del Zángano (H) at km-40. The BA-157 (I), linking the EX-100 and Villar del Rey, crosses very quiet open dehesas that are excellent for passerines and woodland raptors. The EX-325, north from Villar del Rey (J), also crosses good habitat and links with the EX-303.

♿ Most areas, including key watchpoints, may be viewed from or near your car.

CALENDAR

All year: Black-winged and Red Kites, Griffon and Black Vultures, Goshawk, Sparrowhawk, Common Buzzard; Spanish Imperial, Golden and Bonelli's Eagles, Eagle and Long-eared Owls, Kingfisher, Thekla Lark, Southern Grey Shrike, Azure-winged Magpie, Rock Sparrow.

Breeding season: Black Stork, White Stork, Black Kite, Egyptian Vulture, Short-toed and Booted Eagles, Lesser Kestrel, Great Spotted Cuckoo, Scops Owl, passerines.

Winter: Crane, Woodpigeon, Song Thrush, Redwing, Bullfinch.

SOUTHERN BADAJOZ BA3

Status: Includes ZEPA 'Campiña Sur y Embalse de Arroyo Conejos'.

Site description

Much of the undulating terrain of southern Badajoz, to the east of the A-66, is open country and largely devoted to cereal fields and rough grazing, interrupted by olive and encina groves. Stony steppe-like habitat occupies large tracts in some areas, notably around Usagre, southwest of Fuente de Cantos and to the east of Valencia de las Torres/Llerena. A small reservoir, the Embalse de Arroyo Conejos (de Llerena), is of interest and there are a number of ephemeral lagoons that only appear in wet years.

Species

Open-country species, including bustards, sandgrouse and other steppe birds, are present if thinly spread. Mixed flocks of House and Spanish Sparrows are frequent, with Rock Sparrows present near Segura de León and also elsewhere. Lesser Kestrels occupy many of the nestboxes attached to powerline posts in the east of the region. A good selection of raptors will be encountered. Cranes winter in very large numbers, especially east of Llerena, foraging in the fields and dehesas and roosting at the Embalse de Llerena. Islets in the reservoir have breeding Pratincoles, Gull-billed Terns and Little Terns in spring and good numbers of waterfowl occur in winter.

An Iberian Lynx introduction project is underway in the southeast of the area; you will see signs warning of lynx on the road but are unlikely to see the animals themselves.

Timing

Visits can be rewarding all year round.

Access

This area (Map 1) is likely to be visited en route to other sites, notably La Serena (BA9). The N-432 between Zafra and Llerena (A) passes through some

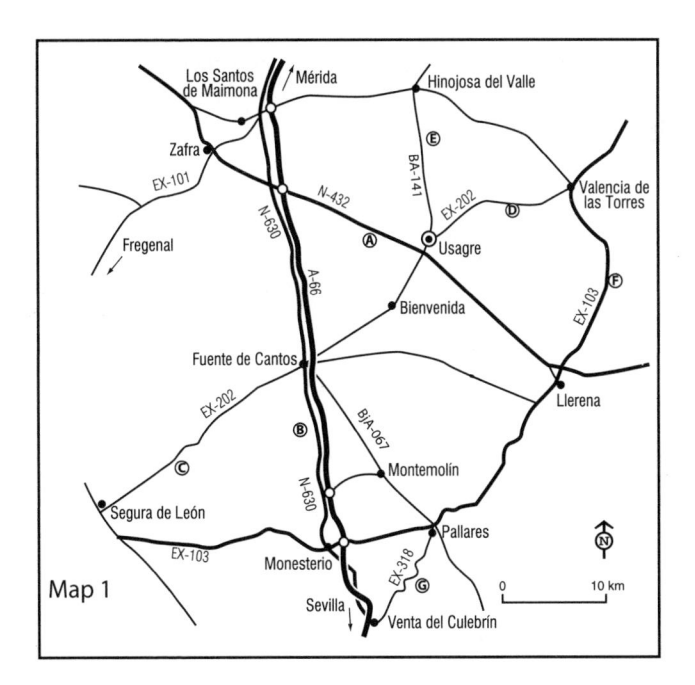

Map 2

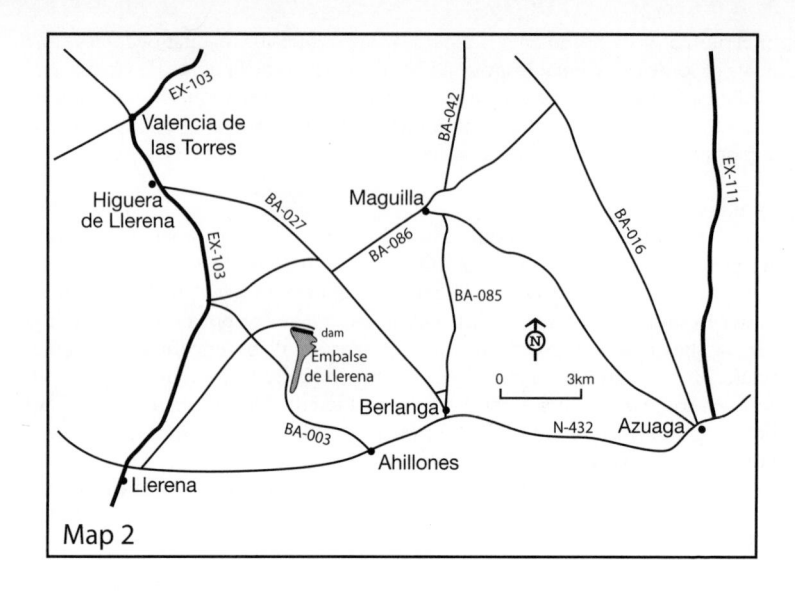

good steppe habitats but, as ever, stopping on busy main roads is strongly discouraged, except where farm tracks provide lay-bys. The minor roads mapped are a selection of those from which the area can be surveyed at greater leisure. Farm tracks are frequent and allow closer inspection of the more promising pastures. In addition, the replacement of the N-630 by the A-66 has made it worth visiting the section of the old road south of Fuente de Cantos (B), which crosses excellent steppe terrain. The N-630 remains as a service road and is kept in good repair and benefits from very light traffic. Scanning the open terrain along sections of the EX-202, notably between Segura de León and Fuente de Cantos (C) and between Usagre and Valencia de las Torres (D), and also along the quiet minor road between Usagre and Hinojosa del Valle (E) often proves worthwhile. The EX-103 between Valencia de las Torres and Llerena (F) also offers wide vistas. The EX-318 (G) winds through hilly terrain and is good for raptors, notably vultures, which no doubt are attracted by the livestock farms in the valleys.

The rolling farmlands east of the EX-103 (Map 2) are sometimes rewarding to visit for open-country species. A good tarmac track heading northeast from the N-432 near Llerena leads to the Embalse de Arroyo Conejos (de Llerena). The BA-003 northwest from Ahillones, also on the N-432, or the first right turn along on the BA-086 to Maguilla (signposted from the EX-103 11km north of Llerena), similarly leads to the reservoir, which can be viewed from the dam on the north shore. Recommended roads to drive in this region are the BA-086 itself, continuing northeast beyond Maguilla to the BA-016, by keeping right at the first fork on the BA-042, and then driving south to Azuaga. Further east still the EX-111, heading north from Azuaga towards Campillo de Llerena, also crosses promising open country.

♿ Most areas may be viewed from or near your car.

CALENDAR

All year: Black-winged Kite, Griffon and Black Vultures; Golden, Spanish Imperial and Bonelli's Eagles, Great and Little Bustards, Stone-curlew, Black-bellied Sandgrouse, Calandra Lark, Zitting Cisticola, Southern Grey Shrike, Azure-winged Magpie, Red-billed Chough, Spanish and Rock Sparrows.

Breeding season: Black Stork, White Stork, Black Kite, Egyptian Vulture, Short-toed and Booted Eagles, Lesser Kestrel, Montagu's Harrier, Great Spotted Cuckoo, Roller, Hoopoe, Short-toed Lark, Tawny Pipit.

Passage periods: Garganey, Spoonbill, waders.

Winter: Greylag Goose, ducks including Red-crested Pochard, Crane, Red Kite, Hen Harrier, Merlin, Lapwing, Golden Plover.

SIERRA DE HORNACHOS   BA4

Status: Local Nature Reserve of ADENEX (in part). ZEPA (12,190ha).

Site description

Viewed from the A-66, between Almendralejo and Villafranca de los Barros, the Sierra de Hornachos appears as an isolated rocky island above a gently rolling expanse of open country. Approaching from the south, from Hinojosa del Valle, the Sierra reveals a long, grey, flat-topped ridge, with some cliff faces and extensive scree slopes where the dominant Cistus scrub seems to have relinquished its grip and slid off the mountain. The red-roofed hamlet of Hornachos reposes against the southern slopes and is surrounded by olive groves. To the east and north there are woodlands of encinas and Cork Oaks, with a well-developed understorey of Cistus, Retama and other shrubs. For a change, the Presa de Campillo, north of Campillo de Llerena, holds back a small lake with good reedbeds. Two other reservoirs in the area, the Embalse de los Molinos and the massive Embalse de Alange, are also worth visiting.

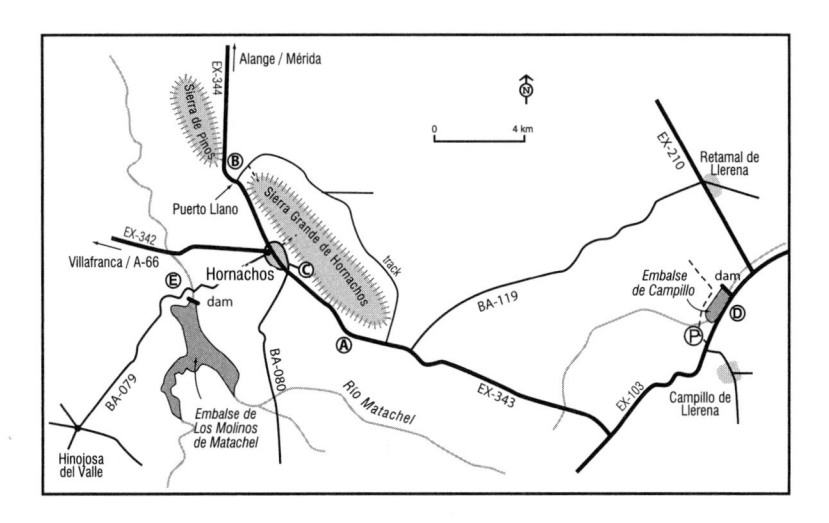

Species

Hornachos provides another chance to see the raptor community of Extremadura with most species represented. The Sierra is a good site for Egyptian Vultures and both Golden and Bonelli's Eagles. Spanish Imperial Eagles also occur and the area regularly attracts dispersing immatures of the species. Woodland passerines are evident. The reservoir and reedbeds at Campillo hold small numbers of waterfowl. Marsh Harriers occur and may nest here. Sparrow flocks around the reservoir include Spanish Sparrows.

Cranes, Great Crested Grebes and waterfowl frequent the Embalses de Los Molinos and de Alange in winter. The dam at Alange houses a large colony of Alpine Swifts and Black Wheatears occur on the adjacent rocky ridge. The area downstream of Alange dam is worth inspecting for aquatic birds: it often attracts herons, egrets, ducks and even Spoonbills.

Timing

Springtime visits are most productive.

Access

Hornachos is readily reached from the west, turning off the A-66 at Villafranca de los Barros and taking the EX-342 to Hornachos, via Ribera del Fresno. Footpaths from the village allow access to the Sierra but the principal species are best located from some distance. Good views are to be had from the roads leading to the village of Hornachos. The EX-343 (A) offers suitable vantage points at many points east of the village. Stop in entrances to the olive groves and other similar 'lay-bys'.

The Hornachos/Alange road (EX-344) climbs up the slope to the west of Hornachos, passing between that Sierra and the Sierra de Pinos, again giving good views of the area. On this latter road, at Puerto Llano (B), a track gives access to the northern slopes. It is possible to drive along this track, which is in good repair, to scan the steep escarpments on the northern side of the Sierra or to explore the woodlands from footpaths. Good stopping places early along this road are the col 0.6km from the entrance and the entrance to a footpath at 1.1km. The latter is marked Mirador de La Magrera. The path is a stiffish but worthwhile climb for only 990m to the mirador, which is an excellent place for viewing the Sierra and its hinterland. The road itself continues all the way around the 'back' of the Sierra but at some distance, eventually joining the EX-343. Take the right fork after 6.5km and turn right again at 8.2km to approach the slopes again. The back road passes through pastures favoured by Bee-eaters, Short-toed Larks and Black-eared Wheatears, among others.

A couple of signposted walking trails ascend the Sierra from the highest reaches of the village. There is an information centre in the centre of the village (Calle Extremadura), open daily except Mondays (C).

There is a stopping point and 'picnic site' on the EX-103 at the south end of the reservoir at Campillo (D), which is accessed by a short track. This track also gives access on foot to marshy riverine vegetation, good for passerines.

The Embalse de los Molinos (E) is readily viewable from the Hornachos/Hinojosa road, which skirts the northwest of the lake and gives elevated views. The river downstream of the dam is also worth inspecting; the bed is a complex of pools, reedbeds and canebrakes and there is some riverine woodland of alders and poplars, which regularly attracts Golden Orioles.

Alternative easy access to Hornachos is from the north, taking the riverside road from Mérida to Alange. The footpath parallel to this road follows the south shore of the Guadiana and is well placed for looking for riparian passerines and Night Herons. There are car parks at either end of the dam at Alange (not mapped) and a footpath leads around the southern side of the rocky outcrop at Alange from here, giving good views of the ridge and across the lake. Continue to Hornachos along the east shore of the Embalse de Alange (BA-005 and EX-344). The reservoir itself may be scanned from where the perimeter roads cross its various arms: the best access is on the southern shore where farm tracks allow close approach to the shoreline.

♿ Footpaths are inaccessible but other areas may be viewed from or near your car.

CALENDAR

All year: Griffon and Black Vultures, Marsh Harrier, Common Buzzard, Spanish Imperial, Golden and Bonelli's Eagles, Eagle Owl, Rock Dove, Thekla Lark, Black Wheatear, Blue Rock Thrush, Azure-winged Magpie, Red-billed Chough, Spanish and Rock Sparrows, Hawfinch, Rock Bunting.

Breeding season: Black Stork, White Stork, Black Kite, Egyptian Vulture, Short-toed and Booted Eagles, Montagu's Harrier, Lesser Kestrel, Great Spotted Cuckoo, Scops Owl,

Red-necked Nightjar, Alpine Swift, Bee-eater, Roller, Hoopoe, Wryneck, Short-toed Lark, Red-rumped Swallow, Rufous-tailed Scrub-robin, Black-eared Wheatear; Great Reed, Reed, Subalpine and Orphean Warblers, Iberian Chiffchaff, Golden Oriole.

Winter: Greylag Goose, Mallard, Wigeon, Gadwall, Teal, Shoveler, Common Pochard, Tufted Duck, Great Crested Grebe, Great Cormorant, Spoonbill, Crane, Alpine Accentor, Bullfinch.

RIVER GUADIANA AT MÉRIDA   BA5

Status: Includes ZEPA 'Embalse de Montijo' (176ha).

Site description
The city of Mérida, on the River Guadiana, is best known for its well-preserved Roman ruins, including the famous bridge, amphitheatre and aqueducts. The old town is attractive but the newer developments south of the river are less pleasing. No doubt the Roman artefacts will outlive them. The city offers the usual haven for rooftop birds, although these are more evident in Cáceres and Trujillo. The river itself has a variable rate of flow, unsurprisingly given the number of reservoirs upstream and the frequent droughts of recent years. The river bed thus has numerous, generally exposed gravel banks with intermediate pools and clumps of rank vegetation, all of them attractive to birds. There are several wooded islets upstream of the Roman bridge and also in the Embalse de Montijo, a long thin reservoir on the river immediately west of the city.

Species
Breeding White Storks, Lesser Kestrels and Common, Pallid and Alpine Swifts enliven the city. The Puente Romano (Roman Bridge) has a colony of some 30 pairs of Alpine Swifts and some Common and Pallid Swifts, which are

present from March to October. Eucalyptus trees and dense scrub on islets just above the Roman bridge support heronries, principally of Cattle Egrets (1,000+ pairs). Both Squacco Herons and Spoonbills breed nearby, at least occasionally. There is a good population of Purple Herons at the Embalse de Montijo, where Great White Egrets first nested in 2016. Purple Swamphens also frequent riverside reedbeds. The riverside scrub attracts finches and other passerines, notably Red Avadavats and Common Waxbills. Numbers of Great Cormorants, Black-headed Gulls and Lesser Black-backed Gulls feed and roost along the river in winter.

A most surprising development here since 2002, annually at first but sporadically later, has been a small passage of Audouin's Gulls across Extremadura in August, along the Guadiana and nearby. Most of the birds involved are juveniles and their rings show that some at least originate from the Ebro Delta colony. Minimum counts of up to 19 birds have been made and the total numbers involved have been somewhat larger in some seasons. They accompany flocks of Black-headed Gulls, which are presumably dispersing to their winter quarters.

Timing
Spring and summer visits are necessary for many species but the area is of interest year-round.

Access
See map for BA7. Mérida is at a crossroads between the A-66 and the A-5 trunk roads. The river and its birds are easily viewed from promenades along both banks and, especially, from the pedestrians-only Roman bridge. A park has been constructed on the two largest islands, just off the east bank, with bridges from the riverside and a ramp from the Roman bridge itself giving access. The island upstream of the bridge allows close observation of the adjacent heronries. Both islands have clumps of reeds and scrub, which are worth searching for passerines. The Embalse de Montijo lies along the River Guadiana, immediately west of the city. It is best viewed from the southern bank, where it is flanked for 4.3km by a road and another riverside park. A track on the opposite bank of the river leads west along a disused railway line and is also useful for finding riverside passerines.

♿ Easily accessed.

CALENDAR

All year: Cattle and Little Egrets, White Stork, Purple Swamphen, Crag Martin, Cetti's Warbler.

Breeding season: Little Bittern; Great White Egret; Night, Squacco and Purple Herons, Spoonbill, Lesser Kestrel, Little Tern; Pallid, Common and Alpine Swifts, Common Waxbill, Red Avadavat.

Winter: Waterfowl, Great Cormorant, Common Coot, Black-headed Gull, Lesser Black-backed Gull.

Passage periods: Audouin's Gull.

CORNALVO RESERVOIR (Embalse de Cornalvo) BA6

Status: *Parque Natural (13,143ha). ZEPA.*

Site description
Cornalvo can claim to be a well-established reservoir: the Romans built the original dam and a smaller one to the northwest. The principal interest of the site is the excellent and very accessible woodland of encinas and Cork Oaks which borders the horn-shaped lake. These woods are a fine example of their kind and shelter a diverse plant and animal community. The southern part of the site is open country.

Species
Breeding species include Black Storks, Lesser Spotted Woodpeckers and a range of raptors including Black Vultures and Black-winged Kites. The open country to the south has some steppe species but these are unlikely to be seen during a brief visit, although Montagu's Harriers are often evident. Several hundred Cranes occur in winter. The reservoirs attract some waterfowl, chiefly in winter.

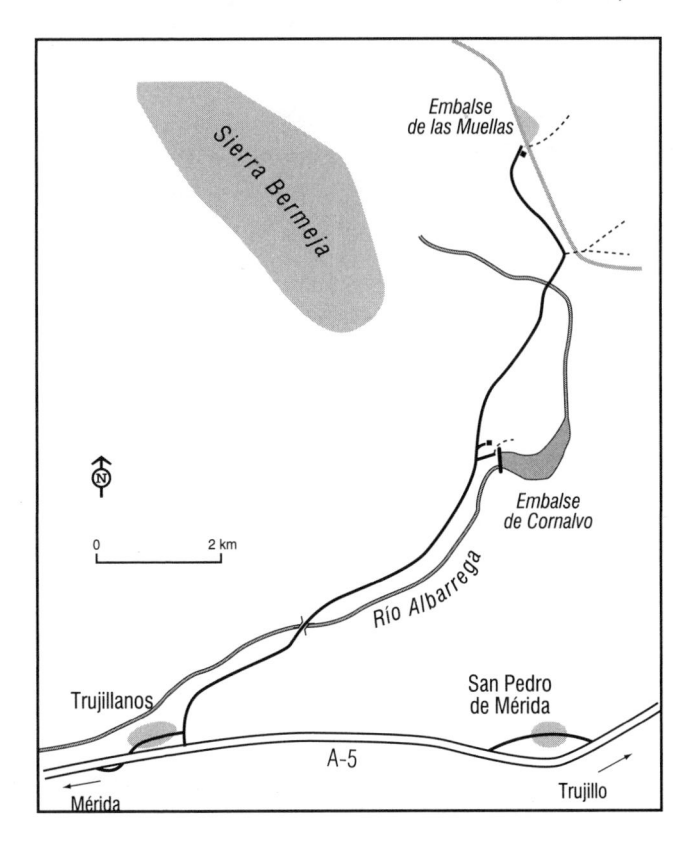

Timing
Spring and early summer for breeding species. Winter for Cranes and waterfowl. The site is a mecca for picnickers from Mérida at weekends or on public holidays and is best avoided then, not least because the access road is very narrow and there are few passing places.

Access
The reserve lies to the north of the A-5, 18km northeast of Mérida and 5km from the town of Trujillanos. It is most obviously signposted from the westbound carriageway of the A5 at km-331. From the eastbound carriageway take the Trujillanos exit and follow the minor road from Trujillanos for 6.5km to the dam. Explore the reservoir fringe and woodlands on foot. The walk around the entire perimeter of the reservoir, crossing the dam, is a pleasant stroll and takes an hour or two. It is also worthwhile continuing north on the access road, past Cornalvo, for a further 6.5km through interesting woodland, chiefly of Cork Oaks, to a second small reservoir, the Embalse de las Muellas, which also attracts some waterfowl and passage waders. Two hides facilitate viewing here and a number of walking trails are signposted, giving access to the woodlands.

♿ The areas around the dams are easily accessed.

CALENDAR

All year: Black-winged and Red Kites, Griffon and Black Vultures, Common Buzzard, Golden and Bonelli's Eagles, Tawny Owl, Lesser Spotted Woodpecker, Thekla Lark, Southern Grey Shrike, Zitting Cisticola, Rock Sparrow. Ringed Plover, Great Spotted Cuckoo, Scops Owl, Red-necked Nightjar, White-rumped Swift, Red-rumped Swallow; Spectacled, Subalpine and Orphean Warblers, Golden Oriole.

Breeding season: Black Stork, White Stork, Black Kite, Short-toed and Booted Eagles, Montagu's Harrier, Little

Winter: Waterfowl, foraging raptors, Common Coot, Crane.

LOS CANCHALES RESERVOIR BA7
(Embalse de Los Canchales)

Status: ZEPA (600ha).

Site description
This reservoir was constructed during the mid-1980s to serve the city of Mérida. Unlike so many such water bodies, some account was taken of the potential wildlife interest of the site when the reservoir was planned. The design provided for shallow feeding areas and the construction of islands to provide secure nesting sites for waterfowl. The result is an interesting reservoir, attractive to birds and birders alike, and well worth a visit. A telescope is highly desirable since the lake is very large.

Species
The reservoir attracts waterfowl all year round but especially in winter when dabbling ducks, including Mallard, Teal, Gadwall, Wigeon, Pintail and Shoveler,

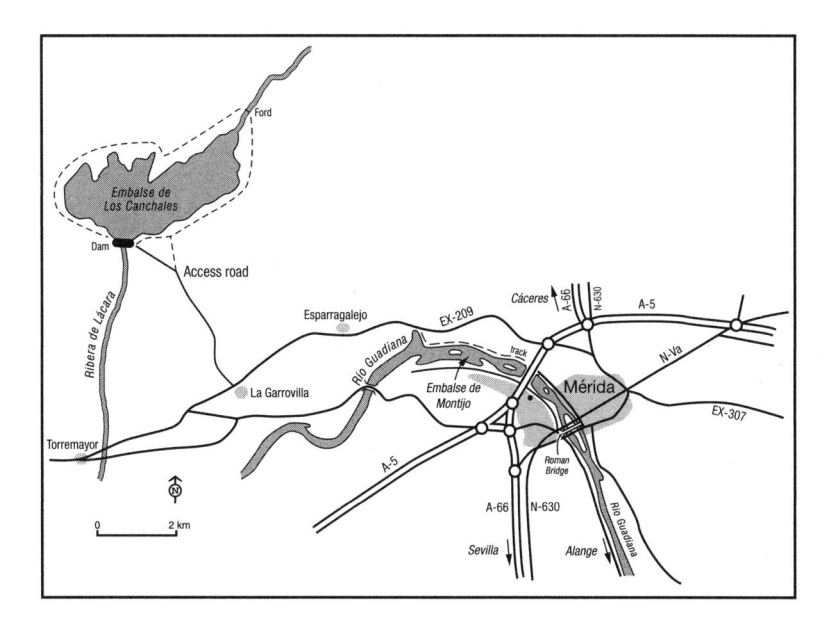

are present in some numbers, as well as Common Pochards, Tufted Ducks and Greylag Geese. This is the best inland site in Spain for Spoonbills, with flocks of over 100 birds present on passage in September. Some Spoonbills are present all year round and a few pairs breed in the heronry nearby. Other breeding aquatic species include Black-winged Stilts and there is a colony of Gull-billed Terns. Wintering and passage waders are another attraction, with a great diversity of species on record, especially during migration periods. Lapwings and Golden Plovers occur in numbers in winter and a wide range of other wader species occur both then and in spring and autumn. Ospreys are frequently present during passage periods and Black-winged Kites nest in the vicinity. There is a large Crane roost in winter, when up to a few thousand make a fine spectacle as they arrive in the late evening to settle in the marshy pastures along the eastern arm of the lake. The olive groves and open woodlands of encinas surrounding the lake attract the usual passerine community. Large flocks of hirundines occur on passage.

This site is something of a hotspot for regionally scarce species, so it is as well to be prepared for the unexpected. Great White Egrets occur annually in winter. Audouin's Gulls occur on passage in some years, generally in August; most records are of small numbers and most are of juveniles – a record of 66 at Los Canchales in August 2004 was exceptional. Scarcer species have recently included Oystercatcher, Sanderling, Knot; Bean and White-fronted Geese, Caspian Tern and White-winged Black Tern as well as real rarities such as Green-winged Teal, Yellow-billed Stork, African Spoonbill and Pectoral Sandpiper.

Timing
The site is always of interest and may turn up some surprises during passage periods. Bird numbers are highest in winter. The site is popular with local

anglers at weekends and gets busy as a picnic site on Sundays and public holidays especially. Hence weekday visits are more peaceful and may well prove more rewarding.

Access
The reservoir lies to the northwest of the city of Mérida. Take the EX-209 west from the N-630 for 15km to La Garrovilla. A sign 'Los Canchales' opposite a workshop (Talleres Guerrero) marks the entrance to a good tarmac road, which leads north for some 6km to the lake (ignore all side turnings). A collection of sand heaps on the right after 1.6km holds a Bee-eater colony. The tarmac road continues to the left where the lake comes into view and leads for 2km to the dam, where there is parking and an information board. It is best to park here and cross the dam on foot, continuing along the west bank along an excellent footpath through Mediterranean woodland. This path skirts the shoreline and gives access to several promontories equipped with blinds, from which the lake may be viewed. The east bank is accessible along a dirt track leading off right from the tarmac road. It is possible to drive around the reservoir to the northern shore (but not all the way round) but this involves negotiating a shallow ford and some damp stretches, which could cause problems to non-4WD vehicles in anything other than dry weather.

CALENDAR

All year: Mallard, Cattle and Little Egrets, Spoonbill, White Stork, Black-winged Kite, Common Coot, Great Bustard, Stone-curlew, Black-winged Stilt, Zitting Cisticola, Southern Grey Shrike.

Breeding season: Black Kite, Marsh Harrier, Collared Pratincole, Gull-billed Tern, Great Reed Warbler.

Winter: Greylag Goose, Mallard, Teal, Gadwall, Wigeon, Pintail, Shoveler, Common Pochard, Tufted Duck, Little and Great Crested Grebes, Crane, Golden Plover, Lapwing, Common Snipe, Black-headed and Lesser Black-backed Gulls, Short-eared Owl.

Passage periods: Osprey, Avocet, Little Ringed Plover, Little Stint, Curlew Sandpiper, Dunlin, Ruff, Black-tailed Godwit, Spotted Redshank, Common Redshank, Greenshank, Green Sandpiper, Wood Sandpiper, Common Sandpiper, Audouin's Gull, Little and Whiskered Terns, Black Tern (has bred), hirundines, passerines, rarities.

SIERRA DE TIROS   BA8

Status: No special protection.

Site description
The Sierra de Tiros forms part of a discrete and abrupt rocky ridge, between the towns of Castuera and Cabeza del Buey. The ridge clearly separates the rolling steppelands of La Serena from the dehesas further south. The Sierra itself is rather barren in its uppermost reaches, with occasional vertical rock faces. Lower down there is well-developed scrub and some woodland, with olive groves near the towns. The Sierra provides points of vantage over both La Serena to the north and the Dehesas de Benquerencia, the savanna-like open woodlands, to the south. The northern side of the Sierra may be viewed from

tracks leading to and beyond the Santuario de Belén, on the fringes of La Serena.

Species

Breeding species include Black Stork, White Stork, Griffon Vultures and both Golden and Bonelli's Eagles, as well as the usual woodland and scrub passerines. Aerial insectivores are often numerous feeding along the ridge. They include Common and Alpine Swifts, Red-rumped Swallows and House and Crag Martins. It is worth scanning the swifts carefully since several White-rumped Swifts have been seen here, some of them even in late October, and they may breed in the area. The steppe species of La Serena (BA9) are present immediately to the north of the Sierra. Montagu's Harriers are abundant near the Santuario de Belén. The trees in the garden of the Santuario itself house a large raucous colony of both Spanish and House Sparrows, including some obviously hybrid individuals. In winter, up to several thousand Cranes cross the area twice daily as they travel to and from their roost at the Embalse del Zújar and their feeding grounds in the open woodlands south of Benquerencia.

Timing

All year, for raptors and steppe species. Summer and autumn for White-rumped Swifts. Winter for Cranes. The Cranes fly south shortly after dawn, traditionally passing through the gaps in the Sierra at Puerto Mejoral, just east of Benquerencia, and further east, through a second interruption in the ridge at Almorchón. They return at dusk along the same routes but more especially then at Almorchón.

Access

See map for BA9. The EX-104 provides easy access to the southern flanks of the Sierra. The road ascends to Benquerencia from Castuera. At Benquerencia the car park on the right when entering the town from Castuera gives fine views over the dehesas. The rocky ridge above the town may be scanned for swifts, hirundines and raptors, notably Bonelli's Eagles; you can walk up to the medieval castle on the skyline for even better views. The 'Crane Gap' at Puerto Mejoral is obvious as the road descends towards La Nava, just east of Benquerencia. ADENEX have constructed a simple watchpoint (Francisco Carbajo Bird Observatory) south of the road at Puerto Mejoral, which provides an excellent vantage point for watching the flypast of Cranes and other birds through the Sierra. The site is marked by a sign (Paso de Grullas) and parking is available next to a telecommunications mast. Follow signs up past the mast to the observatory. Opposite, ADENEX also have an information centre at their Nature School in Puerto Mejoral, with exhibits relating to the wildlife of the Sierra de Tiros and La Serena, although this is only open on special occasions.

The road further east provides plenty of stopping places (entrances to olive groves and farm tracks), from which to scan the Sierra. The 'Crane Gap' at Almorchón is marked by the ruined castle of that name, atop an isolated crag. Cranes fly through this gap in the evenings especially. The minor road to the Santuario de Nuestra Señora de Belén (G), opposite the castle, is the best place to stop here. The same road offers good opportunities to see numbers of Montagu's Harriers in season.

It is also possible to drive past the Santuario to the right and on to a dirt road, which is a main access point for the network of farm tracks serving the southern parts of La Serena and from which the full range of steppe species may be seen (see BA9, F). The principal track follows a railway line and then branches northwest across the plains for a total of some 20km, to meet the EX-103 road 8km north of Castuera. The track is passable to all vehicles in dry weather but the low-lying sectors are apt to become muddy and more challenging in wet weather, when 4WD would be advisable. You can also keep following the railway line to reach Castuera directly.

♿ Most areas may be viewed from or near your car.

CALENDAR

All year: Cattle Egret, Griffon Vulture, Golden and Bonelli's Eagles, Great and Little Bustards, Stone-curlew, Black-bellied and Pin-tailed Sandgrouse, Rock Dove, Eagle Owl, Crag Martin, Black Wheatear, Blue Rock Thrush, Red-billed Chough; House, Spanish and Rock Sparrows, Cirl and Rock Buntings.

Breeding season: Black Stork, White Stork, Black Kite, Egyptian Vulture, Short-toed and Booted Eagles, Montagu's Harrier, Lesser Kestrel, Red-necked Nightjar, Alpine and White-rumped Swifts, Crag Martin, Red-rumped Swallow.

Winter: Crane, Dunnock.

PLAINS OF LA SERENA BA9

Status: ZEPA (153,702ha). These plains are of the greatest interest and importance since they harbour some of the largest concentrations of steppe birds found in Spain, but they have yet to receive the high degree of protected status that they urgently require.

Site description
The undulating steppelands of eastern Badajóz province comprise La Serena, some 100,000ha of rough and often stony pastureland with occasional wheatfields and scattered farmhouses. The region is almost treeless although occasional small clumps of eucalyptus along the roads harbour sparrow colonies and the nests of White Storks. This is one of the best areas of its kind in Spain but the key steppe species are declining, perhaps because of a reduction in habitat quality through overgrazing by sheep. Periodical spraying with insecticides, used to control the massive and undoubtedly spectacular springtime plagues of the Moroccan Locust *Dociostaurus maroccanus*, no longer occur since this large grasshopper has declined considerably.

Species
Breeding steppe species: notably Montagu's Harriers, Red-legged Partridges, Great and Little Bustards, Stone-curlews, Black-bellied and Pin-tailed Sandgrouse, Rollers and Calandra Larks. Collared Pratincoles breed locally and Mallard frequent the occasional ponds. Cranes, Golden Plover and Lapwings occur in winter. Some 30 Dotterel have overwintered in successive recent years, with over 100 found in March 2016. Raptors of a wide range of species occur in small numbers. Birds apart, even the sheep are apt to provide interest during

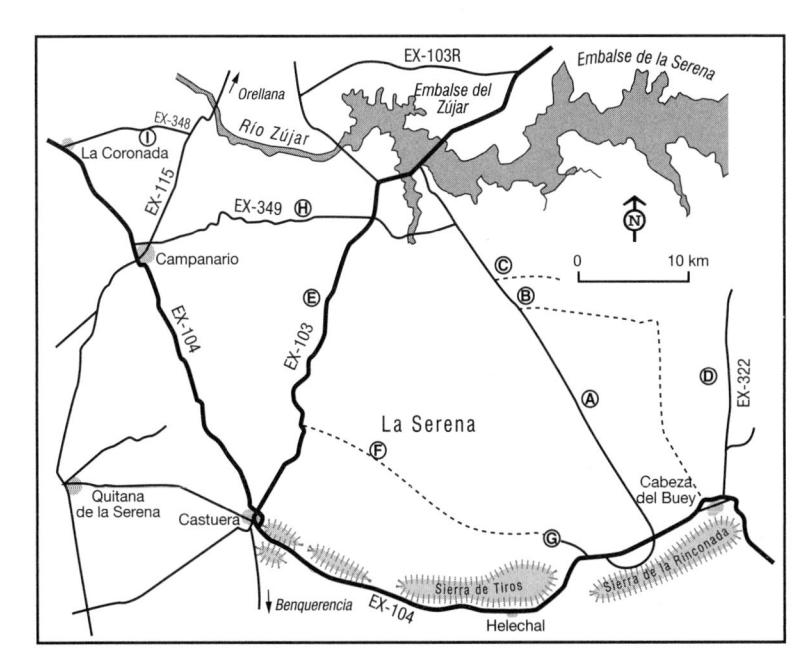

slack periods: the local Merinos have evolved a curious but no doubt effective strategy to cope with periods of intense sunlight, huddling together in tight, circular groups of up to 100 or more, all heads turned towards the centre and tucked between the legs of the animal in front, rugby-scrum fashion.

Timing
Springtime visits reveal displaying bustards and singing larks and other passerines. All times of year will provide birds but the heat of the day in summer can be daunting. Morning and evening visits are recommended and are all but essential in hot weather: the birds are more active then and much more easily located. Accordingly, it is best to start or end your visit by staying somewhere close by, ideally in Castuera, Benquerencia or Cabeza del Buey. At least half a day should be allowed to inspect the area since some species can be elusive. Areas with good densities of the very obvious Calandra Larks are often more generally productive.

Access
Minor roads criss-cross the area and there are numerous tracks leading to farms, which may be negotiated carefully without 4WD in many cases. Please ask for permission to enter private land whenever possible. A telescope is strongly recommended. Bird numbers are high but the area is vast and it may be necessary to cover a lot of ground before all the typical species are located. Even when birds are not immediately obvious, waiting in suitable habitat very often reveals flocks of sandgrouse flying in or bustards moving between feeding grounds. Rather thoughtlessly, many of the farmers keep pigeons, which may be confused with sandgrouse at first glance, so some care with initial identifications is called for.

The precise location of birds varies as land use may shift. Typically the bustards like growing crops in which to nest but they are most obvious when foraging on the open stony pastures. This latter habitat too attracts sandgrouse and Stone-curlews.

The road from Cabeza del Buey to the Embalse del Zújar (A) is often the most productive. This road is not numbered but is good tarmac; locally it is called Carretera de las Golondrinas (Swallow Road), for no obvious reason. The whole road has a number of obvious broad, sandy tracks leading off across the steppes, all of which are worth exploring as far as conditions and your vehicle permit. That leading eastwards from a point 32km north of its junction with the EX-104 (B) is an excellent example of an accessible track that often reveals steppe species. The entrance is opposite two eucalyptus trees and is marked by a large, red 'No Entry' sign but nobody seems to object to birding visitors. The track is particularly reliable for Great Bustards and both sandgrouse species.

Another good track to explore for steppe species also leads eastwards from the EX-104 from a point a further 3.5km from the 'No Entry' track (C). It is inconspicuously signposted 'Finca Pavorosa' but there is a more obvious large yellow box-like structure accompanied by a grey tank on stilts on the north side near the entrance. The road is excellent dirt and continues for several kilometres across undulating terrain. Stop frequently and scan for Great Bustards, which are often numerous here, as well as other steppe birds. In the east, the EX-322 (D) linking Cabeza del Buey to Puebla de Alcocer passes through some good terrain and is another option to explore.

The road between Castuera and the Embalse del Zújar (EX-103) (E) can be productive but tends to be relatively busy. Disused mines alongside this stretch have breeding Red-billed Choughs in spring and Rollers are also characteristic. A track (F) leading to one of the mines, Peña Lobosa, crosses La Serena diagonally, a distance of some 25km 'off-road'. The northern end is off the EX-103 11km north of Castuera. The going is sometimes rough and the track may not be passable without 4WD in wet conditions. The southern end is at the Santuario de Belén (G, see BA8), where there is also good access to the southernmost expanses of La Serena.

In the northwest, the east/west road (EX-349) between the EX-103 and Campanario (H) crosses very stony terrain. Pin-tailed Sandgrouse are characteristic here. Both sandgrouse are also typical of the open grassy steppeland along the east/west road (EX-348) to La Coronada (I), also in the north of La Serena.

♿ Most areas may be viewed from or near your car.

CALENDAR

All year: Red-legged Partridge, Cattle Egret, Griffon Vulture, Red Kite, Common Buzzard, Common Kestrel, Great and Little Bustards, Stone-curlew, Black-bellied and Pin-tailed Sandgrouse, Little Owl, Calandra Lark, Spanish Sparrow.

Breeding season: White Stork, Black Kite, Montagu's Harrier, Lesser Kestrel, Quail, Collared Pratincole, Bee-eater, Roller, Hoopoe, Short-toed Lark, Rufous Bush Chat.

Winter: Hen Harrier, Merlin, Crane, Golden Plover, Dotterel, Lapwing, larks, pipits and finches.

ORELLANA AND OTHER EASTERN RESERVOIRS
BA10

Status: Includes ZEPAs 'Embalse de Orellana & Sierra de Pela', Ramsar site, (42,600ha), 'Embalse de La Serena' (15,889ha), 'Embalse del Zújar' (1,203ha), and 'Puerto Peña – Los Golondrinos' (33,404ha).

Site description
The northern steppelands of La Serena (BA9) and the sierras in eastern Badajoz province, are eroded by several rivers whose dammed valleys comprise a complex of large reservoirs. The Embalses de Orellana, de García de Sola, and de Cíjara (BA11) lie along the course of the Río Guadiana. The Embalses del Zújar and de La Serena have been created along smaller rivers to the south of these. The reservoirs provide large bodies of standing water some of which, notably the Orellana reservoir, attract waterfowl. The waterside vegetation along the reservoirs and feeder streams attracts a range of interesting species. The wooded sierras between the two reservoir complexes, and north of the Embalse de Orellana (Sierra de Pela), also support a wide range of species.

Species
The reservoirs attract Great Crested Grebes, Mallard and other waterfowl in considerable numbers, especially in winter when the Embalse de Orellana may hold up to 20,000 ducks and as many roosting gulls. Gull-billed Terns nest there in summer. The Embalse de Gargáligas attracts large numbers of wildfowl in winter, when it is the best Extremaduran site for Pintail, with remarkable peak counts of over 15,000 birds. Large numbers of Cranes roost regularly at the

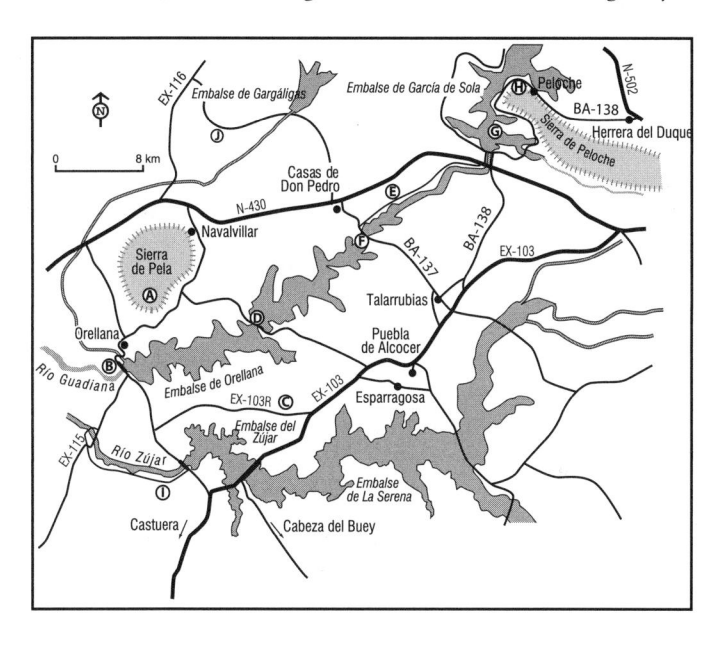

Embalses de Orellana and del Zújar during the winter months. Griffon Vultures nest above the dam of the Embalse de García de Sola, with Crag and House Martins using the dam itself. Black Storks, Egyptian Vultures, Bonelli's Eagles, Eagle Owls and Alpine Swifts all nest nearby. House Martin colonies are typical of all the dams and that of the Embalse del Zújar has hundreds of nests. The adjacent sierras attract a wide range of raptors and woodland species. The town of Talarrubias has noteworthy and photogenic nesting concentrations of White Storks on and around the main church and town square. Puebla de Alcocer boasts a notable colony of Lesser Kestrels.

The Rio Zújar, downstream of the dam of the Embalse del Zújar, has a sizable colony of Cattle Egrets, with some White Stork nests included. Eucalyptus groves nearby hold large colonies of House and Spanish Sparrows. Bee-eaters and Sand Martins have precarious colonies amid the gravel excavations along the river, where Little Ringed Plovers, Common Sandpipers and Little Terns also nest. Red-crested Pochards have bred here, the only location in Extremadura where they do so with any regularity. Breeding passerines include Cetti's, Great Reed and Orphean Warblers and Penduline Tits.

Timing

The region is at its best in winter (November to February) when waterfowl are abundant and dawn and dusk visits to the Embalses del Zújar and de Orellana will reveal concentrations of roosting Cranes. However, breeding season visits are necessary to find the Spanish Sparrows and other nesting birds along the watercourses. Storks, both Black and White, form premigratory gatherings along the Embalse del Zújar and elsewhere in late summer and autumn. An evening visit to the dam at the Embalse de García de Sola will provide good views of Griffon Vultures and other raptors going to roost, and a chance to see or hear Eagle Owls.

Access

Limited access to the reservoirs is available from minor roads and the approaches to the various dams. The region is best explored starting from Navalvillar de Pela (perhaps after exploring the nearby wetlands; see CC11) or Orellana, visiting all or some of the following sites: the margins of the Sierra de Pela (A) for raptors; the Presa (dam) de Orellana (B) for wintering waterfowl; the EX-103R (EX-350) road (C) linking Orellana and Puebla de Alcocer, which crosses steppe habitats; the Puente de Cogulludo (D), a bridge on the minor road crossing the Embalse de Orellana for waterfowl; and the Canal de las Dehesas (E), also on the Embalse de Orellana. This canal has a fringing tarmac road from which the upper reaches of the reservoir may be scanned for waterfowl and waders: it lies north of the reservoir and the canal to either side of the BA-137 but the upstream road is more interesting and leads to the Embalse de García de Sola. The bridge across the Embalse de Orellana on the BA-138 (F) is also worth a visit.

The Embalse de García de Sola (G) attracts some waterfowl but the chief attraction here is the area around the dam, where there is a Griffon Vulture colony and a pleasing range of cliff-nesting species generally. A viewpoint (Mirador de Puerto Peña), a short distance south of the west side of the dam, is ideally placed to view the rocky escarpment opposite; a telescope may be helpful. Further opportunities to find raptors and passerines of rocky hillsides are available from the BA-138 (H), where it skirts the Sierra de Peloche on the eastern side of the Embalse de García de Sola.

The productive region downstream of the dam of the Embalse del Zújar (I) is also worth visiting. Drive down the road which descends to the river at the southern end of the dam. Park at the car park and explore the minor road which follows an irrigation channel (Canal del Zújar) along the south bank of the river. The eucalyptus groves, with their sparrow colonies, are along the eastern part of this road, just below the Presa (dam) del Zújar.

The Embalse de Gargáligas is northeast of Navalvillar de Pela and accessible via the road (J) that follows the Canal de las Dehesas northwards from where the canal passes under the N-430 at Casas de Don Pedro or eastwards from where the canal meets the EX-116. Access is not always freely available, however.

♿ Most areas may be viewed from or near your car. The mirador at Puerto Peña is especially recommended.

CALENDAR

All year: Mallard, Red-crested Pochard, Little and Great Crested Grebes, Cattle Egret, Black-winged and Red Kites, Griffon and Black Vultures, Golden and Bonelli's Eagles, Rock Dove, Moorhen, Common Coot, Stone-curlew, Common Sandpiper, Eagle Owl, Thekla Lark, Crag Martin, Black Wheatear, Blue Rock Thrush, Penduline Tit, Azure-winged Magpie, Southern Grey Shrike, Spanish and Rock Sparrows, Rock Bunting.

Breeding season: Little Bittern, Night Heron, Little Egret, Black Stork, White Stork, Black Kite, Egyptian Vulture, Short-toed and Booted Eagles, Lesser Kestrel, Little Ringed Plover, Gull-billed and Little Terns, Great Spotted Cuckoo, Scops Owl, Bee-eater, Hoopoe, Roller, Alpine Swift, Sand Martin, Red-rumped Swallow, Rufous Bush Chat, Nightingale, Black-eared Wheatear; Cetti's, Reed, Great Reed, Melodious and Orphean Warblers, Golden Oriole.

Winter: Wigeon, Gadwall, Pintail, Shoveler, Common Pochard, Tufted Duck, Black-necked Grebe, Great Cormorant, Hen Harrier, Crane, Golden Plover, Lapwing, Black-headed and Lesser Black-backed Gulls, Alpine Accentor.

CÍJARA GAME RESERVE BA11
(Reserva Regional de Cíjara)

Status: Includes a Reserva Regional de Caza (25,000ha). Not otherwise protected.

Site description

A sparsely populated area of rocky wooded hills around the Cíjara reservoir, in the extreme northeast of the province. It includes a regional, formerly national, deer hunting reserve. The woodlands are of encina with plantations of Stone and Maritime Pines as well as some eucalyptus. There are extensive tracts of matorral, notably Gum Cistus. The reservoir itself has two main arms, extending up narrow valleys. Rocky outcrops, encrusted with yellow lichens, abound both along the reservoir flanks and as the summits of the ridges of the surrounding sierras. Several islands, also rocky, have been created by the reservoir. Water levels in the lake are often low, which results in temporary reedbeds forming along the valleys, attracting amphibians and their predators.

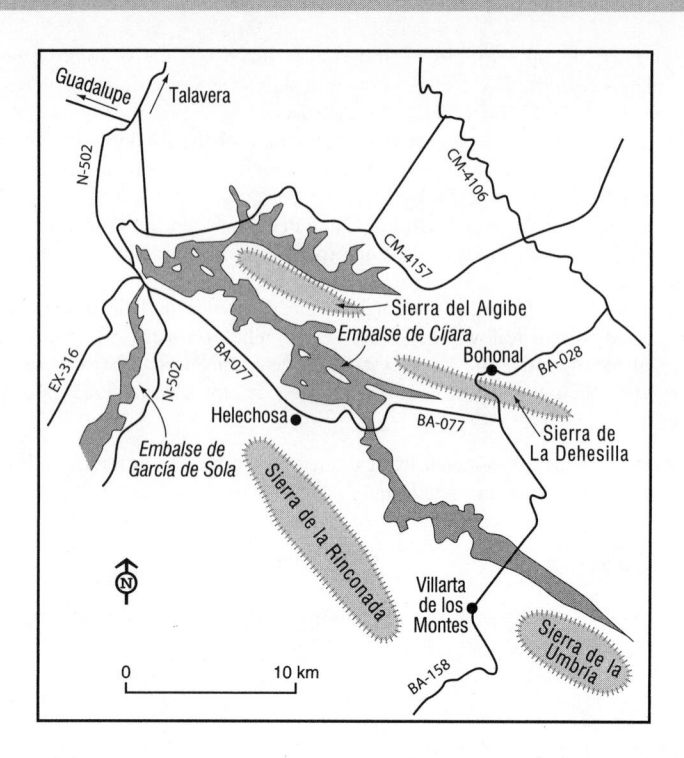

Species
Black Storks nest on the rocky scarps, including those on islands in the reservoir. A good variety of raptors occur, including Griffon, Black and Egyptian Vultures; Spanish Imperial, Golden and Bonelli's Eagles, Hobbies and Eagle Owls. The scrub and woodland house a full range of passerines and others, including Stock Doves and Iberian Green Woodpeckers. Rock Sparrows are very common, especially on and near the various bridges on the perimeter road. There is a good population of Red Deer, which are often seen, unlike some of the other resident mammals, which include Iberian Lynx, Wild Cat and Otter. Fallow and Roe Deer, Mouflon and Wild Boar are also present.

Timing
Springtime visits are likely to prove most productive but the site is of interest all year round.

Access
The area is served by minor roads, from which forest tracks allow closer investigation of the passerine community. A drive around the perimeter road (which leaves Extremadura marginally in the north and east, to enter the provinces of Toledo and Ciudad Real), gives access to all the major habitats. The section between Villarta and Bohonal crosses good woodland and includes a crossing of

the southern arm of the reservoir on a high causeway, which offers stunning views of the valley. Rock Sparrows are common here and both White and Black Storks may often be seen feeding in the valley below. The rocky outcrops of the Sierra de la Dehesilla, above Bohonal, provide views over the region and are a good point to scan for raptors. The northern part of the drive crosses several scenic bridges, again with Rock Sparrows common nearby, and gives views over the islands in the main body of the reservoir.

& Most areas may be viewed from or near your car.

CALENDAR

All year: Cattle Egret, Grey Heron, Griffon and Black Vultures; Spanish Imperial, Golden and Bonelli's Eagles, Stock Dove, Eagle Owl, Iberian Green Woodpecker, Southern Grey Shrike, Azure-winged Magpie, Rock Sparrow.

Breeding season: Black Stork, White Stork, Egyptian Vulture, Hobby, Pied Flycatcher.

RIVER GUADIANA AT BADAJOZ CITY   BA12

Status: ZEPA 'Azud del Río Guadiana'.

Site description
The Azud, a low dam, on the Río Guadiana at Badajoz city maintains constant water levels there, providing residents with fine views and several kilometres of waterside walks alongside what is in effect a broad lagoon. The dam acts as a weir along its entire length and incorporates fish ladders. There is abundant waterside vegetation in the form of clumps of reeds and reedmace and riparian woodland of willows and ash. Several wooded islands, covered in tall eucalyptus with a dense shrubby understorey, provide roosting and nesting opportunities for aquatic birds. The river is badly affected by infestations of Water Hyacinth and an alien waterlily, which are cleared periodically.

Species
This is an excellent site for both nesting and wintering waterbirds. There are small breeding colonies of Cattle Egrets, nesting alongside Little Egrets; Grey, Squacco and Night Herons, Glossy Ibises and Spoonbills. Purple Herons and Little Bitterns nest in the reeds. Purple Swamphens are always present. The breeding passerines include Olivaceous Warblers. Alpine Swifts and numerous House Martins nest on the upstream bridge (Puente Viejo).

Ospreys, waders and a range of passerines occur on passage especially. In winter the resident Mallards are joined by other duck species as well as many hundreds of Cormorants, which nest further downstream. The Cormorants can often be seen fishing cooperatively in large flocks, especially in the early afternoon for some reason. This is one of the very few locations in Spain where Squacco Herons can reliably be found in winter.

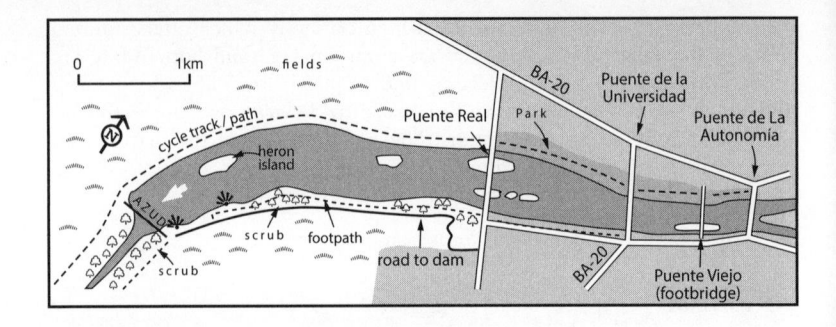

Otters are quite often to be seen here, even close to the bridges.

Timing
Interesting birds are present year-round and there is always a lot of entertaining activity, including large movements of egrets, ibises and cormorants to and from their roosts at dusk and dawn.

Access
Most species can be seen during the course of a short walk (up to 2km) along the riverside walkway between the three road bridges. The river can be viewed from either bank but on the many sunny days the east bank is best in the mornings and the west bank in the evenings.

A recommended longer walk will take you downstream along the southern bank from the Puente Real as far as the dam itself, where birds of many species often linger on the spillway. Walk down to the good sandy waterside track at the bridge. The track continues for some 4km to the dam. Opposite the buildings and jetty of the kayaking club, the large island with a clump of large eucalyptus trees, some of them dead, in the middle of the river is the site of the

Bluethroat

largest mixed heronry. A viewpoint alongside the dam is a short distance further on. If time is short you can drive as far as the Azud on a tarmac road parallel to the above track but you will see more birds if you walk. A track continues south downstream of the Azud where a walkway allows access to the west bank, so that a circular walk (some 10km in total) back to the Puente Real is possible.

♿ Most areas, including the Azud itself, may be viewed from or near your car.

CALENDAR

All year: Mallard, Little Grebe, Great Cormorant, Little Bittern; Night, Squacco and Grey Herons, Cattle and Little Egrets, White Stork, Glossy Ibis, Purple Swamphen, Kingfisher; Cetti's, Sardinian and Dartford Warblers, Zitting Cisticola, Penduline Tit, Southern Grey Shrike, Spanish Sparrow, Common Waxbill, Red Avadavat.

Breeding season: Purple Heron, Spoonbill, Black Kite; Alpine, Pallid and Common Swifts, Bee-eater, Sand Martin, Nightingale, Savi's Warbler (scarce); Reed, Great Reed, Olivaceous and Melodious Warblers, Golden Oriole.

Passage periods: Osprey, Black-winged Stilt; Green, Wood and Common Sandpipers.

Winter: Teal, Gadwall, Shoveler, Common Pochard, Great Crested Grebe, Great White Egret, Marsh Harrier, Black-headed and Lesser Black-backed Gulls, Wryneck, Crag Martin, Water Pipit, Bluethroat, Common Chiffchaff.

LAGOONS AND FARMLANDS OF LA ALBUERA   BA13

Status: ZEPA 'Llanos y Complejo Lagunar de La Albuera (36,367ha). Ramsar site.

Site description

Open country offers steppe-like habitat around La Albuera village, where bustards now patrol the 1811 Peninsular War battlefield. Large tracts are devoted to cereal growing and pastureland but olive groves and vineyards have made increasing inroads into the steppe in recent years. A complex of seasonal lagoons attracts waterbirds.

Species

Open-country species are widespread but can be hard to find during a casual visit. Little Bustards occur but here, as elsewhere, have become very scarce. Great Bustards breed but are considerably more abundant in winter. Other steppic birds present include both sandgrouse species, Stone-curlew and Calandra Lark. Sky Larks and Meadow Pipits are numerous in winter. Montagu's Harriers breed in the cereal crops. Wintering raptors include Red Kites, Hen Harriers, Merlins and Short-eared Owls.

The Laguna Grande is the best of the lagoons and holds a wide range of breeding and wintering waterfowl in wet years, when Marsh Harriers also nest here. Three pairs of Black-necked Grebes nested on the Laguna Grande in 2018, the first recorded breeding in Extremadura. Black-winged Kites are present in the dehesas. Cranes are common in winter when large numbers roost at the lagoons. The cropfields and dehesas surrounding the lagoons also attract a diversity of open-country and woodland species.

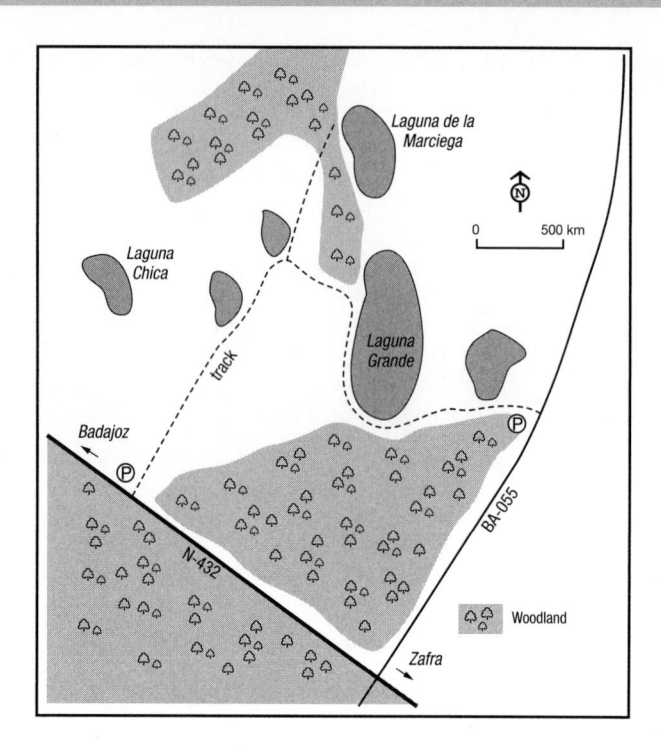

Timing
The lagoons are usually only worth visiting in winter and into spring, provided there has been sufficient rainfall. They dry up totally in summer. Many open-country species, including bustards and larks, are most numerous in winter, but spring visits are also rewarding.

Access
Steppic species are best sought between La Albuera and Valverde de Leganés, where they may be viewed from the main road (BA-006) or from minor tracks off it, for example, those leading eastwards 3.5km and 5.7km from La Albuera (see map for BA1). The minor road west of La Albuera leading to Entrín Bajo, and the cropfields around the lagoons, are also worth visiting for steppe species.

The lagunar complex is readily accessed from the N-432. A signposted turn-off on the north side at km-28.9 indicates a sandy track. It is best to park where the track forks a short distance from the road and continue to the lagoons on foot. There are hides. The Laguna Grande and the other lakes are also accessible from the BA-055; park at the signposted entrance to a sandy track on the west side, at a point 1.7km from the N-432, and continue on foot.

♿ Areas near the main roads are viewable from your car but the tracks serving the lakes are unsuitable for vehicular access.

CALENDAR

All year: Black-winged Kite, Griffon and Black Vultures, Great and Little Bustards, Stone-curlew, Lapwing, Black-bellied and Pin-tailed Sandgrouse, Calandra Lark, Zitting Cisticola, Southern Grey Shrike, Azure-winged Magpie, Spanish and Rock Sparrows.

Breeding season: Mallard, Gadwall, Little and Great Crested Grebes, White Stork, Black Kite, Short-toed and Booted Eagles, Lesser Kestrel, Marsh and Montagu's Harriers, Collared Pratincole, Black-winged Stilt, Whiskered Tern, Great Spotted Cuckoo, Hoopoe, Greater Short-toed Lark, Tawny Pipit, Orphean Warbler.

Passage periods: Garganey, Black Stork, Spoonbill, waders.

Winter: Greylag Goose, Wigeon, Teal, Pintail, Shoveler, Red-crested and Common Pochards, Crane, Red Kite, Hen Harrier, Golden Plover, Short-eared Owl, Sky Lark, Meadow Pipit.

ALQUEVA RESERVOIR INLET   BA14

Status: No special protection.

Site description
The Embalse de Alqueva, a relatively new reservoir (2002) on the Río Guadiana along the Spanish/Portuguese border, has the distinction of being the largest in Europe. It extends over some 250km^2 but only 33km^2 are in Badajoz, the remainder being in Alentejo, Portugal. Despite its size, the lake is highly dissected, having flooded a number of valleys, so much so that its shoreline is over 1,000km long. The Badajoz section described here includes relatively shallow areas, with plenty of drowned trees and many rocky islets revealed when water levels are low.

Species
The shallower lake sectors attract numbers of waterfowl, especially in winter. Little Ringed Plovers, Common Sandpipers and other waders occur, on passage

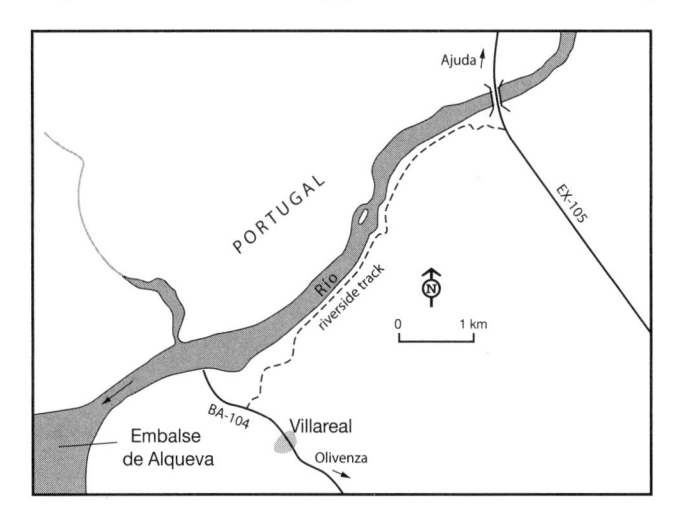

especially, in small numbers. Great Crested Grebes breed and Cormorants nest locally on the dead trees. An Osprey introduction programme conducted as an environmental impact mitigation measure in the Portuguese sector has produced results. Ospreys nested successfully in the Portuguese section in 2016, and in Badajoz in 2017. There were two pairs in residence in 2018 and the population is likely to grow.

Timing
The full birding potential of this site has probably yet to be realised or discovered. It is likely to be most productive in winter and interesting aquatic species are liable to occur during passage periods also. Areas devoted to watersports, such as at Cheles beach further south, are clearly best avoided in summer.

Access
Excellent birding and close views of the Guadiana as it enters the reservoir are available from a good sandy track that follows the waterside for some 6km between the bridge on the EX-105 and the BA-104 near Villareal. The area is rapidly accessible from Badajoz city on the EX-107, which intersects the EX-105 and BA-104 at Olivenza. The EX-105 entrance to the riverside track is 300m before the bridge leading to Ajuda, Portugal. The BA-104 entrance is on the right halfway between Villareal and the river, 600m from Villareal. Both entrances are signposted, indicating an 'Observatorio Ornitológico' (hide). The hide itself is poorly sited and in some disrepair and your car will provide better cover throughout the route.

Easily viewable from your car.

CALENDAR

All year: Egyptian Goose, Mallard, Gadwall, Great Crested Grebe, Great Cormorant, Osprey

Breeding season: Black Kite, Booted and Short-toed Eagles, Little Ringed Plover, Gull-billed Tern.

Passage periods: Garganey, waders.

Winter: Crane, Teal, Pintail, Shoveler, Red-crested and Common Pochards, Red Kite.

CÁCERES PROVINCE

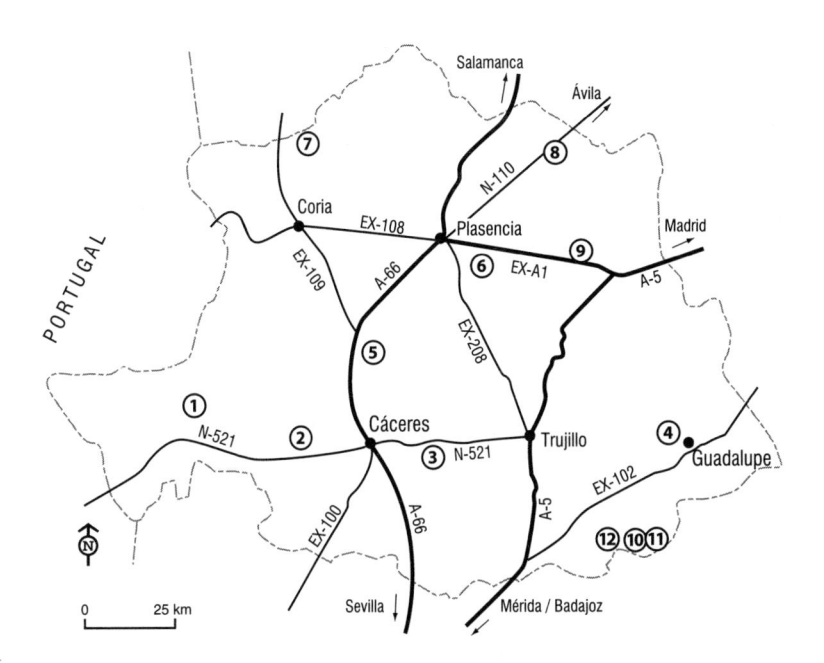

Cáceres province boasts large expanses of wooded sierras, much of the habitat in unspoilt condition. Indeed, some of the large estates have tracts of country that have remained pretty well unaltered since medieval times. The terrain is dissected by significant rivers, notably the Río Tajo (Tagus), which have been dammed, producing extensive reservoirs. Pasturelands in the province (CC1–3, CC5) have a steppe-like character and are the home of large populations of bustards and other steppe birds. The province is also well known for the diversity of breeding raptors, which can be seen well at a number of sites but most famously at Monfragüe (CC6). There are also opportunities to see some mountain species, notably on the flanks of the Sierra de Gredos in the Valle del Jerte (CC8) and in the Sierras de las Villuercas (CC5). The Embalses de Sierra Brava (CC10), de Arrocampo (CC9) and de Alcollarín (CC12) are three of the most important wetlands in Extremadura and the southeastern ricefields (CC11) are also of great interest for aquatic species.

Sites in Cáceres Province
CC1   Plains of Western Cáceres
CC2   Steppes of Malpartida de Cáceres
CC3   Cáceres–Trujillo steppes
CC4   Cuatro Lugares steppes
CC5   Sierras de Las Villuercas
CC6   Monfragüe National Park

CC7   Sierra de Gata and Borbollón reservoir
CC8   Jerte valley
CC9   Arrocampo reservoir and Campo de Arañuelo
CC10 Sierra Brava reservoir
CC11 Guadiana Valley ricefields
CC12 Alcollarín reservoir

Getting there
The A-66 motorway provides north/south access. The N-521 Trujillo/Portugal road crosses the province east/west, branching off the A-5 Madrid/Badajoz motorway at Trujillo; the Trujillo/Cáceres stretch has recently itself been upgraded to a motorway. The EX-108 and the EX-A1 also cross the province east/west in its northern sector, the latter linking Plasencia and the A-5, and providing easy access to Monfragüe and the northern sites.

PLAINS OF WESTERN CÁCERES   CC1

Status: Includes the ZEPAs 'Río Tajo Internacional y Riberos' (20,271ha) along the Tajo valley and 'Llanos de Alcántara y Brozas' (51,200ha).

Site description
Westernmost Cáceres province and the adjacent portion of Badajoz province form a spur of territory projecting westwards into Portugal, the rivers Tajo (Tagus) and Sever marking the border. The northern portion of this region, north of the excellent N-521 Portugal/Cáceres trunk road, has good tracts of encina woodlands. The same area has important expanses of rough pastureland and steppe habitats, especially between the towns of Membrío and Brozas. A number of small reservoirs are also of interest. The Tajo itself has been dammed at Cedillo but the escarpments flanking the reservoir remain an interesting feature here and upstream.

Species
The steppelands have good numbers of both sandgrouse and Stone-curlews. Great and Little Bustards are also present although they can be elusive and are most visible here in winter. Large numbers of White Storks and colonies of Lesser Kestrels frequent the towns and villages. A nestbox installation scheme between Brozas and Villa del Rey has proved a great success with Rollers. Flocks of Cranes, Golden Plovers and Lapwings are present in winter. The reservoirs attract a range of wetland species also especially in winter but also during passage periods. The rocky escarpments of the Tajo valley attract cliff-nesting raptors, including Bonelli's and Golden Eagles, Griffon and Egyptian Vultures and Eagle Owls, and also Black Storks.

Timing
Interesting species are present all year round. Winter for steppe birds. The first Rollers return early in April and some remain into September.

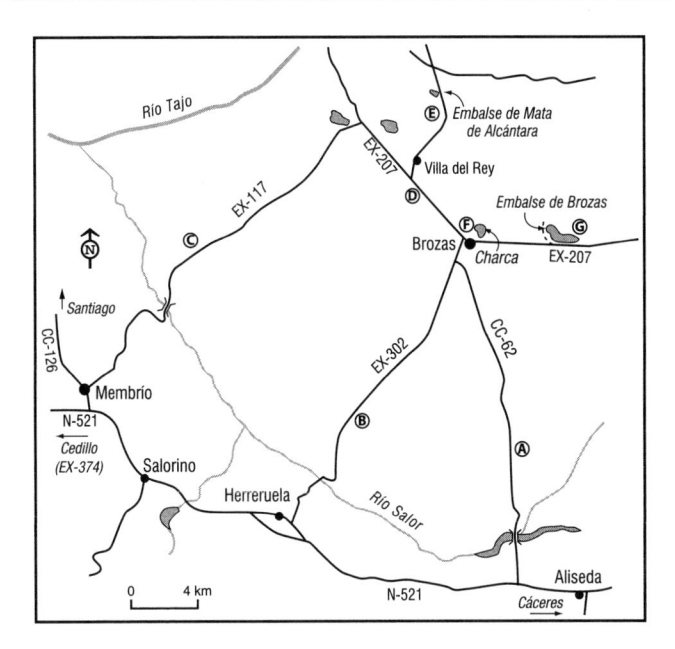

Access

The N-521 provides easy and swift access to the region. From here a zig-zag route can be followed, starting from near Aliseda, taking in Brozas, Herreruela, Membrío, Villa del Rey and back to Brozas (see map). This route traverses all the most interesting areas and it will be necessary to stop frequently and scan the plains. Most of the roads are fairly quiet but the EX-302 Herreruela/Brozas link is often busy with fast traffic. Even here though there are frequent entrances to farm tracks which allow you to pull off the road safely and allow closer access to promising terrain.

Worthwhile stopping places include the picturesque stone bridge across the Río Salor on the CC-62 Aliseda/Brozas road (A), where the thin cover of encinas provides a range of woodland passerine species, including Crested Tits. Here too there are small reedbeds where Great Reed Warblers breed. Further north, the road ascends and crosses interesting areas of steppe land, as does the EX-302 Brozas/Herreruela road (B). Flocks of White Storks, bustards, sandgrouse and Cranes (in winter) can be located here. Closer access is possible along farm tracks: good examples are those on the west side of the road 13km and 16km south of Brozas. The EX-117 north of Membrío (C) crosses vast expanses of open country from km-13 northwards, also requiring frequent scanning to locate the steppe species. White Storks and Montagu's Harriers can be especially abundant here and this quiet road is ideal to enjoy the evocative songs of Quail, Calandra Larks and Corn Buntings in spring.

The EX-207 returning towards Brozas (D) is distinguished by the nest boxes which have been attached to the roadside pylons, which have proved attractive to Rollers. A detour north through Villa del Rey gives access to a reservoir, the Embalse de Mata de Alcántara (E), which attracts waders and waterfowl during passage periods and in winter. Two other small water bodies that always merit inspection are at Brozas. The Charca de Brozas (F) is a small reservoir with shallow marshy margins that lies north of the EX-207 immediately northeast of Brozas. It is not well marked but most of the small roads off the EX-207 at Brozas converge on the Charca. It may be simplest to take the road signposted to Santuario de S. Gregorio immediately east of a petrol station; the Charca perimeter road is the first left turn along this road, 375m from the EX-207. The Brozas reservoir (G) is 4.5km east of Brozas along the EX-207, where there is ample parking. Both are attractive to waders and ducks in winter.

The Tajo valley is not very accessible but there is worthwhile access at Cedillo and Herrera de Alcántara (not mapped). Take the EX-374 northwestwards from the N-521 for some 50km to reach the dam at Cedillo. Scan the area from the dam and from a viewpoint on the right 200m above the dam. Two marked trails in Cedillo village lead to the shore of the reservoir. The shore may also be reached by retracing your route and then taking the EX-376 to Herrera de Alcántara, following the road past the village to the reservoir.

Another possibility is to take the good dirt road that runs north for some 7km to the river from the CC-37 1.5km north of Santiago de Alcántara; the entrance is signposted 'Observatorio' and 'Tajo Internacional'. This road crosses dehesas to a mirador that gives excellent views of the river and its hinterland, with excellent possibilities for seeing vultures, eagles, Black Storks and many other woodland and rupestral species.

&. Most areas are easily viewed from or near your car.

CALENDAR

All year: Little Grebe, Cattle Egret, White Stork, Black winged and Red Kites, Griffon and Black Vultures, Spanish Imperial, and Bonelli's Eagles, Great and Little Bustards, Stone-curlew, Black-bellied and Pin-tailed Sandgrouse, Eagle and Little Owls, Calandra Lark, Crested Tit, Southern Grey Shrike, Rock Sparrow.

Breeding season: Quail, Black Stork, Black Kite, Egyptian Vulture, Short-toed and Booted Eagles, Montagu's Harrier, Lesser Kestrel, Black-winged Stilt, Little Ringed Plover, Great Spotted Cuckoo, Roller, Greater Short-toed Lark, Zitting Cisticola, Great Reed Warbler.

Winter: Wigeon, Gadwall, Teal, Mallard, Shoveler, Common Pochard, Tufted Duck, Great Crested and Black-necked Grebes, Great Cormorant, Red Kite, Hen Harrier, Merlin, Crane, Golden Plover, Lapwing, Little Stint, Temminck's Stint (formerly regular), Ruff, Common Snipe, Common Redshank, Greenshank, Sky Lark, Meadow Pipit.

STEPPES OF MALPARTIDA DE CÁCERES   CC2

Status: Los Barruecos is a Monumento Natural. Otherwise no special protection.

Site description

Undulating grassy pastures and cereal fields with broad vistas characterise the area northeast of the Sierra de San Pedro, west of Malpartida de Cáceres. The region south of the town includes Los Barruecos, a spectacular expanse of immense granite boulders with several small pools nearby: a bizarre enough landscape to have served as a location site for *Game of Thrones*.

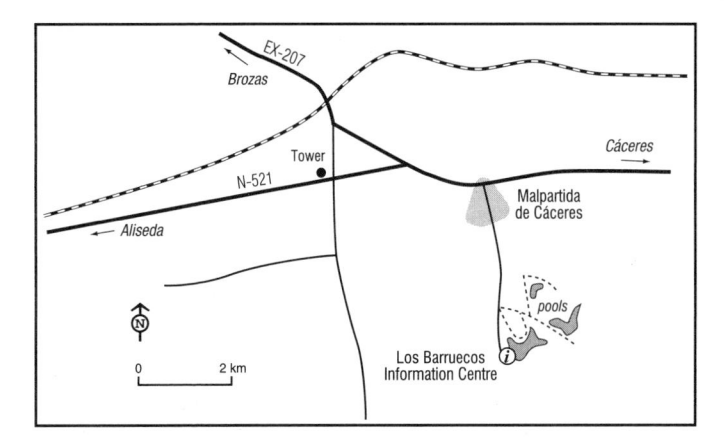

Species
Good numbers of steppe species are present, including Great and Little Bustards, Stone-curlews and sandgrouse. In spring and summer, White Storks are particularly abundant, many of them originating from Los Barruecos where they nest on the huge granite boulders there. Montagu's Harriers and Lesser Kestrels are also very much in evidence in spring and summer. Vultures and other raptors are regular visitors from the sierras to the south. The shallow margins of the pools at Los Barruecos attract a few waders on passage and in winter as well as a diversity of common waterfowl.

Timing
Morning and evening visits in the breeding season are most productive, but interesting species are present at all times.

Access
This area offers another opportunity to see the steppe species of Extremadura from easily accessible tracks. Useful access is available from a broad sandy road which extends south from the N-521 at km 62.5, about 3km west of Malpartida de Cáceres or 14km east of Aliseda. A ruined tower north of the N-521 is just east of the entrance. The road gives access to a network of good sandy tracks across the pastures. Los Barruecos is signposted south from the N-521 in Malpartida.

Most areas are easily viewed from or near your car.

CALENDAR

All year: Cattle Egret, Red Kite, Griffon and Black Vultures, Great and Little Bustards, Stone-curlew, Black-bellied and Pin-tailed Sandgrouse, Hoopoe, Calandra Lark.

Breeding season: White Stork, Black Kite, Egyptian Vulture, Montagu's Harrier, Booted Eagle, Lesser Kestrel (some resident).

Winter: Crane, Lapwing, larks, pipits, finches.

CÁCERES–TRUJILLO STEPPES   CC3

Status: Includes ZEPAs 'Llanos de Cáceres y Sierra de Fuentes' (70,000ha), 'Magasca' (10,846ha) and 'Llanos de Trujillo' (7,757ha).

Site description
An accessible region of steppe-like habitat comprising mainly gently undulating country traditionally given over to sheep grazing on rough pastures with areas of wheat cultivation. The pastures are resplendent with their varied flora in spring. There are many small waterholes dotted around the area for the benefit of the cattle that have been replacing sheep in recent years. These attract a few passage waders, notably Black-winged Stilts. A small reservoir, the Embalse del Salor, is attractive to waterfowl, especially in winter. The vicinity of Trujillo is distinctive: the town is centred on a giant rockery of massive boulders and granitic pavements, interspersed with stony pastures.

The towns are picturesque, with the characteristic rooftop birds much in evidence. Indeed, to sit in the medieval square of Trujillo on a spring or summer evening, having a drink or a meal accompanied by a dynamic backdrop of swirling clouds of swifts, commuting Lesser Kestrels and storks clattering their greetings on the rooftops, is for many an unforgettable highlight of any trip to Extremadura.

Species
This remains one of the prime regions of Extremadura for steppe birds, with notable breeding densities of Montagu's Harriers, Great Bustards, Stone-curlews, Black-bellied Sandgrouse and Pin-tailed Sandgrouse. As such it is also one of the most popular destinations for birding visitors, whose interest in the region helps to counteract such negative pressures as inappropriate use of pesticides, changes in land use and urban encroachment. Developments such as these are implicated in the local decline of Little Bustards, which are still present but are nowhere near as numerous as they once were (see Introduction). Vultures and

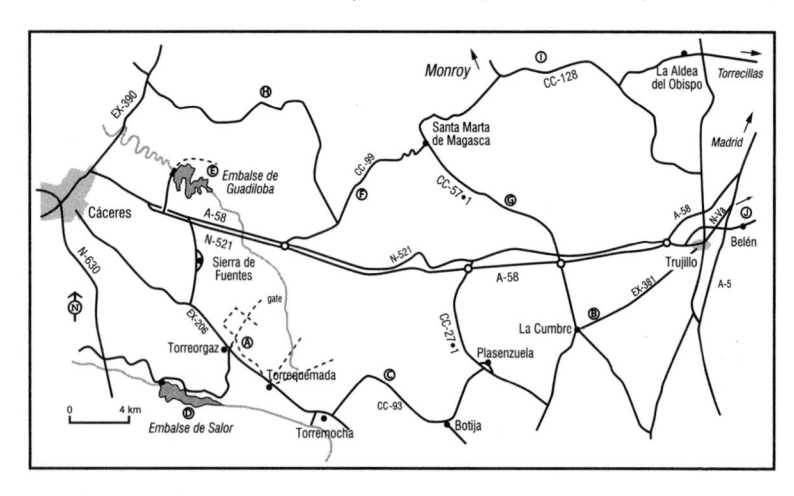

other raptors are obvious year-round. They include Spanish Imperial Eagles, which have recently begun nesting in the area. The villages and, notably, the cities of Cáceres and Trujillo have large and conspicuous breeding populations of White Storks, Lesser Kestrels and swifts. Common Magpies have a conspicuous presence near Trujillo particularly and so attract good numbers of their brood parasite, the Great Spotted Cuckoo. Cranes, plovers and waterfowl are present in winter.

Timing

Birds are obvious all year round but, as with other similar sites, midday excursions in warm weather will be least productive: early mornings and evenings are usually best. Springtime visits are memorable with the harriers visible everywhere quartering the fields. In February, March and into April, scattered white 'sheep' prove to be displaying male Great Bustards on closer inspection and the male Little Bustards stand revealing their neck patterns on their territories. The clamour of the Calandra Larks is the most obvious sound but occasional two-tone 'raspberries' can be traced to the Little Bustard males: these uninspiring calls accompany their territorial display. In late spring and early summer particular attention should be paid to areas of recently cut wheat, which often attract sandgrouse flocks. The steppe birds remain in winter, when they form large flocks; wintering Cranes and waterfowl add to the variety then.

Access

The once-busy N-521 Cáceres/Trujillo road has been replaced by the A-58 motorway. A great benefit of this is that the N-521, which runs roughly parallel to the motorway, now provides an excellent opportunity for viewing the often productive expanses of steppe alongside the road, where stopping was previously out of the question. The N-521 now just serves the local farms (and visiting birders). It is almost traffic-free and you can stop wherever you like to scan likely areas of habitat. A couple of underpasses in the western half of the road give access to the farm tracks across the steppelands south of the motorway. Ask for permission from the local people whenever possible. A telescope is useful for scanning for distant bustards, although the birds generally permit approaches to within 200m. Access to the N-521 is clearly signposted at each end from the roundabouts that also feed the motorway sliproads.

Other excellent places to try include the tracks north of the village of Torreorgaz, on the EX-206 14km southeast of Cáceres (A). Here an encouraging tiled mural of a Great Bustard on the wall of the 'Parador de los Llanos' restaurant, a stone cross on a plinth and the Bar Extremadura are directly opposite the entrance to these tracks, which are sandy but in good repair. The best track extends for 4km northwards from Torreorgaz. Stop at the green gate at the end of the road and scan for steppe species or explore further on foot. Alternative parallel tracks are available at intervals from the sandy road that runs from the above track entrance alongside the EX-206 towards Torrequemada. Tracks north of the EX-206 at, and east of, Torrequemada may also be worth exploring.

Other worthwhile areas include the region on both sides of the Trujillo/La Cumbre road (B) and the Plasenzuela/Botija/Torremocha road (C). North of Torremocha, the latter road gives panoramic views over rolling steppelands and provides excellent vantage points for scanning wide areas. The roads in these areas are all fairly quiet and offer obvious stopping places as well as tracks leading to farms, which may be used for wider exploration.

A detour may be made southwards from Torreorgaz to the Embalse del Salor (D). The entrance to this road lies immediately east of the village. The road follows the northern shore to the dam, which offers good views across the lake. The shallows where the feeder stream enters the river are marshy and often hold waders and the lake itself is attractive to waterfowl, especially in winter. Spoonbills occur on passage. Granite boulders provide islands in the reservoir, which are popular with Grey Herons and waterfowl. The eucalyptus trees fringing the reservoir hold numbers of photogenic White Stork nests. A very large roost of Great Cormorants, sometimes exceeding 2,000 birds, forms downstream of the dam in winter.

A short road signposted Embalse de Guadiloba leads to a dam retaining a small reservoir (E). Access is from the N-521 6km east of Cáceres. The steppelands around the reservoir may be viewed from this access road, which may be followed left past the dam and then around the northern flanks of the reservoir. Both bustards and other steppe species are again normally present here. The reservoir itself attracts waterfowl, including Greylag Geese, in winter.

Also north of the N-521, the western road (CC-99) to Santa Marta de Magasca gives views of the steppelands and the usual farm tracks offering closer access (F). Numerous nestboxes on the roadside power posts accommodate Rollers, Little Owls, Jackdaws and other hole-nesters. The eastern road (G) also passes through some open terrain attractive to Little Bustards. The minor road (H) looping from the western Santa Marta road and north of the Embalse de Guadiloba, ascends through vast expanses of open pastures and is, again, often productive of steppe species, especially Little Bustards and Black-bellied Sandgrouse (G). The road is largely unsurfaced but was in very good repair in 2017. The eastern entrance (tarmacked) is 2.4km from the N-521 along the EX-99.

Further north still, the CC-128 (I) from Monroy westwards towards La Aldea del Obispo crosses large tracts of open country attractive to steppe species including Little Bustards.

The hinterland of the village of Belén (J), just northeast of Trujillo, is very popular with tour groups and is another reliable area for steppe species. Take the N-Va northeast from Trujillo towards the km-250 junction of the A-5 and turn right on the road signposted Belén. Drive through the village and continue towards Torrecillas de la Tiesa. Stop frequently to scan for steppe species for the next 15km.

♿ Most areas are easily viewed from or near your car.

CALENDAR

All year: Egyptian Goose, Little and Great Crested Grebes, Cattle Egret, White Stork, Black-winged and Red Kites, Spanish Imperial Eagle, Common Kestrel, Great and Little Bustards, Stone-curlew, Black-bellied and Pin-tailed Sandgrouse, Little Ringed Plover, Barn and Little Owls, Calandra Lark, Southern Grey Shrike.

Breeding season: Quail, Black Stork, Black Kite, Montagu's Harrier, Lesser Kestrel, Black-winged Stilt, Great Spotted Cuckoo, Roller, Pallid and Common Swifts, Spanish Sparrow.

Winter: Greylag Goose, Wigeon, Gadwall, Teal, Shoveler, Common Pochard, Tufted Duck, Great Cormorant, Hen Harrier, Golden Eagle, Merlin, Crane, Golden Plover, Lapwing, Green Sandpiper, Common Redshank, Lesser Black-backed Gull, larks, pipits and finches.

CUATRO LUGARES STEPPES   CC4

Status: Partly within the ZEPA 'Llanos de Cáceres'. Includes ZEPAs 'Embalse de Alcántara' (7,648ha) and 'Embalse de Talaván' (7,303ha).

Site description
The rivers Tajo (Tagus) and Almonte enclose an area of steppeland to the north of Cáceres city, between two narrow arms of the massive reservoir, the Embalse de Alcántara. Here dry stone walls traverse vast vistas of rough pastureland. By contrast, rather large woodlands of encinas replace the steppes towards the east, from Monroy towards Trujillo. A small reservoir, the Embalse de Talaván, is of limited interest for waterfowl. Much of this formerly celebrated area has lost its attractiveness to open-country species as a result of peripheral planting of olive groves, and a projected 500ha solar farm will make matters worse.

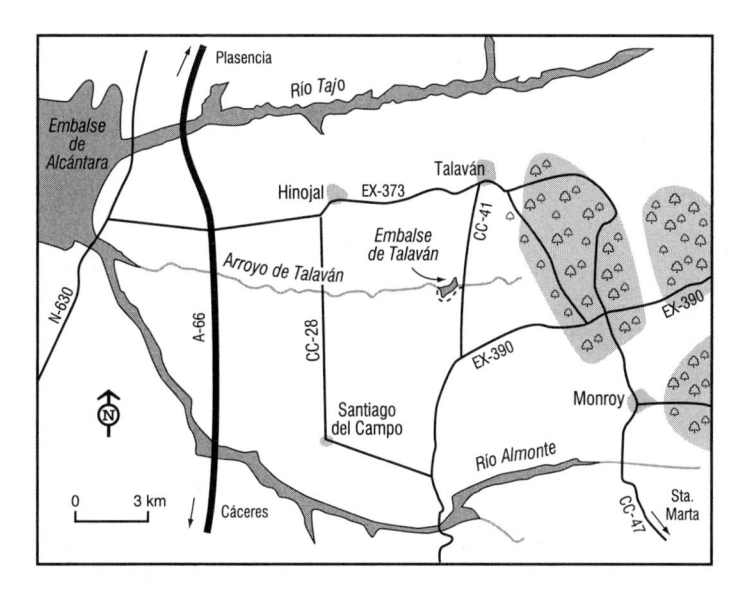

Species
Perhaps the chief attraction of this area is the presence of sandgrouse. Both species occur year-round but the Pin-tailed Sandgrouse is the most abundant. Great Bustards, Little Bustards, Stone-curlews, Quail and Calandra Larks breed in reduced numbers. White Storks from the villages gather here to feed. The Alcántara reservoir attracts large flocks of Cormorants in winter, as well as grebes, waterfowl and thousands of both Black-headed and Lesser Black-backed Gulls. Spanish Sparrows flock in the vicinity. The woodlands have numbers of Azure-winged Magpies and Rock Sparrows among the passerine community.

Timing
Steppe species are present all year but springtime is most generally productive. Visits to the reservoirs are most rewarding in winter.

Access

From the minor roads and farm tracks crossing the area. The areas south of Hinojal and east of Santiago del Campo are the most productive for sandgrouse. Steppe species may also be found south and west of Talaván. The eastern part is dominated by dehesas, of which there are excellent examples southeast of Talaván along the EX-373 and southeast of Monroy along the CC-128 towards Santa Marta. The latter have many large clearings offering wide views ideal for scanning for raptors, which include Black-winged Kites.

The Embalse de Alcántara is readily viewed from the N-630, which ascends the eastern flank of the central basin. This road is now relatively quiet since it was superceded by the A-66, further to the east. An island in the lake has a small nesting colony of Grey Herons and is used as a winter roost by Cormorants. Both Collared Pratincoles and Little Terns have recently nested on another island there.

The Embalse de Talaván is easily reached from the minor road (CC-41) leading south from Talaván, which crosses the eastern arm of the reservoir, where there is a lay-by and adjacent hide overlooking reedbeds. The reservoir and the steppe areas of its hinterland are also viewable from the track along the south bank and the dam at the western end.

♿ Most areas are easily viewed from or near your car.

CALENDAR

All year: Little and Great Crested Grebes, Cattle Egret, Grey Heron, Black-winged and Red Kites, Griffon and Black Vultures, Great and Little Bustards, Stone-curlew, Black-bellied and Pin-tailed Sandgrouse, Calandra Lark, Southern Grey Shrike, Azure-winged Magpie, Spanish Sparrow, Rock Sparrow.

Breeding season: White and Black Storks, Egyptian Vulture, Short-toed and Booted Eagles, Collared Pratincole, Little Ringed Plover, Little Tern, Greater Short-toed Lark.

Winter: Wigeon, Teal, Shoveler, Common Pochard, Tufted Duck, Great Cormorant, Little Egret, Crane, Golden Plover, Lapwing, larks, pipits and finches.

SIERRAS DE LAS VILLUERCAS   CC5

Status: ZEPA (76,336ha).

Site description

Las Villuercas is the mountainous region of southeast Cáceres, in the vicinity of Guadalupe. The parallel rocky ridges that traverse the area enclose narrow and steep-sided valleys, cultivated with olive groves and cherry orchards. Elsewhere, the Mediterranean scrub is interrupted by mixed oak and chestnut woods and plantations of Maritime Pines. In April the heady honey-like scent of Gum Cistus pervades the whole region. The mountains themselves are craggy and have abundant precipitous escarpments, the rocks attractively splashed with patches of moss and yellow lichens. The highest summit, Pico Villuerca, is 1,601m. The local climate is relatively mild in spring and summer, and mists and drizzle envelop the mountains in cool weather. Winters are cold.

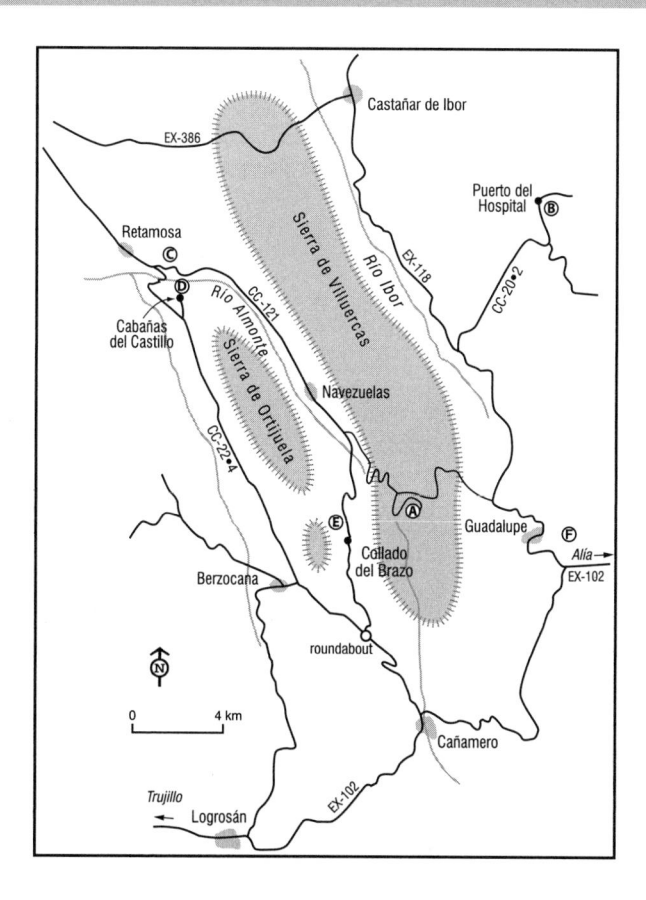

Species

The area offers a good variety of resident and breeding raptors and passerines. It is regarded as one of the best parts of Extremadura for both Peregrines and Red-billed Choughs. White-rumped Swifts have nested recently near Cabañas del Castillo.

Timing

Springtime and early summer visits are most rewarding. Spring arrives relatively late and deciduous trees may only just be coming into leaf in mid-April. Nights especially can be cold and suitable clothing for inclement weather is worth having.

Access

The numerous minor roads that cross the area are not busy and offer many opportunities to stop and view the region. Cliffs should be scanned for the nests of Black Storks and raptors. Forest tracks allow access on foot to explore the woodlands for passerines. Las Villuercas are generally explored by car. A recommended, largely circular, route visiting most of the representative habitats and including some spectacular views on clear days would take in Guadalupe, the summit of Pico Villuerca, the gorge below Retamosa, the village of Cabañas del Castillo, the Collado del Brazo col and Cañamero.

The summit of Pico Villuerca (A), above Guadalupe, is easily accessible by road. The eastern access road is west of the EX-118 just above Guadalupe and is signposted for a 'Zona Militar' military post (now unmanned) and hermitage (Ermita del Humilladero). This road was rather rough tarmac in 2017 and may deteriorate further but otherwise is perfectly drivable. The road winds upwards for some 4km to a col below the summit, where it is worth scanning for raptors and rupestral species and walking along the trails though the scrub. Raptors here may include Golden Eagles and Peregrine Falcons. The rocky outcrops harbour both Blue and Rufous-tailed Rock Thrushes, Black Redstarts of the striking southern form *aterrimus*, Red-billed Choughs and Rock Buntings. The scrub community includes Dartford and Subalpine Warblers. The much shorter western access road (2.2km), linking the summit to the CC-121 south of Navezuelas, has an excellent concrete surface and is accordingly the most comfortable way to access the peak. The peak itself is marked by a cluster of aerials and the unmanned military post. You can drive almost to the very top and then walk on to the summit. The views from here are remarkable, reputedly extending over a 300km radius on the clearest days in winter. The snowy tops of the Sierra de Gredos are obvious to the north. The oak and chestnut woodlands of the lower reaches of the access roads are excellent for Orphean and Bonelli's Warblers in spring and summer as well as being a regular breeding location of Honey-buzzards.

Woodland birds can usefully be sought along the CC-202, which extends eastward from the EX-118 9km north of the entrance to the eastern peak road. This very quiet road (B) climbs for 9.6km through mixed woodlands, chiefly Pyrenean Oak, to a pass (Puerto del Hospital del Obispo). Stop frequently to look for woodland species, which include Lesser Spotted Woodpeckers and Bonelli's Warblers, and to scan for raptors. It is also worth continuing beyond the pass to scan the rocky ridge beyond for raptors, including Golden Eagles and Peregrines.

The bridge below Retamosa (C) faces a gorge whose cliffs accommodate nests of Griffon and Egyptian Vultures. Bonelli's Eagles may also be seen here.

Red-rumped Swallows

Park at the bridge and follow the footpath towards the cliff for a closer look. The river through the gorge has Dippers and Grey Wagtails. The hamlet of Cabañas del Castillo nearby (D) merits the short detour. Parking is available at the entrance to the village, where a mirador offers scenic views westwards. The tall rocky outcrops above the village are reached by walking up the steps past the houses and then bearing right to follow the cliff base. The path leads round the back of the ridge and up to the ruined castle on the summit, where again there are spectacular views eastwards across the valley to the Sierra de la Ortijuela. Griffon Vultures nest nearby and can be seen at close quarters together with other raptors, such as Bonelli's Eagles. This is a regular site for Alpine Accentors in winter and for Black Wheatears all year round. The aerial insect feeders along the ridge often include Alpine Swifts and the White-rumped Swift has also been found here.

Continuing south along the CC-224 it is worth detouring back along the CC-121 from the roundabout at the Puerto de Berzocana, above Cañamero. The roundabout is distinguished by an enormous sculpture of horseshoes. The Collado del Brazo (E) is a col 5.4km from the roundabout, from where the rocky escarpments to the east may be scanned for Golden Eagles, Griffon Vultures and other raptors. The same cliffs can also be viewed from a rough lay-by on a bend of the same road, a further 2.1km north of the Collado.

A Griffon colony may also be seen 8km east of Alía (F) on the EX-102, where a mirador on the north side of the road overlooks a gorge. Other raptors, including Egyptian Vultures and Bonelli's Eagles, may appear here. The river Jarigüela below is fast flowing and may attract Dippers. It can be viewed more closely at km-96, where it is joined by another stream: a picnic site at the confluence also provides access to riparian woodland.

CALENDAR

All year: Red Kite, Griffon and Black Vultures, Goshawk, Sparrowhawk, Common Buzzard; Spanish Imperial, Golden and Bonelli's Eagles, Peregrine Falcon, Eagle and Long-eared Owls, Lesser Spotted Woodpecker, Wood Lark, Crag Martin, Dipper, Dunnock, Black Wheatear, Blue Rock Thrush, Dartford Warbler, Crested Tit, Nuthatch, Red-billed Chough, Raven, Cirl and Rock Buntings.

and Red-necked Nightjars, Alpine and White-rumped Swifts, Hoopoe, Crag Martin, Red-rumped Swallow, Northern Wheatear, Rufous-tailed Rock Thrush; Melodious, Subalpine, Orphean and Bonelli's Warblers.

Winter: Alpine Accentor, Fieldfare, Song Thrush, Redwing, Siskin.

Breeding season: Black Stork, Honey-buzzard, Egyptian Vulture, Short-toed and Booted Eagles, Hobby, European

MONFRAGÜE NATIONAL PARK   CC6

Status: Parque Nacional, designated in 2007 (17,852ha). ZEPA (116,151ha).

Site description

Monfragüe lies in rugged country at the confluence of the rivers Tiétar and Tajo (Tagus). The rivers have eroded deep gorges, flanked by sheer cliff faces.

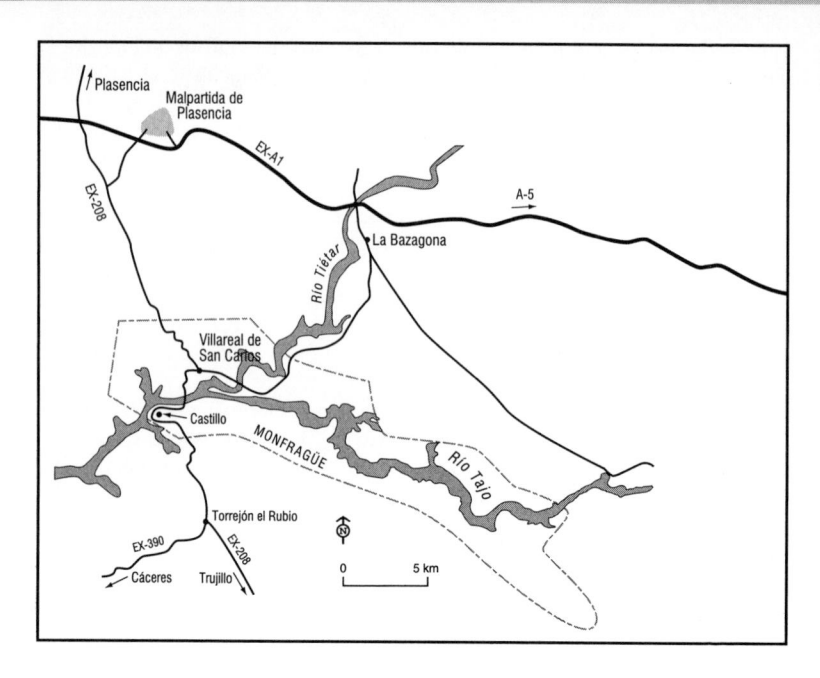

However, their flow was tamed by dams in the late 1960s and the resulting reservoirs (Embalses de Torrejón and de Alcántara) ensure an abundance of standing water year-round. Monfragüe offers large expanses of unspoilt Mediterranean woodland and scrub. Mixed oak forest predominates, with a characteristic understorey of Cistus. Rocky outcrops and ridges divide the area.

The region was under severe threat of development in the 1970s, when there were plans to replace the natural cover with eucalyptus plantations. Some planting was done in the north of the area. However, energetic representations by local conservationists were successful and the site was declared a Parque Natural in 1979. The eucalyptus plantations were largely cleared from 2000 onwards, to make way for reestablishment of native plant cover. The importance of Monfragüe received particular recognition when it was designated as Spain's 14th National Park in 2007.

Monfragüe must be regarded as an essential stop for birders but it plays an increasingly important role in sustaining the interest of many of the general public, who come hoping to see for real what they have enjoyed on television. Such visitors turn up in considerable numbers, especially at holiday periods. Diehard birders will always want to avoid crowds but the continuing upsurge in interest must be welcomed since it augurs well for the broader interests of conservation. It is very heartening now to see how many of the visitors are quite knowledgeable, come suitably equipped with binoculars and compete with each other, often volubly, to identify the birds.

The geographical features of Monfragüe actually make it an ideal site for ecotourism. Many of the target species are very large and easily seen, some of them having conspicuous nests. They are separated from their observers by reservoirs which are wide enough to prevent undue disturbance but sufficiently narrow to give visitors excellent views, even with the naked eye. By and large, people see what they come to see and they see it well. They are helped by the infrastructure, which includes information boards and hides and lay-bys opposite key sites. The watchpoints at the Salto del Gitano and the Portilla del Tiétar are especially popular, even late in the evening, when groups gather to see the resident Eagle Owls, which generally oblige their admirers.

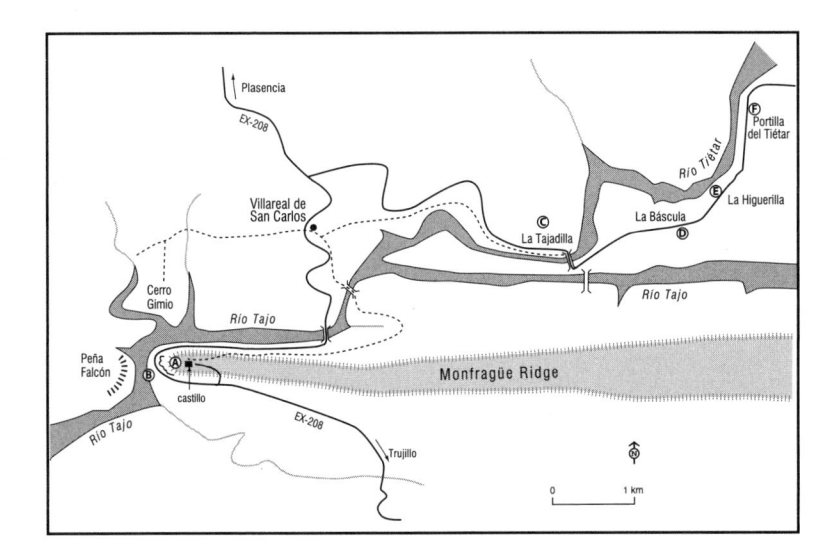

Species

This is arguably the prime site of Extremadura, offering all the characteristic species, except for the steppe birds, in large numbers. The concentration of over 300 pairs of Black Vultures makes this one of the best sites in the world for this spectacular and unforgettable species. All five eagles breed and can be seen with relatively little difficulty; the ten or so pairs of Spanish Imperial Eagles are an obvious attraction. Other breeding raptors include a few Red Kites and very many Black Kites. The cliff faces are populated by easily viewed colonies of Griffon Vultures (600+ pairs), as well as Black Storks (25–30 pairs), Egyptian Vultures (35 pairs), Peregrines and Eagle Owls. White-rumped Swifts may be found more easily here than anywhere else in Extremadura. The woodlands support a high density of Azure-winged Magpies among the representative range of passerines.

The park protects a wide range of fauna, other than birds, as well as an impressive floral diversity. Red Deer are easily seen here but the other mammals, especially the iconic Iberian Lynxes, are more elusive. Otters are quite often reported at dusk and dawn by observers waiting to see Eagle Owls.

Timing

Spring visits (April–May) are recommended since the sight of large raptors and Black Storks on their nests is the major attraction of the park. Visits at other times of year will still produce many of the characteristic species. Eagle Owls may be seen and heard just after sunset and around dawn at both El Salto del Gitano and the Portilla del Tiétar, especially during the breeding season, although they tend to be even more vocal between December and February. White-rumped Swifts are present from May until as late as the end of October.

Access

Monfragüe is easily accessible from Cáceres and Trujillo to the south and Plasencia to the north. A large sector of the park is closed to visitors but the remainder is accessible by car and also on foot, following the several designated footpaths (see map). The visitors' centre at Villareal de San Carlos offers information, an exhibition and the usual car stickers and similar souvenirs. A campsite (often busy) is adjacent.

A panoramic and productive (birdwise) view is to be had from the top of the Castillo de Monfragüe (A). Cars may drive up the road and park under the large Nettle Trees below the castle mount. The views from the top of the tower itself are vertiginous and some care is needed since there are no balustrades and there is a risk of emulating the gypsy who gave his name to the gorge nearby, El Salto del Gitano (Gypsy's Leap). History does not seem to record why the gypsy leapt but at least his mortal remains would soon have been recovered by the inhabitants of the Griffon Vulture colony, which occupy the facing escarpment of Peña Falcón. Over 100 Griffon pairs nest here and over 400 roost on the cliff in autumn. Black Storks and Egyptian Vultures can also be seen on their nests on the same cliff face, from the mirador by the main road at El Salto del Gitano (B). Black Wheatears frequent the scree slopes to the right of the escarpment. The mirador is also a good site to watch and wait for Eagle Owls at dusk or dawn, with a high chance of success in winter especially but also during the breeding season. The castle is a good place to sit and scan the area for raptors: Black Vultures join the Griffons overhead and both Golden and Bonelli's Eagles are regularly seen here. White-rumped Swifts breed nearby and are often seen in the area in late spring and summer, and into the autumn. The woodlands on the south slope are alive with passerines in spring and will repay a patient search. Hawfinches are particularly common here.

The Mirador de la Tajadilla (C) overlooks a small cliff on the opposite side of the reservoir, where some Griffon Vultures and a pair of Egyptian Vultures nest. Bonelli's Eagles nest nearby and may often be seen from this point. The Mirador de la Báscula (D) faces rocky wooded slopes favoured by large raptors. Black Vultures may regularly be seen on their nests here.

The roadside continuing northwards from the Mirador de la Báscula, following the River Tiétar, gives good views across the narrow reservoir to the low cliffs opposite, where several nests of Black Storks are traditionally placed. The Mirador de la Higuerilla (E) gives elevated views over the river. Black Vultures regularly occupy a nest on a solitary, though distant, encina opposite this mirador. A lookout and hide at the Portilla del Tiétar (F) faces the main cliff face, where the nests of a small colony of Griffon Vultures may be clearly seen. Black Vultures nest nearby and are often to be seen perched on the rocks here. Spanish Imperial Eagles have nested on Cork Oaks just to the right of the cliffs in some years. Eagle Owls are regularly visible in the area at dawn and dusk and a pair nests on the cliff face.

♿ Most of the important roadside miradors are readily accessible, although La Higuerilla is reached down a short flight of steps. Walking trails are difficult and the Castillo (A) is a hard climb.

CALENDAR

All year: Grey Heron, Red Kite, Griffon and Black Vultures, Goshawk, Common Buzzard; Spanish Imperial, Golden and Bonelli's Eagles, Peregrine Falcon, Eagle Owl, Lesser Spotted Woodpecker, Crag Martin, Thekla Lark, Black Wheatear, Blue Rock Thrush, Southern Grey Shrike, Azure-winged Magpie, Red-billed Chough, Hawfinch, Rock Bunting.

Breeding season: White Stork, Black Stork, Black Kite, Egyptian Vulture, Short-toed and Booted Eagles, Common Cuckoo, Scops Owl, Red-necked Nightjar, Alpine and White-rumped Swifts, Bee-eater, Hoopoe, Red-rumped Swallow, Black-eared Wheatear; Spectacled, Subalpine and Orphean Warblers, Golden Oriole.

Winter: Great Cormorant, Black-headed Gull, raptors, passerines including Alpine Accentor.

SIERRA DE GATA AND BORBOLLÓN RESERVOIR   CC7

Status: Includes ZEPAs 'Sierra de Gata y Valle de Las Pilas' (18,522ha) and 'Embalse de Borbollón' (946ha). The island in the reservoir is an ADENEX reserve, protecting the heronry.

Site description

The Sierra de Gata is part of a long granitic ridge forming the northwestern boundary of Extremadura with Castilla y León, extending eastwards to the Sierra de Gredos. Woods of pine and Pyrenean Oak cover the southern slopes and there are expanses of heathland on the tops. The Embalse de Borbollón to the south of the Sierra was formed by a dam across the river Arrago. The low, undulating hills surrounding the natural-looking lake include rough pastureland and woods of encinas and Pyrenean Oaks.

Species

The Sierra de Gata is home to a notable colony of Black Vultures (50 pairs) and it is not unusual to see flocks of 30 or more of these massive birds over the ridge in spring. Other raptors include Goshawks, which favour the pine forests on the lower slopes. Honey-buzzards also breed in the area. The breeding species of the summit heathlands include the Tawny Pipit, Northern Wheatear and Ortolan Bunting. Dippers occur along the upper Río Arrago.

The Embalse supports a colony of Grey Herons, which nest on eucalyptus clumps and encinas on the island. Little Ringed Plovers also breed here. Black-winged, Red and Black Kites are among the breeding raptors. Bee-eaters are very common and there is a colony of Spanish Sparrows. Cranes occur in winter and Black-tailed Godwits occur on passage very early in the year, peaking in February.

The pastureland and wheatfields to the southeast, around Guijo de Coria, have a population of steppe species: Great Bustards, Little Bustards, Stone-curlews and Calandra Larks.

Timing

Springtime for most species.

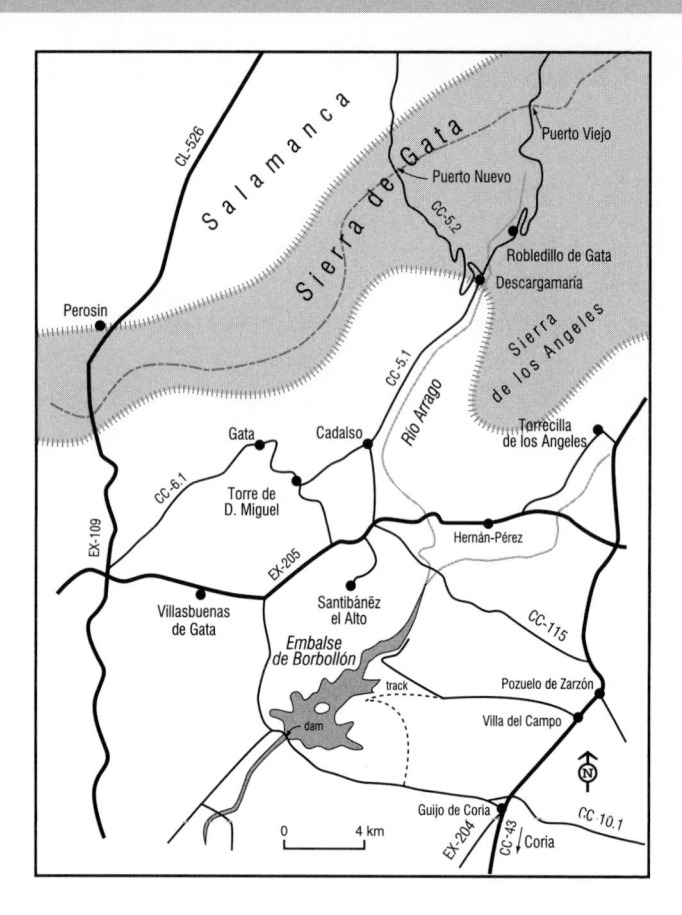

Access

The whole area can be reached easily from the south via the EX-A1 motorway or the adjacent EX-108. Leave at Coria and take the the CC-43 towards Guijo de Coria: a peaceful road through pastureland. The Embalse de Borbollón is signposted from the CC-43 at the crossroads just north of Guijo de Coria. Open-country areas can be scanned for steppe species from the roads that converge on Guijo de Coria: the CC-43 and EX-204 from the south, the CC-10.1 from the east and the minor road from/to the Embalse de Borbollón from the west.

The reservoir is most easily viewed from the dam on the south side (telescope advisable). A minor road, leading from Villa del Campo eastwards to the reservoir inlet and passing through woodlands, is an alternative access route that is also worth exploring. Eucalyptus clumps along this road near the reservoir hold sparrow colonies in spring.

Several roads ascend northwards across the Sierra de Gata. Among these the CC-5.2 provides a highly tortuous transect of the ridge, ascending for 15km from pinewoods through Pyrenean Oak woodlands to a great undulating expanse of heathland at Puerto Nuevo. The heathlands are a spectacular sea of purple in April. This road sees very little traffic and there are ample opportunities for stopping and scanning, and for exploring the heathlands along the broad tracks/firebreaks at the summit.

♿ Most areas are easily viewed from or near your car.

> **CALENDAR**
>
> **All year:** Cattle Egret, Grey Heron, Black-winged and Red Kites, Black Vulture, Goshawk, Bonelli's Eagle, Great and Little Bustards, Stone-curlew, Pin-tailed Sandgrouse, Calandra Lark, Dipper, Spanish Sparrow.
>
> Tawny Pipit, Northern and Black-eared Wheatears, Common Redstart, Ortolan Bunting.
>
> **Winter:** Greylag Goose, Crane, Black-tailed Godwit, Golden Plover, Lapwing, Black-headed Gull, Dunnock, larks, pipits and finches, including Bullfinch.
>
> **Breeding season:** Quail, White Stork, Black Kite, Montagu's Harrier, Little Ringed Plover, Bee-eater, Roller.

JERTE VALLEY (Valle del Jerte) CC8

Status: Includes Reserva Natural (Garganta de los Infiernos, 6,800ha) in the central valley and ZEPA 'Embalse de Gabriel y Galán' (8,401ha) nearby.

Site description

The valley of the Jerte river is a long, straight canyon descending for some 40km from the southwestern slopes of the lofty Sierra de Gredos to the historic city of Plasencia. The region offers a contrast with most of the rest of Extremadura, having a temperate rather than a Mediterranean character, with the birds to match. The high tops, easily accessed at the Puerto de Tornavacas

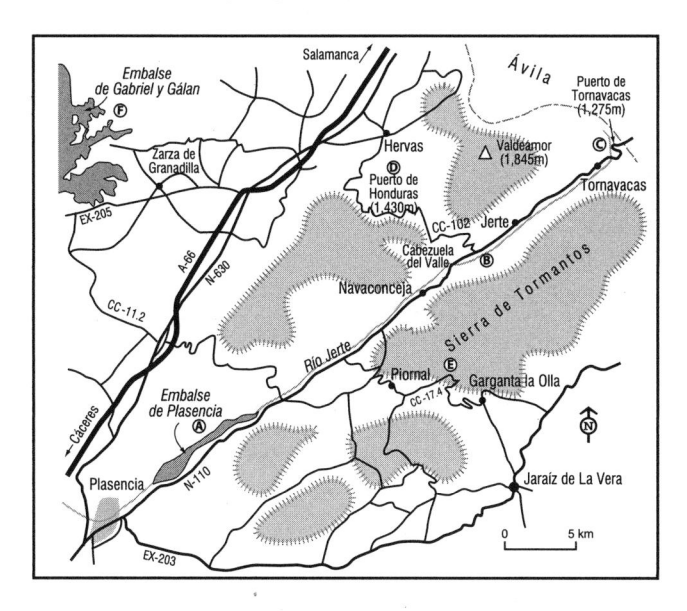

Azure-winged Magpie

(1,275m) and the Puerto de Honduras (1,430m), have a moorland appearance. These upper slopes are rocky but adorned with large tracts of the broom *Cytisus purgans,* whose vanilla-scented flowers provide spectacular sheets of yellow in early summer. Lower down there are extensive woods of Pyrenean Oak as well as Sweet Chestnut. There are also scattered pinewoods, chiefly of Scot's Pine. The lowest slopes are terraced and heavily committed to the cultivation of cherries. The cherry orchards are a local tourist attraction in spring when the trees are in blossom. The whole area has the rocky and often snow-capped summits of the Gredos range, which lie outside Extremadura, as a spectacular backdrop. The Gredos sites are described in the sister volume, *Where to Watch Birds in Northern and Eastern Spain.*

A narrow reservoir, the Embalse de Plasencia (Embalse del Jerte), occupies the lower valley, upstream of Plasencia, which it serves. A much larger reservoir, the Embalse de Gabriel y Galán, some 30km north of Plasencia, is also of ornithological interest.

The tourist information centre at Cabezuela del Valle, on the main road halfway up the valley, can provide further information and a guide to the many walking trails.

Species

The region provides good opportunities to see a range of upland and mountain species, such as Sky and Thekla Larks, Alpine Accentors, Dunnocks, Water Pipits, Dippers, Bluethroats, Northern Wheatears, Rufous-tailed Rock Thrushes, Carrion Crows and Rock Buntings. There are small breeding populations of Water Pipits, Bluethroats and Alpine Accentors at the highest levels, above 2,000m. There is also a southern outpost breeding presence of the Red-backed Shrike around the Puerto de Tornavacas. Spectacled Warblers breed in the expanses of low scrub. The woodlands also have significant breeding populations of Honey-buzzards and Lesser Spotted Woodpeckers, two species that are otherwise very local or absent in the southern half of Spain. The village of Piornal is notable for its large colony of Pallid Swifts: a few hundred pairs nest here at the considerable elevation of 1,175m. The reservoirs attract gatherings of Black Storks in late summer, and waterfowl in winter especially.

Timing
The area requires a full day's birding to do it justice. Visits in good weather in late springtime and in summer are the most productive given that many of the valley's attractions are summer visitors. In fact, the relatively high altitude means that mid-summer birding is pleasant and worthwhile, at a time when the heat lower down is problematic. Visits at other times are still enjoyable, again provided that the mountain mists haven't descended, but be aware that the weather may be very cold in winter, with frosts making driving potentially tricky on the minor roads. The passes are occasionally blocked by snow.

Access
The N-110 Plasencia/Ávila road gives ready access to the region. The road is moderately busy but there are a number of obvious stopping points. A visit should take in the following sites:

A *Embalse de Plasencia (del Jerte)* This long, narrow reservoir is worth a look, especially in late summer when Black Storks often congregate along the shore, and in winter when there may be small numbers of grebes and waterfowl. Good views may be had from the dam (Presa): follow the access road off the N-110, which is signposted 6km north of Plasencia. The reservoir may also be viewed from the service road below the villas that fringe the eastern shore at the top end.

♿ Easily viewed from or near your car.

B *La Garganta de los Infiernos* Don't be put off by the grim name (The Maw of Hell!). This Reserva Natural features a range of marked trails through woodland along the Rio Jerte and its tributaries and up the eastern slopes of the valley. Maps and information are available from the splendid reserve centre. Some of the trails involve steep climbs of eight hours' duration each way. However, much of the bird interest is available without quite such a marathon effort. Honey-buzzards, Lesser Spotted Woodpeckers and Dippers, all of them very local species in southern Spain, breed in the area as well as a pleasing variety of other raptors and woodland passerines. The reserve is entered along a short access road located a short distance north of the road to Puerto de Honduras, between the villages of Cabezuela del Valle and Jerte. The entrance is on the right (if you are driving up the valley) opposite and between a conspicuous pottery workshop and craft centre (alabasteria) and a timberyard. Park signs point the way. A short drive leads across the river and to the information centre, from where the trails originate.

♿ The trails are not easily accessible.

C *Puerto de Tornavacas* This col lies on the boundary between Extremadura and Castilla y León. There are fine views to be had from the ample car park, right down the Valle del Jerte to Plasencia and north down into Ávila province. The col is a natural route for migrant birds, including raptors and passerines and also flocks of Woodpigeons in autumn. These last are the quarry of hunters whose shooting platforms dot the hillside on the northern (Ávila) slopes. The area is peaceful in spring and summer when walks up the slopes may reveal the local speciality species: Northern Wheatear, Rufous-tailed Rock Thrush and Bluethroat. Carrion Crows, a mountain species in southern Spain, are often around the car park. The Red-backed Shrike has established a breeding presence here, chiefly on the Ávila side, since 2000. Dartford Warblers and Common Cuckoos are also characteristic. Raptors overhead frequently include Golden Eagles and Griffon Vultures. The Sierra de Gredos is proving attractive to Lammergeiers from the Cazorla reintroduction programme (see J1), some of which are present for months on end and include the Extremaduran section in their wanderings: you have a chance of seeing one here. Spanish Ibex occur on the ridges.

♿ Easily accessed by car.

D Puerto de Honduras The col is reached along a quiet, winding road that climbs upwards from the Valle del Jerte and then descends to the town of Hervas in the next valley. Starting from the Jerte end, the entrance is signposted some 2km north of Cabezuela del Valle. The road climbs through the terraced cherry orchards, popular with Azure-winged Magpies and other passerines, to reach open woodlands of Pyrenean Oaks and then the moorland above. The upper woodlands are a reliable site for breeding Pied Flycatchers and Ortolan Buntings. Traffic is light (although ponies and other livestock on the road are a minor hazard) and there are abundant stopping places from which to look and listen for woodland species, and also overhead raptors. There are easy walks around the col itself where breeding species include Dunnocks and Rock Buntings. Finches and other passerines often fly over the col. The committed will not be deterred from walking 4km up to the northern summit for a good chance of seeing Northern Wheatears and Rufous-tailed Rock Thrushes. Once over the top of the col, on the Hervas side, there are some fine old oaks around the first hairpin bend. An easy walk of a few hundred metres through these open parkland-like woods leads to the ridge, giving magnificent views to the north and east. This is an excellent spot to scan for raptors and there is also another good chance of finding Rock Thrushes and Northern Wheatears.

The descent to Hervas is through much denser and more mature woodland, chiefly of Pyrenean Oaks and Sweet Chestnuts, with stands of Poplars and a dense undergrowth. There are numerous paths through the woods including the signposted trail 'Ruta Heidi' just above Hervas. The main road itself is so quiet though that it provides ideal access for looking for and listening to the woodland passerines. The town of Hervas provides easy access to or from the A-66/N-630.

♿ The col is easily accessed by car. The trails are steep.

E Piornal/Garganta la Olla A network of small roads serves the villages on the southern side of the Jerte valley, climbing through cherry orchards and chestnut groves. The higher tops have expanses of heathland. The road from Piornal to Garganta la Olla is recommended, especially for woodland and scrub species. The Black Wheatear occurs here.

♿ Worthwhile observation is possible from or near your car.

F Embalse de Gabriel y Galán   This large reservoir is not in the valley but instead lies some 30km north of Plasencia. Access is from the A-66/N-630 and thence west along the EX-205 to Zarza de Granadilla and beyond, where there is roadside access to the eastern shore. There are also good views from the dam, which is on the southern shore, on the EX-205. The lake attracts wintering Cranes and also waterfowl, the latter also in winter especially. Ospreys occur on passage and one or two individuals remain to winter.

♿ The dam is accessible by car.

CALENDAR

All year: Red Kite, Black and Griffon Vultures, Common Buzzard, Goshawk, Sparrowhawk, Golden Eagle, Peregrine Falcon, Eagle and Tawny Owls; Iberian Green, Great Spotted and Lesser Spotted Woodpeckers, Wood and Thekla Larks, Crag Martin, Grey Wagtail, Dipper, Alpine Accentor, Dunnock, Nuthatch, Black Wheatear, Blue Rock Thrush, Dartford Warbler, Firecrest, Azure-winged Magpie, Carrion Crow, Raven, Rock Sparrow, Rock Bunting.

Breeding season: Black Stork, Honey-buzzard, Egyptian Vulture, Short-toed and Booted Eagles, Common Cuckoo, Red-necked and European Nightjars, Scops Owl, Water Pipit, Bluethroat, Northern and Black-eared Wheatears, Rufous-tailed Rock Thrush, Pied Flycatcher, Whitethroat; Melodious, Spectacled, Subalpine, Garden and Bonelli's Warblers, Golden Oriole, Red-backed Shrike, Ortolan Bunting.

Winter: Wildfowl, Great Crested Grebe, Osprey, Crane, Goldcrest, Siskin, Common Crossbill (has bred).

Passage periods: Osprey, Common Sandpiper, Garden and Willow Warblers.

ARROCAMPO RESERVOIR AND CAMPO DE ARAÑUELO   CC9

Status: Includes ZEPA 'Embalse de Arrocampo' (687ha).

Site description
The Campo de Arañuelo is a broad expanse of gently undulating country in the north of Extremadura sandwiched between the rivers Tiétar and Tajo. Here too is an exceptionally interesting reservoir, the Embalse de Arrocampo-Almaraz, perhaps the best wetland in Extremadura for its diversity of nesting waterbirds. The Sierra de Gredos forms an imposing, purple-hued and snow-capped backdrop to the whole region. There are excellent tracts of Cork Oak and encina woodlands, and riverine woodlands, in the north and west of the area. Elsewhere there is much cultivated land, a lot of it given over to tobacco. Indeed, tobacco-drying barns, many of them derelict, are a feature of the region. The nuclear power station of Almaraz is in the southeastern corner.

Species
The reservoir is one of the few sites in Extremadura inhabited by Purple Swamphens and it also plays host to breeding Marsh Harriers and abundant waterfowl, these last especially in winter. The largest colony of Purple Herons in Extremadura is here. Little Bitterns; Grey, Squacco and Night Herons and a few pairs of Great White Egrets and Spoonbills also breed. Great Bitterns are

occasionally reported in winter and may well be regular then. Penduline Tits and Savi's Warblers breed: this is the only regular site in Extremadura where the latter nests. The Bearded Reedling (Bearded Tit) has been recorded here in some recent springs and has nested, this again being the only site in Extremadura where this species might be found.

This is an especially good area for seeing the birds of the dehesas. The raptor population is high, not least because the region borders Monfragüe Natural Park. Griffon and Black Vultures are overhead for much of the day. All five eagles occur, as well as Common Buzzards; Black-winged, Black and Red Kites (these last abundant in winter) and Hobbies. White Storks are abundant and Black Storks frequently feed along the river channels. Passerines include breeding Sand Martins, Tree Sparrows and Hawfinches. Cranes have a winter presence and roost in the irrigated farmland flanking the Tiétar on the approaches to La Bazagona.

Timing
Interesting species are present all year round.

Access
The primary access to the region is the EX-A1 motorway. The stretch between La Bazagona and Casatejada crosses some interesting woodlands but stopping on the motorway is impossible. However, there is periodic access to the service road, the old EX-108 which the EX-A1 replaced, which is virtually traffic-free. The better woodlands may also be explored from the quieter roads to the north of the EX-A1, leading to and beyond Majadas. The road south from La Bazagona to Monfragüe crosses open country and a small river before passing through some exceptionally fine woodlands, boasting unusually large, very old Cork Oaks. The roads to Toril, south of the EX-A1, are especially quiet and productive.

Beyond Toril there is a network of very good dirt roads crossing excellent woodlands and leading to Monfragüe and Serrejón. The entrances at either end

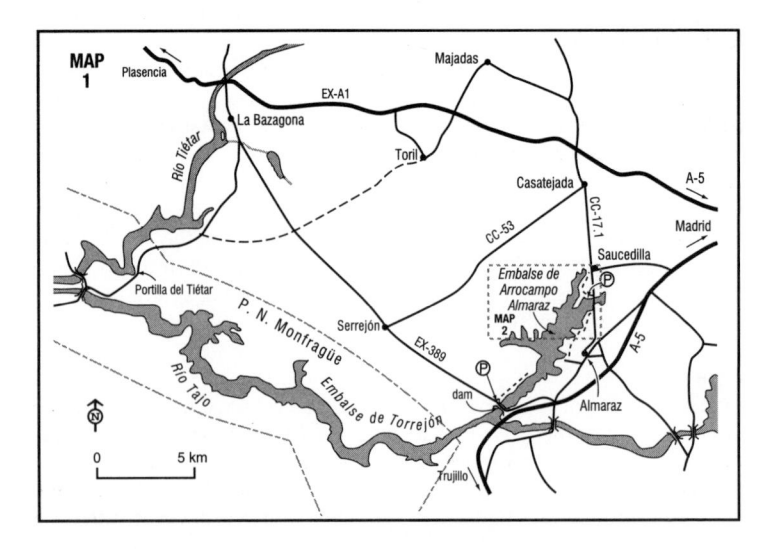

MAP 2

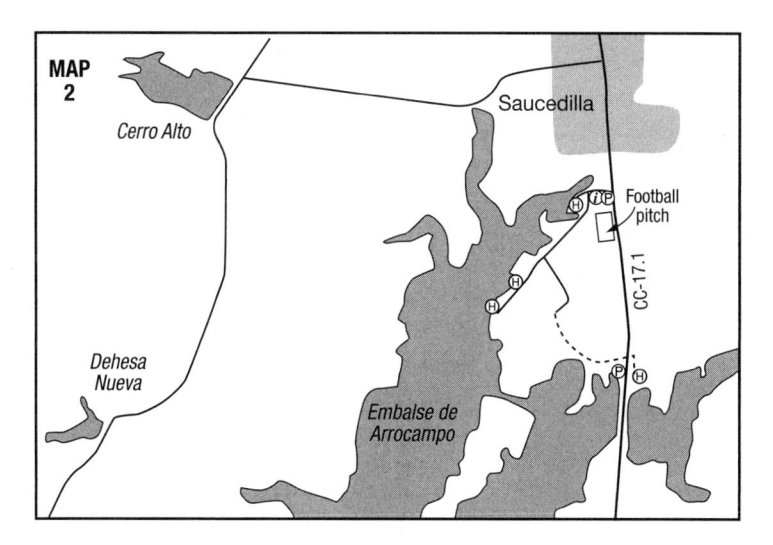

- Cerro Alto
- Saucedilla
- Football pitch
- CC-17.1
- Dehesa Nueva
- Embalse de Arrocampo

are indicated by a sign saying 'Ruta a Toril y Serrejón', adorned with a drawing of a Hoopoe. Stop frequently and scan over the woodlands. Some of the cattle here are bullfight material so it is inadvisable to stray beyond running distance from your vehicle.

The Embalse de Arrocampo-Almaraz is reached by driving south from the EX-A1 on the CC-17.1, past Casatejada and to Saucedilla. Alternatively you can exit the A-5 for Almaraz and travel north to Saucedilla. Saucedilla is the location of a so-called 'Parque Ornitológico' and the point of origin of two birding trails. The trail offering close views of the reservoir and its fringing reedbeds starts just south of Saucedilla, where there is an information centre alongside the local football pitch, on the west side of the CC-17.1. Park here and follow the trail on foot, or drive it if conditions allow. It leads to three lakeside hides. A fourth hide is on the east side of the main road, where it cuts across an arm of the reservoir south of Saucedilla. Park on the west side and explore the shore of the lake and also the margins of the reed-fringed lagoon on the east side of the road on foot; the lagoon is the best site for Purple Swamphens. A footpath (Via Pecuaria) leads from this point for 5km along the eastern shore of the reservoir to Almaraz, giving good views over the lake. You can also continue to and through Almaraz to the dam on the EX-389. There is parking just west of the dam and a footpath follows the dehesa-lined western shore from here.

The second trail originating from Saucedilla crosses fields and dehesas, and visits two small irrigation pools, Cerro Alto and Dehesa Nueva, which also attract waterbirds. This drivable trail offers the chance to see raptors such as Lesser Kestrels and Black-winged Kites, as well as such open-country birds as Stone-curlews and Golden Plovers in winter. The trail is a minor road, partly tarmacked, which is signposted on the west of the CC-17.1 at the northern edge of Saucedilla. Follow this road to access the dehesas, fields and pools, returning the same way or continuing north past Cerro Alto to the CC-53.

♿ The best areas are easily accessed by car.

SIERRA BRAVA RESERVOIR   CC10

Status. Within ZEPA 'Llanos de Zorita and Embalse de Sierra Brava' (18,787ha).

Site description
This fairly small reservoir (1,650ha) was first inundated as recently as 1994 but has become a renowned and most important site for waterfowl. The reservoir is surrounded by low hills offering convenient elevated viewpoints. Encina woodlands fringe most of the area.

Species
The site is of greatest value in winter when often immense numbers of waterfowl are present. Indeed, the duck flocks have exceeded 100,000 birds in some recent winters, making this at times one of the best sites for wintering waterfowl in the whole of Spain. It remains to be seen whether the even more recent (2015) Embalse de Alcollarín (CC12) nearby affects the numbers that winter here. Dabbling ducks, mainly Gadwall, Teal, Mallard, Pintail and Shoveler, predominate but there are also numbers of Greylag Geese and Red-crested and Common Pochards. Great and Little Grebes are also common. The mass of birds is augmented by hundreds of Cormorants and thousands of roosting Black-headed and Lesser Black-backed Gulls. Black-tailed Godwits fly from the ricefields nearby to roost on the islands in winter. Thousands of Cranes feed in the maize stubble and ricefields in winter. Part of the enjoyment of such a varied avian throng comes from scanning the flocks for scarcer or rare species. In recent winters these have included Barnacle, Red-breasted, Pink-footed and Bean Geese, Greater Scaup, Common Shelducks and Ferruginous Ducks. Egyptian Geese are present year-round. The reservoir is fairly quiet at other times of year but the islands in the lake have large numbers of nesting Gull-billed Terns.

Timing
Winter visits (October–February) are by far the most worthwhile.

Access
See map for CC11. The reservoir lies east of the EX-378. At km 16.2, 12km north of Madrigalejo, a tarmac road signposted to 'Presa de Sierra Brava' leads for 3.2km to the dam. There is a 'No Entry' sign but we are assured by SEPRONA (the police wildlife service) that birders are welcome to enter at their own risk. The lake may readily be viewed from the dam itself. You can also continue beyond the dam and follow the road round to the left, from where tracks lead down to the east bank. A telescope is essential.

♿ The dam is easily accessed by car.

Cáceres Province 351

CALENDAR

All year: Egyptian Goose, Gadwall, Mallard, Little and Great Crested Grebes.

Breeding season: Gull-billed and Little Terns, Collared Pratincole.

Winter: Greylag Goose, Wigeon, Teal, Pintail, Shoveler, Red-crested Pochard, Common Pochard, Tufted Duck,

Black-necked Grebe, Great White Egret, Marsh Harrier, Common Coot, Crane, Black-tailed Godwit, Black-headed Gull, Lesser Black-backed Gull, rarities.

Passage periods: Black Stork, Spoonbill, Osprey, Black Tern.

GUADIANA VALLEY RICEFIELDS CC11

Status: No special protection.

Site description

This region owes much of its interest to the ongoing expansion of rice cultivation, which has seen increasing areas converted to paddy fields and the establishment of reservoirs to irrigate them. Water levels vary with the time of year and the management regime so there are nearly always some muddy expanses suitable for waders. However, avoid spring, when the ricefields are dry and birdless (see Timing). Drainage channels are often reed-fringed and attractive to warblers and finches. The region is bordered by open encinares to the east and by rolling open pastures to the north, these last accommodating steppe species.

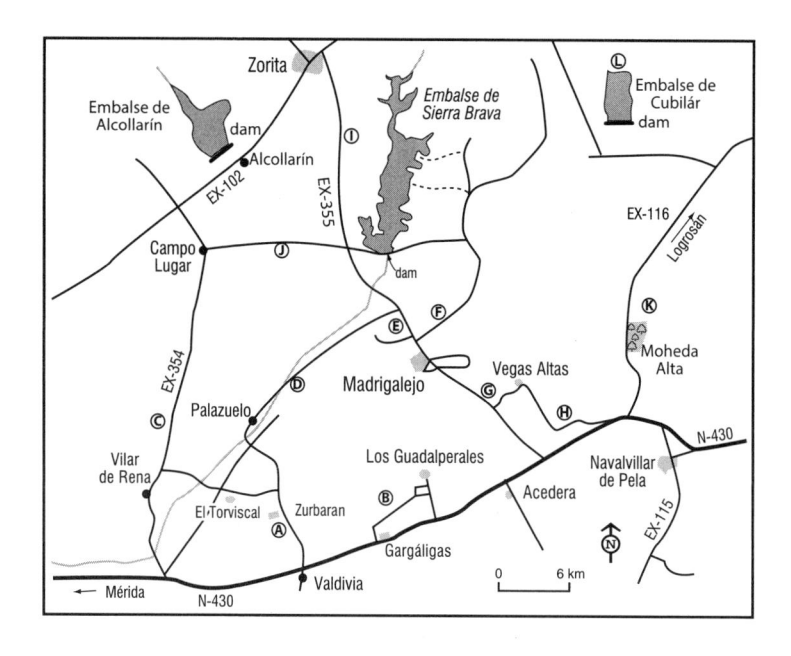

Species

The wetlands attract numbers of waders, notably Black-tailed Godwits, Ruff, Avocets and Black-winged Stilts but also a wide range of other species, especially during passage periods. White Storks, Grey Herons and Cattle Egrets are common. Winter in particular sees numbers of waterfowl, notably Greylag Geese, Mallard, Teal, Shoveler, Common Pochards and Tufted Ducks. The zone is also noteworthy as a major wintering ground for Common Cranes. Large gatherings are regularly to be found in the encinares to the north of Navalvillar de Pela and also foraging for spilled rice in the paddy fields. Migrating Cranes, which follow the Guadiana river, are often to be seen overflying the area in November and February especially.

The ricefields are inevitably a magnet for seed-eaters and large flocks of Spanish and House Sparrows are characteristic. There is also a thriving population of Red Avadavats, which no doubt find echoes of their original Far-Eastern home in the paddies. Common Waxbills are also present. Little Bustards are among the species attracted to the ricefields in winter. Resident raptors include Marsh Harriers and Black-winged Kites, with Hen Harriers common in winter. This is often a particularly good area for Black-winged Kites and roosts of over 30 have been recorded in winter.

The pastures to the north, around Zorita, have good numbers of steppe species, including Great Bustards. They attract flocks of Lapwings, Golden Plovers and Sky Larks in winter. Gull-billed Terns, which nest on the reservoir lakes, forage for insects over the pastures in spring and summer.

Timing

The region offers plenty to see with its wide variety of interesting resident species. However, the birding interest of the ricefields is dependent on whether or not they are flooded and the nature and extent of the management regime. The basic pattern is that the ricefields are prepared for planting in early spring, when they are bone dry. They are planted and flooded in May and remain so until the rice harvest in September/October, after which they are allowed to dry out gradually. In practice therefore they are wet to some degree, and worth visiting, from May until January/February. Waterfowl, herons and waders are most abundant in winter although the less common wader species occur especially during the (late) spring and autumn passage periods. Harvest time is also a good time to visit, the harvesters being followed by hordes of egrets, gulls and White Storks, among others. Common Cranes are only present in winter and their season is a relatively short one, from mid-November to late February. They alone suffice to make a winter visit memorable, however. Spring and summer visits are of course necessary to see the breeding species but, as ever, the heat haze makes visits in the middle of the day in hot weather inadvisable.

Access

A series of major and minor roads traverse the area and it is necessary to discover which paddy fields are holding birds at any particular time. Different areas change in attractiveness to waders especially as water levels change or the harvest progresses. Stopping along the busy N-430 is sometimes tempting but ill-advised, except where farm tracks allow parking right off the road. More relaxed viewing is to be had along the road between Valdivia and Palazuelo (A), the road from Gargáligas through Los Guadalperales (B) and also along the southern part of the Villar de Rena/Campo Lugar road (C). A service road from Palazuelo (D) runs adjacent to a disused railway line, now converted to a 'Vía Natural' — a walkway and cycle track — all the way to the EX-378 (and beyond), passing through many kilometres of rice paddies.

The vicinity of Madrigalejo offers good access to the paddy fields. Three sites are particularly productive:

1. Heading north from Madrigalejo, turn left at km-15.5 (west) just before a clump of eucalyptus trees surrounding a small pool (E). This is the entrance to a network of sandy tracks serving the paddy fields. The car serves as a good hide here. Care should be taken in wet weather since the tracks may become muddy and impassable to anything other than 4WD vehicles. The area may also be readily visited on foot along the Vía Natural, the footpath/cyclepath along a disused railway line, which is most easily identified by the bridge that takes it across the EX-378 here.

2. Also just north of Madrigalejo, turn right (east) into a road running eastwards from the Pension Mayve (F). The rather rough road crosses open fields and encinares before giving access to paddy fields. Common Cranes are often present in winter.

3. Heading south from Madrigalejo take the signposted road left (east) to Vegas Altas (G). The road skirts the town to the east and offers elevated views across the paddy fields. A branch road (H) descends southwards from Vegas Altas to the N-430, giving access to additional large expanses of ricefields.

The steppe species can be seen most readily along the northern half of the Madrigalejo/Zorita road (EX-355, I) and also from the much quieter road to Campo Lugar (J). In winter the road (EX-116) through the encinares for 10km north from Navalvillar de Pela towards Logrosán is excellent for Common Cranes. There are also large expanses of ricefields along this latter road. The small reserve of Mojeda Alta (K) at km-4.8 on the EX-116 offers a watchtower for observing Cranes, which in winter arrive in their thousands to roost on the ricefields beyond. Winter roosts of Marsh and Hen Harriers, and Greylag Geese, also occur in this area. The reserve is signposted on the east side of the road where there is ample parking.

The EX-116 crosses the Canal de las Dehesas, an irrigation channel, some 8km north of Mojeda Alta. The canal is signposted. The tarmac road parallel to the canal on the west side of the EX-116 soon leads to a small and secluded

Calandra Lark

reservoir, the Embalse del Cubilar (L). The 'No Entry' sign at the entrance to the canal road is widely disregarded. The dam here is a good watchpoint across the reservoir, which attracts some numbers of waterfowl in winter. Cranes also roost around the northern edge. The fringing dehesas are also good for Black-winged Kites and other woodland birds.

♿ The best sites are all accessible by car.

CALENDAR

All year: Grey Heron, Black-winged Kite, Common Buzzard, Marsh Harrier, Little and Great Bustards, Stone-curlew, Black-bellied and Pin-tailed Sandgrouse, Calandra Lark, Zitting Cisticola, Cetti's Warbler, Southern Grey Shrike, Common Waxbill, Red Avadavat, Spanish Sparrow.

Breeding season: Little Bittern, Little Egret, Montagu's Harrier, Lesser Kestrel, Black-winged Stilt, Little Ringed Plover, Collared Pratincole, Gull-billed and Little Terns, Roller, Sand Martin, Greater Short-toed Lark, Tawny Pipit,

Black-eared Wheatear; Reed, Great Reed and Melodious Warblers.

Winter: Greylag Goose, ducks, Great White Egret, Spoonbill, Red Kite, Hen Harrier, Golden Plover, Lapwing (a few may breed), Common and Jack Snipes, Black-tailed Godwit, Ruff, Short-eared Owl, Penduline Tit.

Passage periods: Spoonbill, Avocet, Common Redshank, Greenshank, Little and Temminck's Stints, Curlew Sandpiper, Whiskered and Black Terns.

ALCOLLARÍN RESERVOIR   CC12

Status: Within ZEPA 'Llanos de Zorita and Embalse de Sierra Brava' (554ha).

Site description

This new reservoir, inaugurated in January 2015, has rapidly become one of the key birding sites in Extremadura. In large measure this is due to the construction incorporating several features that enhance its interest to wildlife, notably a shallow, gradually sloping natural shoreline, several islands and a second dam ensuring the retention of an area of deeper water even when levels in the main basin are low. The whole site is a delightful place to visit and highly recommended. It is often very peaceful, the predominant sounds in spring being the songs of Hoopoes and Common Cuckoos. The lake is bordered by dehesa and some grassy meadows, with pastures sloping down to the water on the western and northern shores. There are designated areas for picnicking, fishing and bathing. The shoreline everywhere should be inspected for waders and others. The lake retained by the small dam has permanent water, fringing reed and emergent dead trees, where White Storks nest.

Species

The reservoir and its fringing habitats attract a large diversity of species. Its potential has probably not yet been fully discovered but it is undoubtedly very high. In winter it attracts large numbers of similar waterfowl species to those that use the Sierra Brava reservoir and no doubt there are frequent exchanges between the two assemblages. Some waterfowl remain all year round, with

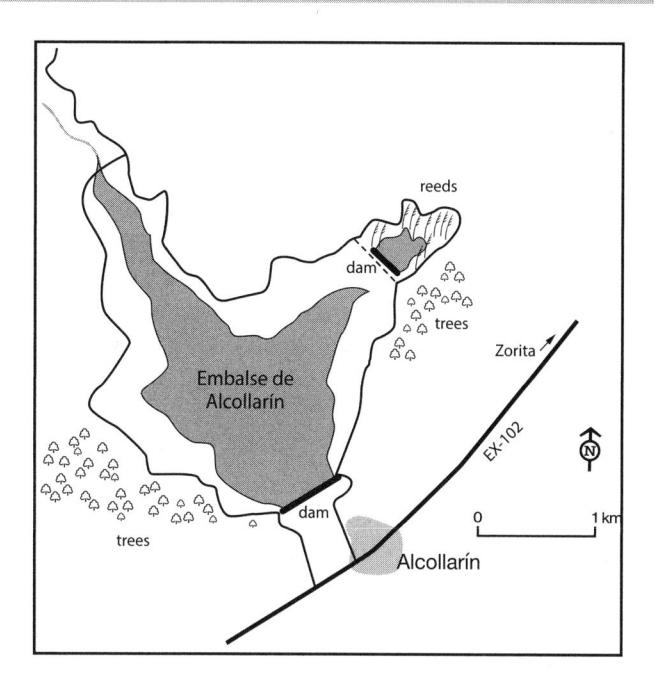

Egyptian Geese, Mallard, Common Coots and Great Crested Grebes breeding. Little Grebes probably nest at the small reservoir. Post-breeding concentrations of 2,000+ Little Grebes and 800+ Great Crested Grebes have been recorded here in August–October. Collared Pratincoles occur on passage and probably in the breeding season. A diversity of waders frequent the shorelines, especially during passage periods. The reservoir margins attract Spoonbills, Grey Herons and egrets, including a few Great White Egrets. Purple Herons occur in spring and may breed. Both Bee-eaters and Spanish Sparrows have colonies nearby and the dehesas have the usual complement of woodland birds. The site has a high potential for attracting and retaining rarities. Scarce or rare waterfowl found here have already included Ruddy Shelducks, a number of Ring-necked Ducks, several Red-knobbed Coots, a pair of Greater Scaup, a White-headed Duck and a redhead Smew. There are also records of a Bonaparte's Gull and of several juvenile Audouin's Gulls.

Timing
Of interest year-round but especially during passage periods and also in winter, when waterfowl numbers are highest. In winter especially, a visit here could ideally be combined with a visit to the nearby Embalse de Sierra Brava (CC10) and the Guadiana ricefields (CC11).

Access
Broad tarmac roads, signposted 'Presa de Alcollarín' connect each end of the dam to the EX-102 immediately south of Alcollarín village. Car parks at the dam offer elevated views of the main basin. A telescope will be necessary. Good dirt roads give access to the shorelines but a locked gate prevents travelling all

the way round. The road from the western end of the dam gives access to woodland and areas of open meadow sloping down to the water. The eastern road reaches the smaller dam after 2.2km, where there is a recreation area with picnic tables.

CALENDAR

All year: Egyptian Goose, Gadwall, Mallard, Little and Great Crested Grebes, Common Coot, Thekla Lark, Cetti's Warbler, Southern Grey Shrike.

Breeding season: Gull-billed Tern, Collared Pratincole, Bee-eater, Hoopoe, Common Cuckoo, Woodchat Shrike, Spanish Sparrow, woodland passerines.

Winter: Greylag Goose, Wigeon, Teal, Pintail, Shoveler, Red-crested and Common Pochards, Tufted Duck, Black-necked Grebe, Great Cormorant, Great White Egret, Marsh Harrier, Crane, Black-tailed Godwit, Black-headed and Lesser Black-backed Gulls, rarities.

Passage periods: Black Stork, Spoonbill, Osprey, waders; Whiskered, Black and Little Terns, rarities.

GIBRALTAR

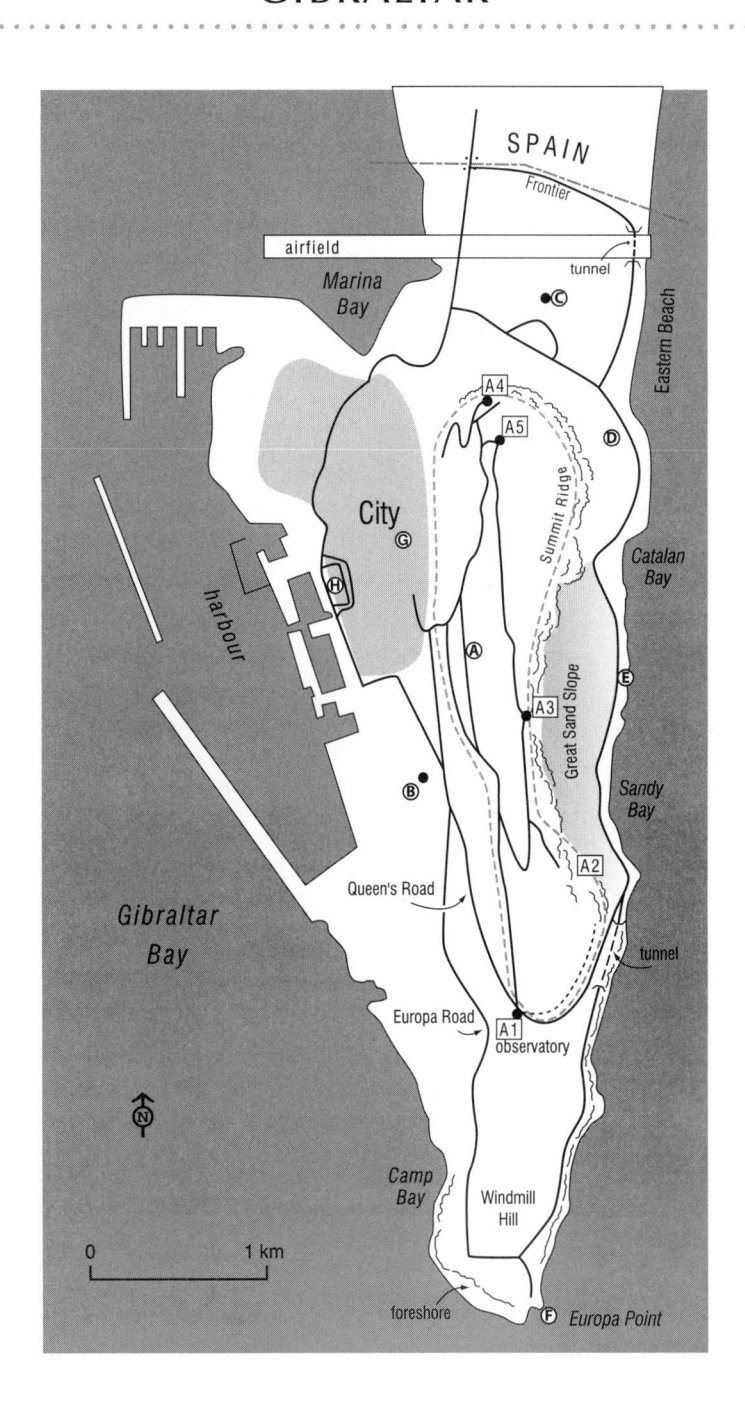

This is a really special place, which you must visit if you can. At least, I think so, probably because I was born there. There can't be many other spots though where you can see a Black Stork sharing a thermal with a Gannet or migrating Zitting Cisticolas accompanying flocks of Honey-buzzards! Gibraltar is exceptionally well placed for migration watching, of raptors and seabirds particularly. The Jews' Gate bird observatory often makes it possible to observe bird ringing in action, with excellent in-the-hand views available of a whole range of species. Its operation is one of the many functions of the Strait of Gibraltar Bird Observatory, the ornithological 'wing' of the Gibraltar Ornithological and Natural History Society (GONHS). The Society is the Gibraltar Partner of Birdlife International. Its diverse activities include raptor counting and other monitoring of visible migration, seabird recording, rehabilitation of sick or injured raptors and other wildlife, and many others related to the recording and conservation of all wildlife in Gibraltar and to ensuring the protection of the natural environment.

Watching migrating raptors and seabirds is essentially a static activity: you wait and the birds come to you. Birders with mobility problems can therefore see much of the best on offer without the need to move around very much. Gibraltar as a whole is not as wheelchair friendly as it might be, although the situation continues to improve. Nevertheless, birders with disabilities can enjoy a very productive visit here without needing to hire a car since all sites are a very short taxi ride away.

Information

A useful resource is the GONHS website (www.gonhs.org). This provides current news about the Society and its activities and it includes the Gibraltar Species List as well as a 'Latest sightings' noticeboard. The Gibraltar Bird Report, edited by me and published annually by the GONHS since 2001, is now online and is a good pointer to what may be expected. The Society may be contacted at GONHS, PO Box 843, Gibraltar; email *info@gonhs.org*.

UK-based visitors may contact the Gibraltar Government Office, who will gladly supply tourist brochures, details of hotels and car hire companies and similar useful information. Their address is: Gibraltar Government Office, 150 The Strand, London, WC2R 1JA (Tel. 0207 836 0777. Website https://www.gibraltar.gov.gi/new/). For tourism information see www.visitgibraltar.gi/. Locally, there are the Gibraltar information centres at the frontier, in the airport terminal building and at Casemates Square in the city.

Where to stay

Gibraltar offers a range of hotels among which the Caleta Hotel, on the east side of the Rock at Catalan Bay, is popular with visiting birders. The Elliot Hotel is centrally placed in the city: its rooftop pool lies directly underneath the main raptor flylines. The Rock Hotel is adjacent to the Botanic Gardens. All three have websites enabling online booking and all have their own car parks – a vital asset in Gibraltar. Camping and caravanning are not permitted but there are numerous campsites in nearby Spain.

The GONHS itself offers accommodation within the Upper Rock Nature Reserve. There are two sites: Bruce's Farm and Jews' Gate Field Centre. Bruce's Farm is a converted army officers' residence, offering twin room accommodation as well as kitchen, bathroom and other facilities, at the modest charge of £20

Barbary Partridges

per person per night. Jews' Gate accommodates up to five persons, at £10 per night, in simple but comfortable quarters. Bird ringers have priority and will find it ideal since it is only five metres from the nearest mist nets. Accommodation at either site should be pre-booked; contact the Society via their website (as above) or by post, email, etc. Suitably qualified persons wishing to help with the ringing or other activities for extended periods, a month or more, are accommodated free of charge: contact GONHS for more details.

Sea trips

Pelagic trips are increasing in popularity worldwide, often catering for those wishing to see cetaceans (whales and dolphins). Gibraltar was something of a pioneer in this field since the 'Dolphin Safari' (Tel. 350 71914, www.dolphinsafari.gi, email dolphin@gibraltar.gi) has provided day trips for dolphin watchers for over 30 years now, operating from Marina Bay. A couple of other tour companies now run a similar service. None of these trips are pelagic in the strict sense: they operate only in coastal waters, principally in Gibraltar Bay. The main attraction is always the dolphins (Common, Striped and, less often, Bottle-nosed), which are more or less guaranteed to appear. The boats do allow a closer look at seabirds such as Cory's, Scopoli's and Balearic Shearwaters, and occasionally such exotica as flying fish, sunfish, whales, turtles and sharks, and a trip on one of them is always a pleasant experience.

Status: Protected area. There is no hunting and all wildlife is protected from deliberate destruction, although the usual pressures from 'development' arise and these are severe in such a small territory. Fortunately much new building is on land recently reclaimed from the sea, which has added some habitat for birds incidentally. The whole of the Upper Rock comprises a Nature Reserve.

Site description

The massive Rock of Gibraltar looms out of the Mediterranean at the western end of the Costa del Sol. Surprisingly, many people still believe that Gibraltar is an island but the Rock is firmly attached to the mainland via a broad sand bar, now mainly under the airport, the border posts and other tarmac. Gibraltar covers an area of some 8km^2. The northern and eastern sides of the Rock are sheer cliff faces, rising to 426m at O'Hara's Battery, the southern end of the summit ridge. The western slopes are more gentle and largely built up on their lower reaches but covered by diverse vegetation higher up. A dense scrub of wild olive, lentisc, buckthorn and spiny broom is interrupted by some rather overgrown firebreaks and other open grassy areas.

An enormous massive lean-to sand bank, the Great Sand Slope, is an unmissable feature of the eastern side of the Rock and visible for miles around. It is the result of an accumulation of windblown sand during the glaciations, when sea levels were lower and a sandy plain extended east of Gibraltar. The Great Sand Slope was covered with corrugated iron sheeting for much of the 20th century, when it served as a water catchment area. The sheeting was removed during the late 1990s and the Slope reseeded with its original flora by the GONHS. The Slope represents an enormously significant enhancement of the natural attractions of Gibraltar, being in effect an extensive (29ha) new grassland habitat, attractive to many bird species, not least the Barbary Partridge.

Gibraltar has a striking climate that contrasts with that of the Costa del Sol. The prevailing winds are westerly (good for raptors) and easterly (good for seabirds in spring and winter): the infamous *levanter*, which often produces a pall of cloud which caps the Rock for days on end. Winter rainfall is considerable, sometimes torrential. Combinations of murk and wetness during passage periods produce falls of migrants.

Species

Gibraltar's claim to ornithological fame is due almost entirely to its strategic position. It is a first-rate viewpoint for raptor migration, which occurs in some form year-round. Here raptors can be seen (and photographed) really well at close quarters, which isn't always possible further along the Strait. A good day in spring or autumn will produce 12–15 raptor species, occasionally more. The southern tip (Europa Point) offers excellent year-round seabird watching, with Balearic Shearwaters and Audouin's Gulls among the regulars, and Lesser Crested Terns if you are lucky. Falls of migrants occur at times, when the vegetated areas harbour a wide variety of small birds: thirty species of warbler have been recorded, for example. The high densities of birds and birders mean that rare or vagrant species are detected annually. Post-2000 sightings have included Common Eider, Rüppell's Vulture, Pallid Harrier; Lesser Spotted, Greater Spotted and Steppe Eagles, Allen's Gallinule, Moroccan White Wagtail, Seeböhm's Wheatear, Yellow-browed Warbler, Mountain Chiffchaff, Red-breasted Flycatcher, Hooded Crow, African Chaffinch, Little Bunting and Indigo Bunting. Lanners occur annually, especially in late summer and autumn, making Gibraltar the best site in Iberia to see this locally rare species. These and other rarity records are included in the 'Status List' Chapter (pp. 368–98).

Relatively few species breed in Gibraltar but it is a good site to see Peregrines, Pallid Swifts and Blue Rock Thrushes. The only wild Barbary Partridges in mainland Europe occur on the Upper Rock and other suitably vegetated areas. The breeding Shags are also an area speciality: the only other colony in the region is at Cabo de Gata (AL1). The Gibraltar Shags have proved amazingly tenacious and the population of at least five pairs is stable despite the threats from pollution and, especially, disturbance. The colony of Yellow-legged Gulls, some 2,000 pairs, is a less welcome feature of Gibraltar since the gulls tend to monopolise open ground for nesting, are aggressive to migrating raptors and cause a nuisance in the city: a culling programme has had some success in limiting the gull population. Gibraltar has a pleasing bird community in winter, when Black Redstarts, Robins and Common Chiffchaffs are present in numbers and Crag Martins roost nightly in some of the Rock's caves. Alpine Accentors occur in some years. Wintering seabirds can be numerous and include Mediterranean Gulls and Sandwich Terns among the regulars. Individual Booted Eagles include Gibraltar in their home range in some winters.

Birds apart, Gibraltar is interesting botanically: the limestone flora, which comprises over 600 vascular plant species, contrasts with that of the surrounding sandstone hills (see *http://floraofgibraltar.myspecies.info/*). Gibraltar Candytuft is one of the more obvious of the special species. The famous apes are Barbary Macaques *Macaca sylvanus*; like the partridges, they too were probably originally introduced from Morocco. They are much easier to find than the partridges; in fact, they will find you if they think you have anything edible to offer. The apes are of uncertain but mainly malevolent temperament and should be treated with discretion. Feeding them is illegal, even though they much prefer junk food to the healthy official rations which they are served daily at public expense.

Timing

Visible migration of soaring birds is obvious from March–May and August–October. The movements mainly involve raptors. Storks and Cranes favour Tarifa, further west (CA2). March has large numbers of Black Kites and Short-toed Eagles especially. April is best for all-round variety. May is dominated by vast flights of Honey-buzzards, especially in the first half of the month. The return passage begins in July with Black Kites. The Honey-buzzards mainly return at the beginning of September. Short-toed and Booted Eagles are prominent in September and early October. A trickle of 'northern' species: Sparrowhawk, Common Buzzard, Hen Harrier, continues into November when there may also be very large movements of Griffon Vultures. Raptors pass over Gibraltar mainly in westerly winds: the easterlies displace the flow westwards. However, numbers of Booted and Short-toed Eagles especially mill about over the Rock in easterly winds in August–October, often flying on a circular course to and from the mainland as they wait for conditions to improve.

Falls of passerine migrants may occur between February and May and again between August and November. Bad weather, especially the low cloud and drizzle or thunder associated with the levanter, can cause thousands of migrants to be grounded on the Rock, although falls as large as this are rare.

Seabird watching is rewarding at all times of year, including in mid-summer. On bright days, however, viewing from Europa Point is best in the late afternoons and evenings. Otherwise the glare from the sun on the water makes life difficult.

Access

Gibraltar Airport is one of the obvious arrival points for visitors to our area and it is becoming increasingly busy. In 2019 frequent scheduled flights were operated by British Airways and Easyjet, together serving London Gatwick, London Heathrow, London Luton, Bristol and Manchester. Other destinations are becoming available. The infamous level crossing for road traffic at the runway is being replaced by a tunnel, although this had not yet been completed in 2018: the access road to Gibraltar will be as mapped once this is in use but pedestrians will still cross the runway on foot.

Some visitors arrive by sea, mainly by ferry from Tangier. The land frontier provides access from Spain for vehicles and pedestrians: don't forget your passports. The border can be badly affected by traffic queues at certain times, mainly when commuters enter the Rock in the early mornings and (particularly) when they leave again in the evenings. Avoid these times if possible if crossing the border in a car. Pedestrians and motorcyclists are not usually delayed. There is a very good and frequent bus service from the border serving Europa Point and passing the cable car, Alameda Botanical Gardens and the southern entrance to the Nature Reserve on the way: ask the driver to drop you off at any of these.

Gibraltar's roads are very congested and parking can be a real problem. You should avoid driving into the city centre at all costs: instead, as you drive south, bear right from the dual carriageway opposite the Gib Oil petrol station and follow what is effectively a bypass, which skirts the conspicuous city walls and continues to the major points of ornithological interest. Several large car parks below the walls, and by the lower cable car station at Grand Parade, offer a chance of parking if you intend to visit the city. Sundays are always quiet and a good choice for a day visit.

Once in Gibraltar, there are a number of key sites to visit. All of Gibraltar can be 'done' in a day but a visit during good westerly winds in the migration seasons usually means staying for hours at either Jews' Gate or the cable car station, with perhaps an evening visit to Europa Point. A visit in winter or during easterly 'fall' conditions in passage periods needs to take in the other sites mentioned.

A UPPER ROCK NATURE RESERVE

The Upper Rock is most readily reached via Queen's Road, which branches off from Europa Road a short distance south of the Rock Hotel. However, you will have to climb the rather steep hill on foot, take a taxi or use the cable car, since foreign-registered cars are no longer admitted to the Nature Reserve. There is a toll booth and an entrance fee that includes entry to the 'obligatory' tourist sites such as the limestone caverns of St Michael's cave and the 'apes' den. However, walkers who do not intend to visit these sites only pay £5. If you are staying at Bruce's Farm or Jews' Gate, or if you are planning to make several visits during the course of a birding holiday, you should contact the Gibraltar Tourist Board in Casemates Square, at the north end of Main Street, who will issue you with a complimentary pass.

All of the Upper Rock is of interest, especially when falls of passerine migrants occur. However, there are a number of notable sites to visit:

A1 JEWS' GATE
GONHS has a bird observatory, bird ringing station and information centre at Jews' Gate, just past the entrance to the Upper Rock Nature Reserve. The centre is often manned by the very helpful local ornithologists and visitors are welcomed. If you are a qualified ringer you may well be able to take part in the bird ringing there. This is the place to go for information on the latest sightings both in Gibraltar and in nearby Spanish provinces. Jews' Gate is the preferred

site for watching raptors in spring: in autumn try the cable car station on the summit ridge (*A3*).

♿ The observatory is only accessible with help: there are several steps. Good watchpoints are also available below the observatory and from the lay-by north of the toll booth.

A2 MEDITERRANEAN STEPS
This is a scenic footpath that takes in the highest point of the Rock and offers spectacular views of the Strait, Spain and Morocco. Park at Jews' Gate and follow the mapped route up the western side to the summit at O'Hara's Battery. The Mediterranean steps descend the eastern side and the path then turns west above Windmill Hill to Jews' Gate once more. It is of course perfectly possible to follow this route in reverse, up the eastern side and down the western side, but the steepness of the eastern climb makes this an option only for the foolhardy or the super-fit. The Mediterranean steps are occasionally frequented by Barbary Partridges and good views of Peregrines and Blue Rock Thrushes are often to be had. Yellow-legged Gulls are abundant, some nesting on the very footpath itself. Alpine Accentors sometimes occur here in winter. The Steps are best visited in the early morning when any grounded night migrants or partridges will still be undisturbed. However, in hot weather they are a pleasant place to be in the late afternoons and evenings, when the shadow of the Rock covers the eastern side.

♿ Impossible.

A3 CABLE CAR TOP STATION
A good site for raptor watching, especially during the southward migration (mainly August–October). The cable car provides a lift there for those who are lazy or just in a hurry (2018 charges: £15.50 return, £18.50 one way, including entrance to the Nature Reserve). The energetic will walk up the Rock from the town: a leisurely progression taking in the summit ridge will require 3–4 hours, obviously more if there are good birds to delay you. The cable car originates from Grand Parade, which is adjacent to the Botanic Gardens and where you may park. The top station has observation platforms and a restaurant and other tourist amenities. Tourists may be numerous and most birders prefer to walk a short distance south along the summit ridge and watch in relative peace from the old concrete gun emplacements there.

Plans to replace the cable car system in its entirety, including the top station, were announced in 2018. The cable car will not be available during construction.

♿ The original cable car is inaccessible to wheelchairs.

A4 PRINCESS CAROLINE'S BATTERY
An alternative site for watching southbound raptor migration in August–October. The decommissioned Gun Battery overlooks the airfield. Scan for raptors approaching Gibraltar from the north. This site often offers excellent views of the birds but it misses those which approach the eastern side of the Rock from the northeast. Walk down to the Battery from the main road or follow signs to the Upper Galleries, from where viewing platforms give

dizzying views down the precipitous north face. The local Peregrines and Blue Rock Thrushes may readily be observed here.

♿ Accessible with help, given the steepness of the roads.

A5 GOVERNOR'S LOOKOUT
This area of scrub, firebreaks and Aleppo Pine groves is attractive to grounded migrants, such as Hoopoes and Bonelli's Warblers, and wintering passerines, such as Firecrests. The firebreaks have a population of Barbary Partridges, which can sometimes be seen here at dawn and in the evenings, when they may call noisily. The area can only be explored on foot. Cars may be parked at Princess Caroline's Battery.

♿ Accessible with help, given the steepness of the roads.

B GIBRALTAR BOTANIC GARDENS
The Botanic Gardens at the Alameda are of considerable and increasing floral interest: the beds of aloes, attracting nectar-seeking Common Chiffchaffs and other birds, are a curiosity in December and January. At other times, the gardens hold grounded migrants, notably Hoopoes, chats and warblers. Park in Grand Parade, next to the bottom cable car station, which is adjacent to the gardens. There is no entrance fee although donations are welcome and are a contribution to the botanical conservation work that is underway here. The garden is under the supervision of Dr Keith Bensusan, who is also General Secretary of the Gibraltar Ornithological and Natural History Society and an invaluable source of local information.

♿ Excellent access by ramps, except at the main gate.

C NORTH FRONT CEMETERY
The public cemetery, below the vertical precipices of the north face of the Rock, is also excellent for grounded migrants during 'fall' conditions. Turn north off Devil's Tower Road into Cemetery Road. Parking spaces are seldom available in this road and it is advisable instead to use the multistorey car park immediately west of the cemetery, on the northern side of the Devil's Tower Road dual carriageway. The cemetery is open to the living during daylight hours but is much disturbed by visitors, who tend to frighten the birds away. Morning visits are therefore best. The eastern and northern parts have most of what vegetation there is and hence attract more birds.

♿ Accessible.

D NORTHEAST TALUS
The steep, rocky vegetated slopes just north of Catalan Bay are a reliable site for Barbary Partridges and Blue Rock Thrushes and may hold migrants at times. Use the car park on the north side of Catalan Bay and observe from the coast road on foot. Scan the slopes and gullies here for the partridges.

♿ Viewable from or near your vehicle.

E EAST SIDE ROAD

This two-way road serves Catalan and Sandy Bays and is linked to Europa Point via Dudley Ward tunnel. A large lay-by just north of Catalan Bay and the hotel offers excellent views of the Great Sand Slope but you will need a telescope to make much of the upper reaches. You can stop here for a while but leave your car at Catalan Bay if you plan to stay longer and walk along the road to Sandy Bay. Barbary Partridges are often seen here as well as resident and migrant passerines. Seabirds are visible offshore, especially in the afternoons when the light is in your favour. The groynes at Sandy Bay, and the rocky coastline, occasionally attract such waders as Whimbrel, Turnstone and Common Sandpiper.

♿ Viewable from or near your vehicle.

F EUROPA POINT

The southern tip of the Rock, marked by the conspicuous lighthouse. Europa Point is one the best seawatching sites in the region, offering excellent views from an elevated position. Raptors and other landbirds can also often be seen arriving here in Europe from Africa in spring. Use the public car park and continue to the watchpoint at the cliff edge, to the right of the lighthouse. A telescope is a great asset but not essential: some fly-bys of Audouin's Gulls and other key species are very close indeed.

It is also possible to follow the coastal promenade westwards to where a wooden flight of steps descends to the rocky and sparsely vegetated foreshore. The lee of a small building here provides some shelter on windy days and the foreshore itself receives passerine migrants in some numbers at times. The rocky shoreline is one of the few places in Gibraltar visited quite frequently by waders: chiefly Whimbrels, Turnstones and Common Sandpipers. The stony ground at the north of Europa Flats, around the conspicuous white mosque, sometimes holds several Thekla Larks in winter. There is also a watchpoint here giving views along the east coast to the Costa del Sol beyond. The World Heritage Site of the Gorham's Cave complex at Governor's Beach, once inhabited by Neanderthals, is also visible from here, as are the Shags that nest on the cliffs nearby.

♿ Good access.

G GIBRALTAR CITY

The city has a large and noisy population of both Pallid and Common Swifts. Sunny evenings from April to July provide excellent opportunities to compare these two confusing species in excellent light. The rooftop nesting population of Yellow-legged Gulls is a relative novelty: the birds are an overspill from cliffs and scree slopes of the Rock. The first nests appeared in the late 1980s and many pairs now attempt to nest in the city each year, if they are allowed to get away with it. It is ironic and unfortunate that city-nesting has become especially popular with the gulls since culling activities made some of their former nest sites on the Upper Rock untenable. Some human town-dwellers find their swooping defence of the nests rather intimidating and the gulls' habit of spraying generous splashes of excreta over the washing on clothes lines has done little for their public image.

♿ Good access. The city centre is pedestrianised.

H COMMONWEALTH PARK

This small park, which opened in 2014, is one of the most pleasing amenities in Gibraltar, attracting plenty of human visitors, young and old, but also an increasing diversity of wildlife. The park occupies a former car park, itself built on land reclaimed from the sea. It boasts the only sizable (4,000m^2) public lawns and the largest freshwater areas in Gibraltar, these last in the form of two ponds. Over 100 trees, a range of Mediterranean species, have been planted as well as many nature-friendly shrubs and herbaceous plants. The park has its resident Blackbirds, and White Wagtails have also nested there. Other species will doubtless follow as the vegetation matures. Many common passage migrants linger there and Black Redstarts and Common Chiffchaffs are among the winterers that occur. More unexpected visitors have included Moorhen, Squacco Heron, Little Egret and Wood Sandpiper, as well as a Cormorant that terrorised the fish in the ponds one winter. The park is always worth a look if you are passing by. It is popular with children of course and early morning visits are likely to be the most productive.

&. Excellent access. Descending from the city walls to the park is enabled by a lift as well as stairs.

CALENDAR

All year: Barbary Partridge, Cory's and Balearic Shearwaters, Gannet, Shag, Common Kestrel, Peregrine Falcon, Great Skua, Audouin's and Yellow-legged Gulls, Collared Dove, Eagle and Little Owls, Wren, Blue Rock Thrush, Blackbird, Sardinian Warbler, Blackcap, Blue and Great Tits, Raven, Spotless Starling, House Sparrow, Chaffinch, Serin, Greenfinch.

Breeding season: European Storm-petrel (offshore only); Common, Pallid and Alpine Swifts, Spotted Flycatcher.

Winter: Common Scoter, Leach's Storm-petrel; Mediterranean, Little, Black-headed and Lesser Black-backed Gulls, Kittiwake, Sandwich Tern, Razorbill, Thekla Lark, Crag Martin, Meadow Pipit, Grey Wagtail, White Wagtail, Alpine Accentor, Robin, Black Redstart, Stonechat, Zitting Cisticola, Common Chiffchaff, Firecrest, Short-toed Treecreeper, Siskin, Rock Bunting.

Passage periods: Quail, Sooty and Levantine Shearwaters, Greater Flamingo, Black Stork, White Stork, Honey-buzzard, Black and Red Kites, Egyptian and Griffon Vultures, Short-toed and Booted Eagles; Marsh, Hen and Montagu's Harriers, Sparrowhawk, Common Buzzard, Osprey, Lesser Kestrel, Merlin, Hobby, Eleonora's and Lanner Falcons, Pomarine and Arctic Skuas; Gull-billed, Caspian, Lesser Crested, Common and Black Terns, Puffin, Turtle Dove, Great Spotted Cuckoo, Scops Owl, European and Red-necked Nightjars; Common, Pallid and Alpine Swifts, Kingfisher, Bee-eater, Hoopoe, Wryneck, Greater Short-toed and Sky Larks; Sand, Crag and House Martins, Barn and Red-rumped Swallows, Tawny and Tree Pipits, Yellow Wagtail, Rufous-tailed Scrub-robin, Nightingale, Bluethroat, Common Redstart, Whinchat, Stonechat, Northern and Black-eared Wheatears, Rufous-tailed Rock Thrush, Ring Ouzel, Song Thrush, Redwing, Zitting Cisticola; Grasshopper, Reed, Olivaceous, Melodious, Dartford, Spectacled, Subalpine, Orphean, Garden, Bonelli's, Wood and Willow Warblers, Whitethroat, Iberian and Common Chiffchaffs, Spotted and Pied Flycatchers, Golden Oriole, Woodchat Shrike, Chaffinch, Brambling, Serin, Greenfinch, Goldfinch, Siskin, Linnet, Ortolan Bunting, rarities.

Records

The Gibraltar Ornithological and Natural History Society keeps a database of records of birds observed in Gibraltar and publishes the annual Gibraltar Bird Report. The Society is most grateful to receive news of your observations, which should be sent to:

The Records Officer, GONHS, PO Box 843, Gibraltar. Tel. (350) 72639. Email: *records@gonhs.org*.

Records of rarities seen in Gibraltar are vetted by the Gibraltar Rarities Committee. The species for which a description is required are indicated on the GONHS website (*www.gonhs.org*) and in the Gibraltar Bird Reports. You should realise that many 'familiar' species, such as Woodpigeon and Moorhen, which are very common in southern Spain, are rarities in Gibraltar: a consequence of the limited range of habitats available there. Any published records are fully acknowledged.

STATUS LIST OF THE BIRDS OF SOUTHERN AND WESTERN SPAIN

The following list outlines the status of all those species which have been recorded in the region as wild birds, whether post-1950 (Category A), pre-1950 (Category B) or exotic but naturalised (Category C). It allows you to put your observations in context. Much more comprehensive information, on distribution, habitats, populations, movements, phenology and occurrence patterns, and analyses of rarity records, is available in *The Birds of the Iberian Peninsula* (De Juana and Garcia 2015).

For those species that occur regularly, the present List shows the main periods when each occurs in the region, excluding marginal dates for which there are only a few records. The sites at which the scarcer species may be located are in the Site Index. Province names are given as abbreviations in parentheses, i.e. AL Almería, BA Badajoz, CA Cádiz, CC Cáceres, CO Córdoba, GR Granada, H Huelva, J Jaén, MA Málaga and SE Sevilla.

The list also includes those additional species that are vagrants or occur very rarely. Records from Andalucía and Extremadura are within the geographical area covered by the Comité de Rarezas de la Sociedad Española de Ornitología (CR-SEO: the Spanish Ornithological Society's Rarities Committee). Rarities seen in Gibraltar are vetted by the Gibraltar Rarities Committee. Only records ratified by either of these bodies up to 2017 are included, with the exception of some more recent sightings, which are pending scrutiny. The listed records are of single birds unless otherwise indicated.

The List sequence and the names used largely follow the most recent (2012) edition of the Spanish List, which in turn is based on the List maintained by the AERC (Association of European Rarities Committees). However, some English names have been changed to reflect British usage (e.g. diver for loon). Qualifiers such as 'Eurasian' have also been omitted where no ambiguity results in a European context.

Weekly updates on news of rarities from throughout Spain are available on a highly recommended website: Rare Birds in Spain (*www.rarebirdspain.net*). If you are lucky enough to come across any rare birds, you are urged to submit the details. Records to CR-SEO may be submitted in English. Email *rarezas@seo.org* attaching your description. Gibraltar records should be sent to the Gibraltar Rarities Committee, GONHS, PO Box 843, Gibraltar (email *cperez@gonhs.org*).

Descriptions are only required by the Spanish Rarities Committee for species that are rarities in Iberia as a whole, marked ★ in the List below. The full list, together with the Rarity Record Submission Form, is available on the SEO/Birdlife website (www.seo.org). Records of other scarce species are still welcomed and may be sent to the same address. Annual reports of rare birds, and other interesting records from around Spain, are published in the Spanish Ornithological Society's journal *Ardeola* or in its magazine *Aves y Naturaleza*, as well in provincial and regional bird reports. All published records are acknowledged.

The Gibraltar Rarities Committee vets records of species that are rarities in

Gibraltar though not in Spain, in addition to Iberian rarities recorded in Gibraltar. See the GONHS website (*www.gonhs.org*). Reports of rarities from Gibraltar are published annually in the *Gibraltar Bird Report*.

Mute Swan *Cygnus olor* Cisne vulgar
Introduced but still very scarce. Wandering individuals may occur anywhere.

Tundra (Bewick's) Swan* *Cygnus columbianus* Cisne chico
One old record from Doñana: in February 1910.

Whooper Swan* *Cygnus cygnus* Cisne cantor
One record from Fuente Obejuna (CO).

(Tundra) Bean Goose* *Anser fabalis* Ánsar campestre
Formerly common in central Spain in winter but now very rare. Occasional in ones and twos in Doñana and Extremadura, November–February. Five at Sierra Brava reservoir (CC) on 13 January 2002.

Pink-footed Goose* *Anser brachyrhynchus* Ánsar piquicorto
Occasional in winter in very small numbers in Doñana. Three records from Extremadura.

Greater White-fronted Goose *Anser albifrons* Ánsar Careto grande
Occurs in the Doñana area with some frequency in very small numbers, October–February. Several recent winter records from Extremadura, from ricefields and reservoirs.

Lesser White-fronted Goose* *Anser erythropus* Ánsar careto chico
Cited as a very rare winter visitor to Doñana, December–February.

Greylag Goose *Anser anser* Ánsar común
Abundant in winter in and near Doñana, where up to 80,000 gather, and very locally elsewhere, although increasing in winter in Extremadura, where over 15,000 have been recorded in recent years. Present from late October to early March mainly, peak numbers occurring in December and January.

Bar-headed Goose *Anser indicus* Ánsar indio
Annual in winter in Doñana. These are believed to be members of the small feral population in Scandinavia, which move south with the Greylags. Several recent records from Extremadura.

Snow Goose* *Anser caerulescens* Ánsar nival
Occasional in winter in Doñana. These are presumably of feral origin.

Canada Goose* *Branta canadensis* Barnacla canadiense
Claimed to be annual in winter in Doñana: October–February. Four records from Extremadura.

Barnacle Goose *Branta leucopsis* Barnacla cariblanca
Annual in winter in Doñana and also recorded in Cádiz and Extremadura; October–February. Records are normally of single birds, but 14 were in Doñana in January 1996.

Brent Goose *Branta bernicla* Barnacla carinegra
Very scarce. Annual in winter in Doñana: December–February. Birds are nearly always of the pale breasted subspecies *B.b.hrota*. Eleven together at Sancti Petri (CA11) on 25 January 1992 is noteworthy. Two Extremaduran records.

Red-breasted Goose* *Branta ruficollis* Barnacla cuellirroja
Several Doñana records. One at the Sierra Brava reservoir (CC) on 12 January 2002.

Egyptian Goose *Alopochen aegyptiaca* Ganso del Nilo
One at Zorita (CC) on 1 September 1993 and two at the Embalse de Arrocampo (CC9) on 3 March 2001 were forerunners of continuing colonisation of Iberia. Small numbers may be encountered increasingly widely in Extremadura and western Andalucía. Successful breeding has occurred at La Serena reservoir (BA) since 2012 and elsewhere.

Ruddy Shelduck *Tadorna ferruginea* Tarro canelo
Formerly a regular visitor, especially to Doñana from August–March, originating from breeding grounds in the Moroccan High Atlas mountains. Rare at present but some still turn up every year on the

Mediterranean coast and in the lower Guadalquivir valley. A flock of seven at the Salinas de Cabo de Gata (AL) on 26 June 1993 is noteworthy. At least 26 Extremaduran records, including a flock of eight at Logrosán (CC) on 28 December 2004.

Common Shelduck *Tadorna tadorna* Tarro blanco
Breeds in small numbers at Doñana and sporadically elsewhere in coastal Andalucía. Also nests at the Valdecañas reservoir (CC). Common November–March in Doñana and the Guadalquivir estuary. Elsewhere present in small numbers in winter.

Mandarin Duck* *Aix galericulata* Pato mandarín
Rare, chiefly in winter, with at least 20 records of 1–3 birds scattered throughout the region.

Eurasian Wigeon *Anas penelope* Silbón europeo
Widespread in winter and on passage: present October–March. Locally abundant, notably in Doñana where peak counts in January may exceed 30,000 birds. Also abundant in the Guadiana ricefields (CC), with peak counts there of 8,000 birds.

American Wigeon* *Anas americana* Silbón americano
At least one record from Doñana in October 1971.

Gadwall *Anas strepera* Ánade friso
Small numbers breed widely in Extremadura and Andalucía: especially in the Guadalquivir valley. Common in winter, when hundreds and locally thousands gather on suitable lakes.

Eurasian Teal *Anas crecca* Cerceta común
Widespread and common in winter and on passage. Locally abundant; peak January counts in Doñana have exceeded 60,000 birds. Up to 50,000 winter in Extremadura, with peak counts of over 30,000 at the Sierra Brava reservoir (CC). A few nest in Doñana in wet years.

Green-winged Teal* *Anas carolinensis* Cerceta americana
Three winter records from Extremadura.

Mallard *Anas platyrhynchos* Ánade azulón
Widespread and common as a breeding species. Numbers are greatly increased by migrants in winter. Up to 90,000 winter in Extremadura, with peak counts of up to 20,000 at the Sierra Brava reservoir (CC). Peak winter counts at Doñana exceed 20,000 birds.

Pintail *Anas acuta* Ánade rabudo
Widespread and locally numerous on passage and in winter: present November–March mainly. Up to 25,000 winter in Doñana in an average season and up to 26,000 in Extremadura: notably at the Embalses de Gargáligas and Sierra Brava (CC). Has bred Doñana and Las Norias (AL).

Garganey *Anas querquedula* Cerceta carretona
Commonly encountered in small flocks during passage periods; February–April and August–October, especially in spring. Small numbers winter, chiefly in Doñana. Breeds rarely.

Blue-winged Teal* *Anas discors* Cerceta aliazul
Eleven records (12 birds) from coastal Andalucía, October–June. One from Extremadura.

Shoveler *Anas clypeata* Cuchara común
Common and locally numerous on passage and in winter, present September–April mainly. Counts at Doñana may exceed 100,000 birds. Up to 50,000 winter in Extremadura, with up to 45,000 counted at the Sierra Brava reservoir (CC) alone. A small population breeds in Doñana and the Guadalquivir valley and mainly sporadically elsewhere.

Marbled Duck *Marmaronetta angustirostris* Cerceta pardilla
A very scarce breeding species, nesting in Doñana, the Brazo del Este (SE) and a few lagoons in southern Andalucía. Most numerous in autumn and winter when flocks of 100+ may occur in this area in good years.

Red-crested Pochard *Netta rufina* Pato colorado
Widespread as a breeding species on suitable lagoons in Andalucía, especially in Doñana. Nests sporadically in Extremadura. Flocks of several hundred form locally in winter. Several thousand winter in Doñana.

Common Pochard *Aythya ferina* Porrón europeo
Some 3,000–5,000 pairs breed in Doñana. Small numbers breed widely elsewhere in Andalucía but very locally in Extremadura. Widespread and numerous in winter when large numbers of migrants arrive from the north. Peak winter counts in Doñana exceed 7,000 birds.

Ring-necked Duck* *Aythya collaris* Porrón de collar
Fifteen records (17 birds) from coastal Andalucía and 11 from Extremadura (23 birds): December–June.

Ferruginous Duck *Aythya nyroca* Porrón pardo
A very scarce and sporadic breeding species. Individuals or small groups may appear at any time on suitable lagoons.

Tufted Duck *Aythya fuligula* Porrón moñudo
Widespread in winter in small numbers. Perhaps most abundant in Extremadura, where over 1,000 occur in some winters. Has bred sporadically.

Greater Scaup *Aythya marila* Porrón bastardo
A rare winter visitor to Extremadura: 16 records. Only four records of single birds in winter from Andalucía.

Lesser Scaup* *Aythya affinis* Porrón bola
Four Andalucían records and two from Extremadura.

Common Eider *Somateria mollissima* Eider
A very scarce and irregular winter visitor to Atlantic coasts with occasional records from the Cádiz region. One Gibraltar record and two from Málaga Bay.

Long-tailed Duck *Clangula hyemalis* Havelda
A rare winter visitor to coastal Andalucía: 19 records (22 birds), seven of them from Almería.

Common Scoter *Melanitta nigra* Negrón común
Present off all coasts mainly October–April, although a few non-breeders remain in summer. Large numbers (up to 5,000) winter off Doñana and the Huelva coast. Exceptional inland. One Extremaduran record.

Surf Scoter* *Melanitta perspicillata* Negrón careto
Two were off Doñana on 6 February 1991 and one was at Odiel during January–April 2008.

Velvet Scoter *Melanitta fusca* Negrón especulado
A rare winter visitor to Atlantic and Mediterranean coasts. One Extremaduran record.

Common Goldeneye *Bucephala clangula* Porrón osculado
A rare winter visitor to Andalucía. At least ten records of single birds, three from Almería.

Smew *Mergus albellus* Serreta Chica
One overwintered at the Alcollarín reservoir (CC) in 2016/17. Also a June record from Granada.

Red-breasted Merganser *Mergus serrator* Serreta mediana
Widespread offshore in winter in small numbers. Most frequent off the Huelva coast, from October–March. Exceptional inland.

Goosander* *Mergus merganser* Serreta grande
Two records: three at the Odiel estuary (H) on 16 December 2001 and one at Pantaneta de Alhama (GR) on the surprising dates of 4–10 May 2005.

Ruddy Duck* *Oxyura jamaicensis* Malvasía canela
Strays formerly appeared occasionally at lakes in Andalucía, where hybridisation threatened the genetic

integrity of the White-headed Duck population. Culls of European feral populations have largely eliminated this threat.

White-headed Duck *Oxyura leucocephala* Malvasía cabeciblanca
A locally common breeding species largely confined to the lagoons of Andalucía where it is progressively more widespread. Numbers have increased and flocks of over 300, exceptionally more, occur in winter in favoured localities such as Las Norias (AL). Still very rare in Extremadura: only three records.

Red-legged Partridge *Alectoris rufa* Perdiz roja
Widespread and abundant. A prime quarry species of hunters, who release large numbers of captive-bred birds. Resident.

Barbary Partridge *Alectoris barbara* Perdiz moruna
Only present in Gibraltar where the stable population of 30–50 pairs inhabits open scrub and the Great Sand Slope. These, the only partridges in Gibraltar, probably include the descendants of birds introduced from Morocco at least 200 years ago. Numbers have recently been reinforced by further introductions.

Common Quail *Coturnix coturnix* Codorníz común
A common and widespread breeding species. Present mainly between March and October but some remain in winter.

Common Pheasant *Phasianus colchicus* Faisán vulgar
A small feral population may be established in southern Cádiz province. Released birds may be encountered elsewhere but few if any naturalised populations exist.

Red-throated Diver *Gavia stellata* Colimbo chico
Very scarce, in winter: December–February, along the Atlantic coast. Rare in the Mediterranean.

Black-throated Diver *Gavia arctica* Colimbo ártico
Rare, in winter: November–March. Coastal.

Great Northern Diver *Gavia immer* Colimbo grande
Very scarce, in winter: November–March on the Atlantic coast, most often reported from Cádiz Bay. Rare in the Mediterranean.

Pied-billed Grebe* *Podylimbus podiceps* Zampullín picogrueso
One Extremaduran record, in May 2010.

Little Grebe *Tachybaptus ruficollis* Zampullín común
Common on reed-fringed lakes. Mainly resident but post-breeding concentrations form locally.

Great Crested Grebe *Podiceps cristatus* Somormujo lavanco
Breeds commonly on larger lakes and on reservoirs. Post-breeding concentrations form locally. More widely dispersed in winter.

Horned (Slavonian) Grebe *Podiceps auritus* Zampullín cuellirrojo
Three Andalucían winter records. Coastal.

Black-necked Grebe *Podiceps nigricollis* Zampullín cuellinegro
Locally common on reed-fringed lakes, chiefly along the Guadalquivir valley. Uncommon in Extremadura but has nested at La Albuera (BA). Resident but numbers are increased by migrants in winter, when some also occur on the coast in sheltered inshore waters.

Black-browed Albatross* *Thalassarche melanophris* Albatros ojeroso
Two records of single birds. Off Conil (CA) in August 1999 and off Málaga in July 2002.

Northern Fulmar *Fulmarus glacialis* Fulmar boreal
Vagrant. Usually recorded as tideline corpses on the Atlantic coast. A wreck in Cádiz Bay in early 2009 involved 13 birds (12 dead).

Cape Petrel* *Daption capense* Petrel damero
One at Gibraltar on 20 June 1979.

Bulwer's Petrel* *Bulweria bulwerii* Petrel de Bulwer
One was netted at Odiel (H) during a wader ringing session on 7 June 2000. It is regular off Portugal in summer and so is likely occur then well off the Andalucían Atlantic coast.

Scopoli's Shearwater *Calonectris d. diomedea* Pardela cenicienta mediterránea
Present in summer off Mediterranean coasts. Large, and at times spectacular, movements occur through the Strait: westwards in late October–November and eastwards in February–March, as Mediterranean populations move to and from winter quarters in the south Atlantic.

Cory's Shearwater *Calonectris diomedea borealis* Pardela cenicienta canaria
Present off Atlantic coasts, mainly March–October. Large feeding concentrations occur in the Strait and Alborán basin in summer.

Great Shearwater *Puffinus gravis* Pardela capirotada
Present off Atlantic coasts and, more rarely, in the Strait and seen occasionally from the shore. Exceptional in the Mediterranean. September–December.

Sooty Shearwater *Puffinus griseus* Pardela sombría
Present off Atlantic coasts and in the Strait and seen occasionally from the shore. Exceptional in the Mediterranean. Chiefly September–November.

Manx Shearwater *Puffinus puffinus* Pardela pichoneta
Rare. There are very few satisfactory records from the Atlantic sector or the Strait. Vagrant in the Mediterranean: there are only two records from Alborán (MA).

Levantine (Yelkouan) Shearwater *Puffinus yelkouan* Pardela mediterránea
Small numbers occur off Mediterranean coasts and in the Strait in summer and autumn.

Balearic Shearwater *Puffinus mauretanicus* Pardela balear
Present offshore all year round. Numbers migrate into the Atlantic through the Strait in June–August, returning from late September–March.

Macaronesian (Little) Shearwater* *Puffinus baroli* Pardela chica
Seven records: five of them at Gibraltar in July–August and two off the Costa del Sol.

Wilson's Storm-petrel* *Oceanites oceanicus* Paíño de Wilson
Probably regular in late summer/early autumn in the western approaches to the Strait and off Atlantic Andalucía, where recently seen in some numbers during 'pelagic' birding trips.

European Storm-petrel *Hydrobates pelagicus* Paíño europeo
Present offshore, mainly April–September. Most often seen in the Strait but common in central Alborán. Infrequently seen from shore. Exceptional inland.

Leach's Storm-petrel *Oceanodroma leucorhoa* Paíño de Leach
Present offshore in winter: November–February, and sometimes visible from land in the Strait and from Atlantic coasts, especially after westerly gales. Scarce in the Mediterranean.

Band-rumped Storm-petrel *Oceanodroma castro* Paíño de Madeira
At least four records (five birds) of storm-driven birds or tideline corpses, including one found inland in Badajoz city, Extremadura.

Red-footed Booby* *Sula sula* Piquero patirrojo
One off Estepona, Málaga, in August 2010.

Masked Booby* *Sula dactylatra* Piquero enmascarado
Two records, probably of the same adult bird, off the Costa del Sol in late 1985.

Brown Booby* *Sula leucogaster* Piquero pardo
Three records off the Mediterranean coast.

Gannet *Morus bassanus* Alcatraz común
Present offshore all year round, although only a few non-breeders remain in summer. Common in winter: October–March. Conspicuous migration of small flocks occurs through the Strait, entering the Mediterranean in September–November and returning in February–May.

Great Cormorant   *Phalacrocorax carbo*   Cormorán grande
Common and locally numerous in winter on lakes and reservoirs as well as on coasts. Breeding colonies occur locally, usually at reservoirs.

Shag   *Phalacrocorax aristotelis*   Cormorán moñudo
A few pairs nest at Gibraltar and around Cabo de Gata (AL). Exceptional elsewhere.

Great White Pelican*   *Pelecanus onocrotalus*   Pelícano vulgar
Recorded from both Extremadura and Andluacia. At least 14 records in total, of single birds and groups of up to three. Origins unknown but at least some have certainly been wild birds.

Pink-backed Pelican*   *Pelecanus rufescens*   Pelícano rosado
Two records of single birds: at the Celemín reservoir (CA) in April 2001 and the Valdesalor reservoir (CC) in May 2004.

Magnificent Frigatebird*   *Fregata magnificens*   Rabihorcado magnífico
A subadult female flew past Torremolinos (MA) on 22 October 2000.

Great (Eurasian) Bittern   *Botaurus stellaris*   Avetoro común
Very scarce but widespread in winter. Small numbers nest at Doñana in wet years.

Little Bittern   *Ixobrychus minutus*   Avetorillo común
Widespread but inconspicuous during April–September, nesting in reedbeds. A few remain in winter.

Night Heron   *Nycticorax nycticorax*   Martinete común
Widespread April–September, nesting in riverine trees, often in mixed colonies with other herons. Many winter in the Guadalquivir valley and some occasionally elsewhere.

Squacco Heron   *Ardeola ralloides*   Garcilla cangrejera
Very local and easily overlooked. Mainly present April–September, chiefly in the Guadalquivir basin. Small numbers remain in winter.

Cattle Egret   *Bubulcus ibis*   Garcilla bueyera
Abundant resident in Extremadura, Western Andalucía and the Guadalquivir valley. Local elsewhere. Less widespread in the breeding season when birds stay closer to the heronries. These colonies are relatively few but can be very large, having up to several thousand nests.

Western Reef Egret*   *Egretta gularis*   Garceta dimorfa
Very scarce but single birds are seen with some regularity in the Guadalquivir salt pans and occasionally at coastal sites elsewhere. About half the records seem to involve hybrids between this species and the Little Egret. One accepted Extremaduran record.

Little Egret   *Egretta garzetta*   Garceta común
Common. Mainly resident. Nests colonially with Cattle Egrets and other herons.

Great Egret   *Egretta alba*   Garceta grande
Formerly rare but now well established and increasing. A few pairs nest in Doñana and locally in Extremadura. Several hundred in total winter along the lower Guadalquivir, at Atlantic coastal wetlands and along the Guadiana valley in Extremadura, occasionally elsewhere.

Grey Heron   *Ardea cinerea*   Garza real
Numerous and widespread on passage and especially in winter. Increasingly common breeding species in western Andalucía and Extremadura, with new colonies being established annually.

Purple Heron   *Ardea purpurea*   Garza imperial
Locally common, mainly in the Guadalquivir valley and locally in Extremadura, breeding in large reedbeds. Present April–October. Widespread on passage. A few occur in winter.

Black Stork   *Ciconia nigra*   Cigüeña negra
Widespread in Extremadura and in the Sierra Morena in Andalucía, typically nesting on cliff ledges. Increasing. Widespread on passage when small flocks converge on the Strait. Mainly present late February–October but a few hundred remain to winter, especially in Doñana.

White Stork   *Ciconia ciconia*   Cigüeña blanca
Abundant and increasing in western Andalucía, the Guadalquivir valley and throughout Extremadura, very local or absent further east. Nests on buildings, trees and pylons. Present all year round but many leave Spain in late July and August, returning once the rains arrive, from November to February. Spectacular movements occur across the Strait.

Marabou Stork*   *Leptoptilos crumenifer*   Marabú africano
At least 13 records (15 birds) from Andalucía and three from Extremadura.

Glossy Ibis   *Plegadis falcinellus*   Morito común
Nests very locally, chiefly at Doñana where several thousand pairs breed in wet years. Large numbers winter in the lower Guadalquivir, with smaller numbers then occurring in the Guadiana valley and La Janda, chiefly at ricefields. Occasional elsewhere, in winter and on passage.

Northern Bald Ibis*   *Geronticus eremita*   Ibis Eremita
Introduced birds have established a small nesting colony at Vejer (CA). This population ranges between Barbate and La Janda, some occasionally reaching the Strait. Category C.

Sacred Ibis*   *Threskiornis aethiopicus*   Ibis Sagrado
Occasional records of escaped or feral birds, chiefly at coastal wetlands. Category C.

Spoonbill   *Platalea leucorodia*   Espátula común
Breeds especially at Doñana and Odiel (H), and also at the Guadiana estuary (H) and Cádiz Bay, where it is resident and increasing. There are several small colonies in Extremadura. Numbers fluctuate, reaching 2,000 or more pairs in good years. Scarce but increasing on passage and in winter elsewhere.

Greater Flamingo   *Phoenicopterus roseus*   Flamenco común
Thousands nest in most years at the Laguna de Fuente de Piedra (MA). Regular and common for much of the year at Doñana (has bred) and the Cabo de Gata salt pans (AL). Feeding flocks may occur at other Andalucían wetlands at any time of year. Occasional in Extremadura.

Lesser Flamingo   *Phoenicopterus minor*   Flamenco Enano   One or two pairs have nested successfully at Fuente de Piedra in some recent seasons. Also occasionally recorded at Doñana and in Almería, usually in ones and twos, but as many as eight have been seen together. Most appear to be of wild origin.

Honey-buzzard   *Pernis apivorus*   Abejero Europeo
Abundant on passage and most readily seen near the Strait. As many as 100,000 fly south there between mid-August and late September, returning from late April to early June. Some 75–150 pairs breed in northern Extremadura, chiefly in the Jerte valley and Las Villuercas.

Black-winged Kite   *Elanus caeruleus*   Elanio común
Widespread and increasing, usually seen on pylons and other exposed perches, or hovering over open country. Most nesting territories are in cereal-growing areas with scattered trees. It can be elusive because of its crepuscular tendencies, especially in summer, and is best sought early in the morning and in the evening. Most abundant in Extremadura where the population exceeds 200 pairs. Increasingly widespread in Andalucía (177 pairs in 2011), especially in the west, but has nested in every province except Almería. Mainly resident but occurs more widely in winter and a few cross the Strait.

Black Kite   *Milvus migrans*   Milano Negro
A very common and locally abundant breeding species in Extremadura and in western and northern Andalucía, characteristically near water. Increasing. Over 150,000 cross the Strait on migration, mainly from February–May and late July–September. A few winter.

Red Kite   *Milvus milvus*   Milano Real
Most numerous as a breeding species in Extremadura with over 200 pairs, especially in Cáceres. Absent from most of Andalucía, except for the Doñana area where 53 pairs were censused in 2015. Most numerous and widespread in winter when numbers arrive from northern Europe: over 10,000 birds may winter in Extremadura alone. Very few cross the Strait.

White-tailed Eagle   *Haliaetus albicilla*   Pigargo europeo
One recent record. An immature at and around Monfragüe, Extremadura, in winter 2012/13.

Bearded Vulture (Lammergeier)   *Gypaetus barbatus*   Quebrantahuesos
Several pairs nest at Cazorla (J) following successful and ongoing reintroductions. Wandering individuals, generally young birds, are reported increasingly widely and may herald eventual recolonisation of former haunts in the Sierra de Gredos, Sierra Nevada and Serranía de Ronda.

Egyptian Vulture   *Neophron percnopterus*   Alimoche común
Local and declining breeding species, nesting on cliffs. Recent censuses (2016) found 151 pairs in Extremadura, chiefly in Cáceres, and only 23 pairs in Andalucía, where Los Alcornocales (CA) is a stronghold. Most are migrants, arriving mainly February–April and departing August–October. Over 100 winter in Cáceres, few elsewhere.

White-backed Vulture   *Gyps africanus*   Buitre dorsiblanco africano
At least three have been recorded on the north side of the Strait of Gibraltar.

Griffon Vulture   *Gyps fulvus*   Buitre leonado
Highly characteristic and common. Large colonies occur on many suitable cliff faces, especially in Cáceres, western Cádiz and eastern Jaén. Adults are largely resident but several thousand chiefly immature birds cross the Strait southwards in October–November, returning April–June.

Rüppell's Vulture   *Gyps rueppellii*   Buitre moteado
A recent non-breeding immigrant from tropical Africa, a few arriving at the Strait with Griffon Vultures in late spring and departing with them in autumn. Some recorded year-round, especially in the vicinity of the Strait. Occasional elsewhere, including in Extremadura.

Black (Cinereous) Vulture   *Aegypius monachus*   Buitre negro
Locally common, increasing breeding species in Extremadura, with over 1,000 pairs mainly in Cáceres province, and several hundred pairs in the Sierra Morena in northern Andalucía. Nests on trees. Resident but some wander more widely in winter, a few crossing the Strait.

Short-toed Snake Eagle   *Circaetus gallicus*   Culebrera europea
Widespread and locally common breeding species. Chiefly a trans-Saharan migrant: arriving mainly between late February and April, departing in September–October, but a few winter. Some thousands cross the Strait on passage.

Bateleur   *Terathopius ecaudatus*   Águila volatinera
A juvenile arrived at Punta Carnero, Strait of Gibraltar (CA), on 5 April 2012.

Marsh Harrier   *Circus aeruginosus*   Aguilucho lagunero
Scarce but locally common breeding species, especially in western Andalucía. Increasing in Extremadura (150+ pairs). Mainly resident but European migrants occur commonly on passage and in winter. Hundreds cross the Strait, mainly in March–April and September–October.

Hen Harrier   *Circus cyaneus*   Aguilucho pálido
Widespread in winter in small numbers: October–March. Very few cross the Strait: in February–April and September–November. A few breed at least occasionally in Extremadura.

Pallid Harrier*   *Circus macrourus*   Aguilucho papialbo
Very scarce migrant, occurring annually at the Strait. A few have overwintered in recent years both in Andalucía and Extremadura, notably at La Janda, Cádiz.

Montagu's Harrier   *Circus pygargus*   Aguilucho cenizo
A common and characteristic breeding species of farmland, especially cereal crops, and steppe. Most numerous in western Andalucía and in Extremadura. Present from mid-March to early October. Hundreds cross the Strait on passage. Melanistic birds are not uncommon.

Goshawk   *Accipiter gentilis*   Azor común
Widespread but scarce in forested areas. Largely resident.

Sparrowhawk *Accipiter nisus* Gavilán común
Widespread but fairly scarce in forested areas. Northern migrants increase numbers in winter. Common migrant: several thousand cross the Strait in March–April and late August–October.

Common Buzzard *Buteo buteo* Busardo ratonero
Widespread and common as a resident breeding species, although largely absent from Almería. Northern migrants increase numbers in winter but very few cross the Strait.

Long-legged Buzzard* *Buteo rufinus* Busardo moro
Rare but regular visitor to the northern shore of the Strait, recorded most years at Gibraltar, the Tarifa area and at La Janda (CA) and occasionally also in the lower Guadalquivir valley and in Extremadura. Most records are in the second half of the year, with a peak in September.

Lesser Spotted Eagle* *Aquila pomarina* Águila pomerana
Very rare but a few individual migrants occur annually at the Strait, chiefly in April and September–October. Occasional elsewhere.

Greater Spotted Eagle* *Aquila clanga* Águila moteada
Very rare but occasional individuals, some apparently *clanga* x *pomarina* hybrids, have wintered in Doñana and at La Janda (CA). Occasional on passage across the Strait.

Steppe Eagle *Aquila nipalensis* Águila esteparia
Three records of single birds: at Tarifa, La Janda and Gibraltar.

Spanish Imperial Eagle *Aquila adalberti* Águila imperial ibérica
A scarce but increasing endemic resident. Most of the global population of some 500 pairs (2018) is in central Spain, Cáceres in Extremadura and the Sierra Morena in Andalucía. Generally resident but juveniles especially disperse more widely in winter, a few crossing the Strait.

Golden Eagle *Aquila chrysaetos* Águila Real
Present in some numbers in all mountain ranges. Over 300 pairs breed in Andalucía and 120 pairs in Extremadura. Resident, but some disperse more widely in winter.

Booted Eagle *Aquila pennata* Águila calzada
A widespread and common breeding species, present mainly from March to October. Small numbers winter mainly in coastal areas. Some thousands cross the Strait.

Bonelli's Eagle *Aquila fasciata* Águila-azor perdicera
Widespread and characteristic of cliffs and broken country. Andalucía (at least 350 pairs) and Extremadura (100 pairs) are strongholds of the Spanish population. Largely resident but disperse more widely into lowland farming areas in winter.

Osprey *Pandion haliaetus* Águila pescadora
Common migrant, in mid-February to May and August to October. Individuals (100+ in total) winter regularly at coastal wetlands as well as at some reservoirs in Extremadura. Some 20 pairs nest, chiefly at Odiel (H) and the Barbate reservoir (CA), following successful reintroduction projects.

Lesser Kestrel *Falco naumanni* Cernícalo primilla
Locally common. Has suffered periodic severe declines in recent decades. Nests colonially on older buildings, notably on church towers and castles. Exceptionally on cliffs. Many villages and towns in Extremadura and the Guadalquivir valley support large colonies. Present mainly from February to October, most wintering in Africa. Post-breeding roosts of up to several hundred birds occur.

Common Kestrel *Falco tinnunculus* Cernícalo vulgar
Widespread and common. Resident, but numbers are increased by migrants in winter and during passage periods: March–May and September–October.

Red-footed Falcon* *Falco vespertinus* Cernícalo patirrojo
Very rare passage migrant with occasional records in April–early June. Rarer still in autumn and exceptional in winter.

Merlin *Falco columbarius* Esmerejón
Scarce but regular and widespread in winter, favouring open country where it hunts larks and finches. A few cross the Strait. Present October–April.

Hobby *Falco subbuteo* Alcotán europeo
A widespread but very local breeding species, most abundant in Huelva and northern Cáceres, present April–October. Regular but scarce migrant at the Strait.

Eleonora's Falcon *Falco eleonorae* Halcón de Eleonora
Single birds and small groups occur regularly in the Strait region between March and October, chiefly in August and September. Occasionally reported from the Almería and Málaga coasts. Exceptional in Extremadura but there were five at La Serena (BA) in late June 2015.

Lanner Falcon* *Falco biarmicus* Halcón borni
Annual on the north side of the Strait, most often at Gibraltar, all during March–October. Many are juveniles in post-breeding dispersal in late summer. One Extremaduran record.

Saker Falcon* *Falco cherrug* Halcón sacré
Very rare migrant. A radiotracked wild bird crossed the Strait southwards in September 2009.

Peregrine Falcon *Falco peregrinus* Halcón peregrino
Widespread, nesting on coastal and inland cliffs and locally on tall buildings, such as on Málaga cathedral. Local birds *F.p.brookei* are resident but dispersive. A few northern Peregrines *F.p.peregrinus* also occur widely in winter, frequenting wetlands and open country.

Water Rail *Rallus aquaticus* Rascón europeo
Widespread and locally common in suitable habitat. Resident populations are increased by migrants in winter.

Spotted Crake *Porzana porzana* Polluela pintoja
Small numbers cross the region on passage and some remain to winter. Some breed Doñana and perhaps elsewhere in the lower Guadalquivir valley.

Sora Crake* *Porzana carolina* Polluela de Carolina
One was shot in Doñana on 30 December 1975.

Little Crake *Porzana parva* Polluela bastarda
Rare migrant and occasional in winter in Andalucía. Six Extremaduran records.

Baillon's Crake *Porzana pusilla* Polluela chica
Very scarce. Some nest in Doñana and the Guadalquivir valley ricefields, occasionally elsewhere. Chiefly a trans-Saharan migrant but some winter in Doñana.

Striped Crake *Porzana marginalis* Polluela culirroja
Two records, from the Sierra Norte in December (SE) and Córdoba in January.

African Crake *Crex egregia* Guión africano
One in the Sevilla farmlands in April 2016 was the first for mainland Spain.

Corn Crake *Crex crex* Guión de codornices
Crosses the region on passage but very rarely recorded.

Common Moorhen *Gallinula chloropus* Gallineta común
A common and widespread breeding species. Mainly resident.

Lesser Moorhen* *Gallinula angulata* Gallineta chica
One was found at Algeciras harbour on 10 March 2003.

Allen's Gallinule* *Porphyrio alleni* Calamón de Allen
At least 16 records: 14 from coastal Andalucía and one each from Gibraltar and Extremadura, all but one between 20 November and 31 March.

Purple Swamphen *Porphyrio porphyrio* Calamón común
Widespread. Locally common in ricefields and reedbeds in Andalucía especially. Also in Extremadura along the Guadiana and locally elsewhere. Resident but dispersive.

Common (Eurasian) Coot *Fulica atra* Focha común
A common and widespread breeding species. Locally abundant. Dispersive, some crossing the Strait to winter in North Africa. Some migrants from northern Europe are also present in winter.

Red-knobbed Coot *Fulica cristata* Focha moruna
A very scarce resident despite repeated reintroductions. Fewer than 50 pairs, most of them in western and central Andalucía, including Doñana. Exceptional in Extremadura.

Kurrichane Buttonquail (Andalusian Hemipode)* *Turnix sylvatica* Torillo andaluz
Probably extinct. No records since 2002.

Common Crane *Grus grus* Grulla común
Abundant in winter in Extremadura, where well over 100,000 occur in some years, and locally in western Andalucía. Some hundreds cross the Strait. Present mainly from late October to early March.

Sandhill Crane* *Grus canadensis* Grulla canadiense
One was with Common Cranes in the Vegas Altas (BA) in winter 2011/12.

Demoiselle Crane* *Grus virgo* Grulla damisela
Only one recent record: one with Common Cranes at Navalvillar de Pela (BA) in February 1996.

Little Bustard *Tetrax tetrax* Sisón común
Mainly resident but in continuing serious decline. Formerly numerous in Extremadura and locally common in Andalucía. Flocks of a few hundred birds form outside the breeding season.

Great Bustard *Otis tarda* Avutarda común
Resident. Locally common in Extremadura, where the population is 5,500–6,500 individuals. A few hundred occur locally in Andalucía, chiefly in Córdoba and Sevilla.

Oystercatcher *Haematopus ostralegus* Ostrero euroasiatico
Present on passage and in winter on all coasts, in small numbers. Exceptional inland but there are several records for Extremadura.

Black-winged Stilt *Himantopus himantopus* Cigüeñuela común
Widespread and locally numerous, breeding commonly in western Andalucía, the Guadalquivir valley, along the Mediterranean coast and in the Extremaduran ricefields. Many are trans-Saharan migrants but others winter in Doñana, along the Huelva coast and, increasingly, inland.

Pied Avocet *Recurvirostra avosetta* Avoceta común
Small numbers breed, chiefly in western Andalucía, the Guadalquivir valley and the Almería coast. Present locally in Andalucía in large numbers on passage and especially in winter. Scarce on passage and in winter in Extremadura. Over 5,000 winter in the Doñana area.

Stone-curlew *Burhinus oedicnemus* Alcaraván común
A common breeding species in steppe habitats. Flocks of up to a few hundred form in winter, some of these probably including migrants from northern Europe.

Cream-coloured Courser* *Cursorius cursor* Corredor sahariano
Very rare but recent breeding records from Almería and Granada (and Albacete) hint at possible colonisation of the arid southeast. Migrants are recorded almost annually in Andalucía, chiefly on coasts. Four Extremaduran records, three of them (five birds) at different sites in May 2017.

Collared Pratincole *Glareola pratincola* Canastera común
Locally common in the Guadalquivir and Guadiana valleys especially. A trans-Saharan migrant, present mainly April–August and departing promptly after breeding. Exceptional in winter.

Little Ringed Plover *Charadrius dubius* Chorlitejo chico
Small numbers breed, especially on stony river beds. Mainly a trans-Saharan migrant, present March–October. Widespread on passage. Small numbers remain in winter.

Common Ringed Plover *Charadrius hiaticula* Chorlitejo grande
Common on passage and in winter. Chiefly coastal but migrants occur inland, including in Extremadura. Most numerous September–April but some non-breeders remain in summer.

Kentish Plover *Charadrius alexandrinus* Chorlitejo patinegro
Widespread, nesting on sandy coasts and salt pans. A few nest locally inland. Mainly resident but the population is increased by migrants during passage periods and in winter.

Dotterel *Charadrius morinellus* Chorlito carambolo
Very scarce migrant in March–April and August–October. Occasional in winter. Most sightings are of small groups but flocks of 100+ have been recorded. Favoured sites include Los Lances beach (CA), the Sierra Nevada (GR), Las Almoladeras (AL) and La Serena (BA).

Pacific Golden Plover* *Pluvialis fulva* Chorlito dorado asiático
Three August records from the Guadalhorce estuary (MA).

American Golden Plover* *Pluvialis dominica* Chorlito dorado americano
Four records: at Doñana in May 2004 and October 2017, and Los Lances (CA) in September–October 2005. One overwintered at the Laguna de Galisteo (CC) in 2016/17.

European Golden Plover *Pluvialis apricaria* Chorlito dorado
Locally numerous in winter, especially in Extremadura. Present November–March mainly.

Grey Plover *Pluvialis squatarola* Chorlito gris
Common in winter and on passage in coastal habitats, rarely inland. Mainly present August–May but some individuals remain all summer.

Sociable Lapwing* *Vanellus gregarius* Chorlito social
At least 19 Andalucían and eight Extremaduran records, chiefly December–February.

Northern Lapwing *Vanellus vanellus* Avefría europea
Small numbers nest, chiefly in Doñana. Common and often abundant in winter, particularly in Extremadura and the Guadalquivir valley, especially when harsh conditions occur further north in Europe. Most numerous from November to early March.

Great Knot* *Calidris tenuirostris* Correlimos grande
One record. One was the Atlantic coast of Huelva/Cádiz in June–July 2017.

Red Knot *Calidris canutus* Correlimos gordo
Widespread on passage in small numbers in coastal habitats, especially in May and September. Rare inland. A few winter.

Sanderling *Calidris alba* Correlimos tridáctilo
Common on passage and in winter on sandy coasts. Rare inland. Present mainly October–April but some non-breeders remain in summer.

Little Stint *Calidris minuta* Correlimos menudo
Common on passage, both at coastal and inland wetlands, mainly in April and August–September. Small numbers are present in winter.

Temminck's Stint *Calidris temminckii* Correlimos de Temminck
A very scarce migrant and winter resident, chiefly at inland wetlands and saltpans.

White-rumped Sandpiper* *Calidris fuscicollis* Correlimos de Bonaparte
Five records from coastal Andalucía.

Pectoral Sandpiper *Calidris melanotos* Correlimos pectoral
Rare migrant, chiefly September–October. At least 19 Andalucían records and nine from Extremadura.

Curlew Sandpiper *Calidris ferruginea* Correlimos zarapitín
Common and sometimes numerous on passage, both at coastal and inland wetlands, including ricefields, mainly in May and August–September. A few individuals winter.

Purple Sandpiper *Calidris maritima* Correlimos oscuro
A few individuals occur in some winters on rocky coastal headlands, such as Punta Secreta (CA) and Punta Calaburras (MA). One exceptional Extremaduran record.

Dunlin *Calidris alpina* Correlimos común
Abundant on passage, mainly in April and September, and in winter, chiefly in coastal habitats but locally common inland. Small numbers of non-breeders remain in summer.

Broad-billed Sandpiper* *Limicola falcinellus* Correlimos falcinelo
Thirteen records from coastal Andalucía, all between 30 April and 17 August.

Buff-breasted Sandpiper* *Tryngites subruficollis* Correlimos canelo
Nine records from coastal Andalucía and three from Extremadura.

Ruff *Philomachus pugnax* Combatiente
Common on passage, mainly in March–April and September, and in winter, both at coastal and inland wetlands.

Jack Snipe *Lymnocryptes minimus* Agachadiza chica
Present on passage and in winter, chiefly at inland wetlands, from October–April.

Common Snipe *Gallinago gallinago* Agachadiza común
Common on passage and in winter, chiefly at inland wetlands, from August–April.

Great Snipe *Gallinago media* Agachadiza real
Four Andalucían records of single birds.

Long-billed Dowitcher* *Limnodromus scolopaceus* Agujeta escolopácea
Four Andalucían records and one from Extremadura. Also seven records (eight birds) of *Limnodromus* sp. from coastal Andalucía, all in late-March–August.

Woodcock *Scolopax rusticola* Chocha perdiz
Widespread in winter but inconspicuous. Present October–March.

Black-tailed Godwit *Limosa limosa* Aguja colinegra
Very common on passage: in December–April and July–September, and in winter, chiefly on the coast. Large numbers pass through Extremadura very early in the year, between December and April, peaking in February when 35,000 have been counted in the Guadiana ricefields. Non-breeders occur in summer in some numbers.

Bar-tailed Godwit *Limosa lapponica* Aguja colipinta
Common on passage, in March and September–October, and in winter in coastal habitats. Rare inland. Scarcer than the Black-tailed Godwit and usually seen in small parties.

Whimbrel *Numenius phaeopus* Zarapito trinador
Common on passage, chiefly on the coast in July–November and March–May. Small numbers winter, most notably in Doñana, and a few non-breeders remain in summer.

Eurasian Curlew *Numenius arquata* Zarapito real
Common on passage: July–October and March–April, and in winter, chiefly on the coast. A few non-breeders remain in summer.

Terek Sandpiper* *Xenus cinereus* Andarríos de Terek
Rare but there are at least 24 recent records from coastal Andalucía. Most are in spring or autumn but several have overwintered.

Common Sandpiper *Actitis hypoleucos* Andarríos chico
Nests very locally in the Guadiana basin in Extremadura and Huelva and along the upper Guadalquivir. Common migrant, both inland and in coastal habitats: July–November and March–May. A few remain to winter.

Spotted Sandpiper* *Actitis macularius* Andarríos maculado
Two spring records from Andalucía.

Green Sandpiper *Tringa ochropus* Andarríos grande
Widespread and common on passage, especially by fresh water: mainly July–October and March–May. Some remain to winter.

Solitary Sandpiper *Tringa solitaria* Andarríos solitario
One at Almanzora (AL) in February/March 2016. Second Spanish record.

Spotted Redshank *Tringa erythropus* Archibebe oscuro
Common on passage, less so in eastern Andalucía: July–October and March–May, and in winter. Often on freshwater marshes as well as in coastal habitats.

Greater Yellowlegs* *Tringa melanoleuca* Archibebe patigualdo grande
Three records from coastal Andalucía.

Common Greenshank *Tringa nebularia* Archibebe claro
Widespread and common on passage, July–October and March–May, mainly on coasts but also at inland wetlands. Hundreds remain in winter along the Atlantic coast.

Lesser Yellowlegs* *Tringa flavipes* Archibebe patigualdo chico
Eleven records from Andalucía, chiefly coastal. Four Extremaduran records.

Marsh Sandpiper *Tringa stagnatilis* Archibebe fino
A very scarce but regular passage migrant: March–April and July–October, usually seen by fresh water. Generally recorded singly. A few winter in Doñana.

Wood Sandpiper *Tringa glareola* Andarríos bastardo
Widespread and common on passage, especially in freshwater habitats, July–October and April–May. Significant numbers winter in Doñana and in small numbers elsewhere.

Common Redshank *Tringa totanus* Archibebe común
A breeding resident in Atlantic coastal marshes and more locally inland. Numbers are greatly increased by migrants in winter and during passage periods: July–October and March–May.

Ruddy Turnstone *Arenaria interpres* Vuelvepiedras
Widespread in small numbers on passage: July–November and February–May, and in winter, chiefly on rocky Atlantic coasts. Very scarce inland, mainly in spring.

Wilson's Phalarope* *Phalaropus tricolor* Falaropo de Wilson
Six records from coastal Andalucían wetlands.

Red-necked Phalarope *Phalaropus lobatus* Falaropo picofino
Very scarce but annual on passage in coastal habitats, most frequently in the Guadalquivir salt pans and Almería. Exceptional inland.

Grey Phalarope *Phalaropus fulicarius* Falaropo picogrueso
Recorded in small numbers irregularly offshore and along Atlantic coasts, especially after westerly gales in late autumn and winter. Exceptional inland.

Pomarine Skua *Stercorarius pomarinus* Págalo pomarino
Regular on passage along coasts, most records coming from the Strait in March–April and August–October. Occasional in winter.

Arctic (Parasitic) Skua *Stercorarius parasiticus* Págalo parásito
Regular on passage: July–November and February–May and some remain in winter, especially in the Strait. Occasional in summer.

Long-tailed Skua* *Stercorarius longicaudus* Págalo rabero
Occasional in the western approaches to the Strait in autumn. Three Atlantic coast records and one from Málaga Bay.

Great Skua *Stercorarius skua* Págalo grande
Regular and sometimes common on passage: September–November and March–June, and in winter, especially in the Strait. Occasional in summer.

Bonaparte's Gull* *Chroicocephalus philadelphia* Gaviota de Bonaparte
Four records from coastal Andalucía.

Black-headed Gull *Chroicocephalus ridibundus* Gaviota reidora
Small breeding colonies occur locally in western Andalucía and Extremadura. Widespread and abundant in winter both inland and on the coast.

Grey-headed Gull* *Chroicocephalus cirrocephalus* Gaviota cabecigrís
Single birds at Doñana in June–August 1971 and at Gibraltar in August 1992.

Slender-billed Gull *Chroicocephalus genei* Gaviota picofina
Small numbers breed locally and often sporadically, notably at Doñana, Fuentedepiedra (MA) and at coastal sites in Almería. Mainly resident but some disperse along coasts.

Mediterranean Gull *Larus melanocephalus* Gaviota cabecinegra
Common on passage, mainly October–November and March–April and in winter on or near coasts. Non-breeders occur in summer in small numbers. Uncommon inland.

Laughing Gull* *Larus atricilla* Gaviota guanaguanare
Nine records from coastal Andalucía and four at Gibraltar.

Franklin's Gull* *Larus pipixcan* Gaviota de Franklin
Ten records from coastal Andalucía. One at Malpartida de Cáceres (CC) in December 2003.

Audouin's Gull *Larus audouinii* Gaviota de Audouin
Near the region some 800 pairs nest on Alborán Island, 800+ pairs in three colonies in Melilla and a small colony has recently become established in Ceuta. Common on passage along Mediterranean coasts and through the Strait, mainly February–April and August–October. Concentrations of summering immature birds occur in the Strait, Málaga Bay and in Almería. Small numbers winter, especially near the Strait. Recorded on passage inland in August, along the Río Guadiana (BA).

Ring-billed Gull *Larus delawarensis* Gaviota de delaware
Very rare, chiefly in winter. There are at least 14 records from the Andalucían coast and two from Gibraltar.

Common Gull *Larus canus* Gaviota cana
A few individuals occur in winter, along Atlantic coasts especially. Very rare inland.

Kelp Gull* *Larus dominicanus* Gaviota cocinera
One record from Isla Cristina (H), an adult in August 2017.

Lesser Black-backed Gull *Larus fuscus* Gaviota sombría
Common on coasts on passage and in winter. A few nest at Atlantic coastal wetlands. Locally abundant inland in winter. Some non-breeders remain in summer. Ringed individuals of the nominate race (Baltic Gull) are occasionally identified in winter, notably in Málaga fishing port.

European Herring Gull *Larus argentatus* Gaviota argéntea
Occasional individuals occur in winter along Atlantic coasts. Rare in the Mediterranean.

Caspian Gull *Larus cachinnans* Gaviota del caspio
One record from Málaga harbour: a first-winter in February 2017, and two from the Cádiz coast.

Yellow-legged Gull *Larus michahellis* Gaviota patiamarilla
Locally abundant, with large nesting colonies on rocky coasts, including c. 4,000 pairs at Gibraltar. Some colonies have also been established on salt marshes, e.g. in Cádiz Bay. Mainly resident but disperses widely along coasts. Scarce inland. Has bred in Extremadura.

Iceland Gull*   *Larus glaucoides*   Gaviota polar
At least 12 winter/spring records from coastal Andalucía and one from Gibraltar. Four occurred during a wider influx in 2009.

Glaucous Gull*   *Larus hyperboreus*   Gavión hiperbóreo
At least 15 winter/early spring records from Atlantic coastal Andalucía/the Strait.

Great Black-backed Gull   *Larus marinus*   Gavión atlántico
Occasional individuals occur in winter, along Atlantic coasts especially. Very rare inland.

Black-legged Kittiwake   *Rissa tridactyla*   Gaviota tridáctila
Occurs October–April in the Strait and off Atlantic coasts. Generally scarce but large numbers may appear inshore after westerly gales, when some enter the Mediterranean.

Sabine's Gull   *Xema sabini*   Gaviota de Sabine
Exceptional on passage. Occasionally recorded in the Strait area in April–May and in late summer and autumn. There are winter records from Ceuta (2) and Punta Umbría (H).

Little Gull   *Hydrocoleus minutus*   Gaviota enana
Regular but usually scarce on passage: August–October and March–May. Common in some winters, especially offshore in the Mediterranean. Rare inland.

Gull-billed Tern   *Gelochelidon nilotica*   Pagaza piconegra
Mainly small colonies breed locally, notably in Doñana, at Fuente de Piedra (MA), at several reservoirs in Extremadura and sporadically elsewhere. Commonly seen hawking insects over fields near the breeding colonies. Seldom seen on coasts. Mainly present April–October.

Caspian Tern   *Hydroprogne caspia*   Pagaza piquirroja
Occurs in small numbers on passage and in winter, July–April, chiefly on SW coasts. Non-breeders summer in the Guadalquivir and Guadiana estuaries. Rare inland.

Royal Tern*   *Sterna maxima*   Charrán real
A migrant from West Africa that reaches the Strait in small numbers, chiefly late July–December. Seldom seen on the Spanish shore but there are records from Odiel (H), Doñana, Tarifa Beach, Punta Carnero (CA), the Guadalhorce estuary (MA), Gibraltar and Punta Entinas (AL), mainly in summer.

Lesser Crested Tern*   *Sterna bengalensis*   Charrán bengalí
Small numbers move east through the Strait chiefly in May, returning west late August–November, but mainly in October. Regular at Tarifa and Gibraltar. Rare elsewhere.

Sandwich Tern   *Sterna sandvicensis*   Charrán patinegro
Common on passage and in winter along coasts. Present mainly August–April but some non-breeders occur in summer.

Elegant Tern*   *Sterna elegans*   Charrán elegante
Occasionally identified on Atlantic coasts, these including some Elegant Tern x Sandwich Tern hybrids. Both pure and mixed pairs have been confirmed nesting in eastern Spain (Valencia).

Roseate Tern   *Sterna dougallii*   Charrán rosado
Migrates off Atlantic coasts but seldom recorded onshore, although a few have appeared on SW beaches during June–September in recent years. Rare in the Mediterranean.

Common Tern   *Sterna hirundo*   Charrán común
Common migrant, especially along coasts, in August–October and March–May. Has bred Cádiz Bay, Doñana, in Almería and in Extremadura. Occasional birds winter.

Arctic Tern   *Sterna paradisaea*   Charrán ártico
Migrates off Atlantic coasts but seldom recorded onshore. Exceptional in the Mediterranean.

Forster's Tern*   *Sterna forsteri*   Charrán de Forster
Five autumn records of single birds: in the Strait and on the SW coast (4).

Sooty Tern* *Onychoprion fuscatus* Charrán sombrío
Two records: at Doñana in May 1963 and the Guadalhorce estuary (MA) in August 1979.

Little Tern *Sternula albifrons* Charrancito
Breeds locally on sandy coasts, especially along the Atlantic shore where it is common, and in Almería. Up to 300 pairs have bred in Extremadura along the major valleys and at reservoirs, but numbers fluctuate greatly. Widespread on passage. Present April–October.

Whiskered Tern *Chlidonias hybrida* Fumarel cariblanco
Locally common but generally declining, nesting at a few inland wetlands in western Andalucía and irregularly in Extremadura. Most are in Doñana but the breeding population here has ranged from zero to over 6,000 pairs, dependent on rainfall. Widespread on passage. Chiefly present March–October but small numbers winter on the lower Guadalquivir.

Black Tern *Chlidonias niger* Fumarel común
Rare and occasional breeder, chiefly at Doñana, formerly more numerous. Widespread on passage, especially in May and August–September, when large numbers sometimes occur along coasts. Present April–October. Occasional in winter.

White-winged Black Tern *Chlidonias leucopterus* Fumarel aliblanco
A very scarce passage migrant in late April and May, more frequent but still rare during the return passage in August–mid-October. Usually occurs after prolonged easterly winds.

Common Guillemot *Uria aalge* Arao común
Present in small numbers in winter, chiefly off the Atlantic coast and very rarely within the Mediterranean, although seldom seen from shore.

Razorbill *Alca torda* Alca común
Common on passage and in winter along coasts. Often fishes inshore. Mainly October–March.

Little Auk *Alle alle* Mérgulo marino
Very rare winter visitor. Most records are of dead beached birds found after winter storms along the Atlantic coast.

Atlantic Puffin *Fratercula arctica* Frailecillo común
Common well offshore on passage and in winter. Most often observed on passage through the Strait: large numbers enter the Mediterranean in November–December, returning March–May.

Black-bellied Sandgrouse *Pterocles orientalis* Ganga ortega
A locally common resident in stony steppe habitats, chiefly in Extremadura and, much less abundantly, in Granada, Jaén and Almería. Scarce and local in western Andalucía.

Pin-tailed Sandgrouse *Pterocles alchata* Ganga común
A locally common resident in steppe habitats, mainly in La Serena (BA) and southern Cáceres, Extremadura. A declining population occurs in western Andalucía, around Doñana.

Rock Dove *Columba livia* Paloma Bravía
Feral birds are common. Wild-type birds are locally resident on rocky coasts and on inland cliffs, especially river gorges.

Stock Dove *Columba oenas* Paloma Zurita
Small numbers breed locally in Extremadura and even more sparsely in Andalucía. More widespread and numerous in winter.

Wood Pigeon *Columba palumbus* Paloma Torcaz
A widespread and common resident. Migrants from the north greatly increase numbers in winter, when large flocks occur. Several million winter in Extremadura.

Collared Dove *Streptopelia decaocto* Tórtola Turca
Widespread and locally common, having completed its colonisation of the entire region in the mid-1990s. Especially typical of city parks and farmsteads.

Turtle Dove *Streptopelia turtur* Tórtola común
Widespread and common April–October. The breeding population is large but has declined considerably since the 1970s, by over 50% in some areas. Trans-Saharan migrant.

Laughing Dove* *Streptopelia senegalensis* Tórtola senegalesa
Rare. Individuals and occasional pairs appear intermittently in Andalucía. One pair nested near Sevilla city in 2018. The origin of all these birds is unknown but it is spreading north in Morocco so some may be genuine vagrants.

Rose-ringed Parakeet *Psittacula krameri* Cotorra de Kramer
Increasing feral populations breed in Sevilla city, coastal Málaga and Almería, the Córdoba Botanic gardens and occasionally elsewhere.

Monk Parakeet *Myopsitta monachus* Cotorra monje
Increasing feral populations breed in parks and gardens along the Málaga coast. Also in the cities of Huelva, Sevilla, Cádiz and Almería.

Great Spotted Cuckoo *Clamator glandarius* Críalo europeo
Most common in Extremadura and eastern Andalucía. More widespread on passage. Some arrive very early, November–February, departing July–August, with juveniles lingering through September. Mainly parasitises the Common Magpie.

Common Cuckoo *Cuculus canorus* Cuco común
Widespread and locally common during the breeding season. Present March–September. Common on passage. Present March–September.

Barn Owl *Tyto alba* Lechuza común
A widespread and common resident.

Scops Owl *Otus scops* Autillo
Widespread and common in open woodlands and often in parks and gardens. Chiefly a trans-Saharan migrant, present March–October. Some remain in winter.

Eagle Owl *Bubo bubo* Bújo real
A widespread and locally common resident, especially in rocky, hilly or mountainous terrain, with scrub and open woodland.

Little Owl *Athene noctua* Mochuelo común
A widespread and common resident, especially in open habitats including cultivated land. Often visible in daytime on telephone poles and other exposed perches.

Tawny Owl *Strix aluco* Cárabo común
A widespread and locally common resident in wooded habitats, especially in the more humid parts of western Andalucía and in Extremadura.

Long-eared Owl *Asio otus* Búho chico
A widespread but very local resident in forested areas, especially isolated pinewoods and woodland edges. Small numbers also occur on passage and in winter.

Short-eared Owl *Asio flammeus* Lechuza campestre
Small numbers occur regularly on passage and in winter, October–April. Winters regularly in the lower Guadalquivir farmlands. Has bred in Extremadura.

Marsh Owl* *Asio capensis* Búho moro
An occasional winter visitor to southwest Andalucía until the end of the 19th century. The sole later record is of a shot bird found at Jerez de la Frontera (CA) on 10 December 1998.

European Nightjar *Caprimulgus europaeus* Chotacabras gris
Locally common in open montane heathland: nesting in the Sierra Nevada (GR), parts of Jaén and the northern fringe of Cáceres. Widespread on passage. Present April–October.

Red-necked Nightjar *Caprimulgus ruficollis* Chotacabras pardo
Widespread and locally common in open pinewoods and other areas with scattered trees, including olive groves. Widespread on passage. Mainly present April–October.

Common Nighthawk* *Chordeiles minor* Añapero yanqui
One was found at Puerto Real (CA) in November 2005.

Chimney Swift* *Chaetura pelagica* Vencejo de chimenea
Two were at Sotogrande (CA) on 27 October 1999, following a small influx into Europe.

Common Swift *Apus apus* Vencejo común
Abundant on passage and numerous in most towns and villages. A trans-Saharan migrant, arriving from late March and departing by the end of September. The southward passage across the Strait in late July and in August is sometimes spectacular.

Pallid Swift *Apus pallidus* Vencejo pálido
Locally abundant, with colonies in most coastal towns and on some coastal cliffs. Numerous inland throughout western Andalucía but more local elsewhere. Local but widespread in Extremadura. A trans-Saharan migrant, mainly present late-February–November.

Alpine Swift *Apus melba* Vencejo real
Widespread and locally common, small colonies inhabiting mountains and coastal cliffs in central and eastern Andalucía. Some nest in buildings in Extremadura, especially on the Guadiana bridges. A trans-Saharan migrant, present March–October.

White-rumped Swift *Apus caffer* Vencejo cafre
Small numbers nest locally, chiefly in Cádiz, the Sierra Morena and across Extremadura. Uses old Red-rumped Swallow nests. A trans-Saharan migrant, mainly present from late March until November. First discovered breeding in Spain in 1966 at the Sierra de la Plata (CA).

Little Swift* *Apus affinis* Vencejo moro
Has bred very locally in western Andalucía since the mid-1990s, with well known colonies at Chipiona harbour and in the Sierra de la Plata (CA). Recorded with increasing frequency elsewhere in Andalucía, with records from Almería, Málaga, Huelva and Jaén. Two Extremaduran records. Some at least are resident.

Common Kingfisher *Alcedo atthis* Martín pescador común
Frequent on passage and in winter, occurring both inland and on rocky coasts. Breeds locally.

Blue-cheeked Bee-eater* *Merops persicus* Abejaruco papirrojo
Two records (four birds), from Andalucía and one each from Extremadura and Gibraltar.

European Bee-eater *Merops apiaster* Abejaruco europeo
A common and characteristic breeding species. A trans-Saharan migrant, present mainly April–September. Large numbers cross the Strait on migration.

Roller *Coracias garrulus* Carraca europea
Widespread on passage and locally common in the breeding season, especially in Extremadura, where nestbox schemes produce local concentrations. Also characteristic of the Guadalquivir valley and eastern Andalucía. A trans-Saharan migrant, present April–September.

Hoopoe *Upupa epops* Abubilla
A widespread and common breeding species and migrant. Most arrive in February–March, departing August–September. Significant numbers winter, especially along the Costa del Sol in Málaga, in Almería and in Doñana.

Wryneck *Jynx torquilla* Torcecuello
A scarce but widespread breeding species in western and central Andalucía, very local elsewhere, favouring open woodlands and orchards. Widespread on passage. Mainly present March–September but small numbers winter locally.

Iberian Green Woodpecker *Picus viridis sharpei* Pito real
A locally common resident, especially across the Sierra Morena and in the Betic mountains.

Great Spotted Woodpecker *Dendrocopus major* Pico picapinos
A common and widespread resident. Particularly associated with conifers in mountain areas but also found in other woodlands.

Lesser Spotted Woodpecker *Dendrocopus minor* Pico menor
A very local resident but apparently increasing. Most likely to be encountered in lusher, often riverine woodlands in northern Extremadura, the western Sierra Morena (H) and Cazorla (J). Increasingly recorded along the Guadalquivir and in corkwoods in Extremadura and may be generally under-reported.

Dupont's Lark *Chersophilus duponti* Alondra de Dupont
A declining resident, confined to steppes in Granada and Almería, where the total population was estimated at only 15 territories in 2018. Some may disperse more widely in winter.

Calandra Lark *Melanocorypha calandra* Calandria común
Locally abundant, characteristic of cereal crops and steppe habitats. Mainly resident but some leave Extremadura in winter. Forms large winter flocks.

Greater Short-toed Lark *Calandrella brachydactyla* Terrera común
Widespread and common in pastures, steppes and fallow areas. A trans-Saharan migrant, present mainly April–September.

Lesser Short-toed Lark *Calandrella rufescens* Terrera marismeña
Locally abundant but largely confined to salt flats in the Guadalquivir estuary (H) and Cádiz Bay, the Almería coastlands and dry steppelands in Granada. Mainly resident. Two Extremaduran records (six birds).

Crested Lark *Galerida cristata* Cogujada común
Widespread and common, generally in lower-lying and more open, often cultivated, country than the Thekla Lark. Mainly resident.

Thekla Lark *Galerida theklae* Cogujada montesina
Common in open, bushy country, typically in hilly terrain. Absent from the Guadalquivir valley. Mainly resident.

Wood Lark *Lullula arborea* Alondra totovía
Common in open woodlands, especially dehesas. Local birds are probably resident but others occur on passage and in winter.

Sky Lark *Alauda arvensis* Alondra común
Abundant on passage and especially on farmland in winter. A small number breed on the summits of the Sierra Nevada and other high tops.

Sand Martin *Riparia riparia* Avión zapador
Nests locally, notably along the Guadalquivir and the Guadiana valleys. Largely absent from southern Andalucía. Widespread and common on passage. Mainly present March–October.

Crag Martin *Ptyonoprogne rupestris* Avión roquero
Widespread and common wherever there are cliffs or rocky areas. Most widespread in winter, when numbers are greatly increased by migrants from further north.

Barn Swallow *Hirundo rustica* Golondrina común
An abundant breeding species and passage migrant. Present February–October mainly but regular in winter in western Andalucía especially, notably in the Guadalquivir valley where roosts of hundreds occur in December and January.

Red-rumped Swallow *Cecropis daurica* Golondrina dáurica
Widespread and locally common, nesting on rocks and under bridges or in roadside culverts especially. Present February–October mainly but some occur in winter, with roosts of up to 100 reported from Sevilla in December and January. Has wintered in Extremadura.

House Martin   *Delichon urbicum*   Avión común
An abundant breeding species and passage migrant. Most nest on buildings but also on dams and cliffs, sometimes in spectacular colonies with hundreds of nests. Present February–October mainly but small numbers occur regularly in winter, especially in western Andalucía.

Richard's Pipit   *Anthus richardi*   Bisbita de Richard
Occasional individuals occur in late autumn and winter in coastal Andalucía.

Tawny Pipit   *Anthus campestris*   Bisbita campestre
Widespread on passage. Nests locally, chiefly in southern Cádiz province and on mountain pastures on south-facing slopes in Jaén, Granada, Almería and northernmost Cáceres. Present April–September.

Olive-backed Pipit*   *Anthus hodgsoni*   Bisbita de Hodgson
Two records of birds ringed in the Doñana area, in October and December.

Tree Pipit   *Anthus trivialis*   Bisbita arbóreo
Large numbers occur on passage, chiefly March–May and September–October. It has recently colonised the Sierra de Gredos and Sierra de Gata just north of Cáceres province.

Meadow Pipit   *Anthus pratensis*   Bisbita común
Widespread and abundant on passage and in winter. Present September–April mainly.

Red-throated Pipit   *Anthus cervinus*   Bisbita gorgirrojo
A very scarce passage migrant and occasional winter resident. Reported mainly from near the Strait and along the Mediterranean coast in March–May and October–November.

Water Pipit   *Anthus spinoletta*   Bisbita alpino
Widespread but scarce on passage and locally common in winter, when it occurs in waterside habitats at low elevations, including the Extremaduran ricefields.

Rock Pipit   *Anthus petrosus*   Bisbita costero
Single birds have appeared at Gibraltar in January and at Doñana, Tarifa and the Antas estuary (AL) in March.

Yellow Wagtail   *Motacilla flava*   Lavandera boyera
M.f.iberiae breeds commonly in the Guadalquivir valley and at coastal wetlands in Andalucía, and more locally elsewhere. Widespread and common on passage, when races *flava*, *flavissima* and *thunbergi* are regular. Race *cinereocapilla* is uncommon and there are two records of *feldegg*. Present February–October but a few winter in the south.

Citrine Wagtail *   *Motacilla citreola*   Lavandera cetrina
Three records from Andalucía.

Grey Wagtail   *Motacilla cinerea*   Lavandera cascadeña
A common resident, frequenting rocky watercourses and hence largely confined to the sierras when breeding. More widespread on passage and in winter.

White Wagtail   *Motacilla alba*   Lavandera blanca
Common on passage and abundant in winter, when numbers frequent roadsides. Large winter roosts occur in cities: over 100,000 have been censused in Sevilla city alone. Present mainly October–March but there is a relatively small breeding population, which especially favours stony riverbeds but also occurs in city parks and near human habitation. The Pied Wagtail *M.a.yarrellii* occurs in very small numbers in winter and on passage. The distinctive Moroccan White Wagtail *M.a.subpersonata* has been found at least seven times, at or near the Strait.

Common Bulbul*   *Pycnonotus barbatus*   Bulbul naranjero
Found nesting at Tarifa in 2013, where at least one pair fledged young. A single male was still there in spring 2019.

Dipper   *Cinclus cinclus*   Mirlo acuático
Very local in mountainous areas, occurring along boulder-strewn, fast-flowing streams, chiefly in eastern Andalucía and in the north and east of Cáceres. Resident.

Wren *Troglodytes troglodytes* Chochín común
A widespread and common resident, absent only from steppes and agricultural lands.

Dunnock *Prunella modularis* Acentor común
Small numbers breed locally in the sierras of northern Extremadura and Las Villuercas (CC) and the Sierra Nevada (GR). Widespread but inconspicuous in winter, occurring in scrub.

Alpine Accentor *Prunella collaris* Acentor alpino
Nests in the Sierra Nevada (GR) above 2,500m and also just north of Cáceres in the Sierra de Gredos. Descends lower in winter, when often reported in Extremadura, in the Sierra de Grazalema (CA) and at Gibraltar, November–February.

Rufous-tailed Scrub-robin *Cercotrichas galactotes* Alzacola rojizo
Locally common in vineyards and olive and citrus groves, as well as in low scrub along watercourses and prickly pear hedges, chiefly in the Guadalquivir valley. Also characteristic of the dry scrublands of Almería. Widespread but local in Extremadura, mainly in Badajoz. A trans-Saharan migrant, present April–September.

European Robin *Erithacus rubecula* Petirrojo europeo
A common resident in more humid woodlands especially, hence often in mountains. Migrants occur widely on passage and it is abundant in winter.

Common Nightingale *Luscinia megarhynchos* Ruiseñor común
Widespread, abundant and characteristic, nesting in dense vegetation especially along watercourses. Common on passage. A trans-Saharan migrant, present March–October.

Bluethroat *Luscinia svecica* Ruiseñor pechiazul
A scarce but regular migrant, in March–May and September–November. Small numbers winter, generally near water, especially in southern Andalucía and along the Guadiana valley. Some nest in broom scrub at the uppermost levels of the Jerte Valley (CC).

Red-flanked Bluetail *Tarsiger cyanurus* Ruiseñor coliazul
One ringing record from Huelva in October. One at Andújar (J) on 5 December 2017.

Black Redstart *Phoenicurus ochruros* Colirrojo tizón
The very dark resident race *aterrimus* is locally common in rocky hills and mountains. Abundant on passage and in winter, when large numbers of the less striking race *gibraltariensis* from northern Europe occur widely at all levels, often around buildings as well as in open areas.

Common Redstart *Phoenicurus phoenicurus* Colirrojo real
Nests very locally in well-wooded, humid mountainous areas, notably in the Sierra Morena, the Serranía de Ronda, northern Cáceres and in the Sierra de San Pedro (BA). Widespread and numerous on passage. A trans-Saharan migrant, present March–October.

Moussier's Redstart* *Phoenicurus moussieri* Colirrojo diademado
At least five records of males from southern Andalucía.

Whinchat *Saxicola rubetra* Tarabilla norteña
Common on passage, crossing the area from late-March–May and August–October.

Stonechat *Saxicola torquatus* Tarabilla común
Abundant, widespread and characteristic. Migrants increase the resident population in winter.

Northern Wheatear *Oenanthe oenanthe* Collalba gris
Nests on highest mountain tops in southern Andalucía and in Extremadura in the Sierra de Gredos (CC). Widespread and common on passage. A trans-Saharan migrant, mainly present late February–November but there are occasional winter records. The Greenland Wheatear *O.o.leucorhoa* occurs regularly on passage, especially in May and October–November.

Seeböhm's Wheatear* *Oenanthe (o) seebohmi* Collalba de Seeböhm
A male was in Gibraltar on 28 April 2005.

Black-eared Wheatear *Oenanthe hispanica* Collalba rubia
Common and widespread inhabiting dry, open habitats. Also common and widespread on passage. A trans-Saharan migrant, present March–October.

Desert Wheatear* *Oenanthe deserti* Collalba desértica
At least six records of single birds from southern Andalucía (5) and Gibraltar (1).

White-crowned Wheatear* *Oenanthe leucopyga* Collalba yebélica
Two recorded in Doñana on 20 May 1977 comprise the only Iberian record.

Black Wheatear *Oenanthe leucura* Collalba negra
Resident. Common in rocky mountainous areas thoughout eastern Andalucía. Local in Extremadura and largely absent from western Andalucía.

Rufous-tailed (Common) Rock Thrush *Monticola saxatilis* Roquero rojo
Nests locally on rocky slopes at high altitudes in the Betic mountains of eastern Andalucía, the Sierra de Gredos (north of Cáceres) and Las Villuercas (CC). More widespread on passage. Present April to October.

Blue Rock Thrush *Monticola solitarius* Roquero solitario
A widespread and common resident, inhabiting coastal and inland cliffs, generally below 1,700m. It is also often seen on castles, churches, ruins and other buildings.

Ring Ouzel *Turdus torquatus* Mirlo capiblanco
Scarce but widespread on passage, in March–April and October–November. Small numbers remain in winter in the southern mountains.

Blackbird *Turdus merula* Mirlo común
An abundant and widespread resident. Migrants from northern Europe also occur in winter.

Eye-browed Thrush* *Turdus obscurus* Zorzal rojigrís
A first-winter bird was in several Cádiz city parks in February–March 2017.

Fieldfare *Turdus pilaris* Zorzal real
Very scarce and irregular on passage and in winter. Present mainly November–February.

Song Thrush *Turdus philomelos* Zorzal común
Abundant on passage and in winter, present October–April. Great numbers are attracted to olive groves in winter. Small numbers nest locally in northernmost Extremadura.

Redwing *Turdus iliacus* Zorzal alirrojo
Regular, sometimes common, on passage and in winter. Present mainly November–February.

Mistle Thrush *Turdus viscivorus* Zorzal charlo
Widespread and common in the southern mountains and in the dehesas and mountains of Extremadura. Mainly resident.

American Robin *Turdus migratorius* Zorzal robín
One in a Granada city park in November 2014 was only the second Spanish record.

Cetti's Warbler *Cettia cetti* Cetia ruiseñor
Widespread and common in the dense vegetation of watercourses and in reedbeds. Resident.

Zitting Cisticola (Fan-tailed Warbler) *Cisticola juncidis* Cisticola buitrón
A common species of open grassy habitats. Mainly resident but some cross the Strait. Higher elevations generally are abandoned in winter. Vulnerable to cold winters.

Grasshopper Warbler *Locustella naevia* Buscarla pintoja
Widespread on passage but easily overlooked. Occurs in April–May and August–October.

Savi's Warbler *Locustella luscinoides* Buscarla unicolor
Very local, found chiefly in Doñana and the lower Guadalquivir, nesting in reedbeds. Also present in wetlands of the upper Guadalquivir in Jaén and at the Arrocampo reservoir (CC). May nest at the Montijo reservoir (BA). A trans-Saharan migrant, present April–September.

Western Olivaceous (Isabelline) Warbler *Iduna opaca* Zarcero bereber
A scarce summer resident in dry scrub, and in bushes, especially tamarisks, along dried up watercourses. Also found in olive groves. Typical of hot, low-lying areas, mainly in coastal and southeastern Andalucía and along the Guadalquivir valley. Nests in the Alagón valley (CC) and probably more widespread in Extremadura. Present May–September.

Moustached Warbler *Acrocephalus melanopogon* Carricerín real
Very rare. Only occasional individuals are recorded. Has bred in Extremadura.

Aquatic Warbler *Acrocephalus paludicola* Carricerín cejudo
Extremely scarce but possibly regular on passage in autumn, August–October.

Sedge Warbler *Acrocephalus schoenobaenus* Carricerín común
Regular on spring passage, much scarcer in autumn. A trans-Saharan migrant, occurring March–May and August–October.

Paddyfield Warbler* *Acrocephalus agricola* Carricero agricola
Four winter records from Andalucía.

Blyth's Reed Warbler* *Acrocephalus dumetorum* Carricero de Blyth
Single September ringing records at Gibraltar in 1973 and Doñana in 2001.

Marsh Warbler* *Acrocephalus palustris* Carricero políglota
A very rare passage migrant. Has been ringed in Doñana in autumn.

Eurasian Reed Warbler *Acrocephalus scirpaceus* Carricero común
Nests locally in colonies in reedbeds. Widespread and common on passage. A trans-Saharan migrant, present early April–October.

Great Reed Warbler *Acrocephalus arundinaceus* Carricero tordal
Nests locally in colonies in reedbeds. Commoner than the Reed Warbler in Extremadura. Widespread on passage. A trans-Saharan migrant, present early April–October.

Icterine Warbler *Hippolais icterina* Zarcero icterino
A rare migrant, recorded at Doñana and Gibraltar, usually after prolonged easterlies in April–May.

Melodious Warbler *Hippolais polyglotta* Zarcero común
Widespread and common, occurring especially in scrub with scattered trees, often near water. Often numerous on passage. A trans-Saharan migrant, present April–October.

Marmora's Warbler* *Sylvia sarda* Curruca sarda
Two autumn records from Gibraltar.

Dartford Warbler *Sylvia undata* Curruca rabilarga
Widespread and common in low scrub, particularly in *Cistus*. Mainly resident.

Tristram's Warbler* *Sylvia deserticola* Curruca de Tristram
One in Gibraltar on 10 April 1988 is the sole Iberian record.

Spectacled Warbler *Sylvia conspicillata* Curruca tomillera
Locally common in extremely low scrub, such as *Salicornia* salt flats, and typical of the Mediterranean coastlands and arid inland regions, as in Almería. More widespread on passage. A trans-Saharan migrant, mainly present March–October, but some remain in winter.

Subalpine Warbler *Sylvia cantillans* Curruca carrasqueña
Common in tall scrub and the understorey of open woodlands, chiefly in hilly or mountainous terrain. A trans-Saharan migrant, present February–October.

Sardinian Warbler *Sylvia melanocephala* Curruca cabecinegra
Widespread and abundant in scrub and dry woodland, as well as in olive groves, orchards and gardens. Mainly resident.

Western Orphean Warbler *Sylvia hortensis* Curruca mirlona
Widespread at low density in open woodland, such as dehesas and olive groves, with or without a shrubby understorey. A trans-Saharan migrant, present from April–October.

Lesser Whitethroat *Sylvia curruca* Curruca zarcerilla
Very rare migrant, chiefly in autumn, recorded in Gibraltar (3) and Andalucía (2).

Common Whitethroat *Sylvia communis* Curruca zarcera
Nests locally in low scrub, notably in the Sierra Morena, the Betic mountains and the northern mountainous fringe of Extremadura. Common on passage. Present March–October.

Garden Warbler *Sylvia borin* Curruca mosquitera
A trans-Saharan migrant. Widespread and common on passage: April–May and August–October.

Blackcap *Sylvia atricapilla* Curruca capirotada
A widespread and common resident, nesting in scrub, gardens and open broadleaved woodlands. Confined in drier regions to riverine woodlands. Local birds are joined by large numbers of migrants in winter, when it is particularly abundant in olive groves.

Arctic Warbler* *Phylloscopus borealis* Mosquitero ártico
Two records: Gibraltar in October 1984 and the Laguna de Medina (CA) in September 2008.

Pallas's Leaf Warbler* *Phylloscopus proregulus* Mosquitero de Pallas
Four records: from Andalucía (3) and Gibraltar.

Yellow-browed Warbler* *Phylloscopus inornatus* Mosquitero bilistado
Rare migrant but has appeared annually in recent years throughout the region, mainly during October–November. Has overwintered.

Hume's Leaf Warbler* *Phylloscopus humei* Mosquitero de Hume
Five winter records: four from Andalucía and one from Extremadura.

Radde's Warbler* *Phylloscopus schwarzi* Mosquitero de Schwarz
One in Doñana on 7 November 1966.

Dusky Warbler* *Phylloscopus fuscatus* Mosquitero sombrío
One in Gibraltar in January 1989 and one ringed at La Janda (CA) in December 2015.

Western Bonelli's Warbler *Phylloscopus bonelli* Mosquitero papialbo
Nests commonly in open mixed woodland, notably in the Betic mountains and also in the Sierra Morena and northern Cáceres. Widespread on passage. Mainly present April–September.

Wood Warbler *Phylloscopus sibilatrix* Mosquitero silbador
A very scarce but regular trans-Saharan migrant, which is widespread on passage across the area in April–May. Rare in autumn.

Common Chiffchaff *Phylloscopus collybita* Mosquitero común
Common on passage and abundant in winter, often occurring in open country far from any trees. Mainly present October–April but also nests in the Sierra Nevada (GR).

Iberian Chiffchaff *Phylloscopus ibericus* Mosquitero ibérico
Very locally distributed. Typical of humid and riverine woodland in the Cádiz hills, the Sierra Morena and the Betic range. Also in a few montane areas in Cáceres. A trans-Saharan migrant, present March–September mainly.

Mountain Chiffchaff *Phylloscopus sindianus* Mosquitero montano
One ringing record from Gibraltar in November 2001.

Willow Warbler *Phylloscopus trochilus* Mosquitero musical
Abundant trans-Saharan migrant, crossing the region March–May and August–October.

Dotterel

Goldcrest *Regulus regulus* Reyezuelo sencillo
Very scarce and perhaps irregular in winter. November–March. May breed in coniferous forests in the mountains of northern Extremadura on the fringes of the Sierra de Gredos.

Firecrest *Regulus ignicapilla* Reyezuelo listado
Locally common in open mixed woodland or mountain pinewoods in the sierras of southern Andalucía, chiefly in Cádiz, Málaga and Granada, the eastern Sierra Morena and northern Extremadura. More widespread in winter.

Spotted Flycatcher *Muscicapa striata* Papamoscas gris
Common in the breeding season, often in open woodlands near water but relatively scarce in Extremadura. Widespread on passage. A trans-Saharan migrant, present April–September.

Red-breasted Flycatcher* *Ficedula parva* Papamoscas papirrojo
Rare migrant, mainly in autumn. Recorded in coastal Andalucía (8) and Gibraltar (3).

Collared Flycatcher* *Ficedula albicollis* Papamoscas collarino
Four spring records from Andalucía.

Pied Flycatcher *Ficedula hypoleuca* Papamoscas cerrojillo
A trans-Saharan migrant, common and widespread on passage. A few nest very locally above 1,000m in montane broadleaved woodlands, in the Sierra Nevada (GR), northern Cáceres and the eastern Sierra Morena. Present April–October.

Bearded Reedling (Parrotbill/Tit) *Panurus biarmicus* Bigotudo
Recorded sporadically in large reedbeds in Extremadura. Nested at the Arrocampo reservoir (CC) in 2011.

Long-tailed Tit *Aegithalos caudatus* Mito
Locally common in mixed and oak woodlands, especially in Cádiz, the Betic mountains, the Sierra Morena and Cáceres. Resident.

Crested Tit *Lophophanes cristatus* Herrerillo capuchino
Common in the Sierra Morena and Betic mountains but local in Extremadura. Occurs in open evergreen woodland, especially of Cork Oaks, as well as in conifers. Resident.

Coal Tit *Periparus ater* Carbonero garrapinos
Locally common in moist coniferous woodlands. Typical of the Betic mountains, eastern Sierra Morena and northernmost Extremadura. Resident.

Blue Tit   *Cyanistes caeruleus*   Herrerillo común
A widespread and common resident. Typically in woodlands, parks and gardens, although it avoids coniferous forests.

Great Tit   *Parus major*   Carbonero común
A widespread and common resident. Typically in woodlands, parks and gardens, although it avoids coniferous forests.

Nuthatch   *Sitta europaea*   Trepador azul
Widespread in mature, open woodland in the Sierra Morena, Serranía de Ronda, the sierras of Extremadura and more locally elsewhere. Resident.

Wallcreeper   *Tichodroma muraria*   Treparriscos
A rare winter visitor, with occasional reports from Gibraltar, El Chorro (MA), Sierra Tejeda (Málaga/Granada), Sierra Nevada (GR), Sierra de Gador (AL), Pico Villuerca (CC) and Monfragüe (CC).

Short-toed Treecreeper   *Certhia brachydactyla*   Agateador común
Widespread and common in all woodland types but favouring mature trees. Mainly resident but more widely dispersed in winter, when many descend from higher altitudes.

Penduline Tit   *Remiz pendulinus*   Pajaro moscón
Inhabits riverine vegetation in the Guadalquivir and Guadiana valleys. More widespread in winter, in reedbeds at principal wetlands and in rank vegetation along watercourses.

Golden Oriole   *Oriolus oriolus*   Oropéndola
Nests commonly in mature open and, especially, riverine woodland, as well as in poplar plantations. Widespread on migration. A trans-Saharan migrant, present April–September.

Black-crowned Tchagra*   *Tchagra senegalus*   Chagra del Senegal
One near Facinas (CA) in July 1995 is the sole Iberian record.

Isabelline Shrike*   *Lanius isabellinus*   Alcaudón isabel
Two records from coastal Andalucía: in November 1999 and in February 2008.

Red-backed Shrike   *Lanius collurio*   Alcaudón dorsirrojo
Very rare migrant, recorded from Andalucía (2), Extremadura (3) and Gibraltar (4). A few nest at the Puerto de Tornavacas at the top of the Jerte valley and nearby (CC).

Southern Grey Shrike   *Lanius meridionalis*   Alcaudón reál
Widespread and common in open woodlands and in scrub with scattered trees, although scarce in Cádiz and along the Costa del Sol. Mainly resident.

Steppe Grey Shrike   *Lanius (meridionalis) pallidirostris*   Alcaudón meridional estepario
Two winter records: at the Andratx estuary (AL) in 2016/17 and at Coria (SE) in January 2017.

Woodchat Shrike   *Lanius senator*   Alcaudón común
Widespread and common, absent chiefly from closed forests and mountains above 1,500m. A trans-Saharan migrant, present mid-March–October.

Masked Shrike*   *Lanius nubicus*   Alcaudón núbico
Four records, all in May: from Gibraltar, Doñana (2) and Cazorla (J).

Jay   *Garrulus glandarius*   Arrendajo
Widespread and common in woodlands, especially where oaks are present. Absent from the Guadalquivir valley and the arid southeast. Resident.

Iberian Azure-winged Magpie   *Cyanopica cooki*   Rabilargo ibérico
A characteristic and locally common species of the evergreen oak woodlands of Extremadura. Also common in open pinewoods in western Andalucía, across the Sierra Morena and in the Genil valley in the central Betic mountains. Resident but wanders in flocks in winter.

Common Magpie *Pica pica* Urraca
Widespread and common in Extremadura and Huelva and also in eastern Andalucía, except along the Mediterranean coast. Scarce elsewhere but colonising new ground locally. Resident.

Alpine (Yellow-billed) Chough *Pyrrhocorax graculus* Chova piquigualda
Small flocks occur infrequently on migration at Gibraltar, in April and May and, once, in September. Also recorded rarely in Andalucía.

Red-billed Chough *Pyrrhocorax pyrrhocorax* Chova piquirroja
Widespread and locally common, principally on cliffs in mountainous country in the Betic mountains and eastern Sierra Morena. Very local in Extremadura. Resident.

Jackdaw *Corvus monedula* Grajilla
Widespread and locally common. Colonies favour bridges, ruined castles and similar buildings as well as cliffs. Resident.

Rook *Corvus frugilegus* Graja
One record from the Málaga coast on 12 May 2017.

Carrion Crow *Corvus corone* Corneja negra
Resident in the mountains of northern Extremadura (CC), the Sierra Nevada (GR) and Cazorla (J). Rare anywhere else in the region but occasional individuals wander in winter.

Hooded Crow *Corvus cornix* Corneja cenicienta
Several individuals have appeared in recent years in Andalucía and at Gibraltar (1), chiefly in winter.

Raven *Corvus corax* Cuervo
Widespread and common, nesting on cliffs but also on pylons and tall trees. Large flocks form on farmland in winter. This is the common black crow of the region. Resident.

Spotless Starling *Sturnus unicolor* Estornino negro
Widespread and common, nesting in holes in trees and on buildings. Mainly resident but some cross the Strait.

Common Starling *Sturnus vulgaris* Estornino pinto
Migrant, wintering in the area in very large flocks. Present October–April. Numbers vary greatly between winters.

Rosy Starling* *Pastor roseus* Estornino rosado
At least five records from Andalucía and one from Extremadura.

House Sparrow *Passer domesticus* Gorrión común
Widespread and common, nesting colonially in trees in open country as well as in towns and villages, where holes in buildings are used. Resident.

Spanish Sparrow *Passer hispaniolensis* Gorrión moruno
Locally abundant in Extremadura, in the lower Guadalquivir and in scattered localities elsewhere in Andalucía. Large colonies in eucalyptus plantations near water are typical. Present all year but some cross the Strait in March–April and September–October.

Tree Sparrow *Passer montanus* Gorrión molinero
Widespread but local. Present all year but some cross the Strait in March–April and September–November.

Rock Sparrow *Petronia petronia* Gorrión chillón
Widespread but local in open woodland in Extremadura and in the rocky hills of Andalucía, generally above 500m. Resident.

Snowfinch *Montifringilla nivalis* Gorrión nival
One record from the Sierra Nevada (GR) in December 1975.

Black-headed Weaver *Ploceus melanocephalus* Tejedor cabecinegro
Small numbers nest at the Brazo del Este (SE). Introduced (Sub-Saharan Africa).

Yellow-crowned Bishop *Euplectes afer* Tejedor amarillo
Small numbers nest in the ricefields and reedbeds of the lower Guadalquivir in Sevilla, notably at the Brazo del Este and Isla Menor. Increasingly numerous in the Guadiana valley (BA). Introduced (Sub-Saharan Africa).

Common Waxbill *Estrilda astrild* Estrilda común
An introduced and naturalised species. Often reported in small groups in reedbeds, mainly near coasts or along major river valleys. Probably resident but groups wander widely. Thousands inhabit the Guadiana valley (BA), as well as wetlands in Cáceres province. Also established on the Guadalquivir in Sevilla and in coastal Málaga and Granada. (Sub-Saharan Africa.)

Red Avadavat *Amandava amandava* Bengalí rojo
Large numbers occur along the Guadiana, the Alagón and other major river valleys in Extremadura. It also occurs on the River Guadalquivir in Sevilla and locally in coastal Málaga and Granada. Wandering birds occasionally appear elsewhere. Introduced (Tropical Asia).

Chaffinch *Fringilla coelebs* Pinzón vulgar
Widespread and common resident in all woodland types. Large numbers also occur on passage and in winter. There are five records of the African Chaffinch *F.(c.) africana* from Gibraltar.

Brambling *Fringilla montifringilla* Pinzón real
Scarce but regular on passage and in winter. Numbers vary considerably from year to year; occurs chiefly November–March.

Serin *Serinus serinus* Verdecillo
Widespread and common. Absent only from treeless areas although winter flocks occur in open country. Local birds are resident but large numbers cross the Strait on passage

Greenfinch *Carduelis chloris* Verderón común
A widespread and common resident. Numerous migrants also occur on passage and in winter.

Citril Finch *Carduelis citrinella* Verderón serrano
Small outpost nesting populations occur in the Sierra Nevada (GR) and Cazorla (J). Also reaches the Jerte valley (CC), especially in winter. Associated with montane conifers.

Goldfinch *Carduelis carduelis* Jilguero
A widespread and common resident. Numerous migrants also occur on passage and in winter.

Siskin *Carduelis spinus* Lúgano
Widespread on passage and in winter in highly variable numbers. In alders and pine/birch woods. Has bred in Cazorla (J) and sporadically elsewhere. Occurs chiefly November–March.

Linnet *Carduelis cannabina* Pardillo común
A widespread and common resident. Numerous migrants also occur on passage and in winter.

Common Redpoll *Carduelis flammea* Pardillo sizerín
One was with a flock of Siskins at Málaga on 11 November 2005.

Red Crossbill *Loxia curvirostra* Piquituerto común
Inhabits pine forests, almost entirely in the southern Andalucían mountains. Has bred in northern Extremadura. Flocks are reported more widely in irruption years.

Trumpeter Finch *Bucanetes githagineus* Camachuelo trompetero
Locally common in the driest parts of Almería: Sierra de Alhamilla, Tabernas and Cabo de Gata, and also in Granada in the Hoya de Guadix. Occasional elsewhere. Mainly resident.

Common Rosefinch* *Carpodacus erythrinus* Camachuelo carminoso
Exceptional, on passage. At least seven records, all of juveniles caught by ringers between 15 September and 7 November. Six were in coastal Andalucía and one in Gibraltar.

Bullfinch *Pyrrhula pyrrhula* Camachuelo común
Very scarce but occurs increasingly widely on passage and in winter.

Hawfinch *Coccothraustes coccothraustes* Picogordo
Locally common in open deciduous woodland, usually near water. Chiefly in southern Andalucía, the Sierra Morena and northern Extremadura. Mainly resident.

Common Yellowthroat *Geothlypis trichas* Mascarita común
An adult male at the Trebujena marismas (CA) on 19 September 2014. First Spanish record.

Song Sparrow *Melospiza melodia* Chingolo cantor
One near Algeciras (CA) on 29 March 2013. First Spanish record.

Snow Bunting *Plectrophenax nivalis* Escribano nival
At least four records from Andalucía and three from Extremadura, all in winter.

Pine Bunting* *Emberiza leucocephalos* Escribano de Gmelin
One record: two at Gibraltar after strong easterlies on 2 May 1987.

Yellowhammer *Emberiza citrinella* Escribano cerillo
Some winter in the Sierra Nevada (GR) and Cazorla (J). Rare elsewhere in winter.

Cirl Bunting *Emberiza cirlus* Escribano soteño
Widespread and common, especially in the Betic mountains and Sierra Morena. More local in Extremadura. Frequents open bushy country near woodland in lower sierras. Mainly resident.

Rock Bunting *Emberiza cia* Escribano montesino
Widespread and common in rocky areas, notably in all the main sierras. Mainly resident but some descend to lower levels in winter.

House Bunting* *Emberiza sahari* Escribano sahariano
Common at Tangier, Morocco, but only recorded four times in Andalucía.

Ortolan Bunting *Emberiza hortulana* Escribano hortelano
Widespread on migration: April–May and August–September. Nests very locally in open woodlands high in the Sierra Nevada (GR) and on northern mountains in Cáceres.

Rustic Bunting* *Emberiza rustica* Escribano rustico
Four records of birds ringed in winter in coastal Andalucía

Little Bunting* *Emberiza pusilla* Escribano pigmeo
At least four autumn/winter records from coastal Andalucía and two from Gibraltar.

Yellow-breasted Bunting* *Emberiza aureola* Escribano aureolado
One caught at Chipiona (CA) in October 1969.

Reed Bunting *Emberiza schoeniclus* Escribano palustre
Winters in small numbers at wetlands in the main river valleys. Mainly present October–March.

Red-headed Bunting* *Emberiza bruniceps* Escribano carirrojo
One was at Sanlúcar de Barrameda (CA) from 20–29 October 1967.

Black-headed Bunting* *Emberiza melanocephala* Escribano cabecinegro
Three Andalucían records of single birds.

Corn Bunting *Emberiza calandra* Triguero
Typical and abundant in open country and farmland. Mainly resident but flocks wander in winter.

Bobolink* *Dolichonyx oryzivorus* Charlatán
One was in Gibraltar from 11–16 May 1984. It was caught and ringed.

APPENDICES

1 SCIENTIFIC NAMES OF PLANT SPECIES MENTIONED

Aleppo Pine	*Pinus halepensis*	Lavender	*Lavandula* spp.
Almond	*Prunus dulcis*	Lentisc	*Pistachia lentiscus*
Black Pine	*Pinus nigra*	Lusitanian Oak	*Quercus faginea*
Agave	*Agave americana*	Maritime Pine	*Pinus pinaster*
Asphodel	*Asphodelus albus*	Myrtle	*Myrtus communis*
Black Poplar	*Populus nigra*	Oleander	*Nerium oleander*
Buckthorn	*Rhamnus alaternus*	Olive	*Olea europaea*
Canarian Oak	*Quercus canariensis*	Orache	*Atriplex halimus*
Cazorla Violet	*Viola cazorlensis*	Prickly Pear	*Opuntia ficus-indica*
Citrus	*Citrus* spp.	Pyrenean Oak	*Quercus pyrenaica*
Common Reed	*Phragmites australis*	Retama	*Lygos sphaerocarpa*
Cork Oak	*Quercus suber*	Rhododendron	*Rhododendron ponticum*
Eucalyptus	*Eucalyptus globulus* etc.	Rosemary	*Rosmarinus officinalis*
Encina	*Quercus rotundifolia*	Sea Daffodil	*Pancratium maritimum*
False Esparto Grass	*Stipa tenacissima*	Smooth-leaved Elm	*Ulmus minor*
Fan Palm	*Chamaerops humilis*	Spanish Fir (Pinsapo)	*Abies pinsapo*
Genista	*Genista* spp.	Spiny Broom	*Calicotome villosa*
Giant Reed	*Arundo donax*	Strawberry Tree	*Arbutus unedo*
Gibraltar Candytuft	*Iberis gibraltarica*	Stone Pine	*Pinus pinea*
Glasswort	*Salicornia* spp.	Sweet Chestnut	*Castanea sativa*
Gum Cistus	*Cistus ladanifer*	Tamarisk	*Tamarix africana*
Hawthorn	*Crataegus monogyna*	Thyme	*Thymus* spp.
Halimium	*Halimium halmifolium*	Tree Heath	*Erica arborea*
Juniper	*Juniperus phoenicea* etc.	White Poplar	*Populus albus*

2 GLOSSARY OF LOCAL GEOGRAPHICAL TERMS

Alcornocal	Cork Oak wood	Matorral	Shrubland
Bahía	Bay	Mirador	Scenic Viewpoint
Cabo	Cape	Montaña	Mountain
Cascada	Waterfall	Peñon	Rock
Dehesa	Grazing woodland	Playa	Beach
Desembocadura	Estuary	Presa	Dam
Desfiladero	Gorge	Puerto	Mountain pass (also seaport)
Desierto	Desert	Punta	Headland
Embalse	Reservoir	Ría	Tidal inlet
Encinar	Holm Oak wood	Ribera	River Valley
Estuario	Estuary	Río	River
Faro	Lighthouse	Salinas	Salt pans
Hermita	Hermitage	Sendero	Footpath/Hiking Trail
Laguna	Lake	Serranía	Mountain range
Marisma	Tidal flats	Sierra	Mountain

FURTHER READING

Brock, P.D. 2017. *A photographic guide to Insects of Southern Europe and the Mediterranean.* Pisces Publications, Berkshire.

De Juana, E. 2006. *Aves raras de España.* Lynx Edicions, Barcelona.

De Juana, E. & Garcia, E. 2015. *The Birds of the Iberian Peninsula.* Bloomsbury, London.

Forsman, D. 1999. *The raptors of Europe and the Middle East. A handbook of field identification.* T. & A.D. Poyser, London.

Forsman, D. 2016. *Flight identification of raptors of Europe, North Africa and the Middle East.* Helm Identification Guides. Bloomsbury, London.

Garcia, E. & Rebane, M. 2017. *Where to watch birds in northern and eastern Spain.* 3nd edition. Bloomsbury, London.

Garzón, J. & Chiclana, F. 2006. *Where to watch birds in Doñana.* Lynx Edicions, Barcelona.

Garzón Gutiérrez, J. & Henares Civantos, I. 2012. *Las Aves de Sierra Nevada.* Consejería de Agricultura, Pesca y Medio Ambiente de la Junta de Andalucía, Granada.

Gutiérrez, R., de Juana, E. & Lorenzo, J.A. 2012. *Lista de las Aves de Españā.* SEO/BirdLife.

Hall, T. 2017. *Wild Plants of Southern Spain. A guide to the Native Plants of Andalucía.* Kew Publishing, Royal Botanic Gardens, Kew.

Madroño, A. *et al.* (Eds). 2005. *Libro Rojo de las Aves de España.* Dirección General para la Biodiversidad–SEO/Birdlife. Madrid.

Martí, R. & del Moral, J.C. (Eds). 2003. *Atlas de las Aves Reproductoras de España.* Dirección General de Conservación de la Naturaleza-Sociedad Española de Ornitología. Madrid

Montero, J.A. *et al.* 2005. *Where to watch birds in Spain. The 100 best sites.* Lynx Edicions, Barcelona.

Porter, R.F. *et al.* 1981. *Flight identification of European raptors.* 3rd edition. T & A.D.Poyser, London.

Svensson, L., Mullarney, K. & Zetterström, D. 2009. *Collins Bird Guide.* 2nd edition. HarperCollins, London.

Thorogood, C. 2016. *Field guide to the Wild Flowers of the Western Mediterranean.* Kew Publishing, Royal Botanic Gardens, Kew.

SITE INDEX

Abdalajís valley	MA4	Bonanza salinas	CA17	Doñana	H5
Adra lakes	AL4	Borbollón reservoir	CC7	East Bank of the lower Guadalquivir	CA17
Aguijón reservoir	BA1	Bornos reservoir	CA18	Eastern Badajóz reservoirs	BA10
Alange reservoir	BA4	Brazo del Este	SE1		
Alcalá de los Gazules	CA6	Brozas reservoir	CC1	El Algarrobo	CA2
Alcántara reservoir	CC4	Cabo de Gata	AL1	El Chorro	MA4
Alcollarín reservoir	CC12	Cabo de Gata salt pans	AL1	El Espigón	H4
Aldeaquemada	J2	Cáceres western plains	CC1	El Picacho	CA6
Algaida pinewoods	CA17	Cáceres-Trujillo steppes	CC3	El Rocío	H5
Alhama de Granada reservoir	GR5	Cádiz Bay	CA12	El Torcal	MA5
Almanzora estuary	AL9	Calaburras Point	MA17	Encinarejo reservoir	J3
Almoraima corkwoods	CA6	Campillo reservoir	BA4	Entremuros zone	SE2
Alpujarras	GR2	Campo de Arañuelo	CC9	Estero de Domingo Rubio	H4
Alpujata migration watchpoint	MA16	Campos de Hernán Perea	J1	Flecha del Rompido	H2
		Cañada de los Pájaros	SE4	Fuente de Piedra	MA1
Alqueva reservoir inlet	BA14	Cape Trafalgar	CA9	Gabriel y Galán reservoir	CC8
Alto del Cabrito	CA2	Castellar de la Frontera	CA6	García de Sola reservoir	BA10
Alto Guadiato farmlands	CO11	Cazalla	CA2	Gargáligas reservoir	BA10
Antas estuary	AL8	Cazorla	J1	Gibraltar	GIB
Arcos de la Frontera	CA18	Celemín reservoir	CA6	Giribaile reservoir	J5
Arcos reservoir	CA18	Charca de Suárez	GR10	Grazalema	CA20
Arrocampo-Almaraz reservoir	CC9	Chipiona coast	CA16	Guadalén reservoir	J5
		Cíjara game reserve	BA11	Guadalhorce estuary	MA8
Arroyo del Agarbe	H5	Cordobilla reservoir	CO6	Guadalquivir river at Córdoba	CO12
Badajoz western sierras	BA2	Cornalvo reservoir	BA6		
Bahía de Cádiz	CA12	Costa Ballena	CA16	Guadalquivir valley farmlands	SE5
Barbate cliffs	CA10	Cuatro Lugares Steppes	CC5	Guadiana estuary	H1
Barbate pinewoods and estuary	CA10	Cuncos reservoir	BA1	Guadiana estuary	H1
		Dehesa de Abajo	SE3	Guadiana Valley ricefields	CC11
Barbate reservoir	CA6	Despeñaperros gorge	J2	Guadiaro estuary	CA8
Baza basin	GR9	Devil's Eye	CA6	Guadiaro river valley	MA13
Bolonia	CA5				

Guadiloba reservoir	CC3	
Guadix basin	GR7	
Hornachos	BA4	
Isla Cristina	H1	
Isla Mayor	SE2	
Isla Mínima	SE2	
Jándula reservoir	J3	
Jerez de los Caballeros woodlands	BA1	
Jerte valley	CC8	
Jimena	CA6	
José Valverde centre	H5	
Juanar	MA9	
La Albuera lagoons and farmlands	BA13	
La Cimbarra waterfall	J2	
La Janda	CA4	
La Nava de San Pedro	J1	
La Ragua pass	GR3	
La Sauceda	CA6	
La Serena plains	BA9	
La Serena reservoir	BA10	
Laguna Amarga	CO2	
Laguna de El Portil	H3	
Laguna de Fuente de Piedra	MA1	
Laguna de Medina	CA15	
Laguna de Padul	GR4	
Laguna de Santiago	CO3	
Laguna de Tarelo	CA17	
Laguna de Tíscar	CO4	
Laguna de Zóñar	CO1	
Laguna del Prado	H1	
Laguna del Rincón	CO3	
Laguna del Taraje	CA13	
Laguna Dulce	CO2	
Laguna Grande de Baeza	J6	
Laguna Salada	CA14	
Laguna Salobral	CO5	
Lagunas de Campillos	MA2	
Lagunas de Espera	CA19	
Lagunas de La Lantejuela	SE6	
Lagunas de Palos y Las Madres	H4	
Lagunas de Puerto Real	CA13	
Lagunas del Puerto de Santa María	CA14	
Las Amoladeras	AL1	
Las Muelas reservoir	BA6	
Las Norias lakes	AL3	
Las Villuercas	CC4	
Llerena reservoir	BA3	
Los Alcornocales	CA6	
Los Barruecos	CC2	
Los Canchales reservoir	BA7	
Los Lances beach	CA1	
Los Molinos reservoir	BA4	
Lucio del Cangrejo	SE2	
Málaga hills	MA7	
Malpartida de Cáceres steppes	CC2	
Malpasillo reservoir	CO7	
Mata de Alcántara reservoir	CC1	
Mérida	BA5	
Mirador de las Águilas	MA16	
Monfragüe	CC6	
Monte Algaida salinas	CA17	
Montijo reservoir	BA5	
Northern Jaén reservoirs	J7	
Northwest Córdoba	CO11	
Odiel estuary	H4	
Ojén Valley	CA2	
Orellana reservoir	BA10	
Palmones estuary	CA7	
Peñon de Zaframagón	CA21	
Piedras estuary	H2	
Pinar del Rey	CA6	
Playa de Composoto	CA12	
Playa de Los Lances	CA1	
Playa de Valdelagrana	CA12	
Puente Nuevo reservoir	CO13	
Puerto de Gáliz	CA6	
Puerto de la Ragua	GR3	
Puerto de las Palomas	CA20	
Puerto del Bujeo	CA2	
Puerto Peña	BA10	
Punta Calaburras	MA17	
Punta del Sabinar	AL2	
Punta Entinas	AL2	
Punta Secreta	CA2	
Punta Umbría	H4	
Rambla de Morales	AL1	
Río Piedras estuary	H2	
River Guadiana at Badajoz city	BA12	
River Guadiana at Mérida	BA5	
Ronda sierras	MA12	
Roquetas de Mar	AL2	
Rumblar reservoir	J3	
Salinas de Cerrillo	AL2	
Salinas Viejas	AL2	
Salor reservoir	CC3	
San Fernando	CA12	
Sancti Petri marshes	CA11	
Santuario de Belén	BA8	
Santuario de la Luz	CA2	

Sierra Alhamilla	AL6	
Sierra Alpujata	MA16	
Sierra Bermeja	MA12	
Sierra Blanca	MA9	
Sierra Blanquilla	MA10	
Sierra Brava reservoir	CC10	
Sierra Crestellina	MA11	
Sierra de Aracena	H7	
Sierra de Baza	GR8	
Sierra de Cabo de Gata	AL1	
Sierra de Camarolos	MA6	
Sierra de Cardeña	CO10	
Sierra de Cazorla	J1	
Sierra de Gata	CC7	
Sierra de Grazalema	CA20	
Sierra de Hornachos	BA4	
Sierra de Hornachuelos	CO9	
Sierra de Huétor	GR6	
Sierra de la Plata	CA5	
Sierra de las Nieves	MA10	
Sierra de Montoro	CO10	
Sierra de Pela	BA9	
Sierra de San Pedro	BA2	
Sierra de Segura	J1	
Sierra de Tiros	BA8	
Sierra de Tolox	MA10	
Sierra Mágina	J5	
Sierra María	AL7	
Sierra Morena (west)	SE7	
Sierra Nevada	GR1	
Sierra Norte	SE7	
Sierra Pelada	H6	
Sierra Tejeda	MA14	
Sierra de Andújar	J3	
Sierras de Las Villuercas	CC4	
Sierras Subbéticas	CO8	
Sotogrande	CA8	
Southern Badajoz	BA3	
Southwest Badajoz reservoirs	BA1	
Southwest Badajoz woodlands	BA1	
Strait of Gibraltar	CA2	
Suárez Wetland	GR10	
Tabernas desert	AL5	
Talaván reservoir	CC4	
Tarifa beach	CA1	
Tarifa hills	CA2	
Teba: sierra and gorge	MA3	
Trebujena Marismas	CA17	
Trujillo	CC3	
Upper Guadalquivir reservoirs	J6	
Valuengo reservoir	BA1	
Vélez estuary	MA15	
Western Cáceres plains	CC1	
Zafarraya	MA14	
Zahara de los Atunes	CA5	
Zújar reservoir	BA10	
Zújar valley	CO12	

INDEX TO SPECIES BY SITE

See also the Status List (pages 368 to 398) for details of very common species and records of rarities. Sites are identified by province name abbreviations. GIB = Gibraltar.

Accentor, Alpine AL7, BA4, BA10, CA2, CA20, CC5, CC6, CC8, CO8, GIB, GR1, H7, J1, J4, MA4, MA5, MA10, MA14.

Avadavat, Red BA5, BA12, CC11, CC9, GR4, GR10, MA8, MA15.

Avocet AL1, AL2, AL3, BA7, CA1, CA4, CA11, CA12, CA15, CA17, CC9, CC11, CO4, H2, H4, H5, J7, MA1, MA8, SE5.

Bee-eater AL1, AL2, AL3, AL5, AL6, AL7, BA4, BA9, BA10, BA12, CA2, CA4, CA5, CA6, CA8, CA10, CA14, CA15, CA17, CA18, CA19, CA20, CC6, CC7, CC12, CO8, CO9, CO10, CO11, CO12, GIB, GR1, GR2, GR3, GR4, GR7, GR8, GR9, H1, H5, H7, J1, J2, J3, J4, J5, MA1, MA3, MA4, MA8, MA10, MA11, MA12, MA13, SE3.

Bittern, Great SE1, CC9.
Little AL3, AL4, AL9, BA5, BA10, BA12, CA8, CA13, CA14, CA15, CA15, CA15, CA17, CC9, CC11, CO6, CO14, GR4, GR10, H3, H4, H5, J6, J7, MA1, MA8, MA15, SE1, SE2.

Blackbird GIB, H5.

Blackcap AL6, AL7, BA1, CA6, CO8, CO10, GIB, GR1, GR2, GR5, H4, H5, H7, J1, J2, J3, J4, MA3, MA10, MA11, MA12, MA13, SE4.

Bluethroat AL3, BA12, CA2, CA4, CA7, CA8, CA14, CA15, CA17, CC8, CO1, GIB, GR4, GR10, H2, H5, J7, MA8, MA16, SE1, SE5.

Brambling CA2, CO9, GIB, GR1, GR3, J2, J3, MA4, MA12.

Bullfinch BA2, BA4, CA2, CC7, CO8, H5, H7, J2, MA6, SE6.

Bunting, Cirl AL1, AL7, BA8, CA2, CA4, CA6, CA17, CA20, CC5, CO8, GR1, GR2, H5, H6, H7, J1, J3, MA5, MA6, MA11, MA12, MA13, SE6.
Corn AL1, AL7, AL9, CA1, CA6, CO3, CO9, GR9, H4, H5, J3, SE4.
Ortolan CA2, CA4, CA5, CA6, CA8, CA17, CC7, CC8, CO11, GIB, H4, H5, MA6, MA11, MA12.
Reed AL2, CA7, CA8, CA17, CO1, GR4, GR10, H4, H5, MA15.
Rock AL1, AL5, AL6, AL7, BA4, BA8, BA10, CA2, CA6, CA17, CA20, CC5, CC6, CC8, CO9, GIB, GR1, GR2, H5, H6, H7, J1, J2, J3, J4, MA3, MA4, MA5, MA6, MA7, MA9, MA10, MA11, MA12, MA13, MA14, SE6.

Bustard, Great BA1, BA3, BA6, BA7, BA8, BA9, BA13, CC1, CC2, CC3, CC4, CC7, CC11, CO11, SE4.
Little AL1, AL5, AL6, AL7, BA1, BA3, BA6, BA8, BA9, BA13, CA2, CA4, CA17, CC1, CC2, CC3, CC4, CC7, CC11, CO5, CO11, GR7, GR9, H5, J5, MA1, MA2, SE4.

Buzzard, Common AL3, AL6, BA1, BA2, BA4, BA6, BA9, CA2, CA4, CA5, CA6, CA19, CC5, CC6, CC8, CC9, CO8, CO9, CO10, CO11, CO12, GIB, GR2, GR3, GR6, H5, H6, H7, J1, J2, J3, J4, J5, MA3, MA5, MA7, MA8, MA8, MA9, MA11, MA12, MA13, MA14, MA16, SE1, SE2, SE3, SE4, SE5, SE6.
Honey See Honey-buzzard
Long-legged CA2, CA4

Chaffinch CO9, GIB, GR2, GR3, GR6, GR8, GR5, J1, J2, J4, MA4, MA11, MA12, MA13.
Chat, Rufous Bush See Scrub-robin, Rufous-tailed
Chiffchaff, Common AL1, AL5, AL6, AL7, AL8, AL9, BA12, CA1, CA6, CA17, CO8, CO9, GIB, GR1, GR2, GR3, GR4, GR10, H4, H5, J1, MA1, MA4, MA5, MA7, MA8, MA10, MA11, MA12, MA13.
Iberian BA4, CA2, CA6, CA20, GIB, GR1, GR5, MA6, MA10, MA13, SE6.
Chough, Red-billed BA3, BA4, BA8, CA20, CA21, CC5, CC6, CO8, CO9, GR1, GR2, GR3, GR6, GR7, J1, J2, J3, J4, MA3, MA4, MA5, MA6,

405

MA9, MA10, MA11, MA12, MA13, MA14.

Cisticola, Zitting AL1, AL2, AL3, BA3, BA6, BA7, BA12, BA13, CA1, CA2, CA4, CA5, CA6, CA8, CA15, CA19, CC1, CC11, CO1, GIB, GR10, H1, H4, H5, H7, MA1, MA4, MA8, MA15, SE4, SE5.

Coot, Common AL2, AL3, AL8, AL9, BA5, BA6, BA7, BA10, CA13, CA17, CA18, CA19, CC10, CC12, CO2, CO3, CO4, CO5, CO7, CO13, GR5, GR10, H3, H4, H5, J7, MA2, MA8, MA15, SE5.
Crested See Coot, Red-knobbed.
Red-knobbed CA13, CA14, CA15, CA19, CO2, GR10, H3, H5, SE3, SE5.

Cormorant, Great AL2, AL3, AL4, AL8, AL9, BA1, BA4, BA5, BA10, BA12, BA14, CA7, CA8, CA10, CA11, CA12, CA17, CC1, CC3, CC4, CC6, CC9, CC12, CO1, CO6, CO7, CO9, CO13, CO14, GR5, H2, H4, H5, J5, MA4, MA8, MA17.

Crake, Baillon's CA14, CA15, H5, SE1.
Little H5, SE1.
Spotted CA8, GR4, GR10, H5, MA15, SE1.

Crane AL1, BA1, BA2, BA3, BA4, BA6, BA7, BA8, BA9, BA10, BA13, BA14, CA2, CA4, CA17, CC1, CC2, CC3, CC4, CC7, CC8, CC9, CC10, CC12, CO6, CO11, CO12, H5, MA1, SE2, SE5.

Crossbill AL7, CA2, CC8, CO8, CO9, GR1, GR2, GR3, GR6, GR8, J1, J2, J3, MA4, MA7, MA9, MA12, MA13, SE6.

Crow, Carrion AL7, CC8, GR1, GR8, J1.

Cuckoo, Common AL7, CA1, CA6, CA17, CC6, CC8, CC12, CO9, CO11, CO12, GR2, GR8, H5, J1, J2, MA10, MA11, MA12, MA13.
Great Spotted AL5, AL6, BA2, BA3, BA4, BA6, BA10, BA13, CA2, CA4, CA5, CA6, CA17, CC1, CC3, CC9, GIB, GR7, GR9, H2, H5, J5.

Curlew AL1, AL2, BA10, CA1, CA4, CA11, CA12, CA17, H1, H4, H5.

Curlew, Stone See Stone-curlew.

Dipper CA20, CC5, CC7, CC8, GR1, GR2, J1, J2, J4, MA10, MA12, MA13, MA14.

Diver, Great Northern CA12, H5.
Red-throated CA12.

Dotterel AL1, BA9, GR1, GR3.

Dove, Collared GIB.
Rock BA4, BA8, BA10, CA10, CO11, J1, MA3.
Stock BA11, CA2.
Turtle AL1, AL3, CA1, CA17, CO8, CO9, CO11, CO12, GIB, GR4, GR5, GR8, H5, H7, J1, J4, J5, MA11, MA12, MA13, SE6.

Duck, Ferruginous AL1, AL3, CA15, CA15, CA15, CO1, GR5, H3, J7, MA8, SE3, SE5.
Marbled AL1, AL3, CA13, CA13, CA14, CA15, CA17, CA19, H5, MA1, MA8, SE1, SE1, SE2, SE3, SE5.
Tufted AL3, AL4, BA1, BA4, BA7, BA10, CA15, CA17, CC1, CC3, CC4, CC10, CC12, CO1, CO2, CO3, GR5, H4, H5, J5, MA8.
White-headed AL1, AL2, AL3, AL4, CA13, CA14, CA15, CA17, CA19, CO1, CO2, CO3, CO5, CO6, CO7, CO12, H3, H4, MA1, MA2, MA8, SE5.

Dunlin AL2, AL3, BA7, CA1, CA4, CA5, CA11, CA12, CA15, CA17, CO14, H1, H4, H5, J7, MA1, SE2, SE5.

Dunnock AL7, BA8, CC5, CC7, CC8, CO8, GR1, H5, H7, J2, MA5.

Eagle, Bonelli's AL1, AL5, AL6, AL7, BA2, BA3, BA4, BA6, BA8, BA10, BA11, CA2, CA4, CA6, CA17, CA19, CA20, CA21, CC1, CC5, CC6, CC7, CO2, CO9, CO10, CO11, CO12, GR1, GR2, GR4, GR6, GR7, H5, J1, J2, J3, J4, MA3, MA4, MA5, MA6, MA9, MA10, MA11, MA12, MA13, MA14, MA16, SE4, SE5, SE6.
Booted AL1, AL6, AL7, BA1, BA2, BA3, BA4, BA6, BA8, BA10, BA13, BA14, CA2, CA4, CA4, CA5, CA6, CA7, CA8, CA10, CA17, CA19, CA20, CC1, CC2, CC5, CC4, CC6, CC8, CC9, CO8, CO9, CO10, CO11, CO12, GIB, GR3, GR6, H5, H6, H7, J1, J2, J3, J4, J5, MA7, MA8, MA9, MA10, MA11, MA12, MA13, MA14, MA16, SE3, SE6.
Golden AL6, AL7, BA2, BA3, BA4, BA6, BA8, BA10, BA11, CA4, CA6, CA19, CA20, CC1, CC3, CC5, CC6, CC8, CO8, CO9, CO10, CO11, CO12, GR1, GR2, GR3, GR6, GR7, GR8, H6, H7, J1, J2, J3, J4, J5, MA3, MA4, MA6, MA9, MA10, MA11, MA12, MA13, MA14, MA16, SE6.
Greater Spotted H5.
Short-toed AL7, BA1, BA2, BA3, BA4, BA6, BA8, BA10, BA13, BA14, CA2, CA4, CA5, CA6, CA10, CA17,

CA19, CA20, CC1, CC5, CC4, CC6, CC8, CC9, C08, C09, C010, C011, C012, GIB, GR2, GR8, H5, H6, H7, J2, J3, J4, J5, MA5, MA7, MA9, MA10, MA11, MA12, MA13, MA14, MA16, SE3, SE4, SE6.
Spanish Imperial BA2, BA3, BA4, BA11, CA2, CA4, CA6, CA17, CA19, CC1, CC3, CC5, CC6, C09, C010, C011, C012, H5, H7, J1, J2, J3.

Egret, Cattle AL1, AL2, AL3, BA1, BA5, BA7, BA8, BA9, BA10, BA11, BA12, CA1, CA2, CA4, CA5, CA6, CA7, CA8, CA10, CA11, CA12, CA15, CA17, CA18, CC1, CC2, CC3, CC4, CC7, C01, C06, C014, GR10, H4, H5, J6, MA8, SE1, SE2.
Great White BA5, BA12, CA11, CA12, CA17, CC9, CC10, CC12, H4, H5, SE1, SE2.
Little AL1, AL2, AL3, BA1, BA5, BA7, BA10, BA12, CA1, CA2, CA4, CA5, CA7, CA8, CA10, CA11, CA12, CA13, CA15, CA17, CA18, CA19, CC4, C013, C014, GR10, H1, H2, H4, H5, J7, MA8, SE1, SE2.

Falcon, Eleonora's AL1, CA2, CA5, CA6, GIB, MA16.
Lanner CA2, GIB.
Peregrine AL1, AL5, AL6, AL7, CA2, CA5, CA6, CA6, CA10, CA11, CA12, CA17, CA20, CC5, CC6, CC8, C07, C08, C010, GIB, GR1, GR2, GR7, GR8, H5, J1, J2, J3, J4, MA3, MA4, MA5, MA6, MA8, MA9, MA10, MA11, MA12, MA13, MA14, MA16, SE5.

Fieldfare AL7, CA2, CA6, CA20, CC5, C08, C09, GR3, H5, J1, J2, J3, J4, MA5, MA6, MA12, MA13.

Finch, Citril CC8, GR1, GR2, GR3, J1
Trumpeter AL1, AL5, AL6, GR7, MA12.

Firecrest CA2, CA6, CC8, C010, GIB, GR1, GR3, GR6, GR8, H4, H5, J1, J2, J3, MA9, MA10, MA11, MA12, MA13, SE6.

Flamingo, Greater AL1, AL2, CA1, CA2, CA5, CA7, CA8, CA10, CA11, CA12, CA14, CA15, CA17, CA19, C01, C03, C04, C05, C07, GIB, H4, H5, MA1, MA1, MA2, MA8, SE2, SE3, SE5.
Lesser MA1.

Flycatcher, Pied AL2, BA11, CA2, CA17, CC8, C08, C09, C010, GIB, GR1, GR2, H4, H5, J2, J3, MA10, MA16.
Spotted AL1, AL5, CA17, C08, C09, C010, GIB, GR1, GR2, GR6, GR8, GR5, H4, H5, J1, J2, J3, J4, MA4, MA10, MA13.

Gadwall AL2, AL3, AL4, BA1, BA4, BA7, BA10, BA12, BA13, BA14, CA13, CA15, CA17, CA18, CA19, CC1, CC3, CC10, CC12, C01, C02, C06, H3, H4, H5, J5, J6, J7, MA8, SE5.

Gallinule, Purple See Swamphen, Purple.

Gannet AL2, AL9, CA1, CA5, CA8, CA9, CA11, CA12, GIB, H1, H2, H4, H5, MA17.

Garganey AL2, AL3, BA3, BA13, BA14, CA4, CA5, CA8, CA15, CA17, C01, GR10, H5, MA2, MA8, SE1, SE2, SE3.

Godwit, Bar-tailed AL2, CA1, CA11, CA12, CA17, H1, H4, H5, MA8.
Black-tailed AL2, AL3, BA7, CA1, CA11, CA12, CA14, CA15, CA17, CC7, CC10, CC11, CC12, H1, H4, H5, J7, MA1, MA8, SE5.

Goldcrest CC8, GR1, GR3.

Goldfinch CA1, C09, GIB, GR8, GR5, J2, J4, MA14.

Goose, Egyptian CC3, CC12, C013, BA14,
Greylag BA1, BA3, BA4, BA7, BA13, CA4, CA11, CA12, CA15, CA15, CA17, CC3, CC7, CC10, CC11, CC12, C03, C05, H4, H5, SE1, SE2, SE3, SE5.

Goshawk AL6, AL7, BA2, CA2, CA6, CA20, CC5, CC6, CC7, CC8, C08, C09, C010, GR1, GR2, GR6, GR7, GR8, H6, H7, J1, J2, J3, MA4, MA7, MA10, MA12, MA13, MA14, MA16.

Grebe, Black-necked AL3, AL4, AL8, AL9, BA10, BA13, CA8, CA11, CA12, CA14, CA15, CA17, CA19, CC1, CC10, CC12, C01, C02, H3, H4, H5, J1, MA1, MA1, MA8, SE3, SE5.
Great Crested AL3, AL4, AL9, BA1, BA4, BA7, BA10, BA12, BA13, BA14, CA8, CA11, CA12, CA17, CA18, CA19, CC1, CC3, CC4, CC8, CC10, CC12, C01, C03, C09, C013, H3, H4, H5, J1, J5, MA1, MA4, MA8, SE6.
Little AL3, AL4, AL8, AL9, BA7, BA10, BA12, BA13, CA4, CA8, CA11, CA12, CA14, CA17, CA18, CA19, CC1, CC3, CC4, CC10, CC12, C01, C02, C03, GR4, GR5, GR10, H3, H4, H5, J1, J5, J6, MA8, SE3.

Greenfinch CA1, CO9, GIB, GR2, GR8, GR5, J2, J4.

Greenshank AL2, AL3, BA7, CA1, CA4, CA11, CA12, CA14, CA15, CA17, CC1, CC11, H1, H5, MA8, MA15, SE5.

Gull, Audouin's AL1, AL2, AL8, AL9, BA5, CA1, CA3, CA4, CA5, CA8, CA9, CA10, GIB, GR10, MA8, MA15, MA17.
Black-headed AL1, AL2, AL3, BA5, BA7, BA10, BA12, CA1, CA9, CA10, CA17, CC6, CC7, CC10, CC12, CO6, CO14, GIB, H1, H4, H5, MA1, MA17, SE5.
Common CA16.
Great Black-backed CA16.
Lesser Black-backed AL1, AL2, AL2, AL8, AL9, BA5, BA7, BA10, BA12, CA1, CA8, CA9, CA10, CA17, CC3, CC10, CC12, CO1, CO4, CO6, CO14, GIB, H1, H4, H5, MA1, MA17, SE2, SE5.
Little AL1, AL2, AL9, CA1, CA5, CA7, CA8, CA11, CA12, CA17, GIB, H2, H4, H5, MA8.
Mediterranean AL1, AL2, CA1, CA5, CA7, CA8, CA9, CA10, CA11, CA12, CA14, CA15, GIB, MA8, MA15, MA17.
Ring-billed CA16.
Slender-billed AL1, AL2, CA17, CO14, H5, MA1.
Yellow-legged AL1, AL2, AL8, CA1, CA8, CA9, CA10, CA17, GIB, H1, H4, H5, MA1, MA17.

Harrier, Hen BA3, BA9, BA10, BA13, CA2, CA4, CA5, CA10, CA11, CA12, CA13, CA14, CA17, CA19, CC1, CC3, CC9, CC11, CO9, CO11, CO12, GIB, GR1, GR4, GR9, H2, H4, H5, J5, MA5, SE1, SE2, SE3, SE4, SE5.
Marsh AL1, AL2, AL3, AL4, BA4, BA7, BA12, BA13, CA2, CA4, CA8, CA13, CA15, CA17, CA19, CC9, CC10, CC11, CC12, CO1, CO2, CO3, CO6, CO7, GIB, GR3, GR4, GR10, H1, H2, H4, H5, J5, J6, J7, MA8, MA16, SE1, SE2, SE3, SE5.
Montagu's BA1, BA3, BA3, BA4, BA6, BA8, BA9, BA13, CA2, CA4, CA5, CA10, CA13, CA14, CA15, CA15, CA17, CC1, CC2, CC3, CC7, CC11, CO2, CO11, CO12, GIB, GR7, GR9, H1, H5, J5, MA1, MA8, MA16, SE1, SE2, SE4, SE5.
Pallid CA4.

Hawfinch BA1, BA4, CA2, CA6, CA20, CC6, CC9, CO8, H5, J2, J3, MA9, MA10, SE6.

Heron, Grey AL1, AL3, AL9, BA11, BA12, CA1, CA2, CA4, CA7, CA11, CA12, CA13, CA14, CA17, CA18, CC4, CC6, CC7, CC9, CC11, CO1, CO6, CO7, CO13, CO14, GR5, GR10, H3, H4, H5, J1, J5, J6, J7, MA1, MA4, MA8, MA13, MA15, SE1, SE2.
Night AL3, BA1, BA5, BA10, BA12, CA2, CA8, CA17, CA18, CC9, CO6, CO14, GR4, GR10, H4, H5, J5, J6, MA8, MA8, SE1, SE2.
Purple AL1, AL3, AL4, AL9, BA12, CA1, CA2, CA4, CA8, CA13, CA14, CA15, CA17, CA18, CC9, CO1, CO6, CO7, CO14, GR4, GR10, H3, H4, H5, J5, J6, J7, MA8, MA15, SE1, SE2.
Squacco AL3, BA5, BA12, CA4, CA17, CC9, GR4, GR10, H3, H4, H5, MA1, MA8, MA15, SE1, SE2.

Hobby AL7, BA11, CA2, CA5, CA17, CA20, CC5, CC9, CO8, GIB, GR1, GR7, H5, J5, MA8, MA16.

Honey-buzzard CA2, CA4, CC8, CO8, GIB, GR3, H5, J1, J3, J4, MA16.

Hoopoe AL1, AL2, AL3, AL5, AL6, AL7, AL8, AL9, BA3, BA4, BA9, BA10, BA13, CA1, CA2, CA4, CA5, CA6, CA8, CA10, CA14, CA15, CA16, CA17, CA20, CC2, CC5, CC6, CC12, CO1, CO2, CO8, CO9, CO10, CO11, CO12, GIB, GR1, GR2, GR4, GR5, GR7, GR8, GR9, GR10, H1, H2, H4, H5, H7, J1, J2, J3, J4, MA1, MA3, MA4, MA8, MA8, MA8, MA9, MA10, MA13, SE6.

Ibis, Bald CA1, CA4, CA10
Glossy AL2, BA12, CA4, CA7, CA11, CA12, CA17, CO3, H4, H5, SE1, SE2, SE3, SE5.

Jackdaw AL5, AL6, AL7, AL8, AL8, CA17, CO7, CO8, CO9, CO11, CO12, GR1, GR2, GR6, GR7, H4, H5, J1, J4, MA3, MA4, MA5, MA6, MA10, MA11, MA12, MA13.

Jay AL7, CO8, CO9, GR1, GR2, GR5, J1, J2, J4, MA4, MA7, MA9, MA13.

Kestrel, Common AL1, AL5, BA9, CA2, CA6, CA17, CC3, CO7, CO8, CO9, CO10, CO11, CO12, GIB, GR1, GR2, GR2, GR5, GR6, GR8, GR10, H4, H5, J1, J2, J3, J4, J5, MA3, MA4, MA5, MA8, MA10, MA11, MA12, MA13, MA14, MA16, SE4.
Lesser AL1, AL5, AL6, BA2, BA3, BA4, BA5, BA8, BA9, BA10, BA13, CA2, CA4,

CA5, CA6, CA10, CA17, CA18, CA20, CA21, CC1, CC2, CC3, CC9, CC11, CO11, CO14, GIB, GR1, GR2, H5, J4, J5, MA3, MA4, MA13, MA16, SE4.

Kingfisher AL9, BA2, BA12, CA1, CA2, CA5, CA8, CA17, CO14, GIB, GR10, H5, H6, J2, J3, MA4, MA13.

Kite, Black BA1, BA2, BA3, BA4, BA6, BA7, BA8, BA9, BA10, BA12, BA13, BA14, CA2, CA4, CA6, CA7, CA10, CA17, CA20, CC1, CC2, CC3, CC6, CC7, CC9, CO8, CO9, CO10, CO11, CO12, GIB, GR3, H4, H5, H7, J1, J2, J3, J4, J5, J7, MA8, MA9, MA10, MA11, MA14, MA16, SE1, SE2, SE3, SE6.
Black-winged BA1, BA2, BA3, BA6, BA7, BA10, BA13, CA2, CA4, CA6, CA17, CC1, CC3, CC4, CC6, CC7, CC9, CC11, CO11, CO12, GR4, H5, H7, J5, SE3, SE4, SE5, SE6.
Black-shouldered See Black-winged.
Red BA1, BA2, BA3, BA6, BA9, BA10, BA13, BA14, CA2, CA4, CA5, CA6, CA15, CA15, CA17, CC1, CC2, CC3, CC5, CC4, CC6, CC7, CC8, CC9, CC11, CO11, CO12, GIB, H5, H7, J1, J3, MA11, MA12, MA16, SE1, SE2, SE3, SE4, SE5, SE6.

Kittiwake AL2, CA1, CA5, CA9, CA11, CA12, GIB.

Knot AL2, AL3, CA1, CA11, CA12, CA17, H2, H4, H5, MA8.

Lammergeier J1.

Lanner See Falcon, Lanner.

Lapwing AL1, AL2, AL3, AL7, BA1, BA3, BA7, BA9, BA10, BA13, CA1, CA2, CA4, CA5, CA11, CA12, CA14, CA17, CO11, CO12, CC1, CC2, CC3, CC4, CC7, CC9, CC11, GR4, H5, J7, J5, MA2, MA8, SE2, SE4.

Lark, Calandra AL1, AL7, BA3, BA9, BA13, CA1, CA2, CA4, CA13, CA17, CC1, CC2, CC3, CC4, CC7, CC11, CO11, CO12, GR4, GR7, GR9, H5, SE4, SE5.
Crested AL1, AL2, AL3, AL5, AL7, CA2, CA17, CO5, CO8, CO9, CO10, CO11, CO12, GIB, GR2, GR4, GR7, GR8, GR9, H4, H5, H7, J2, J3, J4, J5, MA1, MA1, MA2, MA3, MA4, MA13, SE2, SE4.
Dupont's AL1, AL5, AL6, GR9.
Greater Short-toed AL1, AL2, AL7, BA3, BA4, BA9, BA13, CA1, CA2, CA4, CA5, CA7, CA8, CA10, CA17, CC1, CC4, CC11, CO11, CO12, GIB, GR7, GR9, H5, MA1, MA15, SE2, SE4.
Lesser Short-toed AL1, AL2, AL3, CA4, CA12, CA17, GR7, GR9, H4, H5, SE1, SE2.
Sky AL1, BA13, CA1, CA6, CA17, CC1, CO9, CO11, CO12, GIB, GR1, GR2, GR3, GR4, GR7, GR9, H4, H5, MA10, SE4.
Thekla AL1, AL5, AL6, AL7, BA1, BA2, BA4, BA6, BA10, CA1, CA2, CA4, CA6, CA10, CA17, CA20, CC6, CC8, CC12, CO8, CO9, CO10, CO12, GIB, GR1, GR2, GR3, GR7, GR8, GR9, H5, H7, J2, J3, J4, J5, MA3, MA6, MA10, MA11, MA12, MA13, MA14, SE6.
Wood AL7, BA1, CA2, CA4, CA6, CA17, CA20, CC5, CC8, CO8, CO9, CO12, GR1, GR3, H5, J1, J2, J3, MA4, MA9, MA10, MA13, SE6.

Linnet AL1, AL9, CA1, CO9, GIB, GR1, GR2, GR8, H5, J1, J2, J4, MA4, MA14.

Loon See Diver.

Magpie, Common AL8, AL9, CA4, CA17, CO9, CO10, GR7, GR9, GR5, H4, H5, J1, J2, J3, J4, J5.
Azure-winged AL7, BA2, BA3, BA4, BA10, BA11, BA13, CA17, CC4, CC6, CC8, CO9, CO10, CO11, CO12, CO13, GR1, H4, H5, H7, J2, J3, J5, J6, MA6, MA7, SE6.

Mallard AL2, AL3, BA1, BA4, BA7, BA7, BA10, BA12, BA13, BA14, CA4, CA13, CA15, CA17, CA18, CA19, CC1, CC10, CC12, CO1, CO2, CO3, CO4, CO5, CO6, CO7, CO11, CO13, GR5, GR10, H3, H4, H5, J1, J5, J6, J7, MA1, MA2, MA4, MA8, SE5.

Martin, Crag AL1, AL5, AL9, BA5, BA8, BA10, BA12, CA1, CA2, CA5, CA6, CA6, CA10, CA17, CA20, CC5, CC6, CC8, CO8, CO9, GIB, GR2, GR3, GR5, H4, H5, H7, J1, J2, J3, J4, MA3, MA4, MA9, MA10, MA11, MA12, MA13, MA14, SE6.
House CO8, CO9, GIB, GR3, J2, H5.
Sand BA10, CA2, CC9, CC11, CO9, GIB, GR3, GR10, H5, MA13, SE2.

Merganser, Red-breasted AL9, CA4, CA8, CA9, CA11, CA12, H2, H4, H5, MA17.

Merlin BA3, BA9, CA2, CA4, CA10, CA17, CC1, CC3, CC9, CO9, CO11, CO12,

Index to Species By Site 409

GIB, GR1, GR4, GR7, GR9, H5, J5, MA10, SE4, SE5.

Moorhen AL3, AL8, AL9, BA10, CA17, CA19, C02, C03, C04, C011, GR5, GR10, H3, H4, H5, J7, MA2, MA13, MA15, SE5.

Nightingale AL5, AL7, BA10, BA12, CA6, CA8, CA10, CA17, CA18, C01, C02, C03, C07, C08, C09, C010, C012, GIB, GR1, GR7, GR9, GR5, H5, H7, J1, J2, J3, J4, J6, MA3, MA4, MA6, MA7, MA8, MA10, MA11, MA12, MA13, MA15, SE4.

Nightjar, European CA2, CA20, CC5, CC8, C08, GIB, GR1.
Red-necked AL1, BA4, BA6, BA8, CA2, CA4, CA5, CA6, CA8, CA10, CA17, CC5, CC6, CC8, C08, C09, C011, C012, GIB, GR1, GR4, GR7, H5, J3, MA4, MA10, MA13, MA14, MA15.

Nuthatch BA1, CA20, CC5, CC8, C09, C010, H7, J1, J3, MA10, SE6.

Oriole, Golden AL6, AL7, BA4, BA6, BA10, BA12, CA2, CA6, CA10, CA20, CC6, CC8, C07, C08, C09, C010, GIB, GR1, GR2, GR6, GR5, H4, H5, H7, J1, J2, J3, J4, MA4, MA7, MA10, MA13, SE4, SE6.

Osprey AL1, AL3, AL9, BA7, BA12, BA14, CA2, CA4, CA5, CA6, CA7, CA8, CA10, CA11, CA12, CA13, CA17, CA18, CC8, CC9, CC10, CC12, C01, C013, C014, GIB, GR10, H2, H4, H5, J7, J5, MA4, MA8, MA16.

Ouzel, Ring AL7, CA2, CA20, C08, GIB, GR1, GR3, H5, J2, J4, MA6, MA10, MA12, MA13, MA14.

Owl, Barn CA2, CA4, CA15, CA17, CA19, CC3, C09, H5, J4.
Eagle AL1, AL5, AL6, BA2, BA4, BA8, BA10, BA11, CA2, CA4, CA6, CA20, CA21, CC1, CC5, CC6, CC8, C08, C09, C010, GIB, GR1, GR2, GR6, GR7, GR8, H7, J1, J2, J3, J4, J5, MA3, MA4, MA7, MA9, MA10, MA13, MA14, MA16, SE6.
Little AL1, AL5, AL6, BA9, CA2, CA6, CA17, CA21, CC1, CC3, C08, C09, C010, C011, C012, GIB, GR2, GR6, GR7, GR8, H5, H7, J1, J2, J4, MA3, MA4, MA8, MA9, MA10, MA11, MA12, MA13, SE4.
Long-eared BA2, CA17, CC5, C010, GR2, GR7, H5.
Scops AL5, AL7, BA1, BA2, BA4, BA6, BA10, CA2, CA4, CA5, CA8, CA10, CA17, CA20, CC6, CC8, C08, C09, C010, GIB, GR1, GR2, GR6, GR8, H5, H7, J1, J2, J4, MA4, MA7, MA10, MA12, MA13, MA14.
Short-eared BA5, BA13, CA2, CA4, CA5, CA17, CC11, H5, MA8.
Tawny BA6, CA2, CA6, CA17, CC8, C08, C09, C010, GR2, GR6, GR8, H5, H7, J1, J2, J3, J4, MA10, MA16.

Oystercatcher AL1, CA1, CA5, CA11, CA12, CA17, H1, H4, H5.

Parakeet, Monk CA8, CA12, GR10, MA8, MA15.
Rose-ringed MA8.

Partridge, Barbary GIB.
Red-legged AL1, BA9, CA17, CA19, C09, C011, GR8, H5, J1, MA9, MA11, MA12, MA13, SE4.

Peregrine See Falcon, Peregrine.

Phalarope, Grey CA1, CA5, CA9, CA10, CA11, CA12.
Red-necked AL1.

Pintail AL1, AL2, AL3, AL4, BA1, BA7, BA10, BA13, BA14, CA4, CA15, CA17, CC10, CC12, C01, C04, GR5, GR10, H4, H5, J5, J6, J7, MA1, MA2, MA8.

Pipit, Meadow AL1, AL2, AL5, BA13, CA1, CA6, CA17, CC1, C08, C09, C010, C011, C012, GIB, GR7, GR9, H4, H5, J3, MA1, MA5, MA8, MA10, MA11, MA12, MA13, SE4.
Red-throated CA2, CA4, CA6, MA15.
Richard's H5.
Tawny AL5, BA3, BA13, CA2, CA4, CA5, CA6, CA8, CA10, CA17, C08, CC7, CC11, GIB, GR1, GR2, GR3, GR7, GR9, H4, H5, MA10, MA11, MA14.
Tree CA2, CA4, CA17, C08, GIB, H4, H5, MA15.
Water BA12, CA2, CA4, CA8, CA14, CA17, CC8, GR1, GR3, GR4, GR10, H4, H5, MA6, MA8, SE1.

Plover, Golden AL2, AL3, BA1, BA3, BA7, BA9, BA10, BA13, CA1, CA2, CA4, CA5, CA11, CA12, CA14, CA17, CC1, CC3, CC4, CC7, CC9, CC11, C011, C012, H4, H5, SE4, SE5.
Grey AL1, AL2, AL3, AL3, AL9, CA1, CA4, CA5, CA11, CA12, CA17, H1, H4, H5, MA8.
Kentish AL1, AL2, AL3, AL9, CA1, CA4, CA5, CA7, CA7, CA8, CA9, CA10, CA11, CA12, CA15, CA15, CA16,

CA17, H1, H2, H4, H5, MA1, MA8, MA15, SE5.
Little Ringed AL2, AL9, BA1, BA6, BA7, BA10, BA14, CA1, CA6, CA8, CA14, CA15, CA15, CA15, CA17, CC1, CC3, CC4, CC7, CC11, CO1, CO2, CO4, CO12, CO13, CO14, GR4, GR10, H2, H5, H6, H7, MA8, MA15, SE6.
Ringed AL2, AL2, AL3, AL9, CA1, CA5, CA11, CA12, CA15, CA17, H1, H5.

Pochard, Common AL2, AL3, AL4, BA1, BA4, BA7, BA10, BA12, BA13, BA14, CA13, CA15, CA17, CA18, CC1, CC3, CC4, CC10, CC12, CO1, CO2, CO3, CO5, CO6, CO7, CO13, GR5, GR10, H3, H4, H5, J6, J7, J5, MA1, MA2, MA8, SE5.
Red-crested AL2, AL3, AL4, BA3, BA10, BA13, BA14, CA13, CA14, CA15, CA17, CA18, CA19, CC10, CC12, CO1, CO2, CO3, CO4, CO5, CO6, CO13, GR10, H2, H3, H4, H5, J7, MA1, MA1, MA2, MA8, SE2, SE5.

Pratincole, Collared AL1, AL2, AL3, BA1, BA7, BA9, BA13, CA1, CA2, CA4, CA14, CA15, CA17, CC4, CC10, CC11, CC12, H1, H2, H4, H5, MA8, SE1, SE2, SE4.

Puffin AL1, AL2, CA1, CA5, GIB.

Quail AL7, BA9, CA2, CA4, CA5, CA10, CA15, CA17, CA19, CC1, CC3, CC7, CO9, GIB, H5, J1, J5, MA4, MA10, SE4.

Rail, Water AL2, AL3, AL4, CA8, CA11, CA12, CA14, CA15, CC9, CO3, GR4, GR10, H4, H5, MA13, MA15, SE1, SE5.

Raven AL5, AL6, AL7, CA2, CA4, CA5, CA6, CA10, CA17, CA20, CC5, CC8, CO7, CO8, CO9, CO10, CO11, CO12, GIB, GR1, GR2, GR6, GR7, GR9, H4, H5, H7, J1, J2, J3, J4, MA2, MA3, MA4, MA5, MA9, MA10, MA11, MA12, MA13, MA14, SE4, SE5, SE6.
Razorbill AL1, AL2, AL8, AL9, CA1, CA5, CA8, CA9, CA11, CA12, GIB, H4, H5, MA17.

Redshank, Common AL2, AL3, AL9, BA7, CA1, CA4, CA5, CA11, CA12, CA14, CA15, CA17, CC1, CC3, CC11, CO5, CO14, H1, H2, H4, H5, J3, J7, MA8, MA15, SE1, SE2, SE5.
Spotted AL2, AL3, BA7, CA1, CA11, CA12, CA14, CA15, CA17, H1, H5, SE1, SE2, SE5.

Redstart, Black AL1, AL3, AL5, AL7, AL8, AL9, CA1, CA6, CA8, CA20, CO8, CO10, GIB, GR1, GR2, GR7, GR8, J1, J2, J3, J4, MA3, MA4, MA5, MA6, MA10, MA13, MA14, MA15.
Common AL1, CA4, CA17, CC7, CO8, CO9, CO10, GIB, GR1, GR2, H4, H5, H7, J2, MA6, MA10, MA16.

Redwing AL7, BA1, BA2, CA2, CA20, CC5, CO8, CO9, CO10, GIB, GR3, H5, J1, J2, J3, J4, MA4, MA5, MA6, MA7, MA10, MA12, MA13.

Reedling, Bearded CC9.

Robin AL7, CA6, CO9, CO10, GIB, GR2, GR8, GR3, GR5, H5, J2, J3, MA4, MA10, MA11, MA12, MA13.

Roller AL1, AL2, AL5, AL7, BA3, BA4, BA9, BA10, CA2, CA4, CA5, CA17, CC1, CC3, CC7, CC9, CC11, CO8, CO9, CO10, CO11, CO12, GR1, GR7, GR9, H3, H4, H5, J2, J3, MA8, SE4, SE5.

Ruff AL2, AL3, AL9, BA7, CA1, CA4, CA11, CA12, CA14, CA15, CA17, CC1, CC11, H5, MA1, SE1, SE2.

Sanderling AL2, AL9, CA1, CA5, CA11, CA12, CA17, H1, H4, H5, MA15.

Sandgrouse, Black-bellied AL1, AL5, AL6, BA1, BA3, BA8, BA9, BA13, CC1, CC2, CC3, CC4, CC11, CO11, CO12, GR7, GR9, J4, J5, SE4.
Pin-tailed BA8, BA9, BA13, CA17, CC1, CC2, CC3, CC4, CC7, CC11, CO11, CO12, H5, SE4.

Sandpiper, Common AL2, AL3, BA7, BA10, BA12, CA1, CA4, CA14, CA15, CA17, CC8, CO5, CO12, CO14, GR1, GR5, H1, H4, H5, MA1, MA8, MA13, MA15, SE2.
Curlew AL2, AL3, BA7, CA1, CA14, CA15, CA17, CC11, H4, H5, MA1, MA15, SE2.
Green AL2, AL3, BA7, BA12, CA1, CA4, CA14, CA15, CA17, CC3, CO13, CO14, GR5, H1, H5, MA1, MA8, MA15, SE1, SE2, SE5.
Marsh AL1, AL2, AL2, CA17, H5, SE1.
Purple CA3, H4, H5, MA17.
Wood AL2, BA7, BA12, CA1, CA4, CA14, CA15, CA17, CO14, H5, MA1, MA8, MA15, SE1, SE2.

Scoter, Common AL1, AL2, CA5, CA8, CA9, CA11, CA12, CA16, GIB, H1, H2, H4, H5, MA8, MA17.

Scrub-robin, Rufous-tailed AL1, AL5, BA4, BA9, BA10, CA2, CA4, CA5, CA6, CA8, CA10, CA15, CA15, CA17,

CO3, CO9, GIB, H5, MA3, MA13, SE4, SE5.

Serin CA1, CA2, CA6, CO9, GIB, GR1, GR2, GR8, GR5, J1, J2, J4, MA4, MA9.

Shag AL1, GIB.

Shearwater, Balearic AL1, AL8, AL9, CA1, CA5, CA8, CA9, GIB, H2, H4, H4, H5, MA15, MA17.
Cory's AL1, CA1, CA5, CA8, CA9, GIB, H4, MA15, MA17.
Great CA9.
Levantine CA9, GIB, MA17.
Scopoli's See Cory's.
Sooty CA9, GIB, MA17.
Yelkouan See Shearwater, Levantine.

Shelduck, Common AL1, AL2, AL3, CA8, CA11, CA12, CA15, CA17, CO1, CO5, H4, H5, MA1, MA4, SE4.

Shoveler AL2, AL3, AL4, BA1, BA4, BA7, BA10, BA12, BA13, CA4, CA13, CA15, CA17, CA18, CA19, CC1, CC3, CC4, CC10, CC12, CO1, CO3, CO4, CO5, CO6, GR10, H3, H4, H5, J6, J7, MA1, MA2, MA4, MA8, MA8, SE4.

Shrike, Red-backed CC8.

Southern Grey AL1, AL2, AL3, AL6, AL7, AL8, BA2, BA3, BA6, BA7, BA10, BA11, BA12, BA13, CA2, CA4, CA6, CA14, CA15, CA17, CA19, CA20, CC1, CC3, CC4, CC6, CC9, CC11, CC12, CO8, CO9, CO10, CO11, CO12, GR4, GR5, GR7, GR8, H2, H4, H5, H7, J2, J3, MA1, MA3, MA4, MA4, MA5, MA8, MA10, MA11, MA12, MA13, MA13, MA14, SE4, SE5, SE6.
Woodchat AL1, AL2, AL3, AL5, AL6, AL7, CA1, CA2, CA4, CA6, CA15, CA16, CC12, CO8, CO9, CO10, CO11, CO12, GIB, GR1, GR2, GR4, GR5, GR7, GR8, H4, H5, H7, J1, J2, J3, J4, MA4, MA5, MA10, MA11, MA12, MA13, MA14, SE6.

Siskin CA2, CA4, CA17, CC5, CC8, CO8, CO9, GIB, GR1, GR2, H5, J2, MA4, MA7, MA10, MA11, MA12, MA13, SE6.

Skua, Arctic AL1, AL2, CA1, CA5, CA8, CA9, CA11, CA12, GIB, H2, H4, H5, MA8, MA17.
Great AL1, AL2, CA1, CA5, CA8, CA9, CA11, CA12, GIB, H2, H4, H5, MA8, MA17.
Pomarine CA1, CA5, CA8, CA9, GIB, MA8, MA17.

Skylark See Lark, Sky.

Snipe, Common AL2, AL3, AL9, BA7, CA1, CA4, CA11, CA12, CA14, CA15, CA17, CC1, CC11, CO14, GR4, GR10, H4, H5, J7, MA15, SE1, SE2, SE5.
Jack AL3, CA4, CA11, CA12, CA15, CA17, CC11, GR4, GR10, H5, J7, MA15.

Sparrow, House BA8, GIB, SE2.
Rock AL5, AL6, AL7, BA2, BA3, BA4, BA6, BA8, BA10, BA11, BA13, CA20, CC1, CC4, CC8, CO8, CO9, GR1, GR7, H6, H7, MA3, MA6, MA10, MA11, MA13, MA14, SE6.
Spanish AL1, BA3, BA4, BA8, BA9, BA10, BA12, BA13, CA2, CA4, CA17, CC4, CC7, CC11, CC12, CO11, CO12, GR4, H5, H7, SE3, SE4.
Tree AL4, AL9, CA2, CA4, CA17, CC9, GR7, GR10, H5, MA15, SE1, SE2, SE6.

Sparrowhawk BA2, CA2, CA4, CA6, CA17, CC5, CC8, CO9, GIB, GR2, GR3, H5, H7, J1, J2, J3, MA5, MA11, MA12, MA13, MA14, MA16.

Spoonbill AL2, BA3, BA4, BA5, BA7, BA12, CA1, CA2, CA5, CA7, CA11, CA12, CA17, CC9, CC10, CC11, CC12, H1, H2, H3, H4, H5, SE1, SE2, SE3.

Starling, Common CA1, CA2, CA4, CA6, CA17, CO1, H4, H5, MA11, MA12, MA13.
Spotless CA1, CA2, CA4, CA6, CA17, CO9, GIB, H4, H5, H7, J6, MA11, MA12, MA13.

Stilt, Black-winged AL1, AL2, AL3, AL9, BA7, BA12, BA13, CA1, CA4, CA11, CA12, CA14, CA15, CA15, CA17, CC1, CC3, CC9, CC11, CO3, CO4, CO5, GR10, H1, H2, H4, H5, J7, MA1, MA8, SE1, SE2, SE4.

Stint, Little AL2, AL3, AL9, BA7, CC11, CA1, CA11, CA12, CA14, CA15, CA17, CC1, H1, H3, H4, H4, H5, SE1, SE2, SE4.
Temminck's AL2, AL3, CA17, CC1, CC11, H5, SE1, SE2, SE5.

Stonechat AL1, AL2, AL7, CA6, GIB, GR1, GR8, H5, J1, J2, J3, MA8, MA10, MA11, MA12, MA13, MA14.

Stone-curlew AL1, AL2, AL3, AL5, AL6, AL7, BA1, BA3, BA6, BA7, BA8, BA9, BA10, BA13, CC11, CA1, CA2, CA4, CA17, CA19, CC1, CC2, CC3, CC4, CC7, CC9, CO11, CO12, GR7, GR9, GR10, H4, H5, MA1, SE4, SE5.

Stork, Black AL1, BA1, BA2, BA3, BA4, BA6, BA8, BA10, BA11, BA13, CA1, CA2, CA4, CA5, CC1, CC3, CC5, CC4, CC6, CC8, CC9, CC10,

CC12, CO1, CO9, CO10, CO11, CO12, CO13, GIB, H5, H6, H7, J2, J3, J5, MA16, SE1, SE2, SE6.
White AL1, AL3, BA1, BA2, BA3, BA4, BA5, BA6, BA7, BA8, BA9, BA10, BA11, BA12, BA13, CA1, CA2, CA4, CA5, CA6, CA7, CA8, CA10, CA11, CA12, CA17, CA19, CC1, CC2, CC3, CC4, CC6, CC7, CO6, CO9, CO11, CO12, GIB, H1, H4, H5, H7, J5, MA1, MA4, MA16, SE1, SE2, SE3, SE4, SE6.

Storm-petrel, European AL1, GIB, MA17.
Leach's CA5, CA9, GIB, H2.

Swallow, Barn CA15, CO8, CO9, GIB, GR2, GR3, GR5, H5, J1, MA4, MA8, SE4.
Red-rumped AL1, AL5, AL6, AL9, BA1, BA4, BA6, BA8, BA10, CA2, CA4, CA5, CA6, CA8, CA10, CA20, CC5, CC6, CO8, CO9, GIB, GR2, GR3, GR7, GR9, GR10, H5, H6, H7, J1, J2, J3, MA4, MA11, MA12, MA13, MA16, SE6.

Swamphen, Purple AL2, AL3, AL8, AL9, BA12, CA8, CA13, CA14, CA15, CA17, CA18, CA19, CC9, CO1, CO2, CO3, CO6, CO7, GR10, H1, H3, H4, H5, J6, J7, MA1, MA8, MA15, SE1, SE2, SE3, SE5.

Swift, Alpine AL1, AL6, BA4, BA5, BA8, BA10, BA12, CA2, CA6, CA17, CA20, CA21, CC5, CC6, CO8, CO9, GIB, GR1, GR2, GR3, H5, J1, J2, J3, J4, MA3, MA4, MA6, MA9, MA10, MA11, MA12, MA13, MA14, MA16, SE6.
Common AL1, AL9, BA5, BA12, CA2, CA6, CA12, CA18, CC3, CO8, CO9, GIB, GR3, GR5, J2, MA4, MA8, MA11, MA13.

Little CA4, CA5, CA16, SE6.
Pallid AL1, AL9, BA5, BA12, CA2, CA6, CA12, CA16, CA18, CC3, CO8, CO9, GIB, GR1, GR3, GR5, H1, J1, J2, MA4, MA8, MA11, MA13.
White-rumped BA6, BA8, CA2, CA5, CA6, CA20, CC5, CC6, CO9, H6, H7, J2, J3, MA8, MA10, MA11, MA12, MA13, MA16, SE6.

Teal AL2, AL3, AL9, BA4, BA7, BA12, BA13, BA14, CA4, CA13, CA15, CA17, CA19, CC1, CC3, CC4, CC10, CC12, CO1, CO3, CO7, GR5, GR10, H3, H4, H5, J1, J6, J7, MA1, MA8, SE4.

Tern, Arctic CA5.
Black AL1, AL2, AL3, AL4, BA7, CA1, CA5, CA8, CA9, CA10, CA14, CA15, CA17, CC9, CC10, CC11, CC12, CO1, CO14, GIB, H4, H5, J5, MA1, MA8, MA17, SE1, SE2, SE3.
Caspian AL1, AL2, AL9, CA1, CA5, CA8, CA9, CA10, CA11, CA12, CA16, CA17, GIB, H1, H2, H4, H5, MA8, MA15.
Common AL1, AL2, AL3, AL9, CA1, CA5, CA8, CA9, CA10, CA14, CA17, CO14, GIB, H4, H5, MA8, MA17.
Gull-billed AL1, AL2, AL3, BA7, BA10, BA14, CA1, CA5, CA8, CA9, CA10, CA13, CA17, CC9, CC10, CC11, CC12, CO1, CO4, GIB, H5, MA1, MA8, SE1, SE1, SE2, SE3, SE5.
Lesser Crested CA1, CA5, CA8, CA9, GIB, MA15, MA17.
Little AL1, AL2, AL3, AL9, BA5, BA7, BA10, CA1, CA5, CA8, CA9, CA10, CA11, CA12, CA14, CA17, CC4, CC10, CC11, CC12, H1, H2,
H4, H5, MA1, MA8, MA17, SE1, SE5.
Roseate CA5.
Sandwich AL1, AL2, AL3, AL8, AL9, CA1, CA5, CA7, CA8, CA9, CA10, CA11, CA12, CA16, CA17, GIB, H4, H5, MA8, MA17.
Whiskered AL1, AL2, AL3, AL4, BA7, BA13, CA1, CA8, CA13, CA14, CA15, CA17, CC9, CC11, CC12, CO14, H4, H5, J5, MA1, SE1, SE2, SE4.
White-winged Black AL1, AL2, AL4, MA8.

Thrush, Blue Rock AL1, AL5, AL6, AL7, BA4, BA8, BA10, CA2, CA5, CA6, CA10, CA20, CC5, CC6, CC8, CO8, CO9, GIB, GR1, GR2, H6, H7, J1, J2, J3, J4, MA3, MA4, MA5, MA6, MA9, MA10, MA11, MA12, MA13, MA14, MA16, SE6.
Mistle AL7, CO8, CO9, CO10, GR2, H5, J1, J2, J3, J4, MA11, MA12, MA13, SE6.
Rufous-tailed Rock AL7, CA2, CA6, CA20, CC5, CC8, CO8, GIB, GR1, GR3, J1, J2, J4, MA3, MA5, MA6, MA9, MA10, MA14.
Song BA1, BA2, CA6, CC5, CO8, CO9, CO10, GIB, H5, J3, MA6, MA7, MA11, MA12, MA13, SE4.

Tit, Bearded See Reedling, Bearded.
Blue CO9, GIB, GR2, GR3, J2, J2.
Coal AL7, CA20, CO9, GR1, GR3, GR6, J1, J2, MA4, MA6, MA7, MA10.
Crested AL7, BA1, CA2, CA6, CA20, CC1, CC5, CO8, CO9, GR1, GR2, GR6, H5, J1, J2, MA4, MA6, MA7, MA9, MA10, SE6.

Great C09, GIB, GR1, GR2, GR3, J1, J2, J2.
Long-tailed AL7, GR2, J2, MA6, MA7, MA13, SE6.
Penduline AL2, AL3, AL8, BA10, BA12, CA4, CA7, CA8, CA14, CA15, CA15, CA15, CA17, CC6, CC9, CC11, CO14, GR4, GR10, H4, H5, J6, J7, MA8.

Treecreeper, Short-toed AL7, CA6, CA17, CA20, C09, CO10, GIB, GR1, GR2, GR3, GR6, H5, J1, J2, J3, J4, MA9, MA10, MA11, MA12, MA13.

Turnstone AL2, AL3, CA1, CA5, CA11, CA12, CA17, H1, H4, H5, MA17.

Vulture, Black BA1, BA2, BA3, BA4, BA6, BA10, BA11, BA13, CA2, CA4, CA6, CC1, CC2, CC5, CC4, CC6, CC7, CC8, CC9, C09, CO10, H5, H6, H7, J1, J2, J3, MA4, SE6.
Bearded See Lammergeier.
Egyptian BA2, BA3, BA4, BA8, BA10, BA11, CA2, CA4, CA5, CA6, CA10, CA17, CA20, CA21, CC1, CC2, CC5, CC4, CC6, CC8, CC9, C08, C09, GIB, GR7, GR8, H5, H7, J1, J3, MA3, MA4, MA10, MA11, MA12, MA13, MA16, SE6.
Griffon AL7, BA1, BA2, BA3, BA4, BA6, BA8, BA9, BA10, BA11, BA13, CA2, CA4, CA5, CA6, CA7, CA10, CA17, CA20, CA21, CC1, CC2, CC5, CC4, CC6, CC8, CC9, C08, C09, CO10, GIB, GR7, H5, H6, H7, J1, J2, J3, J4, MA3, MA4, MA9, MA10, MA11, MA12, MA13, MA16, SE6.
Rüppell's CA2, CA4.

Wagtail, Grey CA6, CC8, C08, CO10, GIB, GR1, GR2, H4, H5, H6, J1, J2, J3, MA3, MA4, MA8, MA13.
Pied GIB.
White AL1, AL5, CA1, CA6, C08, CO10, CO14, GR5, GR7, GR10, H4, H5, J3, MA4, MA5, MA8, MA11, MA12, MA13, SE4.
Yellow AL2, AL3, CA2, CA4, CA7, CC9, GIB, GR9, GR10, H1, H2, H4, H5, MA1, MA4, MA8, MA8, MA15, SE1, SE2, SE4.

Wallcreeper GR1, MA4, MA14.

Warbler, Bonelli's CA2, CA6, CA10, CA17, CA20, CC5, CC8, C08, GIB, GR2, GR3, GR6, GR8, GR5, H4, H5, H7, J1, J2, MA4, MA6, MA7, MA9, MA10, MA13, MA14, SE6.
Cetti's AL1, AL2, AL8, BA5, BA10, BA12, CA2, CA6, CA8, CA15, CA18, CC11, CC12, CO1, CO3, C06, C07, CO12, GR4, GR5, GR10, H4, H5, H7, J2, J6, MA1, MA2, MA3, MA8, SE4, SE5.
Dartford AL1, AL2, AL6, AL7, AL9, BA12, CA2, CA5, CA6, CA15, CC5, CC8, C08, C09, CO10, GIB, GR1, GR2, GR6, GR7, GR8, H2, H4, H5, H7, J1, J2, J3, MA4, MA8, MA10, MA11, MA12, MA13.
Fan-tailed See Cisticola, Zitting.
Garden CA2, CA17, CC8, GIB, H4, H5, H7.
Grasshopper CA2, CA17, GIB, GR4, H4, H5, MA15.
Great Reed AL2, AL3, AL4, AL8, BA4, BA7, BA10, BA12, CA2, CA4, CA8, CA13, CA14, CA15, CA17, CA18, CA19, CC1, CC9, CC11, CO1, CO3, C06, CO12, GR4, GR10, H4, H5, J6, J7, MA1, MA2, MA8, SE1, SE2, SE5.

Isabelline See Olivaceous.
Melodious AL3, AL5, AL7, BA10, BA12, CA2, CA5, CA6, CA10, CA17, CA20, CC5, CC8, CC11, CO1, CO3, CO4, CO6, C08, C09, GIB, GR1, GR2, GR4, GR5, GR7, GR8, H4, H5, H7, J1, J2, J7, MA1, MA3, MA4, MA5, MA7, MA9, MA10, MA11, MA12, MA13.
Moustached GR10.
Olivaceous AL5, BA12, CA2, CA17, GIB, GR7, H4, H5, MA15, SE4, SE5.
Orphean AL5, AL7, BA1, BA4, BA6, BA10, BA13, CA2, CA6, CA15, CA17, CA20, CC5, CC6, C08, CO12, GIB, GR5, H4, H5, H7, J1, MA5, MA6, MA7, MA8, MA9, MA10, SE6.
Reed AL2, AL3, AL4, AL8, BA12, CA2, CA8, CA13, CA14, CA15CA17, CA18, CA19, CC9, CC11, CO1, CO3, CO6, CO12, GIB, GR4, GR10, H3, H4, H5, J6, J7, MA1, MA2, MA8, MA15, SE1, SE2, SE5.
Sardinian AL1, AL2, AL5, AL6, AL7, AL8, BA12, CA2, CA5, CA6, C08, C09, CO10, CO11, GIB, GR2, GR7, H4, H5, H7, J3, MA4, MA10, MA11, MA12, MA13.
Savi's BA12, CA2, CA15, CA17, CA18, CC9, GR4, GR10, H4, H5, J7, MA15, SE1, SE2.
Sedge CA15, CA17, CC9, GR4, H4, H5, MA15, SE1.
Spectacled AL1, AL5, AL8, BA6, CA2, CA10, CA12, CA17, CC6, CC8, C09, CO12, GIB, GR1, GR7, H4, H5, J1, J2, MA1, MA8, MA10, MA11, MA14, SE2.
Subalpine AL5, AL7, BA4, BA6, CA2, CA6, CA17, CA20, CC5, CC6, CC8, C08, C09, CO10, GIB, GR1, GR2,

GR6, H4, H5, H7, J1, J2, J3, J4, MA5, MA6, MA9, MA10, MA11, MA13, MA14.

Western Olivaceous See Olivaceous.

Willow AL1, CA2, CA17, CC8, CO8, CO9, GIB, H4, H5, J3, MA10.

Wood CA2, GIB.

Waxbill, Common BA5, BA12, CA8, CC9, CO14, GR10, H5, MA8, MA15.

Wheatear, Black AL1, AL5, AL6, AL7, BA4, BA8, BA10, CA20, CC5, CC6, CC8, CO8, CO9, CO12, GR1, GR2, GR7, GR9, J1, J2, J4, MA4, MA5, MA6, MA10, MA11, MA12, MA13, MA14.
  Black-eared AL1, AL2, AL5, AL6, AL7, BA4, BA10, CA2, CA4, CA5, CA6, CA10, CA15, CA17, CA20, CC6, CC7, CC8, CC11, CO8, CO9, CO11, CO12, GIB, GR1, GR2,
GR3, GR7, GR8, GR9, H4, H5, H7, MA3, MA4, MA5, MA6, MA8, MA10, MA11, MA12, MA13, MA14.
  Northern AL2, AL7, CA2, CA4, CA17, CA20, CC5, CC7, CC8, CO8, CO9, CO10, GIB, GR1, GR2, GR3, GR9, H4, H5, J3, MA8, MA10, MA14.

Whimbrel AL1, AL2, AL9, CA1, CA4, CA5, CA11, CA12, CA17, H1, H4, H5, MA17.

Whinchat AL1, AL6, CA2, CA17, GIB, GR1, H4, H5, J3, MA8, MA16.

Whitethroat CA2, CA4, CA5, CA17, CC8, CO9, GIB, GR1, H4, H5, H7, MA10, MA12.

Wigeon AL2, AL3, AL4, BA4, BA7, BA10, BA13, CA4, CA11, CA12, CA15, CA17, CC1, CC3, CC4, CC10, CC12, CO1, CO2, CO3, CO4, CO5, CO6, CO7, GR5, GR10, H4, H5, J7, J5, MA1, MA8, SE5.

Woodcock CA2, CA17, H5, SE6.

Woodlark See Lark, Wood.

Woodpecker, Great Spotted CA6, CC8, CO10, H5, H6, H7, J1, J2, J3, MA13, MA16.
  Green BA11, CA20, CC8, CO8, CO10, GR8, H5, J1, J2, J3, MA9, MA13, MA16.
  Lesser Spotted H7, BA6, CC5, CC6, CC8, J1.

Woodpigeon BA2, CO8, CO9, CO11, GR5, H5, J1, J4, MA10.

Wren AL7, CO9, GIB, GR2, GR3, GR5, GR8, J1, J2, MA3, MA4, MA5, MA6, MA10, MA13.

Wryneck BA4, BA12, CA2, CA4, CA5, CA14, CA15, CA15, CA16, CO8, CO14, GIB, H7, J1, J3, MA13, MA15, SE6.

Yellowhammer GR1, J1, J3.

ACKNOWLEDGEMENTS

This fourth edition has benefited from the most helpful and generous advice of Keith Bensusan, John Cantelo, Francisco Chiclana, Philip Croft, José Montero, Charles Perez, Javier Prieta, Paco Martín and Juan Martín, to all of whom I am most grateful. My thanks too to my wife, Joan Garcia, for her patience and assistance during our many excursions in the field. Andy Paterson kindly read through the final manuscript. Molly Arnold of Bloomsbury Publishing plc ably steered this book through the publication process. The earlier editions, from which this one has evolved, owed a great deal to information and help provided by Steve Abbot, John Bartley, Andy Chapell, Judy Collins, John Cortes, P. F. Cooper, Colin Davies, William Davies, Mary Carmen Fermosell, Bruce Forrester, Joan Garcia, Nacho García Paéz, Jorge Garzón, Richard Gunn, Robert Haigh, Martin Henry, Eduardo de Juana, Eileen Marsh, John Muddeman, Juan Luis Muñoz, Jan Nordblad, Andrés Paterson, Tania Roe, Gene Skelton, Ann Small, Roy Smith, Chris Smout, G. F. Trowmann, Jesús Valiente and Ivor White, and of course from the invaluable coauthorship of Andy Paterson.